P9-AGU-792

DISCARD

The Global Ecology Handbook

4

The Global Ecology Handbook

What <u>You</u> Can Do
about the
Environmental Crisis

The Global Tomorrow Coalition

Edited by
Walter H. Corson

Practical Supplement
to the PBS series
RACE TO SAVE THE PLANET

Beacon Press
Boston

363.7025
G562

Beacon Press
25 Beacon Street
Boston, Massachusetts 02108-2800

Beacon Press books
are published under the auspices of
the Unitarian Universalist Association of Congregations.

© 1990 by The Global Tomorrow Coalition
All rights reserved
Printed in the United States of America

LCN 88-43318
ISBN 0-8070-8501-4

The Global Tomorrow Coalition gratefully acknowledges permission to reprint from the following sources: Concern, Inc.; The Conservation Foundation; Department of Agricultural Economics, Cornell University; Friends of the National Zoo; Institute for Alternative Agriculture; International Union for the Conservation of Nature and Natural Resources; The Johns Hopkins University Press; Center for Advanced Study of International Development, Michigan State University; Organization for Economic Cooperation and Development; Overseas Development Council; Mr. Frank M. Potter; The Royal Swedish Academy of Sciences, AMBIO: A Journal of the Human Environment; Scientific American; Smithsonian Institution Traveling Exhibition Service; Transaction Publishers; United Nations Environment Programme; The Washington Post; World Resources Institute; Worldwatch Institute; Zero Population Growth.

97 96 95 94 93 92 91   8 7 6 5 4

Library of Congress Cataloging-in-Publication Data

The Global ecology handbook : what you can do about the environmental
   crisis / Global Tomorrow Coalition ; edited by Walter J. Corson.
        p.   cm.
     Includes bibliographical references.
     ISBN 0-8070-8500-6. — ISBN 0-8070-8501-4 (pbk.)
     1. Environmental protection—Citizen participation.   I. Corson,
Walter J.   II. Global Tomorrow Coalition.
TD171.7.G56   1990
363.7′025—dc20   88-43318
                                                            CIP

12/94

Brief Table of Contents

Table of Contents

Preface

There is ample evidence of the seriousness of the world's population, resource, and environmental problems—poverty and hunger, deforestation and species loss, soil erosion and desertification, air and water pollution, acid precipitation and ozone layer depletion, as well as the greenhouse effect and climate change. Potential solutions for many of these problems are being identified, and some consensus on how to implement them is emerging. A growing number of people—development experts, environmentalists, business executives, government officials, religious leaders, and journalists—are beginning to recognize that their long-term aims and activities are mutually dependent, not mutually exclusive. They see that poverty and environmental stress are often inextricably linked, that many global environmental problems are rooted in local conditions, that conservation, waste reduction, and recycling can be economically beneficial, and that action at the local level is often the first step toward a global solution.

The Global Tomorrow Coalition has prepared this *Global Ecology Handbook* as a stimulus to action. The decisions we make—as consumers in the marketplace, as professionals in our daily work, as concerned citizens of a powerful nation, and as parents and homemakers helping to form the values of our families—will create the world our children and grandchildren inherit.

A first step toward action is a better understanding of the complex interrelationships among major global problems and some of their potential solutions. The issues covered in the *Handbook* include: foresight capability; population growth; development and environment; food and agriculture; biological diversity; tropical forests; ocean and coastal resources; fresh water; nonfuel minerals; energy; air, atmosphere, and climate; hazardous substances; solid waste management; and global security.

In spite of growing attention to the nature and causes of these global problems, there are political, social, and economic obstacles that continue to retard progress toward solutions. This *Handbook* is designed to help overcome these obstacles by: providing basic facts about global population, resource, and environmental problems; demonstrating how these problems are interrelated; showing how the problems affect the lives of citizens in the United States and other countries; citing successful efforts to alleviate the problems in the United States and around the world; proposing alternative solutions based on the best information available; suggesting how individuals and groups can participate in achieving solutions; and giving sources of further information and assistance.

The overall theme of the book is very close to the message conveyed by the recent report of the World Commission on Environment and Development, *Our Common Future*: the present path of economic and social development is causing serious ecological damage to the planet, is failing to close the North-South gap in per capita income, and is therefore not sustainable over the long term. Yet there are many opportunities for changing course and moving toward new patterns of development that will be sustainable into the foreseeable future.

The first chapter of the *Handbook*, "A Global Awakening," shows how human activities are affecting our increasingly interdependent world, how different global problems are interrelated, and how human impacts in one region can be felt in other parts of the world. Each of the succeeding fourteen chapters focuses on one of the global issues listed above, summarizing the nature and extent of the problem, outlining what can be and is being done to solve it, suggesting specific actions that citizens can take, and listing organizations and printed and audiovisual materials as sources of further information.

The next-to-last chapter, "Toward a Sustainable Future," reviews priority goals and strategies for achieving sustainable development, and summarizes progress in both developing and industrial countries in moving toward more viable policies and actions.

The last chapter, "What You Can Do," offers guidelines and advice about informing yourself and joining with others for action on the issues, communicating your views, working with educators, organizing community activities, shaping public policy, working with the media, and forming international ties.

Each issue chapter also provides specific suggestions of "What You Can Do," and each chapter concludes with a "Further Information" section listing books, articles, periodicals, audiovisual materials, and teaching aids related to the chapter. The Appendix lists names and addresses of organizations that are referred to in the chapters, including Participating and Affiliate Members of the Global Tomorrow Coalition, that can provide information and materials on the topics covered. The Appendix also includes addresses for suppliers of audiovisual materials listed in the Further Information sections.

A major theme of the book is that the issues covered are highly interdependent—for example, poverty contributes to environmental damage, and environmental degradation exacerbates poverty. As a result, a topic that receives primary coverage in one chapter may also be discussed in other chapters. Thus food production receives primary coverage in Chapter 5, but is also considered in the Tropical Forests chapter (as an important cause of deforestation) and the Fresh Water chapter (as a major cause of water depletion and pollution). And population growth is given primary treatment in Chapter 3, but is mentioned as an important cause of problems in nearly every other issue chapter. Where an issue discussed in the text receives primary treatment in a different chapter, a reference to that chapter is included in the text.

Every effort has been made to include the latest information available on the issues, but scientific data and analysis are accumulating so rapidly on certain topics that some facts cited may have been superseded before this *Handbook* reaches the reader. Although the writers have sought to present a balanced view of issues on which there are widely differing opinions (which is the case for a number of issues), some conclusions presented in the book—such as the remaining useful life of fossil fuel and mineral supplies, and the climatic effects of a projected doubling of atmospheric carbon dioxide—are controversial and subject to continuing scientific uncertainty.

References to nations as "developed," "more developed," or "industrial" follow the classification used by the United Nations and include all of North America and Europe, the Soviet Union, Japan, Australia, and New Zealand. All other regions and countries are classified as "developing," "less developed," or the "Third World."

A word about units of measurement. Both English and international metric units are used in the text, and for some data numbers are given in both units. For readers unfamiliar with metric units, one hectare is equivalent to almost 2.5 acres, or roughly 1.3 times the size of a football field; one kilometer is six-tenths of a mile; a square kilometer is equivalent to almost four-tenths of a square mile, and Central Park in New York City covers about 3 square kilometers. One metric ton is fractionally less than one "regular" ton. A temperature change of one degree Fahrenheit is equivalent to a change of 0.55 degree Celsius or Centigrade; a change of one degree Celsius equals a change of 1.8 degrees Fahrenheit. For energy, a kilowatt equals 1,000 watts; a megawatt equals 1,000 kilowatts and can power about 1,000 homes.

The Global Tomorrow Coalition is a national, non-profit alliance of organizations and individuals committed to acting today to assure a more sustainable, equitable global tomorrow. Founded in 1981 in response to the discussion and debate that followed the release of the *Global 2000 Report to the President*, the Coalition now comprises more than 115 organizations with outreach to over 10 million Americans.

The programs of the Coalition include the preparation and dissemination of newsletters and publications like *The Global Ecology Handbook*, drawing on the latest research and analysis of member groups and other organizations; continuing sponsorship of the Globescope Assembly series on the U.S. role and responsibility in dealing with long-term global issues; compilation of curriculum materials and conduct of teacher training workshops; organization of community forums on sustainable development and related issues; and service as clearinghouse on member groups' public policy goals, especially those related to foreign assistance programs and sustainable development.

The activities of the Global Tomorrow Coalition are shaped and guided by its members, who elect its Board of Directors and officers, and are carried out by a small central staff based in Washington, D.C., with the assistance of many volunteers, consultants, and interns, both paid and—more often—unpaid.

The Coalition depends for its tax-deductible financial support on a wide range of sources, including private foundation grants; occasional project support under U.S. government and multilateral organization programs related to sustainable development; donations from individuals and corporations; membership fees and member contributions; income from the sale and rental of publications and audiovisual materials; and provision of in-kind facilities and services.

Without the generosity of this support from multiple sources, and the strong encouragement, guidance, and participation of the Coalition's member organizations, the preparation of *The Global Ecology Handbook* would have been impossible.

RACE TO SAVE THE PLANET

The Global Tomorrow Coalition is working to encourage broad viewership and educational use of **RACE TO SAVE THE PLANET,** a ten-part documentary television series on the same issues discussed in *The Global Ecology Handbook*. The series concept is based on the Worldwatch Institute's annual *State of the World* reports. Produced by WGBH Boston, **RACE TO SAVE THE PLANET** will be broadcast nationwide on PBS in the fall of 1990. Extensive educational materials will be available for this series, including a 13-week undergraduate level telecourse supported by The Annenberg/CPB Project and a separate high school teacher's guide prepared by WGBH. For information on the telecourse, call 1-800-LEARNER. For further information about **RACE TO SAVE THE PLANET,** and suggestions of ways in which you or your organization may plan activities in your community related to these programs, please contact the Global Tomorrow Coalition, 1325 G Street, N.W., Suite 915, Washington, D.C. 20005-3104.

Acknowledgments

The Global Tomorrow Coalition is indebted to many people who helped make *The Global Ecology Handbook* a reality. The book truly is the product of a cooperative, voluntary effort by many organizations and individuals associated with the Coalition.

Many of the chapters draw extensively on publications of the World Resources Institute, especially its *World Resources* series, and on the Worldwatch Institute's *Worldwatch Papers* and annual *State of the World* series. Special recognition and appreciation go to World Resources Institute and its President, Gus Speth, and to Worldwatch Institute and its President, Lester Brown.

Don Lesh, the Coalition's President, contributed sections to several chapters, and provided extensive editorial assistance and basic guidance throughout work on the *Handbook*. Diane Lowrie, Vice President of the Coalition, drafted the final chapter, "What You Can Do," and contributed to the "What You Can Do" sections of all the issue chapters. As consultants, Hal Kane and Linda Starke reviewed and edited many of the chapters and provided valuable advice on content. Walter Corson, Senior Associate in the Coalition's office, wrote a number of chapter drafts, made extensive contributions to all the chapters, prepared the graphics, worked closely with consultants and reviewers, and served as overall coordinator and editor of the volume.

In addition, several other people prepared drafts of chapters. They include: Ann Dorr (Nonfuel Minerals and Solid Waste Management), Hal Kane (Development and the Environment), Charles Loeb (Hazardous Substances and Global Security), Charles McKay (Energy), Mukami Mwiraria (Food and Agriculture), Lynnea Salvo (Tropical Forests), and Elaine Sehrt-Green (Fresh Water). Other individuals who made important contributions to chapter drafts were Jill Antal, Belinda Berg, Ann Corson, Larry Martin, Christina Nichols, Niklaus Steiner, and Mark Valentine.

Each chapter was reviewed by outside writers and experts, who provided valuable criticism, suggestions, and often extensive information that was incorporated in the text. They include: Joan Banfield, Mary Barberis, Gerald Barney, Faith Campbell, John Catena, Ned Dearborn, Eric Draper, Sara Ebenreck, Thomas Gire, Judith Gradwohl, Lindsey Grant, Russell Greenberg, Morgan Gopnik, Carl Haub, Peter Hazlewood, Nicholas Lenssen, Joan Martin-Brown, Alan Miller, David Pimentel, Peter Raven, Michael Renner, Robert Ridky, William Walsh, Michael Weber, Paul Werbos, and Edward Wolf.

There are many others who provided data, research material, advice, and valuable help and support in preparing the book for publication. The Coalition is grateful for their generous assistance and happy to recognize them here: Dirk Bryant, Michael Carrigg, Ryan Cartnal, Robyn Cutler, Craig Downer, Geri Faron, Ruth Flanagan, Jan Hartke, Alison Hilton, Sonya Horowitz, Frazier Kellogg, Pamela Leonard, Douglas Liner, Nicole McCrea, Susie Morris, Mary Paden, Mary Roberts, Susan Terry, and Mark Trexler.

Special acknowledgment and appreciation go to Terry D'Addio of the Coalition's staff for production coordination, Sally Anderson for computer typesetting and page design, Trish Dinkel Crowe for layout, Carol Connett for illustrations, and Robert Pollard and William Lowrie for technical assistance.

While gratefully recognizing the significant contributions and support of all those cited above, the Global Tomorrow Coalition accepts full responsibility for the final form and content of *The Global Ecology Handbook*.

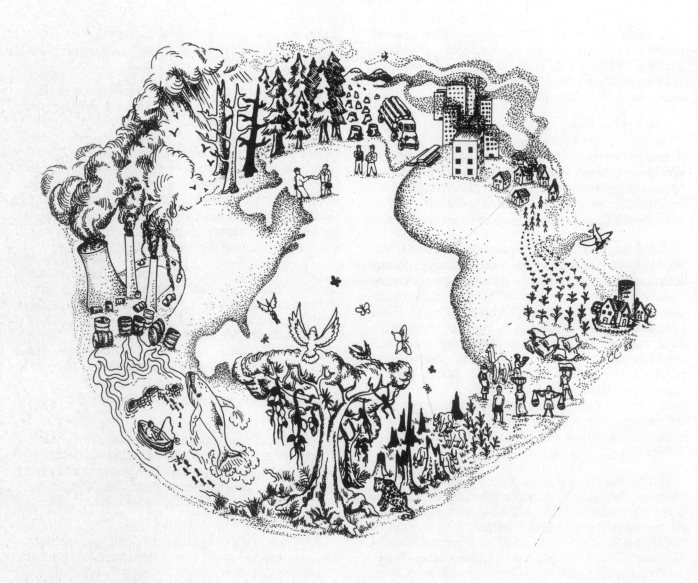

A Global Awakening:
Growing Interdependence, Emerging Challenges

We recognize that poverty, environmental degradation, and population growth are inextricably related and that none of these fundamental problems can be successfully addressed in isolation.

Making Common Cause, A Statement and Action Plan
by U.S.-Based International Development, Environment, and Population NGOs[1]

Mankind is rapidly...altering the basic physiology of the planet.

—Jessica Tuchman Mathews, World Resources Institute[2]

Increasing human demands are damaging the natural resource base—land, water and air—on which all life depends.

—*The State of World Population 1988*[3]

There are many signs that the next general international crisis is going to be about the environment. There have been warnings about environmental abuse for decades, but concerns were separated from high politics and security. Now convergence has begun.

—Flora Lewis, "The Next Big Crisis"[4]

There is something fundamentally wrong in treating the Earth as if it were a business in liquidation.

—Herman Daly, Economist, World Bank[5]

A Global Awakening

Planet Earth is perhaps 4.5 billion years old, and life on Earth has existed for more than 3.5 billion years. Humans have been on Earth for some 2 to 3 million years, living in equilibrium with other life forms. Only within the last 200 years have people begun to affect the global environment significantly, and only in the last 40 years or so has our impact become serious.[6]

As Jessica Tuchman Mathews has put it, humankind is still utterly dependent on nature, but now has for the first time the ability to alter the natural world, rapidly and on a global scale. And George Woodwell has observed that until recently, the environment has been large in proportion to the demands that we have put on it, but now human influences are large.[7]

As the impacts of human influences have grown, so have the risks associated with those impacts. New technologies carry increasing risks, and the scale, frequency, and impact of disasters caused or influenced by human activity are growing. The risks of irreversible damage to the Earth's natural systems are becoming significant.[8]

During the 1980s, evidence of human impacts on the global environment has been accumulating, and in 1988, some of it became highly visible and received wide coverage in the media. *Time* Magazine's essay, "Planet of the Year" referred to 1988 as the year that "the Earth spoke." Its message took many forms, including heat waves, drought, crop losses, forest fires, violent floods and hurricanes, polluted oceans and beaches, and depletion of the ozone layer. As Worldwatch Institute noted in its publication *State of the World 1989*, historians may look back on this period as a turning point for public concern about the environment.[9]

Only in the last two years or so have the citizens and leaders of many countries begun fully to understand the consequences of human impacts on the environment—and their potential threat to our security, economic productivity, health, and quality of life, as well as to those of future generations—and to grasp the need for urgent and concerted action.[10] Only gradually are we starting to perceive the importance of new concepts and challenges and the promise of new opportunities. We are in the midst of what can truly be called a global awakening, the long-term consequences of which are still unclear, but which almost certainly will include historic change in the structures of societies and governments, in levels of multilateral commitment and involvement, in patterns and directions of economic activity, and in the life styles, rights, and responsibilities of the individual.[11]

This book is intended as a working guide, a handbook for the concerned and responsible citizen, as we enter this newly emerging world in which decisions at every level will be shaped increasingly by the hard realities of interdependence and motivated by the goals of global survival and progress.

Growing Global Interdependence

Until recently, ecological and natural resource problems tended to be localized. Soil erosion, for example, historically has been a local problem: civilizations whose agriculture was undermined by erosion and loss of fertility declined in relative isolation or migrated to new lands. But in today's interdependent world economy, food and energy are global commodities: shortages or price changes in one region can have global implications. A country that loses soil fertility may have to import more food, thereby causing pressure on farmlands elsewhere. Oil price increases by major suppliers have had profound economic impacts on oil-importing countries worldwide.[12]

Parallel to increasing economic interdependence is the planet's growing environmental interdependence. The links among the Earth's natural systems of land, water, air, and living matter are often global. Disturbance to any one of them can affect the others in complex and unexpected ways that can be distant in both space and time. For example, deforestation in India and Nepal leads to catastrophic flooding in Bangladesh, emissions of industrial chemicals deplete the Earth's protective ozone layer, and fossil fuel use damages distant forests and contributes to global climate change.[13]

All life is dependent on the planet's land, water, and air, and the quality of the environment influences virtually all aspects of human activity—health and welfare, employment and recreation, cities and villages, industry and agriculture. The environment also affects all groups in society—producers and consumers, rich and poor, women and men, young and old.[14]

The Earth's social, ecological and resource problems are highly interdependent. Issues of poverty, rapid population growth, resource depletion, and environmental degradation are all closely related. Increasing human numbers and widening disparities in economic and political power contribute to resource depletion and environmental damage, and degraded resources and environmental quality affect the lives of people. Environmental degradation can contribute to economic and political instability.[15]

These problems, individually and collectively, threaten our security in ways that many current policies and structures of governance may not effectively address, and may even make worse.[16]

The World Commission on Environment and Development

In response to concern about the growing impacts of human activity on the Earth's natural resources, the World Commission on Environment and Development (WCED) was created in late 1983 as an independent body by the United Nations Environment Program through the U.N. General Assembly. The mission of this diverse group of 22 eminent persons, chaired by Prime Minister Gro Harlem Brundtland of Norway, was to reexamine the critical environment and development problems of the planet and to formulate realistic proposals to solve them. For three years, the WCED conducted inquiries and public hearings around the world, commissioned special studies by experts, and consulted leaders in politics, business, education, science, and development.

In 1987, the Commission released its unanimous report, *Our Common Future*, which documented both successes and failures in global development. The WCED found some good news: human life expectancy is rising, infant mortality is dropping, adult literacy is climbing, scientific and technical innovations are promising, and global food output is growing faster than world population.[17]

At the same time, the WCED report confirmed much bad news: topsoil is eroding and deserts are expanding, forests are dying and disappearing, air pollution is warming the Earth and depleting its protective ozone shield, development programs are failing to close the gap between rich and poor, and industry and agriculture are putting toxic substances into the food chain and groundwater supplies.[18]

The report characterized the situation in this way: "When the century began, neither human numbers nor technology had the power radically to alter planetary systems. As the century closes, not only do vastly increased human numbers and their activities have that power, but major, unintended changes are occurring in the atmosphere, in soils, in waters, among plants and animals, and in the relationships among all of these. The rate of change is outstripping the ability of scientific disciplines and our current capabilities to assess and advise. It is frustrating the attempts of political and economic institutions, which evolved in a different, more fragmented world, to adapt and cope."[19]

Nevertheless, the outstanding contribution of *Our Common Future* has not been the scope and content of its analysis of the global situation, however valuable and well documented, but the forceful introduction into world dialogue and debate of the concept of sustainable development. The WCED report described sustainable development as "...not a fixed state of harmony, but rather a process of change in which the exploitation of resources, the direction of investments, the orientation of technological development, and institutional change are made consistent with future as well as present needs."[20]

By calling for a new and fundamentally different era of sustainable economic growth to combat poverty and improve the human condition worldwide, the WCED has made it possible to engage an extremely wide spectrum of experiences, perspectives, talents, and interests in search for creative and successful solutions to our common problems.

The breadth of that inquiry, the spirit of *Our Common Future*, and the promise of the goal of sustainable development have animated this entire publication.

Crossing Ecological Thresholds

In its *State of the World* report for 1987, the Worldwatch Institute said that human use of land, water, air, forests, and other natural resources that support life on Earth is causing irreversible changes in those resources—such as soil erosion, groundwater contamination, atmospheric carbon dioxide buildup, stratospheric ozone depletion, and extinction of plants and animals. The report maintained that such changes are pushing natural resources over "thresholds" beyond which they cannot absorb such use without permanent damage.[21]

In 1988, Worldwatch Institute's *State of the World* report documented the extent of some of the changes in the Earth's physical condition, including loss of topsoil and forest cover, spread of deserts, disappearance of plant and animal species, depletion and contamination of groundwater, depletion of the ozone layer, and projected climate changes.[22]

The growth of human numbers and their impacts on the Earth's resources have greatly accelerated since World War II. The production of food, energy, and industrial commodities is associated with much of the deterioration of the Earth's life-support system. Between 1950 and 1986, while the world's population doubled, world grain consumption increased 2.6-fold, energy use grew 3.7-fold, economic output quadrupled, and the production of manufactured goods increased sevenfold. During the same period, U.S. production of synthetic organic chemicals, a major source of water and air pollution, increased more than ninefold. Humans now consume, directly or indirectly, about 40 percent of all the food energy potentially available on land.[23]

While there are no precise limits to population growth or resource use beyond which ecological disaster is unavoidable, there are clearly limits to the use of nonrenewable energy, land, water, and other resources in terms of rising costs, diminishing returns, and deteriorating quality rather than in the form of a sudden resource loss.[24] Current trends in the supply and quality of petroleum, cropland, and groundwater graphically foretell such limits.

Indicators of Global Problems and Global Change

The fourteen "issue" chapters in this book document major changes resulting from the rapid growth of human numbers and their impacts on the Earth's resources. This section summarizes those changes and impacts,[25] and the next section considers some of the underlying causes of the problems.

Foresight Capability

Most nations lack the ability to monitor change, to make long-term projections of trends, and to anticipate the impact of interactions among factors such as population size; supplies of food, water, energy, and other resources; and environmental quality. Most nations, including the United States, fail to link the results of any existing projections to current decision-making.

Population Growth

The world's population, currently 5.2 billion, has grown from about 3 billion in 1960, and around 2 billion in 1925. Today it increases by almost 90 million each year, and is likely to reach 10 billion by the year 2025 unless birth control use increases dramatically. At present, less than half of the world's women of reproductive age use birth control measures. If current growth rates continue, many countries in Africa and several in Latin America will double their populations in less than 25 years. Countries with rapid population growth rank low on measures of the physical quality of life and high on measures of human suffering. Rapid growth is often accompanied by severe environmental degradation, including deforestation, desertification, and soil erosion. Given the current rates of degradation in a world of 5 billion people, prospects for environmental protection with a population of 10 billion—only a little more than a generation ahead—are not bright.

Development and Environment

Since World War II, great efforts and many billions of dollars have been devoted internationally to the cause of economic development. The world community has designated successive "decades" of development. Yet, in 1987, developing countries, with 77 percent of the world's population, had an average income per person of only $670 a year, less than 6 percent of the average income of $12,070 in industrial nations. The rich-poor gap is widening; in 1977, developing countries' average per capita income was $490 a year, more than 9 percent of the average income of $5,210 in industrial nations. Third World debt is now more than $1.3 trillion, and in 1988, developing countries' debt payments were more than $30 billion greater than the loans they received. In different ways, development in both rich and poor nations is seriously degrading the planet's land, forests, water, and air, on which all life depends.

Food and Agriculture

Despite aggregate global trends in rising food production, nearly a billion people—almost one-fifth of the world's population—do not consume enough calories for an active working life. Soil erosion is seriously degrading croplands in most of the world's important agricultural regions, the world's grain acreage is diminishing, and grain production per person is declining in Africa, India, and Latin America.

Inefficient or inappropriate crop irrigation is depleting underground water supplies and damaging soils in a number of areas, and agricultural chemicals are polluting groundwater and surface waters and harming people and wildlife in some regions.

Biological Diversity

As tropical forests and other biologically-rich habitats are destroyed or degraded under the pressures of human population growth and economic activity, the normal rates of extinction of plant and animal species are accelerating. We simply do not know the extent of this phenomenon, but experts estimate that we may be losing several thousand species each year, and one-fifth of all species could disappear within the next 20 years. The existing network of parks and reserves established to protect the world's wild plants and animals meets less than one-third of estimated needs.

Tropical Forests

Moist tropical forests—containing at least two-thirds of all plant and animal species, and the source of many foods, medicines, and industrial products—are being destroyed at the rate of at least 11 million hectares each year, an area the size of Pennsylvania. About half of all tropical forests are already gone. Forest loss affects hundreds of millions of people through increased flooding, soil erosion and silting of waterways, drought, shortages of fuelwood and timber, and displacement of societies and cultures.

Ocean and Coastal Resources

The world fish catch is rapidly approaching the annual total—100 million tons—that scientists believe is the maximum sustainable yield. Already, yields from a number of major ocean fisheries have leveled off or are declining, and some have collapsed due to overfishing. Excessive harvesting of whales has depleted a number of species to near extinction. Pollution by oil and municipal and industrial wastes, and development of coastal areas have seriously damaged wetlands, estuaries, mangroves, coral reefs, and other highly productive marine areas.

Fresh Water

In developing nations, only about half the people have access to safe drinking water. Worldwide, more than 10 million deaths each year result from waterborne diseases. In industrial nations, surface and underground water supplies are being polluted by municipal and industrial wastes, and by surface runoff from agricultural and urban areas containing nitrates, pesticides, and other toxic substances. Heavy demands for water from agriculture, municipalities, and industry are rapidly depleting groundwater supplies in parts of Africa, China, India, the United States, and other areas.

Nonfuel Minerals

Global demand for mineral commodities is growing exponentially due to increasing population and rising consumption per person. Some minerals are plentiful, while others are relatively scarce. At current rates of use, estimated life expectancies for mineral supplies vary from almost infinite for salt and magnesium metal to less than 50 years for copper, mercury, zinc, and lead. In some areas, the production of mineral commodities has caused serious environmental damage, including air and water pollution from processing and land degradation from mining.

Energy

Since World War II, the world's energy consumption has quadrupled. At the end of 1986, proven world oil reserves were sufficient to last only about 33 years at 1986 production rates; proven natural gas reserves were projected to last about 59 years. Pollution from growing fossil fuel use is threatening air quality, health, vegetation, and climate stability. The future of nuclear power is clouded by serious safety concerns, toxic waste disposal problems, and escalating costs. Centralized energy systems in many nations are vulnerable to technological failure and sabotage. Inflexible laws and regulations, often bolstered by hidden subsidies, hinder the development of renewable and more environmentally benign energy resources on a competitive basis.

Air, Atmosphere, and Climate

Energy use, industrial production, and deforestation all contribute to air pollution that is harming plants, animals, and human health, and altering the global atmosphere. Sulfur and nitrogen oxides, ozone, and other pollutants are causing acid precipitation and lowering crop yields; air pollutants have damaged more than 30 million hectares of forest in industrial countries. Chlorofluorocarbons and other pollutants have depleted the Earth's protective ozone layer by an average 2 percent worldwide, and by nearly 40 percent during certain months over Antarctica. Atmospheric carbon dioxide levels have increased by 25 percent since before the industrial revolution, and more than 10 percent in the last 30 years. This increase, along with other heat-absorbing air pollutants, is causing a "greenhouse effect" that has apparently raised the Earth's average temperature nearly 1 degree Fahrenheit over the last 100 years, and now threatens to alter the Earth's climate and raise ocean levels, with the potential for economic and ecological disruption on a vast scale.

Hazardous Substances

Since World War II, the production of petrochemicals, pesticides, and other organic chemicals with toxic health effects has skyrocketed. U.S. production of synthetic organic chemicals, many of which are toxic, increased fifteenfold between 1945 and 1985, from 6.7 million metric tons to 102 million. Each year the United States produces more than 260 million tons of hazardous waste; the global total is several times that amount. Exposure to pesticides and other toxic chemicals can occur at worksites such as farms and industrial plants, through residues on food, and through contaminated drinking water. The effects of protracted exposure can include cancer, gastrointestinal and neurological damage, birth defects, and death. Each year, between 400,000 and 2 million pesticide poisonings occur worldwide; most are among farmers in developing countries. Hazardous radioactive wastes and effluents from nuclear power and weapon facilities have contaminated the soil, water, and air around nuclear facilities and are believed to cause cancer. No satisfactory program for safe, long-term nuclear waste disposal has been implemented.

Solid Waste Management

Worldwide, urban residents produce between 1 and 4 pounds of solid waste per person each day. The United States generates at least 140 million tons of municipal solid waste each year; about 80 percent of it is dumped in landfills. In 1978, the United States had about 20,000 landfills, but now only about 6,000 remain unfilled and open. Between now and 1993, some 2,000 landfills will be filled and closed. Solid waste disposal is causing serious pollution of groundwater and surface water in the United States and other industrial countries, and important opportunities for waste reduction and the recycling and recovery of valuable materials are being missed.

Global Security

Since the end of World War II, there have been some 150 armed conflicts, mostly in the Third World, that have taken a total of nearly 20 million lives. Worldwide, military spending is close to 1 trillion dollars a year. Government spending on health and on education both fall short of military spending in most nations, especially in the Third World. The superpowers and their allies possess some 50,000 nuclear warheads, with a combined destructive potential nearly 70 times that needed to destroy all the world's large and medium-sized cities. Scientists believe that a major nuclear war could kill a billion people and put a majority of the world's population at risk of starvation. In addition to the five declared nuclear powers, several other nations have the capability to build nuclear weapons, and the list is growing.

Problem Impacts and Problem Causes

Drawing on analysis of the global problems covered in this book and summarized above, Table 1.1 illustrates some of the relationships between major impacts and important causes of worldwide problems, and indicates the relative importance of these underlying causes. The symbols in

Table 1.1 (see page 7) depict estimates of the likely importance of the problem causes, as follows:[26]

- ●   very important cause
- ◗   moderately important cause
- ○   less important but significant cause
- –   unimportant or insignificant cause

In Table 1.1, the term "unsustainable" as applied to population growth, food production, energy use, and industrial production means that these activities diminish the Earth's long-term capacity to support life, or endanger the natural systems (including land, water, air, and living matter) that support life.[27]

Table 1.1 reflects the following conclusions:

- *Unsustainable population growth* (that is, growth of human population that diminishes the Earth's long-term capacity to support life) is an important underlying cause of all eight problem impacts.[28]

- *Poverty and inequalities* (that is, inequitable access to factors such as land, food, shelter, health care, education, employment, and political power) are important causes of unmet basic human needs, depletion of plant and animal species, land degradation, and conflict and war. They are less important but significant causes of water and air pollution.[29]

- *Unsustainable agriculture* (that is, reliance on methods of food production that emphasize maximum short-term yields while causing environmental damage and long-term loss of natural productivity) is an important or significant cause of all problem impacts.[30]

- *Unsustainable energy use* (use that is inefficient, wastes nonrenewable energy sources, releases harmful effluents, or causes deforestation or other environmental damage) is an important cause of air pollution, depletion of nonrenewable energy sources, and land degradation. It is a less important but significant cause of habitat degradation, water pollution, and conflict.[31]

- *Unsustainable industrial production* (production that is inefficient, wastes primary resources, produces harmful effluents, or causes other environmental damage) is an important cause of energy, mineral, and water depletion, and air and water pollution. It is a less important but significant cause of habitat degradation and conflict.[32]

Based on the estimates in Table 1.1, unsustainable population growth is the single most important causal factor across the entire range of problem impacts listed.

Unsustainable food production is the next most important cause across all impacts, followed by unsustainable industrial production, poverty and inequality, and unsustainable energy use.

For each problem impact, the estimates in Table 1.1 suggest which of the five problem causes are most important. For example, for both unmet needs and species depletion, population growth and poverty are estimated to be very important causes. For land degradation, population growth and food production are very important. For water depletion, food production is a very important cause; and for water pollution, industrial production is very important. For air pollution, both energy use and industrial production are judged to be very important causal factors.

Progress Toward Sustainable Development

In the final chapter of the book, these five causal factors— rapid population growth, poverty and inequality, and unsustainable agriculture, energy use, and industrial production — are included in a series of promising solutions to the global problems discussed in the book's fourteen issue chapters. The final chapter examines the potential of these solutions for alleviating the major problem impacts listed above, and summarizes the progress being made around the world, as reported in the issue chapters, in implementing solutions and moving toward the goal of sustainable development.

The Role of the Individual

This handbook has been produced by the Global Tomorrow Coalition in the belief that true progress toward the solution of long-term global problems, and realization of the potential benefits and opportunities inherent in the new world to which we are awakening, depends on the individual—on *you*, the reader, and the extended circle of family, friends, communities, groups, and institutions which you can influence. In coming years, new international agreements or organizations may be created. New national, regional, and local policies may be instituted. New measures and patterns for economic activity may be defined, and new patterns of education and citizen involvement may be developed. All of this will rely on a foundation of public consensus of which you can be a part.

That is why a section of each chapter in this book, and an entire concluding chapter, are devoted to the theme of "what you can do." However complex and imposing the problems analyzed here may appear, however challenging and demanding the solutions suggested may be, the Global Tomorrow Coalition and its members are convinced that the informed and mobilized individual is where the action starts. "Citizen power" is real, and the time to start exerting that power to achieve a more sustainable national and global future, is now.

Table 1.1
Global Problem Impacts and Problem Causes

Problem Impacts	Problem Causes				
	Unsustainable Population Growth	Poverty and Inequality	Unsustainable Food Production	Unsustainable Energy Use	Unsustainable Industrial Production
Unmet basic human needs for safe water, food, shelter, health care, education, employment, etc.	●	●	○	–	–
Species depletion (extinction of plants and animals), habitat degradation	●	●	◗	○	○
Land degradation: soil erosion, desertification, loss of soil fertility	●	◗	●	◗	–
Depletion of nonrenewable energy and minerals	◗	–	◗	●	◗
Depletion of fresh water (groundwater and surface water)	◗	–	●	–	◗
Water pollution: chemical and bacterial contamination of groundwater and surface water	◗	○	◗	○	●
Air pollution: urban air pollution, acid deposition, ozone layer depletion, greenhouse gas buildup	◗	○	○	●	●
Conflict and war: domestic and international	◗	◗	○	○	○

●	very important cause
◗	moderately important cause
○	less important but significant cause
–	unimportant or insignificant cause

Further Information

A number of organizations have programs or can provide materials related to the wide range of population, resource, environment, and development issues covered in this book.*

Books

Botkin, Daniel B. et al. *Changing the Global Environment: Perspectives on Human Involvement*. San Diego, CA: Academic Press, 1989. 459 pp. Hardback, $49.95. Examines ecological problems and possible solutions sug-

---

* These include: Conservation Foundation, Environmental Defense Fund, Environmental Policy Institute, Friends of the Earth, Global Tomorrow Coalition, National Audubon Society, National Wildlife Federation, Natural Resources Defense Council, Population-Environment Balance, Sierra Club, U.S. Agency for International Development, World Bank, World Resources Institute, Worldwatch Institute, and WorldWIDE—World Women in Development.

gested by new techniques such as remote sensing and computer-based data systems.

Brown, Lester R. et al. *State of the World 1989*. A Worldwatch Institute Report on Progress Toward a Sustainable Society. New York: Norton, 1989. 256 pp. Paperback, $9.95. Published annually since 1984, the volumes contain chapters on most of the issues related to population, resources, environment, and development.

Conservation Foundation, The. *State of the Environment: A View Toward the Nineties*. Washington, DC: The Conservation Foundation, 1987. 614 pp. Paperback, $19.95. A comprehensive review of U.S. progress in improving environmental conditions and managing natural resources.

Council on Environmental Quality. *Environmental Quality 1987-1988*. Annual Report. Washington, DC: U.S. Government Printing Office, 1989. 425 pp. Paperback, $13.00. Includes chapters on municipal solid waste, urban air quality, and water resources, and an extensive appendix on environmental trends in the United States.

Davis, Kingsley et al. (eds.). *Population and Resources in a Changing World*. Current Readings. Stanford, CA: Morrison Institute for Population and Resource Studies, 1989. 532 pp. Paperback, $15.00. Available from Stanford University Bookstore, Stanford, CA 95305. In addition to population, contains selections on energy, fresh water, land, food, biological resources, and the atmosphere.

Ehrlich, Anne H. and Paul R. Ehrlich. *Earth*. New York: Franklin Watts, 1987. 258 pp. Hardback, $19.95. Reviews the origins, nature, and extent of environmental changes brought about by human action.

El-Hinnawi, Essam and Mansur H. Hashmi. *The State of the Environment*. A Publication of the United Nations Environment Program. London: Butterworth Scientific, 1987. 182 pp. Hardback. A valuable review of contemporary and future environmental issues, especially those of global significance.

Freedman, Bill. *Environmental Ecology: The Impacts of Pollution and Other Stresses on Ecosystem Structure and Function*. San Diego: Academic Press, 1989. 424 pp. $39.95.

The Global 2000 Report to the President. Gerald O. Barney, Study Director. Council on Environmental Quality and Department of State. Washington, DC: U.S. Government Printing Office, 1980. Volume I: *Entering the Twenty-First Century*. A summary report. 47 pp. $5.00. Volume II: *The Technical Report*. 766 pp. $14.00. Both volumes published together by Penguin Books. Paperback, $12.95. A survey of the probable changes in the world's population, natural resources, and environment through the end of the century.

The Global Tomorrow Coalition. *Sustainable Development: A Guide to Our Common Future*, the Report of the World Commission on Environment and Development. Washington, DC, 1989. 77 pp. Paperback. Available from the Global Tomorrow Coalition for $2.00 to cover shipping and handling.

Goldsmith, Edward and Nicholas Hildyard (eds.). *The Earth Report: The Essential Guide to Global Ecological Issues*. Los Angeles: Price Stern Sloan, 1988. 240 pp. Paperback, $12.95. Part I includes essays on drinking water, acid rain and forest decline, atmosphere and climate, nuclear energy, and international food aid. Part II is an encyclopedic listing of more than 400 ecological facts, concepts, and key words and names.

Johnston, R.J. and P.J. Taylor (eds.). *A World in Crisis?: Geographical Perspectives*. Cambridge, MA: Basil Blackwell, 1989. 371 pp. Paperback, $17.95. Shows the links between global, regional, and local problems, and demonstrates that these problems have economic, ecological, political, and social components.

Kidron, Michael and Ronald Segal. *The New State of the World Atlas*. Revised Edition. New York: Simon and Schuster, 1987. Paperback, $12.95. Uses text, maps, and other graphics to provide a vivid picture of the economic, political, and environmental state of the world.

Marien, Michael. *Future Survey Annual 1988-89*. Volume 9. A Guide to the Recent Literature of Trends, Forecasts, and Policy Proposals. Bethesda, MD: World Future Society, 1989. 212 pp. Paperback, $25.00. Contains abstracts of recent publications on economics, energy, environment and resources, food and agriculture, science and technology, world futures, and many other topics.

Myers, Norman (ed.). *Gaia: An Atlas of Planet Management*. Garden City, NY: Anchor Books, 1984. 272 pp. Paperback, $18.95. Contains a wealth of data, vivid graphics, and authoritative text based on contributions from more than 100 authorities on the world's critical environmental, political, and social issues.

Myers, Norman. *Not Far Afield: U.S. Interests and the Global Environment*. Washington, DC: World Resources Institute, 1987. 73 pp. Paperback, $10.00. Includes chapters on environment, resource, and population issues; security; Caribbean Basin issues; and policy implications.

National Geographic Society. *Earth '88: Changing Geographic Perspectives*. Washington, DC, 1988. 392 pp. Hardback, $20.00. Proceedings of the Society's centennial symposium with papers on 22 topics including agriculture, biological diversity, climate, deforestation, desertification, energy, oceans, pollution, population, and water.

PBS Environmental Resource Compendium. Can be obtained by sending a money order or check for $10 and a completed mailing label to PBS Elementary/Secondary Services, Dept. GBH, 1320 Braddock Place, Alexandria, VA 22314.

Rambler, Mitchell et al. (eds.). *Global Ecology: Toward a Science of the Biosphere.* San Diego, CA: Academic Press, 1989. 204 pp. Hardback, $24.95. Transcends the boundaries of academic disciplines to show the various facets of the global system within a planetary perspective.

Repetto, Robert (ed.). *The Global Possible: Resources, Development, and the New Century.* A World Resources Institute Book. New Haven: Yale University Press, 1985. 538 pp. $14.95. Gives a broad overview of the state of the world's threatened resources, and proposes realistic and politically practical corrective measures.

Sadik, Nafis. *The State of the World Population 1988.* New York: United Nations Population Fund, 1988. 21 pp.

Speth, James Gustave. *Environmental Pollution: A Long-Term Perspective.* Washington, DC: World Resources Institute, 1988. 24 pp. Paperback, $5.00. Examines four long-term trends in pollution and identifies social and technological transitions needed to deal with pollution in coming decades.

Weber, Susan (ed.). *USA by Numbers: A Statistical Portrait of the United States.* Washington, DC: Zero Population Growth, 1988. 164 pp. Paperback, $8.95. Contains extensive data on U.S. population-linked social, economic, and environmental indicators.

World Commission on Environment and Development. *Our Common Future.* New York: Oxford University Press, 1987. 383 pp. Paperback, $10.95. An authoritative overview of the state of the Earth's natural resources and the environmental impacts of economic and social development; contains a comprehensive series of recommendations for achieving sustainable development.

World Resources Institute and International Institute for Environment and Development. *World Resources 1988-89.* An Assessment of the Resource Base that Supports the Global Economy. New York: Basic Books, 1988. 372 pp. Paperback, $16.95. Contains sections on population and health, human settlements, food and agriculture, forests and rangelands, wildlife and habitat, energy, fresh water, oceans and coasts, atmosphere and climate, global systems and cycles, and policies and institutions. See also *World Resources 1986* and *World Resources 1987.*

Worster, Donald (ed.). *The Ends of the Earth: Perspectives on Modern Environmental History.* New York: Cambridge University Press, 1988. 341 pp. $12.95. Reviews the history of human impacts on the environment and the evolution of attitudes toward the environment.

Articles

Benedick, Richard E. "Population-Environment Linkages and Sustainable Development." *Populi,* Journal of the United Nations Population Fund, Vol. 15, No. 3, 1988, pp. 14-21.

Brown, Lester R. and Christopher Flavin. "The Earth's Vital Signs." In Brown et al., *State of the World 1988* (New York: Norton, 1988), pp. 3-21.

Brown, Lester R. and Sandra Postel. "Thresholds of Change." In Brown et al., *State of the World 1987* (New York: Norton, 1987), pp. 3-19.

Brown, Lester R. et al. "A World at Risk." In Brown et al., *State of the World 1989* (New York: Norton, 1989), pp. 3-20.

Cooper, Kate and Jeff Smoller. "Trends in the Global Environment," *Trends Bulletin,* Trends Analysis Group, Wisconsin Department of Natural Resources (DNR), August 1986. 19 pp. Examines how global environmental trends affect the state of Wisconsin. Available from DNR, P.O. Box 7921, Madison, WI 53707.

Foreign Policy Association. "The Global Environment: Reassessing the Threat," *Great Decisions 1988.* New York, 1988, pp. 62-71. Paperback, $8.00.

Martin-Brown, Joan. "Converging Worlds: The Implications of Environmental Events for the Free Market and Foreign Policy Developments." *The Environmentalist,* Vol. 4, No. 2, 1984, pp. 139-42.

Mathews, Jessica Tuchman. "Redefining Security," *Foreign Affairs,* Spring 1989, pp. 162-77.

"Population, Resources, and Environment," *Ambio: A Journal of the Human Environment,* Vol. 13, No. 3, 1984. Available from Pergamon Press, Fairview Road, Elmsford, NY 10523.

Repetto, Robert. "Population, Resources, Environment: An Uncertain Future," *Population Bulletin,* Vol. 42, No. 2, July 1987. 44 pp. Published by Population Reference Bureau.

Time Magazine. "Planet of the Year: Endangered Earth." January 2, 1989, pp. 24-63.

Periodicals

Ambio: A Journal of the Human Environment. Published eight times a year by Pergamon Press, Maxwell House, Fairview Park, Elmsford, NY 10523. Subscription $43 a year.

The Amicus Journal. Published quarterly by the Natural Resources Defense Council.* Subscription $10 a year.

Audubon Action. Published bimonthly by National Audubon Society.* Subscription $9 a year.

Conservation Exchange. Published quarterly by the National Wildlife Federation.*

EDF Letter. Published bimonthly by Environmental Defense Fund.* Subscription is included in the annual membership fee of $20.

Environment. Published 10 times a year by Heldref Publications, 4000 Albemarle Street, Washington, DC 20016. Subscription $23 a year.

Future Survey. A Monthly Abstract of Books, Articles, and Reports Concerning Forecasts, Trends, and Ideas about the Future. Published monthly by the World Future Society.* Subscription $59 per year; includes *Future Survey Annual* (see listing under Books).

Not Man Apart. Published bimonthly by Friends of the Earth.* Subscription is included in the annual membership fee of $25.

The Renew America Report. Published quarterly by Renew America.* Subscription is included in the annual membership fee of $25.

Sierra. Published bimonthly by the Sierra Club.* Subscription $15 a year.

World Watch. Published six times a year by Worldwatch Institute.* Subscription $20 a year.

Worldwatch Papers. Published monthly by Worldwatch Institute.* Annual subscription $25, which includes the annual publication, *State of the World.*

Films and Other Audiovisual Materials

Fragile Mountain. 1982. Depicts population pressures and environmental degradation in Nepal. 55 min. 16 mm. color film and video. Available from Sandra Nichols Productions, 502 Tideway Drive, Alameda, CA 94501.

Maragoli. 1977. Shows the social and economic dimensions of the population problem as seen by villagers in Kenya. 58 min. 16 mm. color film $505, video $435,

rental, $32. University of California Extension Media Center.

Our Common Future. 1988. Presents the findings of the World Commission on Environment and Development, set up by the United Nations in 1983 to re-examine the planet's critical environment and development problems and formulate realistic proposals to solve them. Each cassette contains a 17-minute and a 12-minute version of the video; the former features Commission Chairperson Gro Harlem Brundtland presenting the report at a London press conference. Available in English or Spanish for purchase ($25.00) or rental ($7.50) from Global Tomorrow Coalition.

Population-Environment Film Catalog. An annotated list of films available for rental from Population Reference Bureau.

Race to Save the Planet. Television series and course, based on the 10-part series produced by WGBH Television in Boston, that explores all the issues posed in *The Global Ecology Handbook* and more. To purchase videocassettes, off-air taping license, duplication rights, or a television course license, call 1-800-LEARNER.

What is the Limit? 1987. Discusses interrelations between human population growth, environmental degradation, resource depletion, habitat destruction, and ethical considerations for the future. Includes a discussion guide, "Where Do We Go from Here?" 23 min. Grade 10 to adult. Video: VHS, Beta, or 3/4 inch, $25. National Audubon Society.

Teaching Aids

Christensen, John W. *Global Science: Energy, Resources, Environment.* 2nd Edition. Dubuque, IA: Kendall-Hunt, 1984. Textbook, 355 pp.; Laboratory Manual, 265 pp.; Teacher's Guide, 356 pp. A new edition is in preparation.

Global Primer: Skills for a Changing World, grades K-8, $29.95; *The New State of the World Atlas,* an activities guide, grades 7-12, $29.95; *Teaching Global Awareness Using the Media,* grades 6-12, $21.95. The Center for Teaching International Relations, University of Denver, Denver, CO 80208.

Global 2000 Countdown Kit. 1982. Contains 14 units based on the major topics in the 1980 *Global 2000 Report to the President,* including population, income, food, fisheries, forests, water, nonfuel minerals, energy, agriculture, climate, species extinctions, and others.

---

* Addresses for these publications are given in the Appendix.

Designed for independent study use with minimal teacher guidance, grades 9-12. Zero Population Growth. $19.95.

Human Needs and Nature's Balance: Population, Resources, and the Environment. Population Learning Series, October 1987. 13 pp. Includes teacher's guide. Population Reference Bureau. $2.00.

Kenya: A Country in Transition. Examines the economic and social challenges facing the world's fastest-growing country. Includes teacher's guide, student handouts, glossary, and reading list. For grades 9-12. Zero Population Growth. $9.95.

The New Global Resource Book. 1989. Approximately 200 pp. $30. A comprehensive listing of global education materials for kindergarten through 12th grade. The American Forum for Global Education.

The New Global Yellow Pages. 1989. Approximately 180 pp. $30. Lists more than 150 organizations, contacts, activities, publications, and services of organizations in global education. The American Forum for Global Education.

12

Foresight Capability

The United States must improve its ability to identify emerging problems and assess alternative responses...the Study found serious inconsistencies in the methods and assumptions employed by the various agencies in making their projections.

—The Global 2000 Report to the President[1]

We recommend that the President issue an executive order establishing a new, government-wide process designed to ensure that all the consequences of proposed decisions including long-term, international, and cross-cutting effects are taken into account. This "foresight" process should apply especially to Presidential decisions.

—Blueprint for the Environment[2]

Major Points

- Formulating effective national policy on matters vital to our future requires (a) the ability, the "foresight capability," to make accurate, long-term projections of global trends and their interactions in areas such as population, natural resources, environmental quality, and other factors of geopolitical significance; and (b) the structure to link the results of such projections directly to current decision-making.

- At present, the United States lacks an integrated foresight system for making such long-term projections or bringing them into the decision process.

- To achieve such a capability, the United States should improve its processes for understanding change and applying that understanding to national decisions.

- The U.S. Congress and the Administration are unlikely to act to improve national foresight capability without broad support and pressure for improved foresight from major sectors of society, and without recognition at the presidential level that this capability can be a tool for more effective governance.

- Many nations and regions around the world have undertaken 21st century studies to explore alternative strategies for achieving sustainable economic development and security. The International Geosphere-Biosphere Program has begun a study of global change, focusing on the long-term future and continued habitability of the Earth.

The Issue: Inadequate Knowledge About Future Impacts of Current Policies and Trends

Humans—our numbers, our technologies, our ceaseless activity in the name of progress—have become so pervasive within recent decades that we have developed a dangerous capability to perturb the very processes of Earth itself, without demonstrating that we can manage our awesome power. When the climate has changed and the seas have risen, it will be too late to decide that we should have done something about the production and release of "greenhouse gases." As a nation, and in cooperation with other nations, we must learn to anticipate problems while they are still manageable. That is the essence of "foresight."

What Is Foresight Capability?

As applied to national decisions, "foresight capability" involves:

- identifying long-range trends such as population growth, resource use, and environmental change, and analyzing how these trends may interact over time; and

- making an understanding of the long-range, multiple impacts of such trends an integral part of current decisions and the policy-making process.

Foresight is not prediction, it is not central planning, and it is not a surrender to mathematical or computer models. Rather, it is a process for bringing better information into the decision-making process, for linking analysis and decision, and for obtaining the best available description of the potential implications of policy choices. Foresight capability is a means for giving substance and formal structure to the ecologists' insight that "everything is connected."[3] As applied to national governments, improvement of foresight capability involves better communication between government agencies, relying where possible on existing government machinery rather than new bureaucratic structures, and assuring the provision of information to senior decision makers in a timely manner.[4]

The Global 2000 Report

In 1980, the U.S. Government published *The Global 2000 Report to the President*, which drew upon the best data and analytical resources of the Executive Branch to project how the world would be if then-current trends in resource use, environment, and population continued over the following two decades. The projections in the study, which was prepared at the request of President Carter, indicated that environmental, resource, and population stresses were intensifying, and that, if current trends and policies continued unchanged, the potential existed for "global problems of alarming proportions by the year 2000."[5]

Foresight Shortcomings Revealed by the Report

The report received international attention and generated widespread debate. No government in the world had ever undertaken such an ambitious effort to foresee the results of present trends. Yet the effort was seriously hampered because, as the *Global 2000 Report* stated, "the executive agencies of the U.S. Government are not now capable of presenting the President with internally consistent projections of world trends in population, resources, and the environment for the next two decades."[6] The report found that "There is no agency of the government that has the responsibility and capacity to conduct the kind of environmental analysis and synthesis required by the *Global 2000* study."[7]

The report's self-identified shortcomings and the lessons it posed could have provided a basis for improving the government's ability to collect and analyze relevant data.

Indeed, the second goal of *Global 2000* was to provide a foundation for more effective long-range planning in the Executive Branch. After 1980, however, a new Administration discounted and then essentially abandoned the report. Budgets for the Council on Environmental Quality and for the small centers of long-range data collection and analysis in several agencies that had participated in *Global 2000* were cut,[8] and no effort was made by the new Administration to implement the recommendations of a follow-up report, *Global Future: Time to Act*, released at the end of the Carter Administration in January 1981.[9]

Some Conclusions from Global Models

The *Global 2000 Report* followed a decade of global modeling efforts that began with *The Limits to Growth* in 1972 and that included at least seven major studies.[10] During the 1970s, the International Institute for Applied Systems Analysis in Austria held a series of meetings to compare results of these efforts. The studies had different objectives, employed a range of modeling techniques, and used models assembled by persons with varied backgrounds. Despite wide differences in goals, methods, and results, participants in the studies agreed on a number of points, including the following:

- No known physical or technical reason will prevent us from meeting the basic needs of all the world's people into the foreseeable future. Failure to meet these needs now owes more to social and political structures, values, norms, and world views than to absolute physical scarcities.

- Population and material growth cannot continue forever on a finite planet.

- Continuing "business as usual" policies in coming decades will not lead to a desirable future—or even to meeting basic human needs. More likely, it will result in an increasing gap between the rich and the poor, worsening problems with resource availability and environmental destruction, and declining economic conditions for most people.

- Because of momentum in physical and social processes, policy changes made soon are likely to have more impact with less effort than the same set of changes made later. By the time a problem is obvious to everyone, it is often too late to solve it.[11]

Current Capabilities and Needs

Recently, the world has seen some striking examples of the need for improved foresight in government. The Third World debt crisis, the effects of the OPEC oil price increases, and the African famine were all predictable and actually foreseen by individual agencies and groups, though their warnings were never given full and serious consideration by policymakers. The consequences of failure to forecast and prepare for these problems have included worldwide economic damage and the loss of thousands of lives.

Foresight in Other Countries

Around the world, many other nations are undertaking what are often called "21st century studies." These long-term, multi-sectoral studies examine alternative futures for a nation in terms of its economy, trade, environment, resources, education, housing, security, and other factors. Nations where such studies have been undertaken include: Canada, Iceland, Japan, Mexico, Netherlands, Norway, People's Republic of China, Peru, Philippines, Poland, South Korea, Sweden, Taiwan, and United Kingdom. Most of these studies are modeled after or influenced by the *Global 2000 Report*, of which more than 1.5 million copies in eight different languages have been distributed.[12]

Foresight at the International Level

Since the global modeling efforts of the 1970s, a number of groups have begun to investigate the interactions of the Earth's physical, chemical, biological, and social systems. In 1986, the General Assembly of the International Council of Scientific Unions established the International Geosphere-Biosphere Program: A Study of Global Change. The program is designed to examine the long-term habitability of the Earth, and to provide information needed to formulate policies that will reverse global environmental decline.[13]

Futures Assessment in the U.S. Government

In addition to the 13 federal agencies that contributed long-range projections to the *Global 2000 Report*, several other government organizations assess future conditions. Among those associated with the Congress are the Office of Technology Assessment, the General Accounting Office, the Congressional Budget Office, the Congressional Research Service's Futures Research Group, a range of congressional committees concerned with long-range trends, and the Congressional Clearinghouse on the Future.

Inadequacies of Government Foresight

Both the development and use of long-range projections within the federal government have been the subject of much debate. Individual agency staff and officials have expressed concern about the adequacy of the government's efforts to improve the data and assumptions on which projections are based, the access of individual agencies to each other's work, and the attention given to the projections in decision-making.

At present, the information systems of federal agencies continue on their separate and individual ways, resisting integration, much less any kind of uniformity. Most of the international information on natural resource use, environmental quality, and population growth is obtained from United Nations agencies and other international organizations, or collected by U.S. officials in embassies and consulates around the world and reported to the Department of State in Washington. Some of the information stops there, but much of it passes on to other departments and agencies for storage. There is no government-wide, centralized index or retrieval system, and the different agencies use highly individual and often mutually incompatible data processing systems and models.

Outside the federal government, private organizations, both profit and nonprofit, have criticized the assumptions and data on which some federal projections are based and the ways in which they are used. They have also complained about inadequate public access to government projections, data, and analytic methods. These same private organizations are themselves dependent on the federal government for much of the data they use in their own foresight efforts.

Foresight at the State and Local Level

There is ample evidence of efforts outside the U.S. government to use projections and trend analysis and to enlist citizen participation in improving decision-making, selecting goals, and choosing among alternative futures

A recent compilation by the Institute for Alternative Futures listed some 60 U.S. organizations, commissions, and other groups concerned with future trends, priorities, and choices at state, regional, and local levels.[14]

Information Needs of Business

In 1984, a report was released on "Corporate Use of Information Regarding Natural Resources and Environmental Quality." It reflected the views of managers and analysts in 30 major U.S. corporations, seven trade associations, and eight private information companies. These executives were generally satisfied with government data available to them, but they felt the federal government should provide an index of resource information, a clearinghouse for information on other countries, better environmental data, more timely release of data, and better governmental projections.[15] Since the policies of large, multinational corporations can have a major impact on the international economy and environment, it is important that the data and methods of analysis used are of high quality.

Improving Foresight Capability

Improvement of foresight requires some standardization and compatibility of federal government information, and a degree of consistency of assumptions underlying data or

projections, so that different agencies can use each other's input data and processed output.[16] Care of federal government information systems and their systematic reform deserve priority attention. The Global Tomorrow Coalition and other organizations have developed the following priorities for improving foresight capabilities:

● Improve the quantity, quality, and timeliness of information reported on population, resources, and environment by collecting agencies;

● Improve the timely processing and dissemination of information to other agencies and to the private sector;

● Expand centralized storage, indexing, and retrieval of information serving both government and the private sector;

● Assure greater cooperation and competence among government agencies in the use of information, especially in making projections, and in the sharing of results within and outside the government; and

● Require periodic reporting by government agencies collectively on their information handling, and particularly on their work on projections, for consideration by the Executive Branch, the Congress, and the private sector. Such reporting should include full publication of results, not selective publication for political purposes.

Recent Initiatives to Improve Foresight

During the 1970s, a number of congressionally sponsored commissions and private groups examined broad areas of national growth policy, resource management, and economic futures.[17] Since the release of the *Global 2000 Report* in 1980 and the follow-up report *Global Future: Time to Act* in 1981, there have been recurrent legislative proposals to improve foresight capability. The most recent bill, introduced in the Congress in 1987, would combine two elements: establishment of a national population policy based on long-term stabilization, and improvement of methods for collecting, analyzing, and employing natural resource, environmental, and demographic data.[18] For a variety of reasons, none of these proposals has been passed by the Congress.[19]

A Proposal to Improve Government Foresight Capability

In November 1988, to provide guidance for the newly-elected Bush Administration on issues of environmental quality, resource management, population growth, and development assistance, a broad cross section of U.S. environmental groups—representing a total of more than 6 million members—prepared and delivered a comprehensive set of recommendations under the rubric of Blueprint for the Environment. The product of more than a year of joint effort, this report comprised over 700 specific action recommendations, directed mainly at the level of Assistant Secretary in each relevant U.S. department and agency.[20]

One of the several recommendations directed to the Executive Office of the President, however, was for the establishment of new structures for improved national foresight in the federal government. The Blueprint for the Environment report urged President Bush to issue an immediate Executive Order "...establishing a government-wide process of 'national foresight' (or strategic analysis) to provide...a foundation for improved decision-making and stronger international leadership on complex, long-term environmental, resource, population, and development issues."

Noting that "No mechanism is currently available to the President to relate day-to-day decisions to long-term trends of national and global importance," this report recommended that the Executive Order should instruct the White House Chief of Staff to ensure that all major proposals for presidential decision include identification of: (a) potential short-term and long-term impacts of recommended actions; (b) likely lateral or cross-cutting effects on other federal programs and goals; and (c) foreseeable international consequences. This function would require "...establishment of a small unit subordinate to the White House Chief of Staff and directly linked to the presidential decision-making process."

The Blueprint report went on to recommend creation of parallel foresight structures at the senior level in each department and agency of the federal government; a concerted effort to assure technical and conceptual compatibility of systems used for data collection and analysis; inclusion in the foresight process of relevant information from other sources such as Congress, corporations, nongovernmental organizations, academic and research institutions, foreign governments, and international agencies; and preparation of periodic public reports—similar in subject matter to the *Global 2000 Report*—on the data and major conclusions derived from this government-wide foresight process.

A schematic diagram of such a structure in the federal government, designed by Frank M. Potter, Jr., was included in the report. (See Figure 2.1 on the following page.)

In a valuable individual study of foresight and national decision processes, Lindsey Grant, a former senior official in the Department of State, proposed a variant of this approach based on a two-tiered network among existing government agencies, linked by an office of an "Ombudsman" in the White House. Grant's proposal would require explicit public identification of the offices and individuals in government with primary working-level responsibility for major long-term issues, and periodic reports, roughly every four years, to identify potential national policy opportunities and provide early warning of threatening issues and problems.[21]

It is important to note that every analyst of improved foresight in the federal government has concluded that, whatever approach is selected, it must (1) serve the decision-making needs of the President, and (2) be established with presidential authority to achieve parallel structures and responses throughout all departments and agencies.

Other Organizations and the Foresight Process

The federal government has no monopoly on the need for foresight. As noted earlier, state and local governments, businesses, technical and scientific associations, and educational institutions have their own foresight processes, and their own needs to influence federal government policies. A range of nongovernmental institutions analyze public policy in relation to present and future trends; these include organizations such as the Brookings Institution, Resources for the Future, the World Resources Institute, the American Society for Public Administration, and the National Planning Association.[22]

While improvement of foresight in government is essential, a case can also be made for the creation of a parallel private foresight institution, either as a new organization or through the cooperation of existing organizations. Such an institution could influence and comment on government forecasts, offer policy suggestions, and make its own forecasts in areas not covered by government predictions.[23]

Benefits of Improved Foresight Capability

The first and most important benefit of improving the federal government's capacity to analyze data on long-term trends and their interactions would be better day-to-day decisions on issues directly affecting the national and global future. The system also could eventually result in an integrated federal information system and foresight capability that would benefit decision-making not only in the federal government, but in state and local governments, corporations, and other nongovernmental organizations. U.S. business, which increasingly depends on international trade and investment, could obtain federal projections of global information that are essential for long-term decision-making. To the extent that there were cooperation between the federal information system and a private foresight institution, the federal system could benefit from the feedback of private experience, data, and technology, and be linked to a range of private sector information systems.

Foresight in Educational Curricula

Techniques for the analysis of trends in population, resources, the environment, and their interrelations should

Figure 2.1
Proposed Structure of National Foresight for Presidential Decision-Making

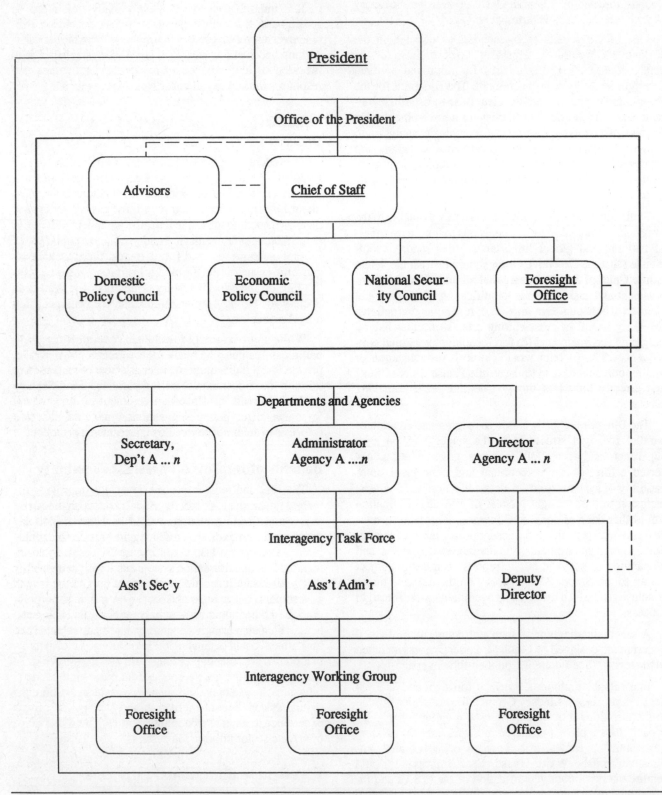

Source: Frank M. Potter, Jr., "Inventing National Foresight," unpublished manuscript, 1988.

also be included in educational curricula. Foresight analysis should become a part of courses and programs in schools, colleges, and universities since future trends ultimately affect every area of human activity. More generally, integrated studies and a holistic, trans-disciplinary approach to global issues should receive greater emphasis throughout the educational system.[24] In the long run, development of a broader constituency for improved foresight capability will depend on changes in educational structures and curricula.

It is difficult to persuade people or nations to make present sacrifices today in order to avoid future catastrophe, or even to agree that the threat of catastrophe exists. If government could improve its foresight processes and open them to the private sector and to an exchange with the rest of the world, government could learn something in the process, and the exchange could lead to greater consensus about the issues and their seriousness. Only with such a consensus is the government likely to persuade others of the need for urgent action.

Foresight and Political Will

Even if government improves its capacity to project future trends and their impacts on society, action to prevent anticipated negative consequences of those trends requires political will, especially when such action itself may have some undesired results. Policy decisions to avert undesirable change usually have significant costs and thus require leadership and courage for their implementation. It is often easier to ignore projections of future crises than to plan how to avoid them, especially if the projections are uncertain, if alternative policies appear costly, and if the call for policy changes implies past policy mistakes or deficiencies.[25]

The Need for a Foresight Constituency

Proposals to improve U.S. national foresight are unlikely to be implemented without widespread public recognition of both the benefits that better foresight capability could provide and the growing risks of decisions made on long-term issues without that understanding. That recognition must be translated into constituent support and pressure for improved national foresight from major sectors of society, including business, labor, education, consumer groups, state and local governments, and a variety of special interests that use federal government data and analyses. Without the support of an informed electorate, the U.S. Congress and the Executive Branch are unlikely to respond to the need for improved foresight capability in national decision-making.[26]

Conclusion

In our rapidly changing world, the ability to make long-range projections of demographic, resource, and environmental trends, to analyze their interrelations, and to take account of the resulting understanding in current policy decisions, is essential in preparing for change at all levels of government and society. The United States, in its own best interests and as a world leader in the age of information, has a responsibility to produce and distribute the best and the most comprehensive knowledge about global trends that will shape the future.

It will require a major effort to develop an effective foresight capability in government. The Global Tomorrow Coalition and its member organizations cannot make a difference on this issue without the determined backing and work of supporters around the country.

What You Can Do

Inform Yourself

There is a long way to go before national foresight becomes an important issue to many people in government. To learn more about the issue, consult some of the sources and organizations listed in the Further Information section. The World Future Society has several publications that can provide good background, and the Institute for Alternative Futures has information about foresight and future studies at the state and local level.[27] Obtain a copy of the article by Cooper and Smoller on trends in the global environment and how they affect individuals, businesses, and institutions in the state of Wisconsin (see Further Information). Find out whether your city or state has conducted a Year 2000 study and, if so, obtain a copy and learn the current status of the report.

Join With Others

Become active in your local civic association. Attend meetings of your city council or county board. Find out which agencies in your local government have responsibility in areas such as population and economic growth, transportation, land use, and waste disposal. Ask whether your community has made projections of how future growth will affect air and water quality, open space, and traffic congestion. Take an active part in local hearings on these issues. Consider making a similar inquiry to agencies of your state.

Review Your Habits and Lifestyle

Practice foresight in your everyday life by considering how decisions you make—for example, about family size and resource consumption—will affect your own future and that of your children. During campaigns for local elective office, attend meetings at which the candidates present their views and ask them how they plan to deal with specific issues affecting the future of your community. Support candidates who are concerned about your region's future quality of life and who have plans to maintain or improve it.

Work With Your Elected Officials

A first step is to encourage elected officials who support improved foresight to continue their work. Chapter 10 of Lindsey Grant's book, *Foresight and National Decisions*, identifies those in Congress who have worked for better foresight.[28] Write and urge them to continue the effort, and send copies to your own Senators and Representatives.

- Write to the President and urge him to develop an effective foresight capability in the White House, and to expand his ability to deal in a timely manner with long-term national and global issues. Send the same message to the President's Chief of Staff (The White House, Washington, DC 20500). If you are active in an environmental organization, population group, or a group concerned with planning or public administration, encourage that group to press for such action in the interests of more efficient national management.

- Write to The Chairman, President's Council on Environmental Quality, 722 Jackson Place, N.W., Washington, D.C. 20006. Ask him to tell you what the Council is doing to increase the federal government's global foresight capability. Ask him how the Council has responded to the *Global 2000 Report* and *Global Future: Time to Act*. Urge that the report be updated periodically with new and better data.

- Write to the Secretary of State, U.S. Department of State, Washington, DC 20520. Ask him how the State Department has taken account of the *Global 2000 Report* and *Global Future: Time to Act*, particularly in the work of its policy planning staff. Urge the State Department to work with the Council on Environmental Quality and other government agencies to develop a stronger foresight capability in the federal government.

- Watch for projections by government agencies and by nongovernmental organizations of future trends in areas such as population growth, energy use, and environmental quality. Look for major differences between forecasts from different sources in the same area, for example, between projections of future energy use with and without improved energy efficiency. Where such a difference occurs, call it to the attention of your congressional representatives and ask them to contact the relevant government agency and request a response

regarding the difference, to call for a hearing where appropriate, and to advise you of their action.

Publicize Your Views

Write letters to the editors of newspapers in your area about the importance of considering the future impacts of current and planned projects affecting the issues covered in this book. Work with local groups to organize meetings, bringing together those active on a wide range of relevant issues, to consider how your community will be affected by future economic and population growth, and which alternative choices may be available.

Raise Awareness Through Education

Find out whether schools, colleges, and libraries in your area have courses, programs, and resources in areas such as future studies, trend analysis, and systems analysis; if not, offer to help develop them or supply useful references and contacts.

Consider the International Connections

Some 40 countries and regions around the world have undertaken long-term studies to examine alternative futures. (See the section on Foresight in Other Countries.) Consider learning about these studies and their relevance to your own state or region. For further information, contact the Institute for 21st Century Studies.

Further Information

Several organizations can provide information about their programs and resources related to foresight capability.*

Books

Barney, Gerald O. and Sheryl Wilkins (eds.). *Managing a Nation: The Software Sourcebook.* Arlington, VA: Institute for 21st Century Studies, 1986. 85 pp. Paperback, $30.00. Contains information on currently available software that can be applied in the management of any nation. Includes sections on global and multisectoral models, rural and urban development, energy, water, agriculture, forests, population, environment and ecology, transportation, security, and politics.

Blueprint for the Environment. Advice to the President-Elect from America's Environmental Community. Washington, DC, 1988. Executive Summary, 32 pp. Paperback, single copies free from Natural Resources Council of America. Book-length version, 352 pp.

---

* These include Californians for Population Stabilization, Congressional Clearinghouse on the Future, The Futures Group, Global Tomorrow Coalition, Institute for Alternative Futures, Institute for 21st Century Studies, Population-Environment Balance, World Future Society, and Zero Population Growth.

Paperback, $13.95 plus $1.50 for shipping, available from Howe Brothers, Salt Lake City, UT 84106.

Cornish, Edward et al. *The Study of the Future.* Bethesda, MD: World Future Society, 1977. 320 pp. Paperback, $9.50. A general introduction to futurism. Discusses forecasting methods, ways to introduce future-oriented thinking into organizations, and likely consequences of current revolutionary social change.

Global Future: Time to Act. Report to the President on Global Resources, Environment and Population. Council on Environmental Quality and U.S. Department of State. Washington, DC: Government Printing Office, 1981. 242 pp. $7.50. An overview of the potential policy and action implications of the *Global 2000 Report.*

The Global 2000 Report to the President. Gerald O. Barney, Study Director. Council on Environmental Quality and U.S. Department of State. Three volumes. Volume I: *Entering the Twenty-first Century.* A summary report. 47 pp. $5.00. Volume II: *The Technical Report.* 766 pp. $14.00. Volume III: *The Government's Global Model.* 401 pp. $9.50. Volumes I and II are published together by Penguin Books, 766 pp. Paperback, $12.95. A summary version is available that includes Volume I and parts of Volume II, edited by Gerald O. Barney and published by Pergamon Press, 360 pp. Paperback, $10.95. Volume I is also available in a new edition published by Seven Locks Press, 64 pp. Paperback, $8.95.

Grant, Lindsey. *Foresight and National Decisions.* Lanham, MD: University Press of America, 1988. 310 pp. Paperback, $14.75. Defines foresight, examines the connections between national foresight processes, state and local planning, planning needs in business, and international foresight processes. Reviews existing foresight proposals and offers concrete suggestions on how to improve the national foresight process.

Meadows, Dennis L. *Alternatives to Growth: A Search for Sustainable Futures.* Cambridge, MA: Ballinger, 1977. 405 pp. Paperback, $9.95.

Meadows, Donella et al. *Groping in the Dark: The First Decade of Global Modelling.* New York: Wiley, 1982. 311 pp. Paperback, $27.50. Reviews the results from seven major global models developed during the 1970s.

Articles, Pamphlets, and Brochures

Barney, Gerald O. "Improving the Government's Capacity to Analyze and Predict Conditions and Trends of Global Population, Resources and Environment." March 1982. Available from the Institute for 21st Century Studies.

Committee on Energy and Commerce, U.S. House of Representatives. "The Strategic Future: Anticipating Tomorrow's Crises." Washington, DC, August 1981.

Cooper, Kate and Jeff Smoller. "Trends in the Global Environment." Madison, WI: Wisconsin Department of Natural Resources, August 1986. 19 pp. Available for $1.00 from the Global Tomorrow Coalition. With text and graphics, shows how global trends may affect people in the United States and in Wisconsin.

Dearborn, Ned. "Radar for the Ship of State." Presentation to the American Association for the Advancement of Science, January 1982. 21 pp. Available for $2.00 from the Global Tomorrow Coalition. Revised and updated version published as "Global 2000: Radar for the Ship of State," *FUTURES*, Vol. 15, No. 2, April 1983.

"Foresight in Government." An Interview with Senator Albert Gore, Jr. *The Futurist*, January-February, 1986, pp. 21-23.

Grant, Lindsey. "Foresight: Addressing Tomorrow's Problems Today." *The Futurist*, January-February 1989, pp. 14-17.

Grant, Lindsey. *Thinking Ahead: Foresight in the Political Process.* Washington, DC: The Environmental Fund, 1983. 76 pp. $3.75. Available from Population-Environment Balance.

Lesh, Donald R., Executive Director, Global Tomorrow Coalition. Congressional Testimony on the Global Resources, Environment, and Population Act, Subcommittee on Census and Population, Committee on Post Office and Civil Service, U.S. House of Representatives, April 12, 1988. Available for $1.00 from Global Tomorrow Coalition.

Peterson, Russell W., Past Board Chair, Global Tomorrow Coalition. Congressional testimony on foresight capability, July 26, 1984. 8 pp. Available for $1.00 from the Global Tomorrow Coalition.

Potter, Frank M. Jr. "Inventing National Foresight." Undated (1988). 9 pp. Available for $1.00 from Global Tomorrow Coalition.

Wagner, Cynthia G. and Blake M. Cornish. "Leading the Way to Tomorrow: Government with Foresight." *The Futurist*, July-August 1989, pp. 35-38.

Periodicals

The Futurist. A bimonthly journal of forecasts, ideas, and trends about the future. $25 per year for individuals. *Future Survey.* A monthly abstract of books, articles, and reports concerning forecasts, trends, and ideas about the future. $59 per year for individuals. Both periodicals are published by the World Future Society, 4916 St. Elmo Avenue, Bethesda, MD 20814.

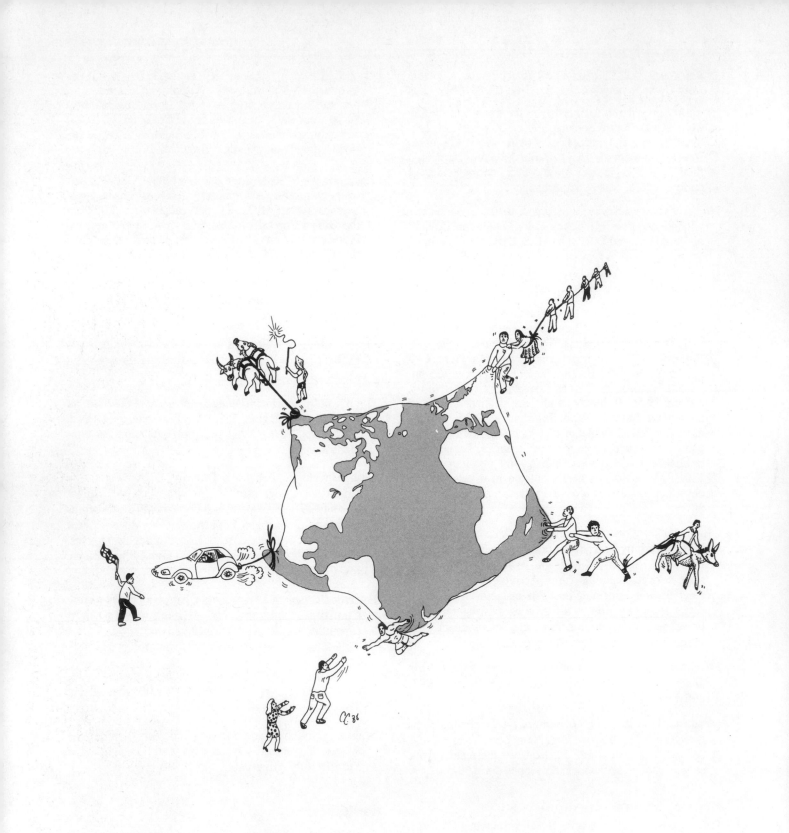

Population Growth

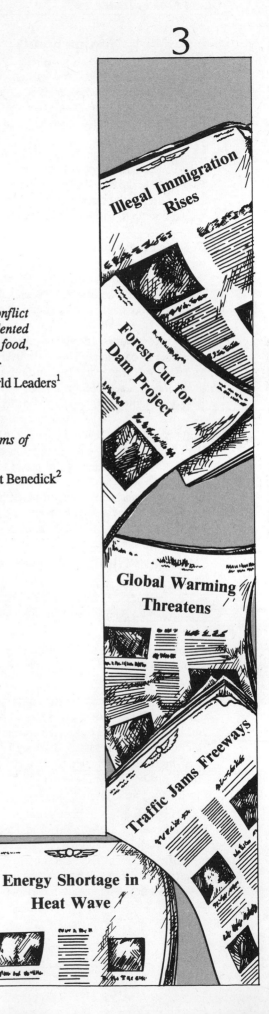

Degradation of the world's environment, income inequality, and the potential for conflict exist today because of over-consumption and over-population. If this unprecedented population growth continues, future generations of children will not have adequate food, housing, medical care, education, earth resources, and employment opportunities.

—Statement on Population Stabilization by World Leaders[1]

Population and environmental policies should be integrated because the problems of poverty, population, natural resources and the environment are closely related.

—Ambassador Richard Elliot Benedick[2]

Major Points

- World population, now 5.2 billion people, grows by almost 90 million each year. It is projected to exceed 6 billion before the year 2000, and reach 10 billion by 2025 unless birth control use increases dramatically.

- More than 90 percent of the projected increase in the world's population between now and the year 2025 will take place in the developing nations in Africa, Asia, and Latin America; at present growth rates, these countries would double in population in just 33 years.

- Countries with rapid population growth rates rank low on measures of the physical quality of life and high on measures of human suffering. Rapid growth is often accompanied by severe environmental degradation, including deforestation, desertification, and soil erosion.

- The United States is one of the fastest-growing industrialized countries. The current U.S. population of 249 million grows by 2.2 million each year and is projected to reach 268 million by the year 2000, more than double the 1940 population of 132 million.

- Eight countries in East Asia and Latin America lowered their fertility rates by more than 50 percent between 1960 and 1987. Such reductions appear to involve the wide availability of varied contraceptive methods, public education stressing responsible parenthood, broad-based economic and social development, and government commitment to population stabilization.

The Issue: Global Population Explosion

In today's world, thanks to improved nutrition and medical care, more babies survive their first few years of life and people live longer. While this is good news, it is a major cause of our rapidly growing population. Whereas infant and childhood deaths and short life spans used to limit population growth, we must use family planning and contraceptive methods today.

From the beginning of recorded history until about 1800, world population grew slowly to about 1 billion people, and it took about 125 years to add the second billion. The third billion was added in about 35 years, by 1960, and the fourth billion was reached 14 years later, in 1974. The fifth billion was added only 13 years later, in 1987.[3] (See Figure 3.1 on the following page.)

Looking ahead, world population is projected to exceed 6 billion before the year 2000. And according to a recent report by the United Nations Population Fund, total population is likely to reach 10 billion by 2025 and grow to 14 billion by the end of the next century unless birth control use increases dramatically around the world within the next few decades. About 90 percent of the growth predicted before the end of the twenty-first century will be in the Third World. Over the next 10 years, the expected addition of a billion people will be nearly equivalent to today's population of Africa and Latin America combined.[4]

World population is now increasing at an annual rate of 1.8 percent. If this rate continues, the number of people on Earth will double in 39 years. If the population of Kenya keeps growing at its current rate of 4.1 percent per year, it will double in only 17 years. In contrast, U.S. population, currently increasing at 0.7 percent, will take 98 years to double, and Sweden, growing at only 0.2 percent, will need nearly 370 years to double its population. Figure 3.2 (on page 26) shows population doubling times for major regions of the world; as of 1989 they ranged from 24 years for Africa to 269 years for Europe.[5]

Causes of Rapid Population Growth

Scientists believe the human species dates back at least 3 million years. For more than 99 percent of this time, or until the dawn of agriculture around 8000 B.C., human population probably numbered less than 10 million. The development of agriculture allowed the evolution of communities that could support more people. World population grew slowly to about 300 million by 1 A.D. and to about 800 million by 1750.

Several explanations have been suggested for the population explosion that began in the 1700s and accelerated during the industrial revolution. One theory, first proposed by Thomas Malthus in the eighteenth century, is that populations are limited by available food, and expand in response to increased food supplies. Although the world's agricultural output clearly has risen, especially since World War II, Malthus' theory is weakened by the fact that much of the increased output has been exported and unavailable for consumption in the world's poorest regions, where population growth has been greatest. Furthermore, anthropological studies show that traditional societies generally do not fully exploit potential food supplies, and rely on a wide range of means to limit population growth.

A second explanation, which does largely account for the population explosion, is that improvements in medicine and sanitation have reduced disease and increased longevity. A third explanation for recent population growth is that economic and social development has undermined peoples' sense of identity and security based on traditional ties to extended family, clan, and community; has heightened their felt need for children as a source of security; and has weakened traditional means for controlling births.[6]

Figure 3.1
World Population Growth, 1 A.D. to 2100

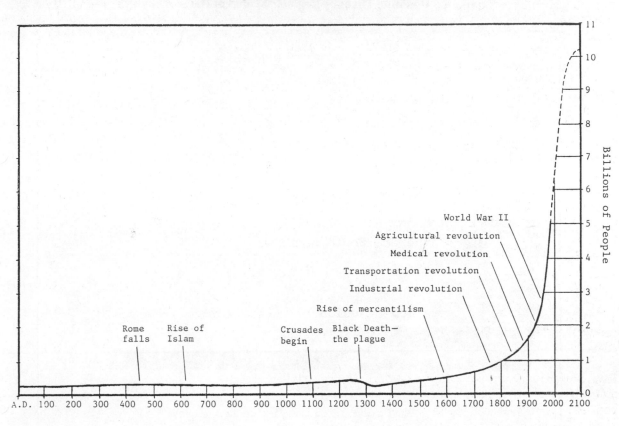

Sources: Population data from Population Reference Bureau, various publications; historical data from Richard D. Lamm, *Hard Choices* (Denver, CO, May 1985), p. 34.

Consequences of Rapid Population Growth

Why is it important to limit our numbers? Of the more than 5 billion people alive today, too many lack adequate food, water, shelter, education, and employment. Ironically, high fertility, traditionally associated with prosperity, prestige, and security for the future, now jeopardizes chances for many to achieve health and security.

Rich and poor countries alike are affected by population growth, though the populations of industrial countries are growing more slowly than those of developing ones. At present growth rates, the population of economically developed countries would double in 120 years, whereas the Third World, with over three quarters of the world's people, would double its numbers in about 33 years.[7] This rapid doubling time reflects the fact that 37 percent of the developing world's population is under the age of 15 and entering their most productive childbearing years. In the Third World countries (excluding China), 40 percent of the people are under 15; in some African countries, nearly half are in this age-group.[8]

The world's projected population growth will call for a commensurate increase in efforts to meet needs for food, water, shelter, jobs, and education. It is estimated that by the year 2000, the world will need 600 million new jobs, teachers for 300 million additional children, and family planning services for 400 million women.[9] In the poorest countries, massive efforts are needed to keep social and economic conditions from deteriorating further; any real advances in well-being and the quality of life are likely to be negated by further population growth. In the industrialized world, many countries lack adequate domestic supplies of basic materials needed to support their current populations.

Figure 3.2
Population Doubling Times by Region, at Current Rates of Growth

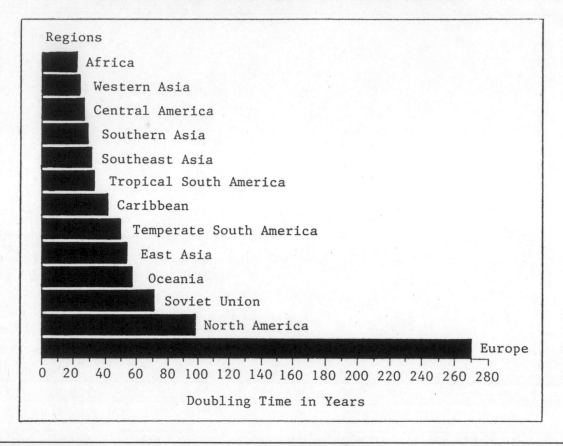

Source: Population Reference Bureau, *1989 World Population Data Sheet* (Washington, DC, 1989).

Population Growth and the Quality of Life

Rapid population growth can affect both the overall quality of life and the degree of human suffering in a nation or region.

The Quality of Life Index

In the late 1970s, a composite index of the physical quality of life (PQLI) was developed by the Overseas Development Council, based on life expectancy, infant mortality, and literacy.[10] Countries with low birth rates (number of births per 1,000 population in a given year) consistently rank high on the PQLI, while those with high birth rates have lower PQLI ratings.[11]

The Human Suffering Index

More recently, The Population Crisis Committee compiled an index of human suffering for each of 130 countries, based on 10 measures of human welfare. When the index is compared with annual rates of population increase, there is a high correlation between level of suffering and rate of increase.[12]

The suffering index and rate of population increase are plotted in Figure 3.3 (on the following page), which shows how the 130 countries are distributed among four levels of human suffering. Points representing the more populous countries are labeled. The data show that:

- The 30 countries falling in the *extreme* suffering range are all in Africa and Asia and have an average annual population increase of 2.8 percent.

- The 44 countries in the *high* suffering range are, with one exception (Papua-New Guinea), in Africa, Asia, and Latin America and also have an average annual population increase of 2.8 percent.[13]

Figure 3.3
Population Growth and Human Suffering

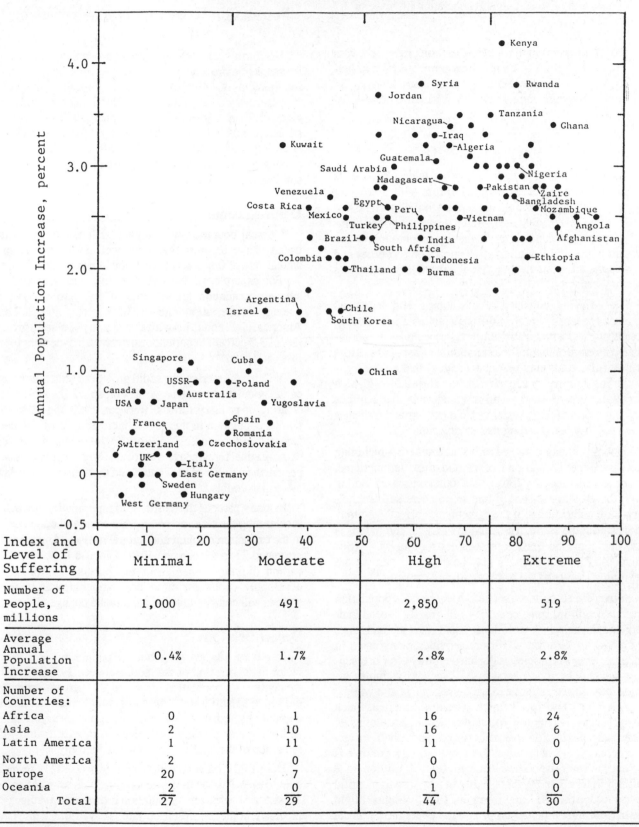

Index and Level of Suffering	Minimal	Moderate	High	Extreme
Number of People, millions	1,000	491	2,850	519
Average Annual Population Increase	0.4%	1.7%	2.8%	2.8%
Number of Countries:				
Africa	0	1	16	24
Asia	2	10	16	6
Latin America	1	11	11	0
North America	2	0	0	0
Europe	20	7	0	0
Oceania	2	0	1	0
Total	27	29	44	30

Source: Sharon L. Camp and J. Joseph Speidel, *The International Human Suffering Index* (Washington, DC: Population Crisis Committee, 1987).

- The 29 countries in the *moderate* range are, with the exception of Mauritius, in Asia, Latin America, and Europe and have an average population increase of 1.7 percent.

- The 27 countries in the *minimal* suffering range are, with three exceptions (Japan, Singapore, and Trinidad-Tobago), in Europe, North America, and Oceania and have an average annual population increase of only 0.4 percent.[14]

Carrying Capacity

Rapid population growth can push a region beyond its economic and natural resource limits—its "carrying capacity" or long-term ability to support the people who live there without degrading the region's resources.[15] The number of people, the nature and quantity of production and consumption, and the cumulative impact on resources and environment are all factors that determine a given area's carrying capacity.[16]

The carrying capacity of developing and industrial countries generally involves different factors. In developing countries, widespread malnutrition, especially if accompanied by environmental deterioration such as rapid soil loss or desertification, may be one indication that a country is exceeding its carrying capacity. In the industrial world, high per capita use of energy and raw materials that requires major imports of these resources to meet demand indicates that a country is exceeding its carrying capacity.

Globally, changes in the Earth's atmosphere—including increasing concentrations of carbon dioxide, methane, chlorofluorocarbons (CFCs), and other gases stemming from the growing impact of human numbers and technology—are contributing to the "greenhouse effect" and may be warming the Earth. The buildup of CFCs and halons is also depleting the Earth's protective ozone layer. [17] (See Chapter 12.) Such changes may be evidence that we are beginning to exceed the Earth's carrying capacity.[18]

Some experts believe that before the world's population reaches 6 billion, some countries will grossly exceed their carrying capacity and experience rapid rises in death rates and drops in birth rates. History provides many cases in which regions and societies declined rapidly after life-support systems became overstressed. Ancient examples include the Tigris-Euphrates fertile crescent in the Middle East, and the Mayan civilization in Central America, which after 17 centuries of growth collapsed within decades when increasing population and soil erosion made their society and economy unsustainable.[19] A more modern example is Ireland, where population doubled from 4 million to 8 million in less than 50 years, partly in response to plentiful crops of potatoes. Beginning in 1845, fungus blight destroyed several potato crops and the population was cut in half through starvation, disease, and large-scale emigration.[20]

Population Growth and Environmental Degradation

The Earth's carrying capacity is burdened not only by the demands of more than 5 billion people, but also by the food requirements of 4 billion cattle, sheep, pigs, goats, and other livestock, as well as more than 9 billion chickens and other domestic fowl.[21] In addition to the pressures from growing human populations, the clearing of land for cattle ranching is an important factor in tropical deforestation, and overgrazing by livestock is a major cause of desertification and soil loss in many countries.[22]

Deforestation

Tropical countries with high population growth rates usually have deforestation rates well above the average annual rate of 0.6 percent for tropical areas. (See Chapter 7.) For example, in West Africa, five countries with an average population growth rate of 2.9 percent have an average deforestation rate of 3.4 percent; and in Latin America, eight countries with an average population growth rate of 3.0 percent experience an average 3.1-percent forest loss each year.[23]

In Central America, the rapid growth of both human and cattle populations has been accompanied by large-scale clearing of primary forests. Between 1950 and 1985, the human population in this region increased by 183 percent, from 9.2 million to 26.1 million,[24] while about 40 percent of the original forest cover was lost.[25] Over roughly the same period, the number of beef cattle and the area of pastureland more than doubled.[26]

In many parts of the Third World, population growth is accelerating the clearing of forested areas and contributing to the extinction of plant and animal species. And since an estimated 23 to 43 percent of the increase in atmospheric carbon dioxide comes from the burning of forests in developing countries, rapid population growth in these regions will contribute to global climate change.[27]

Desertification

In Africa, the growing human population is a major factor in degradation of the land. Between 1950 and 1985, the continent's population increased 149 percent, from 222 million to 553 million.[28] Pressures from human and animal populations translate into deforestation, overcultivation, and overgrazing, which often lead to desertification—the reduction of the land's biological productivity.[29]

Desertification is now an "ongoing process" in some 22 countries in Africa. In the seven Sahelian-zone nations of West Africa, where the population growth rate averages 2.7 percent and the deforestation rate is seven times the Third

World average, desertification is described as "rampant,"[30] and has affected nearly 90 percent of the productive drylands in the region.[31]

Soil Erosion

Estimates of soil loss due to erosion by water and wind have been made for a number of regions. (See Chapter 5.) Estimated erosion rates are high—in some areas many times the rate for U.S. croplands—for parts of Central and Southern Africa, where population growth rates are between 2.8 and 3.2 percent a year; in Central America, with population growth rates between 2.8 and 3.2 percent and population densities between 200 and 650 persons per square mile; and in parts of South and East Asia with population densities between 240 and 640 persons per square mile.[32]

Migration and Growth of Urban Areas

Although in past centuries migration may have served as an escape valve for overpopulation, today there are relatively few habitable empty spaces to settle and few countries that accept immigrants in significant numbers. In the past century, most of the globe's fertile lands have been settled. Now, when population grows in a region, joblessness and overcrowding can result. Relief may take the form of voluntary or forced migration—to a city or other area within a nation, or to another nation.

The proportion of the world's population living in urban areas grew from 29 percent in 1950 to 42 percent in 1985, and is projected to reach 60 percent by 2020.[33] Migration to cities, the crucial force in the initial phases of urbanization, is contributing to the marked increase of urbanization in the developing world.[34] Almost 90 percent of Third World population increase over the next few decades will be in urban areas.[35] Ten of the world's 12 largest cities will be in developing countries by the year 2000; the populations of each one will range from 13 to 26 million.[36]

Urban growth is resulting more from rural poverty than urban prosperity; rapid rural population growth, along with inequitable land distribution, poor income prospects, and inadequate government investment in agriculture, all combine to make even urban slums seem more appealing than rural life.[37] With Third World unemployment rates ranging as high as 30 to 50 percent, the search for jobs is a major incentive for migration.[38] And as urban areas drain the surrounding countryside of resources—for example, as city dwellers destroy forests in their search for firewood in place of unaffordable oil—rural residents may be forced to migrate.[39] Unfortunately, migrants arriving in many Third World cities find large areas of shantytowns and slums characterized by high unemployment, pollution, disease, social disorder, political unrest, and, in many cases, violence.[40]

Rapid urban growth has led to the concentration of political power in cities and the favoring of urban areas in national development plans. This bias toward urban areas' power—demonstrated by subsidies for food and other goods sold in cities, discouragement of agricultural investment, and overvalued exchange rates that lower the costs of imports—leads to increased rural deprivation and flight to the cities. Recently, however, escalating debt burdens have been preventing Third World governments from favoring urban areas and forcing them to limit urban subsidies and services.[41]

In industrial countries, large-scale urbanization occurred relatively slowly and in conjunction with economic growth, allowing urban growth to be accommodated. In the Third World, however, rapid population growth and urbanization without adequate economic growth leaves governments unable to meet the needs of burgeoning urban areas.[42] As a result, urban conditions in much of the Third World are harsh. At least a third of Bombay lives in slums,[43] and Mexico City is surrounded by shanties and garbage dumps. Most developing nations' cities are similarly circled by squatter settlements that lack space, safe water, sanitation, waste collection, lighting, adequate housing, and other essentials for decent living.[44]

These conditions lead to the spread of diseases, including typhoid, cholera, malaria, and hepatitis; to increases in alienation and violence; and to a high level of vulnerability to natural and industrial disasters.[45] Table 3.1 (on the following page) shows the explosive growth that has occurred in selected Third World cities since 1950, along with population projections for the year 2000.[46] In summary, urbanization in developing countries is generally not promoting economic development as it did earlier in the industrialized nations. In many ways, urbanization is impeding the struggle toward sustainable development in the Third World.[47]

Population Pressures and Political Instability

Rapid population growth, especially when it occurs in regions with sharp ethnic differences, can put great stress on political institutions and complicate the problems of governance. In a study of 120 countries, the Population Crisis Committee found that only a few countries with severe demographic pressures managed to maintain stable constitutional governments and achieve good records on political and civil rights. Of the 31 countries rated highest on political instability, nearly all had serious population pressures. Conversely, most nations that were relatively stable and democratic tended to have lower levels of demographic pressure.[48]

Table 3.1
Rapid Population Growth in Third World Citites

Population in millions

City	1950	Most Recent Figures	UN Projection For 2000
Mexico City	3.05	17.3 (1985)	25.8
São Paulo	2.7	15.9 (1985)	24.0
Bombay	3.0 (1951)	10.1 (1985)	16.0
Jakarta	1.45	7.9 (1985)	13.2
Cairo	2.5	7.7 (1985)	11.1
Delhi	1.4 (1951)	7.4 (1985)	13.2
Manila	1.78	7.0 (1985)	11.1
Lagos	0.27 (1952)	3.6 (1985)	8.3
Bogota	0.61	4.5 (1985)	6.5
Nairobi	0.14	0.83 (1979)	5.3
Dar es Salaam	0.15 (1960)	0.9 (1981)	4.6
Gter. Khartoum	0.18	1.05 (1978)	4.1
Amman	0.03	0.78 (1978)	1.5
Nouakchott	0.0058	0.25 (1982)	1.1
Manaus	0.11	0.51 (1980)	1.1
Santa Cruz	0.059	0.26 (1976)	1.0

Source: Nafis Sadik, "The State of World Population 1988" (New York: United Nations Population Fund, 1988), p. 21.

Population Growth and Conflict

There are many links between population growth and social conflict. When growing populations compete for limited or inequitably distributed resources such as land, food, water, or income, conflict may occur. When population growth outstrips economic growth, declining per capita income and living standards can lead to social unrest and civil war. And when deteriorating natural resources can no longer support growing populations, conflicts can arise as people are forced to move to seek a livelihood. In Africa, many of these "ecological refugees," driven by desertification, have crossed national borders and come into conflict with people living in the areas they try to enter.[49]

In general, when large groups move voluntarily or are displaced, they increase the competition for jobs and resources in their new location, and they may be resented. Tensions resulting from migration and ethnic differences can develop in any country, but the most serious conflicts often occur in densely populated regions. As more people compete for scarce resources and seek equality and autonomy, hostilities often erupt. Overpopulation increases crime, and in some places has aggravated ethnic or tribal discord to the point of civil war.[50]

As population growth has accelerated in the twentieth century, civil wars have increased sharply, and are now by far the major form of warfare. Of the nearly 20 million war-related deaths since the end of World War II, over 15 million have occurred in conflicts primarily domestic in nature.[51] Since 1945, nearly 8.7 million war-related deaths have occurred in countries with current population densities

greater than 500 persons per square mile, five times the world average. The area that is now Bangladesh, with the highest population density of any major country (1,969 persons per square mile) has registered 1.5 million war-related deaths since 1945.[52]

Summary of World Population Situation

While some assessments of the Earth's resources and human abilities have led observers to see optimistic prospects for unlimited future development,[53] and while population growth by itself does not necessarily lower the quality of life or result in environmental degradation, in many developing countries population growth is clearly undermining the capacity to meet basic economic and social needs and to protect natural resources. There is ample evidence that many social, economic, political, and environmental problems are worsened by rapid population growth. Malnutrition and disease, unemployment, political unrest, water and fuelwood shortages, deforestation, drought and flooding, desertification, soil loss, air and water pollution, depletion of ocean fisheries, and loss of plant and animal species are all exacerbated by rapid increase in human numbers.[54]

Population Growth in the United States

Contrary to the common misconception that the United States has attained zero population growth, U.S. population, now 249 million, is still growing by about 2.2 million each year and is likely to continue growing well into the 21st century. U.S. population is projected to reach 268 million by the year 2000, more than double the 1940 population of 132 million. In the United States, the birth rate is much higher than the death rate, primarily because women born during the postwar baby boom are now in their childbearing years. In contrast, in Western Europe birth rates are only slightly higher than death rates.[55]

At its present annual growth rate of 0.9 percent (including immigration), U.S. population is growing faster than all but six of the 24 industrialized countries comprising the Organization for Economic Cooperation and Development; only Turkey is growing more rapidly than the United States.[56]

During the next 100 years, U.S. population is projected to grow by nearly 70 million, an increase equivalent to the combined current populations of California, New York, Texas, and New Jersey.[57] The United States is undeniably "a nation of immigrants," founded by people from other countries of the world. Clearly, immigrants have made and will continue to make major contributions to all aspects of American society. But as environmental and other population-related problems facing the United States continue to intensify, so does the debate about future U.S. immigration policy.[58]

Large numbers of legal immigrants and refugees, about 600,000 yearly, enter the United States along with substantial, though uncertain, numbers of illegal immigrants.[59] The U.S. Census Bureau estimated that in 1986, somewhere between 3 million and 5 million illegal immigrants lived in the United States.[60] The portion of U.S. population growth attributable to immigration has risen sharply, from 7.4 percent for the period 1940-44 to 28.4 percent for 1980-85.[61]

The flow of immigrants into the United States probably has both positive and negative effects on the U.S. economy. One U.S. government report concluded that "illegal alien workers appear to displace (or take jobs away from) native or legal workers."[62] Another study showed that while having little negative effect on employment at the national level, immigration has adversely affected job markets in the few regions where most immigrants settle.[63] In the future, immigration to the United States is likely to increase due to extreme poverty and high unemployment rates in developing countries and the prospect of jobs and a better standard of living in the United States.

Arguments that the United States will experience social, economic, and military disadvantages because of declining fertility[64] generally disregard the flow of immigrants to the United States, the increasing role of technology and automation and the corresponding decreasing need for large numbers of people in industry and the military, as well as the environmental benefits of limiting population growth.[65]

Consequences of U.S. Population Growth

Continued U.S. population growth from natural increase and from immigration has major implications for employment, inflation, resource supplies, economic growth, environmental quality, agricultural production, and the provision of social services. U.S. population growth has major impacts on the entire world because the United States, with only 5 percent of the world's population, consumes 25 percent of the world's energy resources[66] and produces a comparable amount of its pollution. An average U.S. citizen consumes more than 12 times as much energy as an average citizen in a developing country.[67]

Agriculture

Domestic population growth will continue to affect U.S. farming and agricultural exports. Because of cropland conversion, the United States may have fewer farm products to export in future years. It is estimated that each year more than 1.2 million hectares (3 million acres) of agricultural land are irreversibly converted to highways, housing, shopping centers, and industrial development. The erosion of some 5 billion tons of topsoil each year from U.S. crop and rangeland, partially because of imprudent farm practices adopted to meet demand pressure, is also diminishing U.S. agricultural capacity.[68]

Water Supplies

U.S. population grew by more than 50 percent between 1950 and 1980, but during the same period the withdrawal of water from streams, reservoirs, lakes, and underground aquifers increased by 150 percent, from 180 billion to 450 billion gallons per day.[69] Water shortages have already occurred in several of the most rapidly growing areas of the country, particularly the South and Southwest. The U.S. Water Resources Council has projected that by the end of the century water supplies will be severely inadequate in 17 of the 106 water supply regions in the United States. Even areas now considered to have plentiful water supplies could develop shortages as industrial centers expand and population grows.[70]

Stress in Urban Areas

Many urban areas in the United States experience population-related stress that adversely affects the social, economic, and environmental well-being of their residents. In an "urban stress test," Zero Population Growth assembled data for 184 U.S. cities. The study identified 13 cities as "red zones," most with rapid population change, rising birth rates, crowding, high rates of violent crime, economic hardship, low educational attainment, hazardous waste, air and water pollution, poor water availability, and inadequate sewage treatment. The red-zone cities with the highest population-related stress ratings were: Miami, Florida;

Pomona, California; Newark, New Jersey; Los Angeles, California; Gary, Indiana; and Jersey City, New Jersey. Cities with the lowest stress ratings were: Fargo, North Dakota; Madison, Wisconsin; Casper, Wyoming; Boise, Idaho; Ann Arbor, Michigan; and Lincoln, Nebraska. Communities with the lowest ratings were generally smaller and less crowded, and had low crime rates, low levels of hazardous waste, and high percentages of high school graduates.[71]

Summary of U.S. Population Situation

Continued U.S. population growth is likely to affect the quality of American life adversely, especially in urban areas, by adding to the burden on transportation, water, sewage and waste management systems; on education and social welfare facilities; and on basic community services such as police and fire protection.[72] To address some of these issues, the "Global Resources, Environment, and Population Act of 1987" was introduced in both houses of the U.S. Congress. The legislation would: (1) declare that there are economic, social, governmental, and environmental advantages to achieving U.S. population stabilization; (2) require the President's Council on Environmental Quality to coordinate the collection of demographic, natural resource, and environmental information for planning and decision-making; (3) require that all federal population-related policies and programs be consistent with attaining U.S. population stabilization; (4) make available to state and local governments and others information useful for their own planning; (5) require preparation of a Presidential report every three years evaluating short- and long-term impacts of population growth and its environmental and resource impacts.[73]

Limiting Population Growth

The consequences of rapid population growth for a country's economy, social fabric, and political stability are increasingly recognized. Between the time of the third United Nations population conference in 1974 and the conference in Mexico City ten years later, a near-unanimous consensus developed that family planning is essential for economic development, and that greater efforts are needed to promote it. Third World leaders are speaking out on the urgency of slowing population growth.

In 1985, a statement supporting stabilization of the world's population was signed by heads of state of countries with over half the world's population and presented to the United Nations. By early 1988, the statement had been signed by 48 heads of government.[74]

In spite of growing international support for family planning programs, the United States, traditionally a leader of international family planning efforts, since 1984 has sharply reduced its monetary and political commitment to family planning throughout the world in order to avoid contributing to abortion-related activities in the nations where abortion is legal.[75]

U.S. funding for international population assistance dropped 20 percent between 1985 and 1987, from $288 million to $230 million. This change in policy impairs the ability of established family planning services to function, and it will likely lead to an increase in births, abortion-related complications, and maternal deaths throughout the developing world.[76]

Benefits of Family Planning

The imminent dangers of not limiting our numbers are obvious, but there are also benefits for all if we do reduce birth rates. Family planning is a primary health care strategy. Success in encouraging family planning helps lower infant mortality (see Figure 3.4 on the following page), and aids the survival of other children in the family.[77] Programs that help women to have fewer, safer, and better-spaced births can reduce maternal mortality and illegal abortions. (Maternal death rates are especially high among malnourished women who have lower immunity to disease. About half the world's abortions are estimated to be illegal; abortion-related deaths are frequent after illegal abortions, but rare after legal ones.) In addition, educating people about reproductive health issues and promoting the use of condoms can curb the spread of AIDS, a critical health and economic problem for many nations.[78] At the family level, there can be economic, social, and psychological benefits associated with small family size.[79]

And reducing birth rates enables both families and national governments to increase their per-child expenditures on health and education and thus build a more productive labor force.[80]

Lower fertility rates also ease pressures on natural resources, the environment, and most aspects of the Earth's life-support systems.[81] Finally, a 1986 study by the National Research Council concluded that "slower population growth would be beneficial to economic development for most developing countries."[82]

Measures to Reduce Population Growth

Since the causes of rapid population growth involve a variety of biological, psychological, social, cultural, economic, and political factors, a number of measures may be important for reducing growth rates. The stimulus of rapid growth varies both within and between countries; for a given region some measures will be more appropriate than others. Some of these measures are listed below.

● Increase public understanding of how rapid population growth limits chances for meeting basic human needs, and about the advantages of limiting family size and spacing births. Reductions in maternal death rates and increases in the level of spending per child on health and

Figure 3.4
Contraception and Infant Mortality

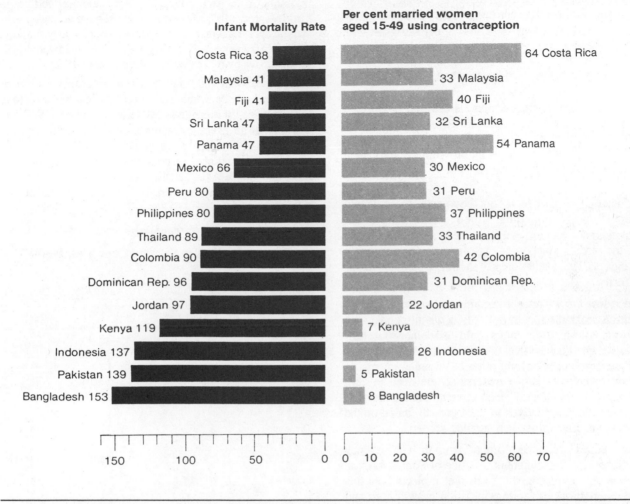

Infant Mortality Rate	Per cent married women aged 15-49 using contraception
Costa Rica 38	64 Costa Rica
Malaysia 41	33 Malaysia
Fiji 41	40 Fiji
Sri Lanka 47	32 Sri Lanka
Panama 47	54 Panama
Mexico 66	30 Mexico
Peru 80	31 Peru
Philippines 80	37 Philippines
Thailand 89	33 Thailand
Colombia 90	42 Colombia
Dominican Rep. 96	31 Dominican Rep.
Jordan 97	22 Jordan
Kenya 119	7 Kenya
Indonesia 137	26 Indonesia
Pakistan 139	5 Pakistan
Bangladesh 153	8 Bangladesh

Source: World Fertility Survey, as published in Nafis Sadik, "The State of World Population 1988" (New York: United Nations Population Fund, 1988), p. 18.

education are among the significant benefits that can be cited.

- Achieve widespread availability and use of effective, affordable family planning services through increased funding. The World Bank has estimated that in order to provide family planning services to all those in need, some $8 billion a year must be spent by the end of the century. The Worldwatch Institute estimates that stabilizing world population at about 8 billion by the year 2050 would require a further $25 billion a year of related expenditures on social improvements and financial incentives by the year 2000.[83] Today, most countries favor family planning, but they lack the funds and other resources to make it widely available. As a result, only about 45 percent of the world's women of reproductive age use

birth control measures, and about 250 million women in the developing world outside China say they wish to plan their families but do not have access to, or information about, the family planning techniques they want.[84]

- Reduce infant and child mortality through improved nutrition, sanitation, maternal and child health care, and education of women.

- Provide social and economic incentives and disincentives to encourage smaller families, such as special tax or social security benefits to parents of small families.[85]

- Promote wide participation in social and economic development, including full integration of women into educational, social, economic, and political activities. In general, the higher the status of women in a country, the fewer children they have.[86] Employment and education opportunities for young women are the most significant factors influencing their age at first marriage. The more education women have, the more opportunities they have besides child-bearing, and the fewer children they choose to bear.[87]

● Develop systems to provide security for older citizens.

● For countries receiving immigrants, develop measures to limit immigration to sustainable levels; for countries sending immigrants, contribute to the long-term reduction of emigration pressures by providing development and family planning assistance.

● Educate political leaders about the consequences of rapid population growth; obtain their public support for policies and programs that would reduce growth. In view of the recent reduction of U.S. funding for population programs, U.S. leaders clearly should be included in this effort.

To reduce growth rates to sustainable levels, population education programs should not only acquaint people with family planning methods, they should also demonstrate how population growth will affect the availability of water, food, fuel, housing, education, and other basic requirements. In addition to helping people understand the need to curb population growth quickly, each government can develop incentives and disincentives to limit growth, based on the values and interests of its own people.

The Futures Group, with support from the U.S. Agency for International Development's Office of Population, has developed a computerized system that projects how different assumptions of future population growth rates and fertility rates for a nation would affect future economic, social, and environmental trends such as requirements for food, water, jobs, housing, education, and health care facilities, and rates of deforestation, desertification, and soil erosion. The system, known as RAPID—Resources for Awareness of Population Impact on Development—demonstrates the heavy burden that rapid population growth places on all aspects of development, and the advantages of achieving lower rates of fertility and growth. RAPID presentations have been made to and have impressed top government officials and national leaders in some 44 countries, including more than 20 in Africa.[88]

Successful Programs

In spite of the difficulties involved in achieving a rapid reduction in total fertility rate (the average number of children per woman), some countries have been successful. In Asia, the countries of Taiwan, South Korea, and China, along with the city-state of Singapore, have made remark-

able progress. Thailand, Malaysia, Indonesia, and Turkey have also made substantial progress.[89]

In Latin America, the countries of Cuba, Chile, Colombia, and Costa Rica have achieved major reductions in fertility rates, while Mexico and Brazil have achieved substantial declines. Only one Middle Eastern country, Turkey, and two African ones, Tunisia and Egypt, have lowered their fertility rates by more than 20 percent since 1960.

The most dramatic reductions have occurred in East Asia and Latin America, where eight countries lowered their total fertility rates by more than 50 percent between 1960 and 1987, from an average for the group of 6.1 children per woman in 1960 to only 2.3 in 1987. (See Table 3.2.) In contrast, for a number of countries in Africa, the Middle East, and South Asia, fertility remains extremely high—above six children per woman.[90] (See Table 3.3 on the following page.)

Table 3.2

Fertility Declines in Selected Countries, 1960-87

Country	Total Fertility Rates		Change
	1960	1987	
	(average number of children per woman)		(percent)
Singapore	6.3	1.6	−75
Taiwan	6.5	1.8	−72
South Korea	6.0	2.1	−65
Cuba	4.7	1.8	−62
China	5.5	2.4	−56
Chile	5.3	2.4	−55
Colombia	6.8	3.1	−54
Costa Rica	7.4	3.5	−53
Thailand	6.6	3.5	−47
Mexico	7.2	4.0	−44
Brazil	6.2	3.5	−44
Malaysia	6.9	3.9	−43
Indonesia	5.6	3.3	−41
Turkey	6.8	4.0	−41
Tunisia	7.3	4.5	−38

Source: 1960 data from Ansley Coale, "Recent Trends in Fertility in Less Developed Countries," *Science*, August 26, 1983; 1987 data from Population Reference Bureau, 1987 *World Population Data Sheet* (Washington, DC, 1987) as published in Jodi L. Jacobson, *Planning the Global Family*, Worldwatch Paper 80 (Washington, DC: Worldwatch Institute, December 1987), p. 8.

Experience has shown that in countries where fertility has been reduced significantly, two mutually reinforcing strategies have been effective. First, family planning services providing basic health care and offering a wide choice of contraceptive methods must be made easily available at the community level, and people must be encouraged to use them through public education campaigns stressing responsible parenthood. A recent study by the Population Crisis Committee shows a high correlation between good access to means of birth control and substantial decline in fertility rates.[91]

Table 3.3
Countries with High Fertility, 1987

Country	Total Fertility Rate	Population Growth Rate
	(average number of children per woman)	(percent)
Kenya	8.0	3.9
Afghanistan	7.6	2.6
Jordan	7.4	3.7
Tanzania	7.1	3.5
Zambia	7.0	3.5
Saudi Arabia	6.9	3.1
Ethiopia	6.7	2.3
Senegal	6.7	2.8
Nigeria	6.6	2.8
Pakistan	6.6	2.9
Sudan	6.5	2.8
Zimbabwe	6.5	3.5
Iran	6.3	3.2
Bangladesh	6.2	2.7
Zaire	6.1	3.1

Source: Population Reference Bureau, 1987 *World Population Data Sheet* (Washington, DC, 1987) as published in Jodi L. Jacobson, *Planning the Global Family*, Worldwatch Paper 80 (Washington, DC: Worldwatch Institute, December 1987), p. 10.

Second, broad-based economic and social development is needed to alleviate those underlying conditions—low status of women, illiteracy, low wages—that contribute to high fertility rates.[92] The success of these strategies requires genuine political commitment to population stabilization and significant investments of national resources, both human and financial.[93]

It is difficult to overemphasize the importance of family planning in efforts to achieve meaningful economic and social development and cope with the range of environmental and resource problems facing our planet. Just as rapid population growth exacerbates most global problems, lower growth rates can make the problems more manageable and increase the chances for improving the quality of human life, especially in the Third World.

What You Can Do

Battling the all-encompassing issue of population growth may seem overwhelming to you as an individual, but don't be discouraged! An effective plan of action can focus on a variety of areas, as the relationship between population and resources is a vital and far-reaching one.

Inform Yourself

Before beginning to act, increase your awareness of the population growth crisis so you can speak and write on the issue with authority.

- Consult local, national, and international organizations, including those sources listed in the "Further Informa-

tion" section, to become more knowledgeable about the population crisis. Subscribe to their newsletters to keep up to date on the latest developments.

- Find out which government and private agencies in your community are concerned with family planning services, health and welfare, services for women, and economic development; learn how actions taken by the agencies may affect population growth in your region.

- If you live in a rural area, visit a large city like New York or Los Angeles and observe firsthand the effects of concentrated population on health, behavior, and the quality of life.

- See how population growth is affecting your own community. Review regional demographic trends and assess their relationship to air and water quality, traffic congestion, and the quality of municipal services. Find out whether your community has a planning ordinance to regulate population growth and development. Get involved in town meetings and your local civic association and planning board, and work on issues related to population which affect you and your family's lifestyle directly.

Join With Others

Form a study group with other concerned citizens in your area, and work together to raise local awareness of the regional and international effects of rapid population growth. Support Planned Parenthood, Zero Population Growth, the National Audubon Society, or other organizations concerned with population issues by joining as a member, contributing to fund raising, or volunteering your services.

Review Your Habits and Lifestyle

Although you may not realize it, the careless actions you may take as an individual inevitably contribute to the severe effects of rapid population growth.

- Look for ways to conserve and protect the resources that the global population shares. For example, try recycling paper, glass and cans; carpooling, walking, or riding a bike instead of driving; turning down the heat or air conditioner; turning off extra lights and running faucets; eating less meat.

- When planning your family, be aware of the impact each human being (especially one living in an affluent country) has on the Earth's resources. Consider adopting a second or third child.

- Remember, you alone decide the impact you make in your home, community, school or workplace as a student, citizen, voter, and consumer.

Work With Your Elected Officials

Contact your local, state, and national officials and find out their positions and voting records on population legislation. In brief, clear letters and phone calls, tell your representatives where you and your organization stand on the issues and urge them to:

» support a national population policy, perhaps along the lines of the Global Resources, Environment, and Population Act of 1987;

» increase funding for domestic family planning programs and population-related research;[94]

» encourage development assistance programs and loans directed at the education and empowerment of women;

» provide more funding for international development assistance, particularly for integrated family health care and family planning programs in developing countries;[95]

» increase U.S. support for the International Planned Parenthood Federation and the United Nation's Fund for Population Activities.

Publicize Your Views

Strive to inform your community leaders, policymakers, and the general public about the adverse impacts of rapid population growth on the quality of life and on communities, nations, and the global environment. Stress the urgent need to meet the rapidly growing demand for family- planning services in developing countries.

- Write letters to the editors of local newspapers; help your organization create a public service message for local television and radio stations.

- Support efforts to allow advertising of contraceptives on radio and television.

- Organize educational workshops for the local PTA, Key Club, Rotary, and other community groups.

- Sponsor a showing and discussion of National Audubon Society's film "What is the Limit" or of Better World Society's video "Increase and Multiply" (see listing under Further Information).

Raise Awareness Through Education

Find out if schools and libraries in your area have courses, programs, films, and other educational resources related to population growth.

- Meet with teachers of biology, ecology, health, and social studies and urge them to incorporate population issues in class discussion.

- Help provide educators with materials; offer to speak to classes or organize projects on population-related topics. Suggest exercises that draw attention to the consumptive lifestyles of industrial nations and stress possible ways to reduce individual consumption.

- Acquire educational posters from national population organizations and donate them to local schools for library displays. (See Further Information for listings of educational materials.)

Consider the International Connections

Try to travel or study in a Third World country. Before you go, learn about the country's population—its size, growth rate, and age structure. Visit a large urban center such as Mexico City. Observe the range of living standards and the quality of life, and consider how they are related to the city's size and growth rate. Compare what you witnessed with the direction your own community is taking, and work on the critical issue of population growth with renewed vigor.

Further Information

A number of organizations can provide information about population issues.[*]

Books

Fornos, Werner. *Gaining People, Losing Ground: A Blueprint for Stabilizing World Population.* Washington, DC: The Population Institute, 1987. 121 pp. Paperback, $6.45. Examines the links between rapid population growth, maternal and child mortality, and environmental destruction.

---

* These include the Planned Parenthood Federation of America, Population Communication, Population Crisis Committee, Population Institute, Population Reference Bureau, United Nations Population Fund, and Zero Population Growth.

Gupte, Pranay. *The Crowded Earth: People and the Politics of Population*. New York: Norton, 1984. 349 pp. Hardback, $17.95. Assesses the effects of population growth both in terms of individual suffering and national disaster for countries that cannot accommodate rapidly growing numbers.

National Research Council. *Population Growth and Economic Development: Policy Questions*. Washington, DC: National Academy Press, 1986. 108 pp. Paperback, $10.00. Concludes that slower population growth would be beneficial to economic development for most developing countries.

Reining, Priscilla and Irene Tinker (eds.). *Population: Dynamics, Ethics, and Policy*. Washington, DC: American Association for the Advancement of Science, 1975. 184 pp. Paperback, $3.50. A collection of articles from the AAAS journal, *Science*.

Sadik, Nafis. *The State of World Population 1988*. New York: United Nations Population Fund, 1988. 21 pp.

Wattenberg, Ben J. *The Birth Dearth*. New York: Pharos Books, 1987. 182 pp. Hardback, $16.95. Asserts that recent declines in fertility in the West will weaken Western nations' liberties, values, and economies.

Weber, Susan (ed.). *USA by Numbers: A Statistical Portrait of the United States*. Washington, DC: Zero Population Growth, 1988. 164 pp. Paperback, $8.95 plus $1.50 postage. Includes statistics and brief discussions about states and cities, immigration, fertility, contraceptive use, adolescent sexuality, abortion, the labor force, income distribution, water and air pollution, waste management, land, and wildlife.

Articles, Charts, and Pamphlets

Brown, Lester R. "Analyzing the Demographic Trap." In Lester R. Brown et al., *State of the World 1987*. New York: Norton, 1987, pp. 20-37. An expanded version is available as *Our Demographically Divided World*, Worldwatch Paper 74. Washington, DC: Worldwatch Institute, December 1986. 54 pp. $4.00.

Camp, Sharon L. (ed.). *Population Pressures—Threat to Democracy*. Chart and text, 18x24 inches. Washington, DC: Population Crisis Committee, 1989. Compares the magnitude of demographic pressures with the performance of national political institutions in 120 countries.

Camp, Sharon L. and J. Joseph Speidel. *The International Human Suffering Index*. Chart and text, 18x24 inches. Washington, DC: Population Crisis Committee, 1987. $5.00. Rates living conditions in 130 countries and allows a comparison of rates of population increase and human suffering.

Jacobson, Jodi L. "Planning the Global Family." In Lester R. Brown et al., *State of the World 1988*. New York: Norton, 1988, pp. 151-69. An expanded version is available as Worldwatch Paper 80. Washington, DC: Worldwatch Institute, December 1987. 54 pp. $4.00.

Keyfitz, Nathan. "The Growing Human Population." *Scientific American*, September 1989, pp. 119-26. Poses the dilemma of how to achieve economic and social progress in poor nations where population growth hastens environmental degradation and threatens development itself.

Merrick, Thomas W. et al. "World Population in Transition." *Population Bulletin*, Vol. 41, No. 2, April 1986. 52 pp. Published by the Population Reference Bureau, Washington, DC.

Population Crisis Committee. "Access to Birth Control: A World Assessment." *Population Briefing Paper* No. 19, October, 1987. Includes companion 18x24 inch chart, "World Access to Birth Control." The study rates the availability of family planning information and services for 95 developing countries and 15 developed ones.

Population Crisis Committee. *Nongovernmental Organizations in International Population and Family Planning*. Briefing Paper No. 21. December 1988. 20 pp. Free. A listing and description of more than 100 private organizations active on population issues of national and international importance.

Population Reference Bureau. *Metro U.S.A. Data Sheet*. Washington, DC: 1987. Chart, 18x35 inches. $3.00. Gives population estimates and selected demographic indicators for the metropolitan areas of the United States.

Population Reference Bureau. *Population Handbook*. 2nd Edition. Washington, DC, 1988. 72 pp. $5.00. A guide to population dynamics for journalists, policymakers, teachers, students, and others.

Population Reference Bureau. *1989 World Population Data Sheet*. Washington, DC: 1989. Chart, 18x24 inches. $3.00. Gives demographic data and estimates for the countries and regions of the world.

Population Reference Bureau. *The United States Population Data Sheet*. 8th Edition. Washington, DC, 1989. Chart, 18x24 inches. $3.00. Gives demographic data by state and region.

World Resources Institute (WRI) and the International Institute for Environment and Development (IIED). "Population." In WRI and IIED, *World Resources 1986*, pp. 9-26. "Population and Health." In WRI and IIED, *World Resources 1987*, pp. 7-24. "Population and Health." In WRI and IIED, *World Resources Report 1988-89*, pp. 15-33. New York: Basic Books, 1986,

1987, 1988. Looks at implications of rapid population growth for natural resources, trends in fertility reduction efforts, the state of the world's health, and AIDS.

Zero Population Growth. *ZPG's Urban Stress Test.* Washington, DC: 1988. $4.95. Wall chart and text analyze population-related stress in 192 U.S. cities, with rankings based on a variety of environmental and socioeconomic indicators.

Films and Other Audiovisual Materials

The Population Reference Bureau rents through the mail over 50 video tapes, films, and slide/tape programs on population dynamics, the environment, and related topics. For a free list, send a self-addressed, stamped envelope.

Silent Explosion 1986. 20 min. Video (VHS, Beta) 3/4" ($25.00). Population Institute.

What is the Limit? 1987. A probing discussion of the inter-relationships between human population growth, environmental degradation, resource depletion, habitat destruction and the ethical considerations for the future. Directed at audiences of high school and up, the video is intended as a trigger film to be followed by a discussion based on the accompanying guide "Where Do We Go From Here?" 23 min. Video ($25.00). National Audubon Society Population Program, 801 Pennsylvania Avenue, S.E., Washington, DC 20003.

Teaching Aids

The American Forum: Education in a Global Age (formerly Global Perspectives in Education). *Annotated Bibliography on Resources for Teaching About Population.* $2.20 including postage, from The American Forum.

Global Tomorrow Coalition. *Population Education Packet.* Includes background information, lesson plans, and activities emphasizing interactive learning aimed at upper elementary and intermediate students; can be upgraded for high school. $7.00 including postage. GTC Education Services, 1325 G Street, N.W., Suite 915, Washington, DC 20005-3104.

Population Reference Bureau. Distributes a variety of teaching materials, including:
  1) Population Learning Series Modules. Teaching units featuring student readings, graphs, tables, and classroom activities, $4.00 each. Modules available include:
    a) World Population: Facts in Focus, 1988.
    b) U.S. Population: Charting the Change, 1988.
    c) Human Needs and Nature's Balance: Population, Resources, and the Environment, 1987.
    d) Global Population Trends: Challenges Facing World Leaders, 1987.
  2) Population Reference Bureau Teaching Kits
    a) An Introduction to Population Dynamics, 1984. $10.00.
    b) The World's Women, 1986. $7.50.
    c) Population in Perspective: A Guide to World Population Issues for Model UN Participants, 1986. $5.00.

The World Bank. *Life Expectancy at Birth* and *Population Growth Rate.* Poster kits including poster map with data, text, and charts; six color photos; and teaching guide. $6.50 each.

The World Bank. *Improving Indonesia's Cities.* A case study of rapid urbanization and its effects on a country's economy. Includes 36 student pamphlets and case study books, teaching guide, and sound filmstrip with script. $60.00 complete or $9.95 for sound filmstrip with script. World Bank Publications, Dept. 0552, Washington, DC 20073-0552.

Zero Population Growth. *EdVentures in Population.* 1984. An introduction to population concepts through 16 activities for upper elementary and secondary school students. Includes data, charts, masters, a poster, follow-up activities, glossary of terms, and teacher's guide. $19.95.

Zero Population Growth. *Global 2000 Countdown Kit.* 1984. A comprehensive teaching package on the Earth's population, environment, and resources, for secondary school use. Contains 14 modules that develop skills in mathematics, science, social studies, and language arts. $19.95.

Zero Population Growth. *USA by Numbers Teaching Kit.* 1988. For high school and college use; relevant to science, mathematics, geography, and social studies. Includes the book *USA by Numbers* (see above), teacher's guide, 14 classroom activity modules, glossary, and resource list. $19.95.

Development and the Environment

The Third World is littered with the rusting good intentions of projects that did not achieve social and economic success; environmental problems are now building even more impressive monuments to failure in the form of sediment-choked reservoirs and desertified landscapes.

— Walter Reid et al., *Bankrolling Successes*[1]

The advanced Western countries completed their escape from poverty to relative wealth during the nineteenth and twentieth centuries. There was no sudden change in their economic output, but only a continuation of year-to-year growth at a rate that somewhat exceeded the rate of population growth.

— Nathan Rosenberg and L.E. Birdzell, Jr.[2]

A drought is a lack of water, but not necessarily a disaster. Whether or not a drought becomes a disaster depends on how people have been managing their land before the drought. . . . Given the overcultivation and land misuse forced on many people in the tropical Third World, one would expect drought to be a major disaster worldwide. It is. During the 1960s, drought affected 18.5 million people each year; during the 1970s, 24.4 million. And in the one year of 1985, at least 30 million suffered in Africa alone.

— Lloyd Timberlake, *Africa in Crisis*[3]

By means of a determined effort in the developing world, the proportion of children protected by immunization has been levered from under 10 percent to over 50 percent in the last eight years. Common illnesses like measles, tetanus, and whooping cough . . . are now on the retreat worldwide. Vaccines are now saving at least 1.5 million children annually. And the incidence of polio . . . has been reduced by 25 percent in the last decade and could be eradicated completely in the next.

— United Nations Children's Fund[4]

We are living in an historic transitional period in which awareness of the conflict between human activities and environmental constraints is literally exploding. . . Never before in our history have we had so much knowledge, technology and resources. Never before have we had such great capacities. The time and the opportunity has come to break out of the negative trends of the past.

— Gro Harlem Brundtland, Prime Minister of Norway[5]

Major Points

- Today, nearly a billion people live in "absolute" poverty—lacking enough income to obtain a diet adequate to prevent stunted growth and serious health risks. About 1.3 billion people do not have safe water to drink; 900 million adults cannot read or write; and average incomes in most of Africa and Latin America have declined between 10 percent and 25 percent in the 1980s.[6]

- The gap between rich and poor nations is large and widening. In 1977, developing countries' average income per person was about 9 percent of that in industrial nations; by 1987, developing countries' average per capita income was less than 6 percent of that for industrial nations.

- Developing countries now pay over $30 billion a year more to industrial countries in the form of interest and repayment on loans than they receive from industrial countries in the form of development assistance. Thus, there is a net flow of money away from developing countries to industrial countries.

- In many developing countries, poverty, foreign debt, the need to export for foreign currency, misconceived government policies, and overpopulation cause natural resources to be overused, misused, and degraded.

- Industrial countries are also in the process of development. Their current uses of energy and resources and production of wastes are causing unacceptable environmental damage and are not sustainable at current levels, and they are not meeting the basic needs of segments of their populations.

- New scientific knowledge about agriculture and ecosystems, along with modern technology and traditional knowledge, once they are applied and extended, offer the potential for making development much more sustainable than it is at present.

- Successful development comes from the integration of many factors, including consultation with local people, equitable distribution of benefits, protection of the environment, limitation of population growth, provision of education and health care, and assurance of equitable trade patterns. All must be approached together; piecemeal development has failed.

The Issue: Unsustainable Development

"A decent provision for the poor is the true test of civilization," declared Samuel Johnson at the time of Britain's commercial and industrial revolution in the 18th century.[7] After World War II, perhaps for the first time, a concerted international effort was made to improve the status of the world's poorest people. Today, however, alarm bells are sounding. The standards of living and the health of national economies in most developing countries have worsened, sometimes dramatically, and there is little optimism for improvement in the near future.

In Ghana and Zambia, for example, infant mortality rates were halved between 1950 and 1970, and per capita income tripled. But today, average incomes are actually lower than they were in the 1960s, and indicators of social well-being have declined. In the 1950s and 1960s, many South American economies grew rapidly. By the 1980s, however, that growth proved to be based on unsound policies, and much of Latin America had become mired in debt and financial crisis, with improvement in economic and social indicators slowed in some countries and reversed in others.[8]

The industrial countries have tried to improve this situation by providing loans and technical assistance to help poor countries improve their agriculture, housing, schools, health-care systems, transportation, and industry. However, many experts now believe that most development projects have failed to achieve their objectives.[9] Such projects have generally been planned and carried out without the involvement of local people and organizations. Development planners often evaluate projects in terms of economic indicators that do not reflect the importance of local cultures, traditions, and environmental factors. Indeed, many development projects have actually reduced the quality of life and degraded natural systems on which life depends.

This loss of natural resources may be the greatest threat of all. It undercuts future economic potential. In some cases, damage is irreversible. All countries depend on air, water, soil, and forests as bases for human survival, as well as for economic and social development.

Environmental degradation and poverty are mutually reinforcing. Growing populations left poor by inequitable and misconceived economic policies are forced to overuse and degrade resources in order to survive, and those damaged resources will no longer support economic growth. Historically, this pattern has undermined the development potential of societies and reduced their stability.[10] The basic investment in healthy people and healthy ecology that all economies need in order to grow is lacking today in many countries.

The Development Stalemate

According to the dictionary, "development" means expansion or realization of potential, or gradual advance to a fuller, greater, or better state. Yet this definition leaves questions unanswered. Who benefits from development? Does development always involve economic growth? Who controls the development process? What are the costs of development? Who pays those costs?

On a personal level, it is easy to consider development as simply making things better. Institutionally, however, development has to be defined before development policies can be implemented. Different institutions follow different definitions, and carry out different policies. The U.S. Agency for International Development has published lists of U.S. foreign aid objectives. They have included "alleviate poverty," "cut infant mortality," "improve housing and urban infrastructure," "improve policy environments in host countries," "modernize agricultural policies in developing countries," "find markets for American farm products," "dispose of U.S. agricultural surpluses," and "reduce population growth rates." They have also included "win friends for the United States among governments of developing countries," "win friends among their people," "counter diplomatic initiatives of the Soviet Union," and many others.[11]

Of the development assistance flowing from industrial countries, less than 15 percent goes for agricultural development, even though the livelihoods of most of the poor come from agriculture. Less than 11 percent goes toward education and less than 5 percent to health and family planning combined. Less than 25 percent of the assistance goes to the forty least developed countries. Further, more than half of the money from assistance is now tied to the purchase of goods and services from donor countries.[12]

Defining the Third World

To many Americans, countries that are nominally lumped under the heading "Third World" are difficult to distinguish from one another, or at best, can only be told apart by geographic region. The term "Third World" is most often used to describe a country's level of economic development, and for many people, it connotes abject poverty. Yet Third World countries range from "newly-industrialized," upper-middle income nations such as Korea and Singapore, to less developed, low-income countries like Chad and Afghanistan.[13]

The term "Third World" comes from a French expression, *le tiers monde*, used in the early 1950s to describe nations shaking off the shackles of colonial rule. Later, the term took on new meanings when many Third World countries refused to align themselves with the industrial democracies of the "First World" or the socialist bloc countries of the "Second World."[14] Today, these distinctions are becoming blurred as most countries move toward incentive-based economic systems, where individuals benefit in proportion to what they contribute to national economic growth.[15]

Of the 103 countries that comprise the Third World, the World Bank considers roughly 39 to be less developed countries with per capita GNP below $425; another 35 are lower-middle income countries, such as Liberia and Thailand, with per capita GNP between $426 and $1600;

and another 19 are labelled upper-middle income countries with a per capita GNP between $1600 and $7500.[16] For comparison, the U.S. GNP per person is about $17,500.

Measuring Development

It is difficult to gauge development, and often impossible to accurately compare a country's level of development with that of other countries. A common indicator of economic development is Gross National Product (GNP). As a measure of economic well-being, however, GNP has many limitations. It primarily measures the monetary value of the production of goods, and does not adequately include activities such as food production for family consumption, the unpaid labor of women, and the local collection of firewood and water. Especially in rural areas, these unmeasured activities are important elements of development. In addition, GNP statistics usually fail to reflect government expenditures on health care and education, which improve national well-being. For example, Sri Lanka's per capita GNP is $400, while South Africa's is $2,000. However, Sri Lanka's mortality rate for children under five is half that of South Africa, a difference unreflected in the GNP statistics.[17]

Further, GNP does not reveal how wealth is distributed within a nation. Kenya's GNP per person is about double that of Bangladesh; but the living standards of the poorest 40 percent of the people in the two countries are roughly equal. Brazil's per capita GNP is twice that of Thailand, but in Thailand the poorest 40 percent of the people have a higher standard of living than those of Brazil.[18] Finally, GNP does not reflect environmental costs of economic production. It does not take account of groundwater contamination, damage to forests, damage to marine ecosystems, or the drawing down of water tables to dangerously low levels.

With GNP and other economic and social indicators, errors and omissions in data are often larger than the changes they are being used to measure.[19] And the sources of data are often questionable. Most developing countries lack the institutional capacity needed to collect accurate social and economic information. Many governments cannot even keep accurate data on their own activities. But these governments commonly report on the status of their countries anyway, even though the data is not reliable. As a result, statistical indicators are best used as rough gauges of national status, not as precise measures of living standards.

The interpretation of indicators is also complex: there is no totally objective way to judge "adequate" standards of living. Different people perceive standards of living differently. Different value systems put different weights on health care, free time, housing conditions, disposable incomes, life expectancies, job opportunities, democratic forms of government, education, rates of population growth, and conservation of resources.[20]

Poverty and the Rich-Poor Gap

For almost 900 million people in the Third World, The United Nations Children's Fund (UNICEF) estimates that development has halted, and their societies are back-pedaling into poverty. Average incomes in most of Africa and Latin America have declined between 10 percent and 25 percent in the 1980s. In the 37 poorest countries, per capita expenditures during the 1980s have dropped 50 percent for health care and 25 percent for education. After years of gain, the average weight for young children has again begun to decline in many countries. Each year, a half million women die from pregnancy-related causes, 99 percent of them in the Third World. Of the 15 million children under the age of five that die each year, the overwhelming majority are in the Third World.[21]

The gap between rich and poor countries is large, and it is growing larger. In 1987, developing countries, with 77 percent of the world's population, had an average income per person of only $670 per year, less than 6 percent of the average income of $12,070 in industrial nations. Ten years earlier, developing countries' average per capita income was $490 a year, more than 9 percent of the average income of $5,210 in industrial nations.[22]

Industrialized countries have less than 25 percent of the world's population, but consume 75 percent of its energy, 79 percent of all commercial fuels, 85 percent of all wood products, and 72 percent of all steel produced.[23] Since 1980, the number of malnourished people in developing countries has risen by 30 percent. It is estimated that one-fifth of the world's population—nearly a billion people—goes hungry. Nations with a per capita GNP of $400 or less account for 80 percent of the world's malnourished.[24] (See Chapter 5.)

The gap between the rich and the poor within individual countries is also great, and often growing. With a per capita GNP of $1,810, Brazil is considered an upper-middle income country, but the top 5 percent of its people get a huge proportion of the national income. According to official statistics, 86 million Brazilians suffer from malnutrition. In northeastern Brazil, where about 35 million people live, children weigh about 16 percent less and are 20 percent shorter than other Brazilian children. In addition to an inequitable distribution of the national wealth, Brazil also is burdened with the requirement to pay about $1 billion a month in debt interest and repayments to the industrial world.[25]

In Brazil and the United States alike, there is a large discrepancy between the incomes of the richest of the people and the poorest. In Brazil, using 1982 figures, the richest 20 percent of the population held 64 percent of the national income, while the poorest 20 percent controlled just over 2 percent of the national income. In the United States, using 1985 figures, the richest 20 percent controlled 44 percent of the national income, while the poorest 20 percent held just under 5 percent. In Brazil, the average income of the richest 20 percent was twenty-eight times greater than the average income of the poorest 20 percent. In the United States, the average income of the richest 20 percent was almost ten times greater than the income of the poorest 20 percent.[26]

In the poor countries of Latin America, Africa, and the Caribbean, debt repayments now consume, on average, more than 25 percent of all export earnings; for sub-Saharan Africa, the figure is closer to 50 percent. For most countries, annual repayments of interest and principal on outstanding loans exceed what they currently receive in aid and loans. These same regions also have experienced an annual drop in their share of world trade throughout the 1980s, and their labor productivity has declined.[27]

Life expectancy figures in the developing world have been rising for the last 25 years. However, a recent study of 73 debtor nations showed that through the year 1982, the higher the per capita debt of a country the lower the increase in life expectancy of its people.[28]

Trade Imbalances

Going back to colonial times, trade and the international division of labor were based on the exchange of manufactured goods from the industrialized countries for the raw agricultural and mineral commodities of the colonies. Under that system, the impact of the industrialized countries' steady economic growth on the colonies was profound.

During the colonial period, Africa and much of the Third World were forced to become primary producers of commodities like cocoa, peanuts, coffee, and cotton. From initially being forced into cash crop agriculture by their colonial masters, African peasants became wedded to cash crops as their only ticket into the global economy. After World War II, the production of cash crops in Africa expanded rapidly. Colonial populations had become dependent on manufactured goods imported from the West. The War had interrupted shipping of these goods and created a large pent-up demand for petty manufactures such as tools, pots, cotton cloth, and lanterns. To earn enough cash to afford such items, more and more farmers were encouraged to switch from staple food crops to commodities that could be exported.[29]

As more peasant farmers in Africa produced these commodities, supply began to outpace demand. And as other countries entered the markets, prices for commodities fell, yet farmers who had already invested in cash crops had few alternatives but to continue to cultivate these crops. They could not switch to producing staples for local consumption, because there was no market due to surpluses from agricultural overproduction in the United States and Western Europe, which were supplied to Africa as food aid.[30] Ironically, it was often easier to ship food across the Atlantic than it was to ship it from the countryside to the markets of the provincial or national capital.

Prospects in many Third World rural areas deteriorated, and many peasant families migrated to urban areas, which were already overcrowded. Governments tried to feed these migrants through food aid or by purchasing surplus food from the United States and Europe. Even rural areas were becoming partially dependent upon food imports. In the 1970s, food imports to Africa rose by 600 percent. By 1985, two-fifths of Africa's food had to be imported.

With swelling urban populations, African governments were compelled to subsidize the costs of obtaining food so as to forestall riots. Food production continued to decline and imports rose. For payment, governments needed more foreign exchange from the sale of cash crops. The only logical choice for raising this foreign exchange was to encourage peasants to further expand cash crop production, which depressed the market once again. When African governments could no longer find the money to import food, many resorted to loans from foreign banks. Between 1973 and 1987, Africa's debt increased from $14 billion to $125 billion.

Financial resources needed by many Third World governments for development went instead to pay interest and principal on loans. Further, the emphasis on production for export has caused significant environmental degradation. Crops were grown even if they were not suited to local conditions. Heavy use of expensive fertilizers, pesticides, and irrigation boosted production but depleted farmers' financial resources and made the soil infertile. Food crops were grown on marginal lands where wind and soil erosion could not be controlled. In many areas, non-native cattle were introduced that required more forage than native livestock, and were susceptible to local diseases. The cattle depleted water supplies and stripped the land of vegetation, aggravating erosion and degrading the land's carrying capacity.

Much of this was encouraged by government subsidies and tax exemptions, and such practices continue today. Agricultural subsidies are meant to stimulate agricultural output, thereby contributing to economic development, generating foreign exchange, increasing food supplies, and improving farmers' incomes. Subsidies depress the prices of such agricultural inputs as pesticides, fertilizers, irrigated water, and machinery, which encourage farmers to use them. In Indonesia, for example, pesticides were sold to farmers at prices discounted by as much as 82 percent. But such excessive use of inputs often raises productivity less than the costs of the inputs, eliminates the incentive to manage resources wisely, and causes severe environmental damage in return for minimal increases in agricultural production. Further, subsidies are costly to national budgets. Indonesia's costs in the early 1980s for pesticide subsidies alone were $100 million per year.[31]

The United States subsidizes food production and assists its farmers by buying their excess crops. The cost of U.S. farm support in 1986 was $25.8 billion.[32] As illustrated earlier, much of that food is sent to Third World countries as food aid; it floods international markets, and greatly lowers the prices that Third World farmers can get for both their food crops and export commodities. The exchange earnings of developing countries are reduced, commodity prices are highly unstable, and Third World farmers have little incentive to produce food.

Brazilian cattle-subsidies have amounted to $1.4 billion between 1965 and 1983. Through huge tax credits, tax postponements, several other highly beneficial tax measures, and loans, they have caused large amounts of land to be devoted to cattle ranching. By 1983, the ranches were responsible for 30 percent of the total deforestation detected by the Landsat monitoring program between 1973 and 1983. The subsidized ranches also significantly increased the gap between the rich and the poor in Brazil. From the perspective of the Brazilian people, the return on a typical cattle ranch over a 15-year program is a 55 percent *loss*. From the perspective of a private investor, the return is a 249 percent *gain*.[33]

Third World Debt

Economists often trace the current debt crisis back to the 1970s, when oil prices rose dramatically. OPEC countries deposited their new wealth in transnational banks, and Third World countries borrowed the "petro-dollars" to finance modernization programs and to pay higher fuel costs.[34]

Between 1973 and 1983, development assistance loans to Latin America increased tenfold, from $35 billion to $350 billion.[35] Funds were readily available, at easy terms, and discipline among both borrowers and lenders broke down. During this period, U.S. interest rates began to soar because of the burgeoning federal deficit and tight monetary policies. Many of the loans to Third World countries were based on variable interest rates, and in the early 1980s those rates rose.

Instead of being spent for sound economic development programs, however, much of the borrowed money was spent in other areas. For instance, it financed the purchase of costly military hardware. Beginning in the early 1970s, military spending rose more than 10 percent a year in most Latin American countries. In Africa, military spending rose an average of 18 percent a year. According to the Stockholm International Peace Research Institute, about 20 percent of Third World debt stems from the purchase of arms.[36] Most Third World countries do not manufacture military hardware but import it from industrial countries, so its purchase does not create jobs or income in the Third World, but only depletes already scarce cash reserves.

Many development loans have been used to finance large-scale projects that never achieved success. For example, a nuclear power plant in the Philippines financed by loans was built near a volcano and cannot operate as a result. It costs the Filipino people $500,000 a day in interest.[37]

Parts of some loans to the Third World have been returned to banks in industrial countries to earn interest for Third World elites. Parts also went for the purchase, by Third World elites, of expensive goods from industrial countries, which does nothing to stimulate the developing economies — even though this practice is encouraged by many government policies that overvalue national currency. Other loans essentially have been lost through mismanagement of national railroads, utilities, and other enterprises.[38]

Without this sound investment in development, Third World countries could not meet their loan payments and became dependent on additional financial help from outside institutions including the World Bank, the International Monetary Fund (IMF), commercial banks in the United States, Europe, and Japan, and bilateral development agencies. The World Bank and the IMF are part of the United Nations system and were established in 1944.

The World Bank makes loans to governments for development projects such as hydroelectric power, irrigation, roads, health clinics, and energy and agricultural improvement. The IMF makes short-term loans to countries with immediate balance of payment problems. The IMF attaches austerity programs to its loans to ensure that recipients will be able to repay the loans later.

Basically, IMF austerity policies are designed to force countries to increase revenues and reduce expenditures, a desirable goal. But by increasing revenues, the IMF means increasing exports that bring in foreign exchange. Since most Third World countries supported by IMF loans produce similar goods for export, world markets are flooded with their exports, and their prices fluctuate dangerously. Between 1973 and 1988, the world price for a kilogram of sugar has varied from as high as 65.5 cents to as low as 8.9 cents and has changed by as much as 44 cents in a single year — for a variety of reasons, not just IMF austerity measures.[39] Between 1984 and 1986, the United States saved about $65 billion on payments for raw materials imported from the Third World because of price declines.[40] In 1986 alone, declining materials prices reduced the revenues of sub-Saharan Africa by $19 billion — about four times the amount that the region was promised in emergency aid during the same year.[41]

By reducing expenditures, the IMF means reducing or eliminating subsidies on fuel and staple foods and making deep cuts in government programs. These often include programs that affect health care, education, environmental protection, and other areas vital for maintaining economic growth. Often, the military and other institutions important to politically powerful local elites are spared these cuts.

The total debt of the developing world is now over $1.3 trillion.[42] For Latin America, about 70 percent of the debt is owed to private banks. In Africa, most debt is owed to

multilateral development banks and bilateral development agencies. Ten years ago, the net flow of funds from industrialized countries to the Third World was about $40 billion. But since 1983, the flow has been reversed. Taking into account loans, aid, and repayments of interest and principal, the developing nations now return to donors over $20 billion a year more than they receive in new aid and loans. (See Figure 4.1.) In 1988, developing countries' debt payments were more than $30 billion greater than the loans they received.[43]

Figure 4.1
Net Resource Transfers to Developing Countries, 1973-87

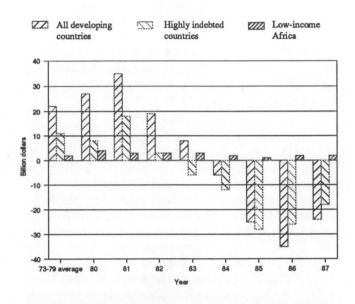

Note: Net resource transfers are defined as disbursements of medium- and long-term external loans minus interest and amortization payments on medium- and long-term external debt.

Source: The World Bank, *World Development Report 1988* (New York: Oxford University Press, 1988), as published in Ralph H. Smuckler et al., *New Challenges, New Opportunities: U.S. Cooperation for International Growth and Development in the 1990s* (East Lansing, MI, Michigan State University, 1988), p. 4.

In 1982, Mexico's finance minister went to Washington to inform the U.S. government that Mexico was unable to meet the payments on its $80 billion debt. It was then that the debt problem assumed crisis proportions, creating fears that the integrity of the entire international economic system was at stake. In 1982, in an effort to curb inflation, the U.S. Federal Reserve Board raised interest rates to nearly 20 percent.[44] The world economy slid into a recession and commodity prices plummeted. Most debtor nations were stunned by the sudden loss of income and the dramatic increase in the cost of the interest on their outstanding loans. By 1983, 42 countries were behind in their debt payments.[45]

Seven countries account for almost half of the Third World debt—Brazil, Mexico, Argentina, Venezuela, South

Figure 4.2
Debt-GNP Ratios in Developing Countries,
1975 to 1987

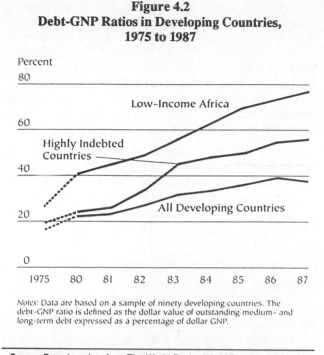

Percent

Notes: Data are based on a sample of ninety developing countries. The debt-GNP ratio is defined as the dollar value of outstanding medium- and long-term debt expressed as a percentage of dollar GNP.

Source: Based on data from The World Bank, *World Development Report, 1988*, op.cit., as published in Smuckler et al., *New Challenges, New Opportunities*, op. cit., p.13.

Korea, The Philippines, and Indonesia. If any of these countries were to default, the international banking system would be endangered. Of particular concern to the United States are the four Latin American debtor countries, which owe U.S. banks nearly $50 billion.[46] U.S. trade with Latin America has been disastrously affected by the debt crisis. The U.S. trade balance with Latin America shifted from a $1.3 billion surplus in 1980 to a $14.1 billion deficit by 1987.[47] On a global scale, the seventeen most indebted nations cut imports by $72 billion between 1981 and 1986, with the United States one of the biggest losers. With the loss of trade with Latin America, the United States has also lost jobs, perhaps more than 2 million.[48]

By contrast, over the first four years of the debt crisis, the profits of the six largest U.S. commercial banks rose by almost 60 percent.

Parallels are being drawn between Latin America's current economic nightmare and the experience of Central Europe in the 1920s, when countries were torn between the extremes of hyperinflation and economic stagnation. Brazil's annual inflation rate for 1988 was near 1,000 percent and could reach 2,000 percent in 1989.[49]

Reflecting the shaky nature of many debtor nations' economies, there is a growing secondary debt market where devalued Third World debt is traded between banks. Currently, Brazil's sells for 35 cents on the dollar, while Peruvian debt sells for only 5 cents on the dollar.[50] The rate at which debt is sold indicates the degree of return that banks believe is likely from a loan. In addition, debt-equity ex-

changes, in which a debt claim is exchanged for stock in a local investment at an advantageous price, have grown to more than a $5 billion market in just over two years.[51] In 1988, U.S. banks set aside $10 billion in reserves against possible losses on loans to Third World debtors. The Bank of Boston has taken the unprecedented step of writing off 20 percent of its outstanding loans to the Third World.

In 1985, then U.S. Treasury Secretary James Baker moved to relieve pressure on the international banking system by calling for closer cooperation between creditors and debtors, more lending by all creditors, and a larger role for the World Bank.[52] However, by 1987, the results of the "Baker Plan" were mixed at best. Production in Latin America was still 5 percent below 1980 levels, and the average inflation rate had risen to 180 percent. In the meantime, creditors continued to reduce their exposure in the region and remained unwilling to extend new loans.

Many banks are comfortable with the Baker Plan's vague approach to the debt problem. The international banking system is still intact, and banks have gained valuable time to improve their balance sheets. Critics point out that five years of belt-tightening has failed to revive the moribund economies of the Third World, and has led to increasing civil unrest in several countries. Debt reschedulings are becoming increasingly difficult, and lenders are still reluctant to make new loans.[53]

Most recently, Treasury Secretary Nicholas Brady unveiled a new U.S. approach to the debt crisis, which asks banks to exchange their existing loans for bonds with a lower interest rate or lower face value. Under the new U.S. plan, the World Bank and IMF would guarantee the principal and interest on new loans. To do this, the United States would have to increase its support for the World Bank and IMF considerably, with the U.S. taxpayer ultimately bearing the burden.

Following adoption of the new U.S. approach, in July 1989 agreement was reached on a debt-reduction package for Mexico that will lower the face value of $54 billion worth of loans by 35 percent, or reduce interest rates to 6.3 percent from more than 12 percent. By the end of 1992, Mexico's debt will be $10 billion to $12 billion less than it would have been under previous arrangements.[54]

An innovative approach to relieving debt and environmental pressures in developing nations is the "Debt-for-Nature Exchange," used successfully with Bolivia, Ecuador, Costa Rica, and the Philippines. First proposed in 1984 by Thomas Lovejoy, the concept initially was met with skepticism. However, three years later, Conservation International (CI) and the Bolivian government signed an agreement concluding a debt-for-nature swap centered around the Beni Biosphere Reserve, home to 13 endangered species, 500 species of birds, and the nomadic Chimane Indians. Using Citicorp as its financial agent in the secondary financial market, CI used a $100,000 grant to purchase

$650,000 in Bolivian debt. Under the terms of the agreement, CI agreed to cancel the debt, and in return the Bolivian government promised to give added long-term protection to the Beni Reserve and to increase the protected areas next to the Reserve. The government also agreed to set up an operating fund of $250,000 in local currency to manage the Beni Reserve and attendant buffer zones.

The World Wildlife Fund has been a principal U.S. actor in subsequent debt-for-nature exchanges in the Philippines, Ecuador, and Costa Rica. So far, the total amount of debt involved in these exchanges is about $18 million — only a tiny fraction of the billions of dollars owed by each of these countries. However, crushing debt pressures have left little or no money for environmental and conservation work in most developing countries, and debt-for-nature exchanges provide modest amounts of capital so that national governments can take effective action to protect their natural heritage.

Environmental Degradation

Natural resources are the basis of economic development; environmental protection and economic development are inseparable. It may appear that the economies of industrial countries have grown detached from agriculture, or have stopped depending on it. In fact, there is no such thing as a post-agricultural society or economy. Economies without ready access to adequate natural resources, and appropriate use of them, are unlikely to be sound and secure.[55]

Soil

In the Third World, economic development and transfer of technology from the industrialized countries have changed the ways people use the soil. The use of pesticides and fertilizers has grown tremendously. A growing percentage of agricultural lands depend on irrigation. In many areas, "technology-intensive" agriculture has replaced traditional "ecology-intensive" agriculture. New seed varieties have raised productivity for some farmers, but also have increased dependence on chemicals and machines. The agricultural revolution has locked some peasant farmers into a vicious cycle of dependence on costly chemical inputs to raise cash crops that are undervalued due to depressed global commodity markets. Intensive, mechanized agriculture depletes the soil, and chemical fertilizers and pesticides are altering agricultural ecosystems.

In industrial countries, mechanized, chemical-intensive agriculture is causing soil erosion, declines in soil productivity, and the contamination of ground water supplies. The United States has a higher rate of soil erosion per hectare of cultivated land than Peru, although erosion rates in many developing countries are higher than in the United States.[56]

Worldwide, 6 million hectares of land are permanently degraded each year to desert-like conditions. Twenty-one million additional hectares provide no economic return because of desertification, which is one direct cause of famine. Soil degradation and loss can have large economic costs. In Canada, a government report estimated that soil degradation costs farmers $1 billion a year. Such costs are leading countries such as the Soviet Union to reconsider their plans to expand agricultural production onto marginal lands.[57]

Methods of food distribution are as important as methods of production. Currently, many countries are highly dependent on food imports, and, in emergencies, on food aid from reserves stored in a few countries. Recent famines have been caused as much by social conflicts and faulty distribution as by inadequate production. (For further discussion of these issues, see Chapter 5.)

Forests

Many struggling tropical countries, reeling from growing foreign debt, burgeoning populations, and rising expectations from their citizens are treating their forests as materials they can export for cash.[58] Tropical hardwoods generate $8 billion each year in much-needed foreign exchange. Each year, some five million hectares of tropical forest are cut for timber, paper pulp, and other wood products. Too often, when logs are removed the entire forest is cleared. Trees not intended for harvest often are damaged or killed by heavy logging machinery, which also compacts the soil and makes forest regeneration much more difficult.[59] Currently, logging is depleting the world's forests much faster than they are being reforested or can regenerate. (See Chapter 7.) In essence, tropical countries are consuming their natural resource capital rather than using the income from that capital.[60]

Every year, an estimated 8 million hectares of undisturbed tropical forest — an area the size of South Carolina — are cleared for cultivation. Without extensive use of fertilizers, deforested soils lose their fertility within a few years. Farmers are then forced to clear more forest.[61]

The clearing of large amounts of forest harms a country's potential for development. Forests perform many functions important to long-term national productivity. In addition to providing homes for hundreds of millions of people, forests protect watersheds, regulate water flow and weather patterns, prevent erosion and sedimentation of rivers, provide foods, fibers, and wood for building, fuelwood for cooking, lighting, and heating, and animal fodder for livestock.[62] (See Chapter 7.)

Brazil is an excellent example of how poverty and inappropriate government policies have led to massive deforestation in the Amazon. Beginning in the mid-1960s, large numbers of landless peasants began moving into the region in search of land and employment. The government attracted them to the region by providing generous tax and credit incentives.

The settlers found that the soils in the region were fragile and could withstand intensive cultivation and livestock management for only a few years. By the 1970s, deforestation in parts of the Amazon, particularly Rondonia, had reached alarming proportions. By 1988, almost 25 percent of Rondonia's tropical forest had been cleared. Despite the vast amounts of forest cleared and the numerous agricultural and ranching operations in the region, Amazonia is still desperately poor and contributes only 3 percent of Brazil's national income.[63]

In 1988 alone, Brazil may have burned as many as 20 million hectares of forest and scrub, an area the size of Nebraska, to clear the land for farming and cattle ranching. The fires caused massive air pollution and probably accounted for about one-tenth of all carbon dioxide emissions from human activities during 1988.[64] (See Chapter 7.)

The plight of Bangladesh illustrates some of the dire consequences of deforestation. In the neighboring upland countries of Nepal and India, and in the hilly areas of Bangladesh itself, much of the forest has been cleared to meet domestic timber needs and to provide fuelwood and building wood for peasants. Lying downstream from these areas, Bangladesh has always experienced extensive flooding, with about 20 percent of its land area under water during an average year.[65]

But in recent years, flooding has increased enormously. In 1988, about two-thirds of the country was under water, including large portions of major cities. Many believe that such catastrophic flooding is caused by the loss of vegetative cover in the upland areas that used to absorb rainfall and moderate the flooding. Rainfall from the Himalayan Mountains also erodes billions of tons of high-quality topsoil that fill river beds, making them shallower and even more prone to flooding.[66]

Before it is exploited, forest land should be classified as to its suitability for agriculture, timber production, wildlife protection, and other uses. Forests unsuited for sustainable agriculture should not be cleared for cropland or cattle ranching, but should be used for other purposes, including watershed protection, sustainable production of forest products, species conservation, and recreation.[67]

Oceans and Coasts

Municipal sewage, industrial wastes, and the misuse of coastal ecosystems have replaced oil contamination as the most serious threats to the health of the world's oceans. Sprawling coastal cities have grown faster than the capacity of sewage treatment plants. Coastal cities often dump their sewage directly into the sea, contaminating marine ecosystems. Industrial pollutants discharged in the oceans reduce their capacity to support life by depleting oxygen levels and damaging fragile food webs. Chemical fertilizers and pesticides from farming and effluents from inland cities and industries travel down rivers into the sea, leading to further marine contamination. As human populations have continued to grow throughout the world, development of coastal areas has damaged many of the Earth's most productive marine habitats, including bays, mangrove forests, and coral reefs.[68] (See Chapter 8.)

Damage from municipal sewage can be prevented. For example, Singapore recently installed modern sanitation facilities for 97 percent of its population, recycling the waste water for industrial use. As a result, the quality of its coastal waters has improved dramatically. In the Mediterranean, after the adoption in 1975 of a UNEP-sponsored action plan for the sea's protection, the waters are cleaner today than predicted ten years ago, the shellfish are less contaminated by mercury, the beaches are cleaner, and tourism is thriving.[69]

But sewage treatment facilities are expensive; instead of paying for them, most coastal municipalities discharge untreated sewage into the sea, even though the practice threatens public health.[70] In Poland, the Vistula River receives so much pollution that the water is too polluted and corrosive even for industrial use.[71] The Vistula is just one of several rivers that flow into the Baltic Sea. The seven countries that border the Baltic dump about 15,000 tons of heavy metals, 70,000 tons of phosphorus, a million tons of nitrogen, and 50,000 tons of oil and highly toxic PCBs every year. Much of the Baltic Sea is biologically dead, and extinction threatens a number of its marine species.[72] The economic and ecological costs of the pollution to the Baltic nations are enormous, and could undermine future economic development.

Oceans are vulnerable to more than just pollution. Overfishing threatens many of the world's major fisheries. Depletion of the Earth's protective ozone layer could damage vital parts of the marine food chain and disrupt planetary life support systems.[73] Climatic warming from the greenhouse effect could raise sea levels and flood coastal regions.

The oceans are massive ecosystems that cover more than 70 percent of the Earth's surface and play critical roles in maintaining its life-support systems, moderating its climate, and providing protein, transportation, energy, and employment.[74] When marine resources are damaged, effects that seem to be localized can have unexpected consequences in distant regions and on seemingly isolated resources. Many effects do not go away with time. Two-thirds of the 1.5 million metric tons of DDT produced before its use was restricted, for instance, may still remain in sediments on the ocean bottom.[75]

The management of ocean systems is not the responsibility of any one country, but all countries together. Fish and pollution do not respect national boundaries, and uncoordinated plans of national management are likely to fail. According to the World Commission on Environment and Development, "the effects of urban, industrial, and agricultural growth ... pass through currents of water and air from nation to nation, and through complex food chains from species to species, distributing the burdens of development, if not the benefits, to both rich and poor."[76]

Cities

For many countries, cities serve as gateways to the global business networks on which national economies increasingly depend. Only by concentrating many of their resources in these cities can nations attract international investment, profit from new technologies, and claim a place in world trade.[77] Many impoverished rural people have flocked to the cities, hoping to find opportunities lacking in rural areas.

Particularly in the developing world, cities are growing at unprecedented rates. In 1950, 29 percent of the world's people lived in urban areas; by 1985, it was 41 percent. In 1950, urban areas contained more than 700 million people; in 1985, urban populations totaled almost 2 billion. Mexico City's population was only 3 million in 1950, but had grown to 16 million by 1982. In the year 2000, the United Nations projects that it will reach 26 million. Sao Paulo is projected to climb from a 1950 figure of 2.7 million to a 2000 figure of 24 million.[78]

Figure 4.3
Percent of Population in Urban Areas, 1950-2000

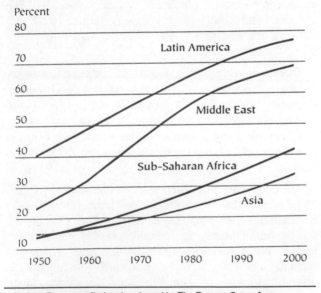
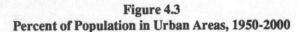

Source: Data compiled and projected by The Futures Group from United Nations and World Bank Sources, as published in Smuckler et al., *New Challenges, New Opportunities*, op. cit., p. 21.

Such urban growth can be seen as development, but it is development with a cost. In the Third World, few governments can provide adequate housing, clean water, sanita-tion, schools, transport, other services to such growing populations. Squatter's settlements spring up, buildings become decrepit, and pollution degrades air quality and contaminates water supplies. Diseases spread easily among people living too close together, weakened by malnutrition, and subsisting without sanitation, health care, or health education. In some Third World cities, urban air pollution is a serious health threat. For example, the mayor of Mexico City has warned of a "collective hysteria" regarding air pollution there.[79]

In the industrial world, the costs may be less painful, but they must be taken seriously. Air pollution is often a significant health hazard. Noise pollution lowers the quality of life in cities. Stress pervades many workplaces, and people feel rushed, sometimes even if they are not in a hurry.

Cities must manage huge amounts of waste, and they rely increasingly on chemical treatment of municipal wastes. John Todd, who has pioneered ecologically sound waste treatment methods, says "the waste-treatment industry is getting scary. To meet regulations on some chemicals, we use others that aren't regulated. We use chlorine to meet ammonia standards, and in the process make chloroform and chloramine, which don't have standards. To get rid of phosphate, we precipitate it out with aluminum. Aluminum is toxic in all parts of the environment, but we haul it out of sewage-treatment plants and dump it onto the land by the ton. We use high concentrations of copper salts, which are not natural in ecosystems, to get rid of algae, which are. Every time the restrictions on one pollutant get stronger, the chemicals to remove it get stronger."[80]

The large quantities of energy and natural resources that cities use, and the pollution they produce, often have profound impacts on distant lands because of their massive scale. Energy consumption per person in industrial market economies is more than 80 times greater than in sub-Saharan Africa.[81] Such intense energy use is altering the global ecosystem. For example, urban emissions enter the atmosphere and are transported over long distances. In the atmosphere, the chemicals combine with moisture, and when they return to earth as rain, snow, and fog they acidify vegetation, soils, and surface water.[82] Acid deposition is killing trees, damaging crops, degrading soil, killing aquatic life, corroding stone and metal, and affecting human health.[83] (See Chapter 12.)

A 1981 OECD study estimated the total crop damage from sulfur dioxide in 11 European countries at $500 million per year. The chemistry of lakes in Sweden has changed more since 1950 than it did between the end of the last ice age and 1950. In the United States, a National Academy of Sciences study estimated the total cost of acid damage in the Eastern United States at $5 billion annually. In Greece, the deterioration of Athenian monuments during the past 25 years has probably been greater than in the previous 2,400 years.[84]

Emissions from urban areas, particularly from motor vehicles, are major contributors to the greenhouse effect. Global warming could cause disastrous consequences, including flooding of coastal cities, alteration of global weather patterns, and disruption of world agriculture. (See Chapter 12.)

Industry

One of the most visible results of development is the enormous growth of industry. The world manufactures seven times as many goods and produces three times as many minerals today as it did in 1950. All aspects of manufacturing affect the environment, including raw materials exploration and extraction, the production process, energy consumption, waste generation, and the use and disposal of products by consumers. Manufacturing is growing most rapidly in developing countries, where industries tend to be pollution- and resource-intensive.[85]

Industry must expand to support growing populations and to raise living standards. Yet industry must not reduce the quality of life, destroy the environment, or undermine the economic potential of future populations by exhausting natural resources. Air and water have traditionally been regarded as "free goods," but in reality they are not free. The economic and environmental costs of their contamination are often not apparent until the capacity of the environment to absorb pollution has been exceeded. After that point, they cannot be avoided, whether they appear as damage costs to human health, property, and ecosystems, or as cleanup payments by manufacturers.[86]

The world's chemical industry has grown rapidly, and now accounts for about 10 percent of world trade, with 70,000 to 80,000 chemicals currently in production. These products are an important part of modern civilization and have made major contributions to today's high standards of living. Yet they also have severe environmental impacts, harming human health and damaging critical natural resources.

Every year, 1,000 to 2,000 new chemicals enter the commercial market, and many have not been adequately tested for toxicity. Even in the United States, the data required to evaluate health effects are available for only 10 percent of pesticides and 18 percent of the chemicals used in commercial products and processes covered by the Toxic Substances Control Act.[87]

In developing countries, pesticides kill some 10,000 people and acutely poison about 400,000 more each year.[88] Chemicals travel through the food chain and spread to areas far from where they are originally used. Pest species rapidly develop resistance to pesticides. The number of resistant species worldwide has climbed sharply, requiring the development of new pesticides to counter the resistant pests. (See Chapter 5.) The heavy use of chemical insecticides is not a sound basis for development.[89]

When industrial countries place restrictions on the use of chemicals they manufacture, developing countries are often not notified, despite their extensive imports of chemicals. Even if those notifications were made, developing countries usually lack institutions capable of receiving and assessing such information. Instead, industrial corporations find profitable markets in developing countries by selling restricted chemicals.

Poverty and Population Growth

Poverty is both a contributor to and a result of population growth and environmental degradation. In this cyclical relationship, the poor often have few alternatives but to exploit resources in an unsustainable fashion. Hungry families attempt to grow food on any land available, regardless of the fragility of the soil or the suitability of crops. When a family is hungry and cold, it will burn wood for cooking and heat, and when the wood supply is exhausted, it will burn dung and crop stems that would otherwise have been used as fertilizer. Often the poor do not have the ability to let land lie fallow, or to undertake reforestation.

The fertility rates of Third World countries are very high—an average of 6.7 children per family in sub-Saharan Africa, for example, and part of the pressure to have large families is economic.[90] More pairs of hands and more strong backs enhance the ability of the family to survive as a viable economic unit. However, exponential population growth imposes pressures on the land and on all resources, and, as a result, large families become a much greater burden than an asset. Agricultural areas are no longer capable of sustaining their rapidly growing populations. In turn, as migrants seek work in urban areas, shanty towns and slums spring up due to inadequate housing and overwhelmed social services. As these volatile urban areas grow, a government's political stability and longevity are threatened. To survive, urban dwellers move into shadowy economic alternatives. In many countries, large sectors of the population are considered wholly outside the national economy. Some observers in Brazil estimate that 50 percent of the labor force is underemployed. It is estimated that 60 percent of the Peruvian economy is part of the underground "informal" sector. Informal activity accounts for 43 percent of all housing in Lima and 93 percent of the urban transportation fleet.[91]

The population in the Nepalese village of Lele has increased by 50 percent over the past ten years from 6,000 to 9,000. With a larger population, demand for fuelwood shot up as did unemployment. Hills that had once been covered in lush forest are now stripped bare. When at one time it took only 30 minutes to collect firewood for the evening meal, it now takes close to five hours. Accompanying the deforestation of the hillsides has been a dramatic increase in landslides and flash floods. In 1987 alone, 500 people were killed in landslides, and every year 250 million cubic meters of topsoil is washed into the Bay of Bengal.

The barren slopes surrounding Lele have proved tempting to unemployed villagers who have created a quarrying industry. With 67 huge craters ringing the village, the probability of landslides has increased and the prospects for successfully reforesting the area have decreased. The loss of forest cover has led to a drier climate for the region. Crop yields are down by 75 percent. Reforestation is an absolute necessity. However, last year, of the 200,000 seedlings planted, only 5,000 survived because not enough was done to protect the plants.[92]

Brazil has long experienced large waves of migration from poorer rural areas of the country to its growing cities. Despite the fact that, after 50 years of such migration, the lot of most migrants is destitution in the slums of Sao Paulo, Rio de Janeiro, Brasilia, and other cities, it continues unabated. However, there is a new twist to the patterns of migration. The Brazilian government has actively promoted Amazonia as an alternative destination for poor, landless migrants with the hope that they will depart from the already overcrowded and increasingly volatile urban areas. Since the 1970s, the bulk of the migrants into the Amazon Basin (5 to 10 million) have come from the southern provinces of Parana and Sao Paulo, which are the wealthiest regions of Brazil, but, nevertheless, have been unable to provide these people with a satisfactory living. To many Brazilians, the Amazon Basin looms as the incarnation of the fabled El Dorado, promising riches for some and a new start for others. The reality, however, reveals not an economic or agricultural mecca but a region that is rapidly losing irreplaceable tropical forest for only a marginal economic return.[93]

Other Factors in Development

In addition to economic and environmental factors in development, many social and cultural factors affect development programs and projects.

The Role of Women

In poor areas of the world, women are responsible for most food production. In sub-Saharan Africa, women account for four-fifths of all food.[94] The extensive role of women in the production of food for local consumption gives them an understanding of natural ecosystems. They are likely to know what crops are appropriate for different land areas, when and how to plant, and how to manage agricultural production, as well as which foods are most needed by their families.

The power to manage, however, lies in the hands of men in most rural societies. They are the legal owners of the land, tools and machines, vehicles, roads, and anything that requires a permit or an ownership deed. They are the politicians, and they have social status. Men tend to make decisions about food production to enhance their status and power, with little regard to the feasibility and sustainability of growing crops. In that system, the relations between women and men regarding agriculture are complex, and sustainable food production often depends on a delicate balance between their priorities.

A major cause of failure in development projects stems from ignorance of the importance of these relations between men and women, and from failure to consider the knowledge, needs, and status of women. Because most of their work is unpaid, the work of women is not included in bookkeeping ledgers, and is therefore considered nonproductive. Despite the major role of women in food production, most census figures show less than 20 percent of women in the agricultural workforce.[95] Yet development planning that ignores their contributions is uninformed planning, and may do more harm than good.

Further, women are a major link between several factors in development. They produce food, and they provide food for their families, and they also have responsibility for family planning. They make use of the environment for agriculture, and they affect the environment by their decisions and their teachings. All of these factors are linked together by women, and the status of women affects each of them.[96] Women who gain education and status in society produce fewer children.[97] That reduces population growth, and puts less of a strain on the environment. Women who have input in agricultural decisions can help assure sustainable production of food, in forms that suit the needs of their families. Women who have adequate health care have the opportunity to produce healthy families. Women who have adequate livelihoods have the opportunity to use resources sustainably, and also have less incentive to produce large families.

In the Third World, however, the needs of women are often neglected. Their health is jeopardized by bearing large numbers of children, and they are at high risk of dying in childbirth; maternity kills half a million women a year, and most of those deaths would be avoided if births were spaced more widely and adequate health services were available. Many Third World women are poor and lack safe water, sanitation, clean places to live, and health care. Their children are susceptible to disease and malnutrition. For women who wish to improve their condition by lowering family size, family planning services are often unavailable; an estimated 300 million couples do not want any more children but are not using any method of family planning, mostly due to inadequate access to services.[98] In many areas, a bias in favor of male education makes it difficult for women to attend school. Education of women, however, is one of the strongest influences on lowering future population growth rates.[99] (See Chapter 3.)

The neglect of women also hinders many development projects. For example, World Bank reforestation projects in India and Kenya failed when directed at men, but succeeded once women were taught to tend the trees.[100] In Ghana and Togo, about 95,000 of the 100,000 people involved in shore-based fishing are women; they process, transport, and market the fish. But a development project in Ghana tried to establish a fish industry with distribution carried out by men. The project failed. Women, however, saved the fish industry in that area when they developed a quick-gutting and brining system to process tiggerfish, after those fish began to dominate catches.[101] In Gambia, development projects had taken land away from women, until a recent project helped women to gain land rights, along with accompanying seed, equipment, and childcare centers. The result was a sixfold increase in rice production that benefits 15,000 people in 40 villages.[102] To succeed, development projects must take account of how indigenous societies manage their resources, and must consider the different roles that men and women play in those societies.

Cultural Tradition

A great many development projects fail because they do not adequately include local people. Indigenous cultures are often quite different from the expectations of development planners in the cities of the industrial or the developing world. Projects that violate cultural rules, norms, desires, or abilities are unlikely to succeed. But many projects are planned almost exclusively around economic variables with little consideration of social and cultural factors. Financial and demographic statistics do not adequately describe particular regions and peoples; at best they tell only a partial story.

Many projects are planned and implemented without the participation of the local people who will be most affected. Planners often assume that they know what is best for indigenous populations, and that projects can successfully be carried out without the approval or involvement of those people. Many unsuccessful projects have been based on such assumptions. Development projects must meet the needs of local people, and include measures to ensure that the people are involved in their implementation and benefit from them.

For example, the building of the Kariba Dam in Zambia caused many of the Gwemba Tonga tribe to be displaced and to change from agriculture based on river flooding to dryland agriculture. Immediately after, the tribe experienced increased malnutrition and disease as well as increases in alcoholism, divorce, and accusations of witchcraft. After a time, many members of the tribe did make the shift to the new form of agriculture, and they had sound livelihoods. However, villagers' diets did not return to normal, diseases that had not existed in that area for many years appeared, and the incidences of others, like tuber-

culosis, increased. Housing deteriorated and remained poor, as did furnishings, and local transport. There were more beer halls, and the tastes of the men drinking in them had shifted from the village beer that women once made and sold to a government-brewed beer that brought little income to the tribe.[103]

Health and Nutrition

In 1985, more than 730 million people did not get enough food to lead productive working lives.[104] This is a precarious base for a sound national economy, and an obstacle to development. Human stamina and creativity are essential for economic growth, but malnutrition weakens human potential and resistance to disease. As long as people are deprived of adequate food and health care, their capacity for development will be reduced.

Development, in turn, has great potential for improving health, even when it does not include a medical component. The greatest reductions in mortality and disease come from improvements in nutrition, hygiene, sanitation, water supplies, and housing. In the Third World, the number of water taps is a better indicator of community health than the number of hospital beds.[105]

Areas with inadequate sanitation often have high rates of malaria, diarrhea, worm infestations, and other diseases. Some 1.7 billion people lack access to clean water, and 1.2 billion people do not have adequate sanitation. Development projects can either improve or worsen sanitation systems. Large dams and irrigation systems can harm sanitation facilities by causing unclean water to stagnate near human communities. Crowded squatters' settlements often expose people to sewage and unclean water. In contrast, new sanitation systems, along with urban water drainage systems, sewers, and pipes for drinking water have improved urban water management and reduced the incidence of water-borne disease.[106]

Development projects can help improve public health, and they can also worsen it. Development that encourages production of the inexpensive foods that the poor traditionally eat, such as coarse grains and root crops, can improve nutrition. Medical care, such as immunization against childbirth infections, oral rehydration therapy against diarrhea, and maternal health care can greatly benefit public health. On the negative side, industrial development projects can increase illness from air pollution and disability from industrial accidents. Projects that cause misuse or overuse of pesticides can cause poisoning. For instance, the World Health Organization (WHO) estimates that approximately a million people are poisoned annually by pesticides, and pesticide poisoning is the leading cause of death in the Pacific Coast cotton-growing region of Nicaragua.[107]

Today, the development of pharmaceuticals, vaccines, and techniques for disease management is primarily directed at diseases common in industrial countries, because those countries provide the most profitable markets. Research is urgently needed on the tropical diseases that affect so many people in the Third World. There is also a critical need for health education. Teaching people how to avoid hazardous water and disease-carrying materials can save and improve many lives.[108]

Diarrheal diseases stemming from malnutrition and lack of clean water supplies are responsible for the deaths of approximately 3.5 million children in the Third World each year. Infectious diseases account for a similar number of deaths annually. In the developing world, 80 percent of all illness is attributed to lack of access to clean water supplies, and half of all hospital beds are allocated to patients suffering from water-related diseases.[109]

Sustainable Development[110]

The Gaia hypothesis, named after the Greek Goddess of the Earth, states that "the evolution of the species of living organisms is so closely coupled with the evolution of their physical and chemical environment that together they constitute a single and indivisible evolutionary process."[111] This is a valuable way to think of development. People are altering the Earth, and the Earth, in turn, affects people. They cannot be divorced from each other.

Many have assumed that environmental protection and economic development are incompatible, and that the interests of industrial nations are incompatible with the needs of Third World countries. This is not the case. In fact, industrial and developing nations depend on each other. Protection of the environment and promotion of economic development are closely interrelated goals. To achieve them, we must align development objectives with the capacity to support sustainable development. Development cannot be sustained by a deteriorating environmental resource base, and the environment cannot be protected when projects consistently fail to consider the costs of environmental destruction, and fail to allocate resources to prevent it. For national economies to grow and be profitable, natural resources must be maintained.

This can be accomplished through "sustainable development"—a policy that meets the needs of people today without destroying the resources that will be needed in the future. It is a policy based on long-range planning and the recognition that, to maintain access to the resources that make our daily lives possible, we must recognize the limits of those resources.[112]

Today, we have the means to attain sustainable development. New technologies and scientific knowledge provide new potential for improving communication, producing food, and maintaining the biosphere. Our new knowledge gives us the ability to make human affairs compatible with natural laws. We can build a secure future that goes beyond military security—one that is also environmentally and economically secure.

Sustainable development is, by its nature, integrated. For example, it integrates concern for protecting the natural resource base with concern for reducing poverty so that people will not be forced to degrade soils and forests to survive. It integrates the need for efficient, sustainable energy use to conserve energy resources with the need for unpolluted cities and healthy global ecosystems. It integrates the value of human health with the importance of human resources to productive national economies. Sustainable development is based on the recognition that:

- Environmental health and economic development are linked to each other, and that environment and economics must be integrated from the start in decision-making processes.

- Environmental stresses are interconnected. For example, the clearing of trees not only means the degradation of forests, but also accelerates soil erosion and silts rivers and lakes.

- Economic and environmental problems are linked to many social and political factors. For example, the rapid population growth that has such a profound effect on development and on the environment in many nations is driven in part by the inferior status of women in these societies.

- Ecosystems, pollution, and economic factors do not respect national borders, making international communication and cooperation critical.

Such integration must also be reflected by institutional change in the agencies and organizations that create policies affecting development. They must bind together development and environment, as well as social and political factors, with industry, agriculture, and trade. They also must bind countries together.

Traditionally, those responsible for managing natural resources and protecting the environment have been organizationally separate from those responsible for managing the economy. But separate institutions and nations acting in isolation can no longer cope with interlocking issues. The interdependence of economic and environmental issues will not change. Institutions must change if they are to regain their effectiveness in the face of these new realities.

The World Commission on Environment and Development, composed of members from twenty-one countries, has been one of the strongest advocates of sustainable development. It has called for sustainable development to be incorporated into the agendas of economic policy discussions and other key discussions at national and international levels, and for environmental issues to be accorded the same value as economic and political topics. It has said that individual governmental departments must be made accountable for the economic and environmental sustainability of their policies.

Perhaps for the first time, such ideas are gathering momentum. The World Commission's significant position in the international community and its close ties to the United Nations have given weight to its recommendations. Other institutions and individuals are calling for similar or identical measures. The World Bank has recently expanded its environmental staff, and has made an effort to incorporate environmental concerns in all its projects, though it has a history of neglecting such factors and has a long way to go to assure their inclusion. The United Nations is firmly committed to sustainable development and protection of the environment, and may play a crucial leadership role in the application and promotion of sustainable strategies, although it, too, has institutional deficiencies to overcome.

Priorities for Sustainable Development

There are many priorities for sustainable development, and they vary according to the needs and opportunities of individual regions. Some priorities, however, are universally important. They include introducing sustainable agricultural techniques, improving energy efficiency, developing renewable energy sources, limiting population growth, developing efficient and clean technology, reducing consumption, and increasing the capacity for foresight in policy development. (For an analysis of sustainable development priorities, see Chapter 16.)

Much is now known about how to achieve these goals. If education and communication can spread this knowledge to those who need it, and if the knowledge can be integrated into regional cultures and institutions, there can be global progress toward sustainable development. But that progress will depend on worldwide commitment and political will.

Success Stories

During recent decades, a variety of development projects have been tried and tested. There have been many successes as well as many failures. A few relatively successful projects from both developing and industrial countries are described below.

"Social Forestry" in Java

Deforestation in Java has been particularly severe. The government took measures to control the clearing of forests, but without success. Traditional methods of policing forest boundaries failed to prevent people from entering the forest for food, fuelwood, and fodder, nor did government policies requiring limited forest use give farmers enough incentive to avoid overexploitation of the forest lands.

In 1984, the Ford Foundation staff in Indonesia, with government backing, helped create a Social Forestry Working Group to find ecologically sound and socially acceptable ways to manage Java's forests. The Group's researchers lived in local communities that would be affected, to understand their cultures and their patterns of forest use.

The researchers found that antagonism between villagers and government agencies was preventing forestry policies from being carried out. The Working Group helped create village-level farmers groups that could work with government officials, and take part in decisions about local forest management. The process allowed the farmers to gain a range of benefits from the forest management programs.

Using agroforestry techniques that combine trees and food crops, the farmers planted tree species far enough apart to allow space for a variety of crops on the same field. Farmers were encouraged to plant trees that yield fruit, fuelwood, fodder, and timber. Agroforestry maintains soil fertility, because crops such as legumes replace nutrients taken from the soil by other crops.

By including local farmers in the design and implementation of the program, their attitudes were greatly improved, and their special knowledge of the region was utilized. Because the farmers received multiple benefits from the program, they had incentives to work for its success.

By 1987, 61 new sites had been established under the program, many on severely degraded land. The Ford Foundation renewed the original grant and provided a new one to support a variety of studies to improve the program.[113]

Sustainable Agriculture

In agriculture, natural methods such as integrated pest management can replace many chemical insecticides, and natural organic substances can replace chemical fertilizers. The benefits of such natural alternatives make them worthwhile wherever science or traditional knowledge can supply them. But their development and successful application depend on scientific research and on government policies that encourage traditional, non-chemical methods.

In the United States, many reports indicate that farmers are increasingly taking up natural and low-input, sustainable practices. The *Kiplinger Agriculture Letter* noted recently that farmers are "showing a lot more interest in biological pest controls, crop rotations . . . the use of few chemicals." A Wisconsin group called the On-Farm Research Network demonstrated that corn whose nitrogen came from plowed-down alfalfa and manure produced the same yields as corn that used commercial nitrogen, but gave a higher net profit. An Iowa farmer told the Congressional Agriculture Subcommittee that he had increased profits $95 per acre on corn and $45 per acre on soybeans by using natural methods instead of high-chemical methods.[114]

Throughout the world, and throughout history, farms have been uniquely self-reliant production systems. They have supplied almost everything they have needed from within their own borders.[115] Early South American cultures in the Andes, for example, had remarkable success on difficult lands in producing food without chemical or external inputs of any kind. Remote forest cultures today practice ingenious forms of agriculture tailored to their individual regions, as they have done for centuries. (For further discussion of sustainable agriculture, see Chapter 5.)

Industrial Efficiency

Industries must be made more efficient and they must minimize hazardous and non-hazardous waste. Indeed, this has been happening. Energy efficiency has improved to the extent that today's industrial production uses the same or less energy and raw materials than industry did two decades ago. As it has become profitable for firms to improve their efficiency, they have developed innovative technologies and improved their competitiveness in world markets. In several industrialized countries, pollution control has even become a thriving branch of industry in its own right.[116]

Recycling of industrial materials has high potential for increasing sustainability: recovering a single run of the Sunday *New York Times*, for instance, would save 75,000 trees. In the United States, since 1981, more than half the 300 billion aluminum cans sold have been returned for recycling. Within about six weeks, those cans are back on supermarket shelves, and American consumers have saved over $1 billion as a result.[117] In 1984, 988,000 acres of trees were spared in nine nations by recycling.[118] The electroplating industry uses electrolysis to recycle gold, silver, tin, copper, zinc, solder alloy, and cadmium. The motor industry recycles polyvinyl chloride in scraps of car-seat fabric. Previously, the scraps had been incinerated, causing vinyl chloride, a potent carcinogen, to be released into the atmosphere.[119]

Great progress can be made in industry by reducing the amount of materials used in production. The 3M Corporation started a reduction program in 1975, and by 1984, it had eliminated 10,000 tons of water pollutants, 90,000 tons of air pollutants, and 140,000 tons of sludge from its discharges. It had also saved $192 million in costs in less than ten years.[120]

Water Pollution Control in the Puget Sound Waterways

In 1985, the Puget Sound water-quality agency developed a master plan for cleaning the heavily polluted, 3,200-square-mile body of water. The state legislature levied an 8 cents per pack surtax on cigarettes to help pay the bill, enough to contribute an estimated $25 million.

The plan causes industrial waste to be monitored carefully, and industrial discharges to be reduced. It also targets urban and agricultural runoff for reduction. Two counties have regulated land clearing and the installation and inspection of septic tanks. Zoning of construction is more stringent in a critical watershed area, where a single-family house must now sit on at least two acres of land. Farmers must fence cattle away from streams, and the number of livestock and poultry per acre is controlled.

The Puget Sound group has an educational program that teaches area residents everything from the history of the Sound to what not to put down the kitchen sink. Pollution control is promoted as everyone's task. High school students take water samples, and island dwellers have been taught what to do if they spot an oil spill.[121]

Solar-Powered, Community-Oriented Homes

In the late 1970s, after years of work to create a residential subdivision designed to better meet the needs of residents and to sustain the environment more effectively, builder Michael Corbett completed his project. It consists of solar-heated, passively-cooled houses grouped together in units of eight. Grassy areas common to all the homes provide community areas and play space, and area devoted to streets is kept to a minimum. The community is designed to be largely self-sufficient to encourage walking and the use of bicycles, and to minimize the need for cars. As a result, energy use and transportation costs are reduced in a state where transportation accounts for half of all energy used, and children are safer from car accidents. And because of the community-oriented nature of the subdivision, the crime rate is 90 percent lower than in the city's other residential neighborhoods.

The average house uses less than two-thirds the energy of a conventional house. Many of the houses employ passive solar design features to improve both winter heating and summer cooling, including massive insulation, concrete slab floors for thermal mass, solar greenhouses, southern orientations, and roof overhangs. The houses require no air conditioning even though summer temperatures often top 100 degrees Fahrenheit. Solar water heating supplies about three-quarters of each home's hot water during the year. Some of the homes meet 70 to 80 percent of their energy needs from solar sources.

To conserve scarce water, a unique natural drainage system retains 85 percent of the rainfall in streams and ponds, maintains soil moisture, and minimizes lawn watering and soil erosion. Family and community gardens are evident. Composting of organic wastes fertilizes the gardens, and reduces the need for refuse collection.

Creating the subdivision, however, was a difficult task for Corbett. He spent years trying to convince partners, lenders, realtors, and the Davis Planning Department of the project's viability at a time when solar energy was still regarded as highly experimental, and when it was assumed that subdivisions were always built on traditional square grids. Corbett frequently risked financial failure, and made an enormous personal investment in the project, before achieving his goal.[122]

International Organizations and Institutions

Development assistance is primarily a transaction between two sovereign governments, or between governments and international institutions that are parts of the United Nations system. The U.S. government provides development assistance directly to friends and allies through the Agency for International Development, which is considered a bilateral agency. Most other industrialized countries also have bilateral aid agencies. Indirectly, the United States also provides development assistance through contributions to multilateral institutions such as the World Bank, the IMF, regional development banks, and various U.N. agencies such as the United Nations Development Program (UNDP) and the Food and Agriculture Organization (FAO). Collec-

tively, the bilateral and multilateral agencies are referred to as the "donor community."

U.S. bilateral aid takes the form of development assistance grants and loans to developing countries; food aid through the PL-480 program, which directs the government to buy commodities produced in the United States and ship them to developing countries; and other assistance programs including trade, narcotics control, the Peace Corps, and migration and refugee assistance. A common thread throughout all U.S. assistance programs is their underlying political motivation. The United States is active in providing foreign aid when it is deemed to be in our best political, economic, and strategic interests. Assistance, with its potential to win friends and to increase economic development, is an increasingly desirable tool in U.S. foreign policy.

In addition to development assistance, the United States also provides many countries with "security assistance." Security assistance includes military aid, which takes the form of military training for personnel, sales credits, grants, and loans to enable countries to upgrade their military equipment. Security assistance also entails cash transfers designed to bolster the economies and ease balance-of-pay-

Figure 4.4
U.S. Foreign Aid Compared to Personal Consumption Expenditures, 1986 ($ Billions)

U.S. foreign aid, particularly the part aimed at furthering development in low-income countries, is relatively small in comparison to a variety of U.S. personal consumption expenditures.

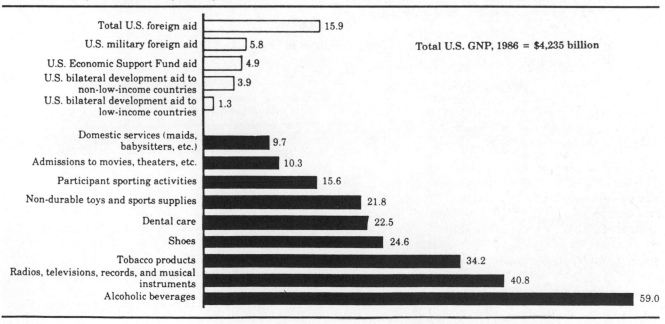

Source: Published by Permission of Transaction Publishers from *Growth, Exports, and Jobs in a Changing World Economy: Agenda 1988*, (U.S. – Third World Policy Perspectives, Series No. 9), by John W. Sewell et al. Copyright © 1988 by Overseas Development Council.

ments problems for allies of particular strategic importance.[123]

As a percentage of GNP, the U.S. foreign assistance budget is a meager 0.19 percent — just over $13 billion for 1988. Almost $2.5 billion was allocated for development assistance; $1.5 billion for food aid; $3.2 billion for economic security; $1.5 billion for contributions to multilateral institutions; and $4.8 billion for military aid. These figures support the contention that, throughout the 1980s, there has been a strong trend toward increasing military and economic security assistance at the expense of development assistance. (See Figure 4.5.) Over half the economic security and military aid goes to the Middle East, primarily to Egypt and Israel.

The primary source of multilateral aid is through the Multilateral Development Banks (MDBs), which include the World Bank, Inter-American Development Bank, Asian Development Bank, and the African Development Bank. The MDBs make loans to developing countries at variable rates. The poorest of the poor are eligible for loans from the International Development Association of the World Bank (IDA) at very low interest with extended time frames in which to make repayment.[124] The MDBs have come under harsh criticism for failing to include environmental concerns in their decisions, for not assisting the poorest people, for not addressing cultural factors in development, and for failing to include local communities adequately in project design and implementation. MDB assistance tends to be directive, even paternalistic in nature, rather than responsive

Figure 4.5
Composition of U.S. Foreign Aid Appropriations, 1977-86

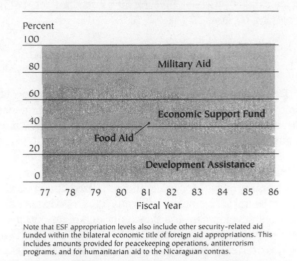

Note that ESF appropriation levels also include other security-related aid funded within the bilateral economic title of foreign aid appropriations. This includes amounts provided for peacekeeping operations, antiterrorism programs, and for humanitarian aid to the Nicaraguan contras.

Source: *Trends in Foreign Aid, 1977-86*, study prepared by the Congressional Research Service for the Select Committee on Hunger, U.S. House of Representatives, November 1986, as published in Smuckler et al., *New Challenges, New Opportunities*, op. cit., p. 35.

to the messages of the communities of Third World countries.

The MDBs, and all multilateral development organizations, are relatively new, created after World War II. In the

Figure 4.6

Net Official Assistance from OECD/DAC Countries in 1987 — in $ Billion and As % of GNP]

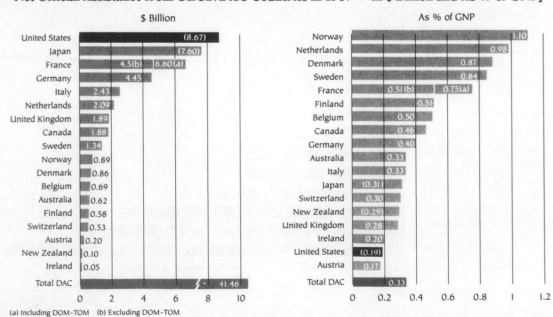

(a) Including DOM-TOM   (b) Excluding DOM-TOM

Source: Organization for Economic Cooperation and Development, Development Assistance Committee, 1988, as published in Smuckler et al., *New Challenges, New Opportunities*, op. cit., p. 35.

Table 4.1
Grassroots Organizations in Selected Developing Countries, Late Eighties

Country	Description
Bangladesh	1,200 independent development organizations formed since 1971, particularly active in health and income generation with large landless population.
Brazil	Enormous growth in community action since democratization in early eighties: 100,000 Christian Base Communities with 3 million members; 1,300 neighborhood associations in São Paulo; landless peasant groups proliferating; 1,041 independent development organizations.
Burkina Faso	Naam grassroots peasant movement has 2,500 groups participating in dry-season self-help; similar movements forming in Senegal, Mauritania, Mali, Niger, and Togo.
India	Strong Gandhian self-help tradition promotes social welfare, appropriate technology, and tree planting; local groups number in at least the tens of thousands, independent development organizations estimated at 12,000.
Indonesia	600 independent development groups work in environmental protection alone; peasant irrigation groups multiplying.
Kenya	16,232 women's groups with 637,000 members registered in 1984, quadruple the 1980 number (1988 estimates range up to 25,000); many start as savings clubs.
Mexico	Massive urban grassroots movement active in squatter settlements of major cities; at least 250 independent development organizations.
Peru	Vital women's self-help movement in Lima's impoverished shantytowns, with 1,500 community kitchens; 300 independent development organizations.
Philippines	3,000–5,000 Christian Base Communities form focal points for local action.
Sri Lanka	Rapidly growing Sarvodaya Shramadana village awakening movement includes over 8,000 villages, one-third of total in country; 3 million people involved in range of efforts, particularly work parties, education, preventive health care, and cooperative crafts projects.
Zimbabwe	Small-farmer groups throughout country have estimated membership of 400,000, 80 percent women; active women's community gardens multiplying.

Source: Alan B. Durning, *Action at the Grassroots: Fighting Poverty and Environmental Decline,* Worldwatch Paper 88 (Washington, DC: Worldwatch Institute, January 1989), p. 10.

last couple of decades, they have been joined by an ever-increasing number of nongovernmental organizations (NGOs) concerned with development and more recently, environment as well. These organizations often have close ties to indigenous populations and access to information or perspectives lacking to larger, more bureaucratic institutions. NGOs include such groups as farming cooperatives, squatter-neighborhood associations, savings clubs, technology institutions, village forestry associations, religious communities, and nongovernmental development organizations.

The growing realization that development must stem from the indigenous, local people whom it is meant to serve has fostered a greater role for NGOs. Both the NGOs of Third World countries and their counterparts in industrial countries focus their efforts on grass roots implementation strategies and are keenly aware of the need for institution-building at the village level. Often NGOs can serve as a link between communities and large development organizations. They sometimes champion such concerns as the roles of cultural tradition, the environment, women, and the poor in development. Their development projects are sometimes innovative and well tailored to specific local conditions. However, NGOs are often weak in the areas of administration and financial and personnel management. Currently 10 to 15 percent of development assistance funds spent by Western industrial countries goes to NGOs for projects in the Third World.

Money devoted to international development also benefits private contractors, universities, consulting firms, banks, lobbyists, and businesses of all sorts. Development assistance has become a multi-billion dollar industry in the United States alone. A United Nations publication,

Development Business, carried the following advertisement:

> Suppliers of a wide range of goods and services are pursuing and winning big contracts in the developing countries—where project funding has reached epic proportions ... not just for infrastructure and basic industries, but for agricultural projects, oil and gas, urban development sites, water supply and sanitation systems, education and health care services. To win your fair share of this huge and highly lucrative market, you need reliable information direct from the source.... This year [1986] more than $25 billion in loans is projected by just the World Bank and the IDB alone.[125]

In the world of international development assistance, the stakes are high for both the suppliers and the recipients of aid.

What You Can Do

Development is a process of change under way not only in the Third World, but also in industrial countries like the United States. Development is not contained within a country, but links countries together and affects foreign territories. Actions of individuals in the United States and other industrial countries affect their own personal circumstances, the status of their own nations, and the status of other countries.

Inform Yourself

Before beginning to act, increase your awareness of international development, the environment, and the factors that affect both, so that you can speak and write about them with authority.

- Contact organizations concerned with economic and social development, the environment, diverse human cultures, population, poverty, agriculture, food relief, health care, renewable energy, and other topics related to the links between development and the environment. Subscribe to their newsletters.

- Learn how local, national, and international decisions by government and private agencies affect development; learn what processes those organizations go through to arrive at decisions, and what information they rely on; learn what their goals are, and what successes or failures they have had.

- If you have the opportunity, travel to areas of the country or the world that have developed differently from your area, or that have faced different problems in achieving development.

- See how the development of your region affects its economy and your community. Consider the ways that the development of neighboring regions and distant countries also affect your community.

- Learn how the foods you eat and the common household materials you use depend on resources from distant lands. Learn more about the products that the poor produce in other countries and how the poor contribute to their national economies.

- Try to recognize the connections between the various elements of international development (such as population, energy, agriculture, foreign trade). Try to see how development must progress as a whole, rather than looking at each element separately.

Join With Others

Support organizations concerned with international development and the environment; go to their meetings; volunteer your services. Help such groups to become more involved in development issues. Learn about pending legislation that affects development and environment.

Develop a network of contacts. Attend meetings of your city council or county board and encourage them to consider issues that affect the future development of your region and other parts of the world.

- Work with local business forums — Chamber of Commerce, Young Presidents Association, Rotary Clubs, etc., — to actively engage the business community in a dialogue on sustainable development. Encourage creation of a statewide forum of chief executive officers (CEOs) on the connection between economic growth and the wise, long-term management of natural and human resources.

- Create environmental or sustainable development awards for responsible business practices and good corporate citizens.

- Ask businesses to review their foreign environmental and pollution control practices to see if they comply with local regulations. Businesses must be conscientious about their impact on the environment, here and abroad.

- Work with corporate planners to develop a definition of sustainable development for the business community.

- Work with shareholders and board members of corporations engaged in overseas development projects to see that their projects meet sustainable development criteria.

- Form a state-wide coalition on sustainable development with leaders from diverse sectors — business, finance, environment, government, education, peace, health, population, labor, etc. — to create a long-term sustainable development plan for your state. Consider how the coalition can exert leadership internationally.

Review Your Habits and Lifestyle

- Look for ways to conserve and protect the resources that the global population shares. Try to recycle paper, glass, and cans; carpool, walk, or ride a bike instead of driving; turn off extra lights; eat less meat.

- Try to reduce your use of chemicals and industrial products. Reduced consumption will conserve the resources used in all stages of the production of those products.

- Consider installing energy-conserving appliances and equipment in your home, such as efficient heating, cooling, and lighting systems; solar panels, extra insulation, and energy-efficient windows.

- Avoid using products that directly reduce the natural resource base or the biological diversity of industrial or developing countries.

Work With Your Elected Officials

Contact your local, state, and national officials and find out their positions and voting records on development and environment legislation. In brief, clear letters and phone calls, tell your representatives where you and the organizations you represent stand on the issues, and urge them to:

- encourage the World Bank, International Monetary Fund, and other multilateral agencies to incorporate environmental considerations into all decisions.

- increase national attention to international environmental and developmental concerns.

- increase national support for the United Nations.

- allocate more resources to developing renewable energy sources and improving energy efficiency.

Publicize Your Views

Inform your community leaders, policymakers, and the general public about the urgency of poverty in the Third World and its effects on industrial countries. Stress the irreversible nature of the human injury, environmental damage, and economic loss that result from poverty, and the potential of industrial countries for reducing poverty.

- Write letters to the editors of local newspapers about development and environment issues; help your organization create a public service message for local television and radio stations.

- Help spread the views of people who agree with you on development issues, for example by supporting letter-writing campaigns to government representatives.

- Try to write articles for magazines, newspapers, and newsletters about subjects related to development and the environment.

- Meet with editorial boards of local newspapers about the issues of sustainable development and how they affect your community.

Raise Awareness Through Education

Find out whether schools and libraries in your area have courses, programs, films, and other educational resources related to development and the environment. If not, offer to help develop or supply them. Offer to speak to classes or organize projects on issues related to development and environment.

- Find out whether your state has a Global Issue Mandate. If it does not, work with teachers and local school administrators to get one adopted. See that curriculum materials on global sustainable development, environment, and population are available to elementary and secondary school teachers.

- Work with school administrators, teachers, and PTA groups to provide teacher in-service training sessions that include units on environmental and sustainable development topics.

- At the college level, work with professors of various departments — economics, biological and physical sciences, liberal arts, etc. — to sponsor a sustainable development forum to discuss how that concept relates to their respective departments and course offerings. Encourage multi-department events.

- Work to create a Foreign Student Speakers Bureau. Visiting foreign students are an untapped community resource. Contact heads of university departments to see how these students can work with organizations in your community. Invite foreign students to speak on sustainable development issues as they relate to what is happening in their native countries. Encourage them to write articles for newsletters and participate in local projects.

- Start a development education project with your school or organization. Contact groups such as the American Forum (see Appendix) for a resource list of materials available.

Consider the International Connections

Try to travel or study in a Third World country. Before you go, learn about the country's history, culture, environment, and state of development. Observe the range of living standards and the quality of life, and consider how they are related to the country's economic and social development and to its environment. Compare the communities you visit to your own community. Learn which U.S. organizations have links with organizations in the country you visit, and how they operate. Become involved in exchange programs and sister-city initiatives.

Further Information

A number of organizations can provide information about development and the environment.*

Books

Berger, John. *Restoring the Earth: How Americans are Working to Renew Our Damaged Environment*. New York: Anchor Press/Doubleday, 1987. 241 pp. Paperback, $9.95.

Brown, Lester et al. *State of the World 1989*. New York: Norton, 1989. 256 pp. Paperback, $9.95. Worldwatch Institute.

Carr, Marilyn. *The AT Reader: Theory and Practice in Appropriate Technology*. London: Intermediate Technology Publications, 1985. 468 pp. Paperback, $19.50. Available from Intermediate Technology Group of North America, Croton-on-Hudson, NY 10520. A collection of more than 200 extracts from articles on appropriate technology. Describes its history and current status, and covers its use in agriculture, health, housing, transportation, manufacturing, mining, recycling, education, and communication.

Darrow, Ken and Mike Saxenian. *Appropriate Technology Sourcebook: Guide to Pratical Books for Village and Small Community Technology*. 800 pp. Paperback,

$17.95. Available from Appropriate Technology Project, Volunteers in Asia, P.O. Box 4543, Stanford, CA 94305.

Conroy, Czech et al. *The Greening of Aid: Sustainable Livelihoods in Practice*. London: Earthscan, 1988. 302 pp. Paperback, #8.95.

The Debt Crisis Network. *From Debt to Development: Alternatives to the International Debt Crisis*. Washington: Institute for Policy Studies, 1985. 65 pp. Paperback.

Dixon, John A. et al. *Economic Analysis of the Environmental Impacts of Development Projects*. London: Earthscan, 1988. 134 pp. Paperback, $27.50. Available from Winrock International Agribookstore.

El-Hinnawi, Essam and Mansur Hashmi. *The State of the Environment*. A publication of the United Nations Environment Program. London: Butterworths, 1987. 182 pp. Hardcover.

Global Tomorrow Coalition. *Sustainable Development: A Guide to Our Common Future*. Washington, DC: Global Tomorrow Coalition, 1989. 77 pp. Paperback.

Goldsmith, Edward and Hildyard, Nicholas (eds.). *The Earth Report: The Essential Guide to Global Ecological Issues*. Los Angeles: Price Stern Sloan, 1988. 240 pp. Paperback, $12.95.

Hellinger, Stephen et al. *Aid for Just Development*. Boulder, CO: Rienner, 1988. 231 pp. Paperback. Available from The Development Group for Alternative Policies.

The Hunger Project. *Ending Hunger: An Idea Whose Time Has Come*. New York: Praeger, 1985. 430 pp. Paperback.

Maguire, Andrew et al. *Bordering on Trouble: Resources & Politics in Latin America*. Bethesda, MD: Adler & Adler, 1986. 448 pp. Paperback, $14.95.

Meyers, Norman (ed.). *Gaia: An Atlas of Planet Management*. New York: Anchor Press/Doubleday, 1984. 272 pp. Paperback, $18.95.

Overseas Development Council. *Growth, Exports, & Jobs in a Changing World Economy: Agenda 1988*. New Brunswick, NJ: Transaction Books, 1988. 274 pp. Paperback.

---

* These include: Appropriate Technology International, The Development Group for Alternative Policies (The Development GAP), Environmental Defense Fund, Environmental Policy Institute, Friends of the Earth, Global Tomorrow Coalition, Institute for Development Anthropology, Institute for Food and Development Policy (Food First), Institute for Policy Studies, Inter-American Development Bank, National Audubon Society, National Parks and Conservation Association, National Wildlife Federation, Natural Resources Defense Council, Organization for Economic Cooperation and Development, Panos Institute, Population Crisis Committee, Sierra Club, Society for International Development, United Nations, United States Agency for International Development, World Bank, World Resources Institute, World Wildlife Fund, Worldwatch Institute.

Reid, Walter et al. *Bankrolling Successes: A Portfolio of Sustainable Development Projects*. Washington, DC: Environmental Policy Institute & National Wildlife Federation, 1988. 48 pp. Paperback.

Schumacher, E.F. *Small is Beautiful: Economics as if People Mattered*. New York: Harper & Row, 1973. 324 pp. Paperback, $9.95.

Sierra Club. *Bankrolling Disasters: International Development Banks and the Global Environment*. San Francisco: Sierra Club, 1986. 32 pp. Paperback, $3.00.

Spero, Joan. *The Politics of International Economic Relations*. New York: St. Martin's Press, 1985. 447 pp. Paperback, $13.95.

Stewart, Frances. *Macropolicies for Appropriate Technology in Developing Countries*. Boulder, CO: Westview Press, 1987. 315 pp.

Timberlake, Lloyd. *Africa in Crisis: the Causes, the Cures of Environmental Bankruptcy*. London: Earthscan, 1985. 232 pp. Paperback.

United Nations Children's Fund (UNICEF). *The State Of The World's Children 1989*. New York: Oxford University Press, 1989. 116 pp. Paperback, $6.25.

U.S. Agency for International Development. *Development and the National Interest: U.S. Economic Assistance into the 21st Century*. Washington, DC, 1989. 159 pp. Paperback.

The World Bank. *World Development Report 1988*. New York: Oxford University Press, 1988. 307 pp. Paperback.

The World Bank. *Population Growth and Policies in Sub-Saharan Africa*. Washington, DC: World Bank, 1986. 102 pp. Paperback.

World Commission on Environment and Development. *Our Common Future*. New York: Oxford University Press, 1987. 400 pp. Paperback.

World Resources Institute and International Institute for Environment and Development. *World Resources 1988-89: An Assessment of the Resource Base that Supports the Global Economy*. New York: Basic Books, 1988. 372 pp. Paperback, $16.95.

Articles

Bock, David, "The Bank's Role in Resolving the Debt Crisis," *Finance & Development: A Quarterly Publication of the International Monetary Fund and the World Bank*, June, 1988, p. 6.

Blaikie, Piers. "The Use of Natural Resources in Developing and Developed Countries." In R.J. Johnston and P.J. Taylor (eds.), *A World in Crisis?: Geographical*

Prespectives (Cambridge, MA: Basil Blackwell, 1989), Chapter 5.

Fujioka, Masao, "Alleviating Poverty Through Development: The Asian Experience," *Development: Journal of the Society for International Development*, 1988 2/3, p. 26.

Hyman, Eric L. "The Identification of Appropriate Technologies for Rural Development," *Impact Assessment Bulletin*, Vol. 5, No. 3, 1987, pp. 33-55. Available from Appropriate Technology International.

Killick, Tony, "Africa's Commodity Dilemma," *Africa Recovery: Tracking the Continent's Development*, United Nations, August 1988, p. 5.

Kinley, David, "Turning a New Leaf: Brazil Seeks Ecological Balance in the Amazon," *World Development*, United Nations Development Programme, March 1989, p. 4.

Mellor, John, "The Intertwining of Environmental Problems and Poverty," *Environment*, November 1988, p. 8.

Periodicals

Development. Journal of the Society for International Development (SID). Published for SID members. For information contact SID Washington Chapter (See Appendix for address).

The Ecologist. Published bi-monthly by Ecosystems Ltd., Worthyvale Manor Farm, Camelford, Cornwall, UK. $36 per year.

Environment. Published monthly by Heldref Publications, 4000 Albemarle Street, NW, Washington, DC, 20077-5010.

Impact Assessment Bulletin. Published quarterly by the International Association for Impact Assessment, c/o Maurice Voland, P.O. Box 70, Belhaven NC 27810. Subscription is included in the annual membership fee of $30.

NGO Networker. Published quarterly by World Resources Institute (See Appendix for address).

Panoscope. Published six times a year by Panos Institute (See Appendix for address).

Rocky Mountain Institute Newsletter. Published quarterly by Rocky Mountain Institute (See Appendix for address).

World Development. Published by the United Nations Development Program, Division of Information (See Appendix for address).

World Development Forum. A report of facts, trends, and opinion in international development, published twice monthly by The Hunger Project, 1300 19th Street, N.W., Suite 407, Washington, DC 20036.

Films and Other Audiovisual Materials

New Alchemy: A Rediscovery of Promise. Portrays the New Alchemy Institute, a pioneering appropriate technology group, and its work on solar aquaculture, bioshelters, wind power, and organic agriculture. 28 min. version, 16 mm film ($550), video ($250), rental ($50); 58 min. version, 16 mm film ($850), video ($450), rental ($85). Grade 7-adult. Bullfrog Films.

Conserving Our Environment. Examines some of the ways people are polluting the land, sea, and air, and suggests various solutions for protecting the environment. 16 min. 16mm film ($355), video ($250), rental ($75). Coronet/MTI Film & Video.

Land Use and Misuse. Illustrates the sources and hazards of water pollution. Prevention methods such as better sewage treatment, garbage recycling and wise consumerism are discussed. 20 min. 16mm film ($310), video ($220), rental ($50). Coronet/MTI Film & Video.

Rivers of Life. Against tremendous natural obstacles, the people of Bangladesh struggle to raise their standards of living. 10 min. Video ($19.95). World Bank Publications.

The Neighborhood of Coehlos. People living in the slums of Recife in northeast Brazil work with the government's urban development program to improve their everyday lives. 28 min. Video ($39.50). World Bank Publications.

Dandora. Slum-dwelling families on the outskirts of Nairobi, the capital, build their own houses clustered around roads and sources of clean water. 20 min. Video ($30). World Bank Publications.

Seeds of Progress. The poorest of Mexico's rural families work with agencies of the Mexican government to increase farm output, extend electricity, and build roads, schools, and health centers. 28 min. Video ($39.50). World Bank Publications.

Teaching Aids

Teaching guides come with each videocassette offered by World Bank Publications listed above.

The Developing World. Students learn the characteristics of developing and industrial countries, what kinds of activities help improve standards of living, how these activities affect people's lives, and what the results of development have been. Includes Student book, 104pp., Sound Filmstrips, and Teaching Guide, 116 pp. with 30 worksheets. $75. World Bank Publications.

Small-Scale Industries in Kenya. Tells the story of Karari Ngugi, a carpenter who wants to expand his furniture business, and of the people who help him. Shows the difficulties Kenya faces in improving living conditions for its people and the important role small-scale industries play in this process. Includes Student Pamphlet, 8pp., Student Book, 48 pp., Sound Filmstrip, and Teaching Guide, 52 pp. with 12 worksheets. $60. World Bank Publications.

Food and Agriculture

Enormous gains have been made in increasing global food production over the past few decades. But, because of rapid population growth and inequitable distribution of food, hunger still threatens millions throughout the developing world.

—*World Resources 1988-89*[1]

Hunger is a massive problem. We will enter the 1990s with at least 700 million hungry people by even the most minimal definition of food adequacy. A more humane definition would include a full one billion people.

—John W. Mellor, Director, International Food Policy Research Institute[2]

Declining food output per unit area will leave a lengthening list of countries unable to feed themselves from their own farmlands by the year 2000.

—Norman Myers, Consultant in Environment and Development[3]

Unless national governments are prepared to wage the war against hunger on a far broader front, it may not be possible to arrest the decline in per capita food production that is undermining the future of so many poor countries.

—Lester R. Brown, President, Worldwatch Institute[4]

Integrated pest management can cut pesticide requirements by half or more while providing adequate protection for crops.

—*World Resources 1988-89*[5]

Major Points

- Each year an estimated 40 to 60 million people die from hunger and hunger-related disease. Nearly a billion people—almost one-fifth of the world's population—do not consume enough calories for an active working life.

- Although world grain production has tripled since the end of World War II, it has fallen about 14 percent since 1984, and since 1981 the area planted in grain has dropped 8 percent. Grain production per person has declined 12 percent in Latin America since 1981, 24 percent in India since 1983, and 22 percent in Africa since 1967.

- Many developing countries with rapid population growth may face food shortages by the year 2000. If Africa continues its rapid population growth and maintains its current level of fertilizer use and other agricultural inputs, Africa's food production will support only about half of its population in the year 2000.

- The use of unsustainable agricultural methods is causing serious degradation and erosion of croplands in most of the world's important agricultural regions, and erosion of fertile topsoil is increasing as more marginal land is farmed. In the United States, agricultural land may be losing more than 3 billion tons of topsoil each year.

- Growing use of chemical fertilizers, pesticides, and herbicides is polluting groundwater and surface waters and harming people and wildlife. Each year there are hundreds of thousands of accidental pesticide poisonings worldwide, and about 45,000 in the United States.

- Farmers in a number of countries are beginning to use low-input, regenerative agricultural methods that reduce expensive material inputs of pesticides and inorganic fertilizers, and increase the use of biological pest controls, organic fertilizers, crop management, and other sustainable farming methods.

The Issue: Growing Food Demand, Limits to Increasing Supply

Although there are sharp regional differences, on a global basis food production is growing somewhat faster than human population. Yet hundreds of millions of people are malnourished because they are too poor to afford an adequate diet, because of environmental conditions such as drought, or, especially in affluent areas, because their diets are poorly balanced. This chapter covers the following topics:

- recent trends and projections in world food supply;

- causes of malnutrition and hunger;

- limits to increasing food supply;

- priorities for improving food supply and achieving sustainable agricultural production; and

- examples of progress toward sustainable agriculture.

World Food Supply

Between 1950 and 1984, world grain production grew substantially faster than population, increasing 2.6-fold and boosting output per person by 40 percent. In some regions, the increase per person was much greater. In China, per capita output increased more than 80 percent between 1950 and 1984, and in Western Europe, where population growth was relatively slow, per capita output rose more than 130 percent during the same period.

This remarkable growth in world grain production resulted from a variety of factors, including expansion of grain acreage between 1950 and 1981; increased use of irrigation, agricultural chemicals, and high-yield crop varieties; and improved management techniques.

Since 1984, however, world grain output per person has declined each year, dropping 14 percent over the last 4 years. And total grain output fell sharply in 1987 and 1988, declining nearly 10 percent below the all-time high of 1986. (See Figures 5.1 and 5.2 on the following page.)

This drop was caused in part by India's 1987 drought, and by serious drought conditions in the United States, Canada, and China during the summer of 1988. The drop in grain output caused world grain stocks to fall from a 101-day supply in 1986—an all-time high—to an estimated 54-day supply in 1989, the lowest since the late 1940s.[6]

During the mid-1980s, grain output leveled off in several populous countries, including China, India, Indonesia, and Mexico. In Japan, South Korea, and Taiwan, grain production has been declining for many years.

On a per person basis, grain output in Latin America has declined more than 10 percent since 1981; output in India has fallen over 20 percent since 1983; and in Africa, per capita production has declined more than 20 percent since 1967. (See Table 5.1 on page 70.) In all three regions, population growth has been a major factor in the drop in per capita production.[7]

In sub-Saharan Africa, agriculture has been undercut by overcultivation of marginal cropland, overgrazing, deforestation, desertification, wind and water erosion, two decades of unusually low rainfall, and ineffective agricultural policies, all compounded by social unrest and, in some countries, civil war. Many African governments lack economic resources and programs to help small farmers on

Figure 5.1
World Grain Production, 1950-88

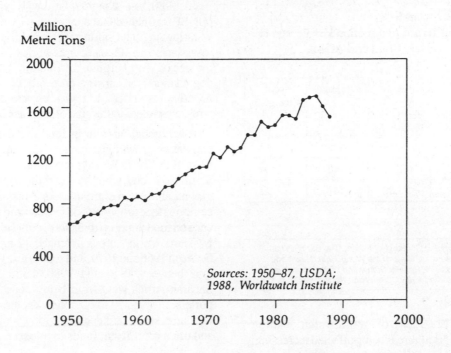

Million
Metric Tons

Sources: 1950–87, USDA;
1988, Worldwatch Institute

Figure 5.2
World Grain Production Per Capita, 1950-88

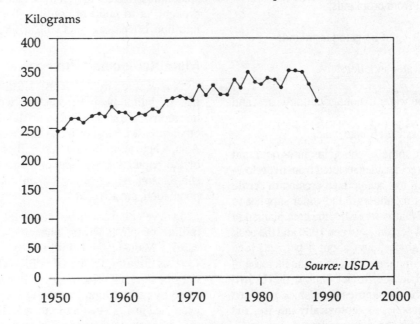

Kilograms

Source: USDA

Source: Lester R. Brown, *The Changing World Food Prospect: The Nineties and Beyond*, Worldwatch Paper 85 (Washington, DC: Worldwatch Institute, October 1988), pp. 9, 43.

marginal land, and 13 percent of Africa's cropland is used for export crops, even though these do not earn enough foreign currency to pay for food imports.[8]

Table 5.1
Decline in Annual Grain Production Per Person in Latin America, India and Africa

	Year of Peak Production	Kilograms Per Person in Peak Year	Kilograms Per Person in 1987	Percent Change
Lat. Amer.	1981	258	228*	-12
India	1983	185	140	-24
Africa	1967	180	141	-22

*1986

Sources: Lester R. Brown and Christopher Flavin, "The Earth's Vital Signs," in Lester R. Brown et al, *State of the World 1988 (New York: Norton, 1988), p. 11; Lester R. Brown, "Sustaining World Agriculture," in* Brown et al., *State of the World 1987* (New York: Norton, 1987), p. 135.

Lester Brown, President of Worldwatch Institute, believes that world agriculture is being affected increasingly by environmental trends and resource constraints. In addition to the growth of human numbers, Brown cites several factors that may be contributing to downturns in grain production, including:

- erosion of fertile soil from croplands;

- waterlogging and salting of irrigated farmland;

- depletion of groundwater supplies;

- diversion of irrigation water to nonagricultural uses; and

- adverse effects of possible climate change.[9]

Brown suggests that some of the gains in agricultural production of the last three decades resulted from overplowing erodible land, which has accelerated erosion of fertile topsoil, and from depleting underground water supplies to irrigate crops. He notes that worldwide, the area planted in grain increased about 24 percent between 1950 and the peak year of 1981, but has since fallen about 8 percent. (See Figure 5.3 on the following page.) Much of the increase in grain acreage occurred in the Soviet Union, the United States, and China; all three countries have since provided evidence that the expansion was ecologically unwise, and have consequently reduced the acreage planted in grain.[10]

Among the causes for the decline in grain acreage, Brown includes abandonment of eroded land, as in the Soviet Union; retirement of eroded land under conservation programs, recently a common practice in the United States; and diversion of cropland to nonagricultural uses, a major trend in Asia.[11]

In 1980, the United Nations Environment Program (UNEP) estimated that degradation of cropland by erosion, waterlogging, and salting was resulting in the irreversible loss of some 6 million hectares worldwide each year, an area the size of West Virginia. In addition, UNEP estimated that degradation was causing crop productivity to approach a negative net economic return on another 20 million hectares, equivalent to the size of Nebraska.[12]

Even though growth in food production has exceeded population growth throughout much of the world, in many parts of the Third World production has not kept pace with *demand* for food, which is determined largely by population growth and income growth. As a result, most developing countries face growing food deficits and increasing dependence on food imports from North America. Net food imports by Third World nations averaged 12 million tons per year between 1966 and 1970, and rose to an average of 38 million tons between 1976 and 1980. Latin America and sub-Saharan Africa went from being net exporters to net importers. In some 50 to 60 countries, the majority in Africa, the outlook is bleak: declining food production per person, and often insufficient funds to import enough food.[13]

Malnutrition and Hunger

There are at least two kinds of malnutrition. One is inadequate nutrition, often associated with poverty, and sometimes related to environmental conditions such as drought, or to political conflict. The other kind is faulty nutrition, often associated with affluence.[14]

Malnutrition and Poverty

A new study by the World Health Organization found that 1.3 billion people around the world suffer from malnutrition or are seriously ill with diarrheal, respiratory, and other diseases. Most of those affected are in south and east Asia, where an estimated 500 million, or about 40 percent of the population are malnourished or seriously sick. In sub-Saharan Africa, about 160 million, or 30 percent of the population, are affected.[15]

In developing countries, each year an estimated 40 to 60 million people die from hunger and hunger-related diseases.[16] Malnourished children are especially prone to infectious diseases. Of the 14 million children under 5 years old who die each year, most perish from infections complicated by malnutrition.[17] Even in the United States, it was estimated in 1985 that more than 12 million children and 8 million adults were malnourished through hunger.[18]

Malnutrition caused by chronic lack of nutrients can result in listlessness, impaired mental functions, damage to the immune system, and susceptibility to infectious diseases

Figure 5.3
World Grain Harvested Area, 1950-1988

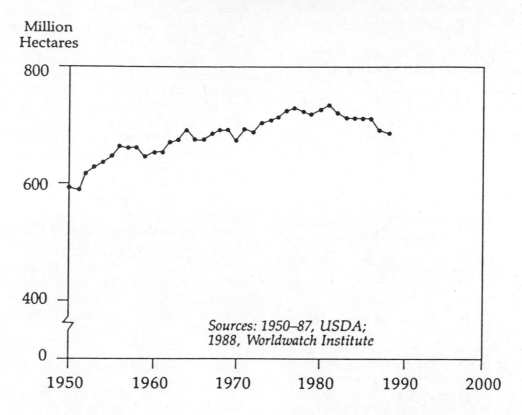

Source: Lester R. Brown, *The Changing World Food Prospect: The Nineties and Beyond,* Worldwatch Paper 85 (Washington, DC: Worldwatch Institute, October 1988), p.18.

such as tetanus, diphtheria, measles, polio, and tuberculosis. Malnutrition can also cause deficiency diseases such as kwashiorkor, anemia, pellagra, beri-beri, and rickets (caused by deficiencies of protein, iron, niacin, thiamine, and vitamin D, respectively).[19]

A 1988 report estimated that there are 950 million people in the Third World—nearly a fifth of the world's population—who do not consume enough calories for an "active working life." Of these, nearly half received too few calories to avoid stunted growth and serious disease. In 1985, the average consumption of calories in 46 nations was less than that needed for productive work. Of the 46, most were low-income countries in sub-Saharan Africa and Asia. In India, according to government reports, 37 percent of the people cannot purchase enough food to sustain themselves.[20]

In the developing world, there are not only great disparities in wealth and nutrition between nations, but within nations as well. In some Third World countries, the top one-fifth of the population may have 10 to 20 times more income than the bottom fifth. The poorest people usually are the hungriest. In cities, recent migrants and the unemployed suffer the most; in rural areas, the landless often have the poorest diets. In Bangladesh, for example, landless families averaged 1,925 calories per person per day, compared with 2,375 calories for families owning more than 7 hectares. The United Nations-recommended daily intake is 2,350 calories per person. An average industrial-nation citizen consumes about 40 percent more calories than the U.N. recommendation, while the average Third World resident receives about 10 percent less.[21] In general, people living in the poorest countries receive only one-half to two-thirds as many calories as those in industrial nations. (See Table 5.2 on the following page.)

Famine and Drought

In normally dry regions, an extended period of subnormal rainfall can lead to severe malnutrition and even starvation. In sub-Saharan Africa, after years of relatively plentiful rainfall in the 1950s and 1960s, precipitation has been below normal every year since 1970. Severe drought conditions occurred in 1983 and 1984, and rainfall deficits

Table 5.2
Daily Caloric Supply Per Person for Selected Countries

	Calories
Mozambique	1,617
Low-income countries except India & China	2,100
India	2,126
China	2,620
Middle-income countries	2,719
Industrial countries	3,357
United States	3,682
East Germany	3,769

Source: World Bank, *World Development Report 1988* (New York: Oxford University Press, 1988), pp. 278-79.

continued through 1987.[22] Drought conditions contributed to severe famine in sub-Saharan Africa in 1975 and again in 1984-85, and parts of the region face the threat of famine again today. In 1970, Africa was essentially self-sufficient in food, but by 1985, the United Nations Food and Agriculture Organization (FAO) considered more than 30 African nations to be food-deficit nations. In the 1984-85 famine, drought affected 22 African nations, and by 1985, 11 million people had been displaced and many livestock herds were decimated. An estimated 1 million deaths occurred in Ethiopia and Somalia during this period, even though millions of lives were saved by more than $1 billion in food aid from around the world.[23]

Famine is a problem not only in Africa. In 1988, India's drought entered its fourth year, reaching nearly 285 million people in rural areas. About 93 million of the severely affected were agricultural workers and small-scale farmers. One-eighth of India's 470 districts face extreme water, food, fodder and employment losses. Four million hectares of crops have been lost as a result of the drought.[24]

Malnutrition and Affluence

In affluent countries, malnutrition takes the form of excessive consumption of animal products, fats, and sugars, resulting in a wide range of diseases.[25] There is extensive medical evidence that a high-fat, low-fiber, animal-centered diet is a direct cause of many serious health problems, including heart disease, stroke, hypertension, cancer, obesity, diabetes, osteoporosis, arthritis, and kidney and gallbladder disease. An animal-centered diet is also a major or contributing factor in hypoglycemia, multiple sclerosis, stomach ulcers, appendicitis, anemia, asthma, salmonellosis, and sterility. Of the 2.1 million Americans who died

in 1987, dietary factors were associated with at least two-thirds of the deaths. In 1987, health care costs in the United States reached $500 billion.[26]

Heart attacks and strokes are now the major causes of death in industrialized nations, accounting for more than half of all deaths. In Eastern Europe, growing affluence and improving food supplies during the 1970s were accompanied by rising mortality rates. Between 1972 and 1982, death rates from cardiovascular disease in Hungary rose 33 percent, and in Poland and Romania, heart attack deaths among men jumped more than 50 percent. Experts attribute the increases to a diet rich in animal fats. In contrast, in the United States and Japan, education programs on the links between diet and disease have helped reduce the consumption of animal fats. As a result, between 1972 and 1982 the cardiovascular death rate dropped nearly 30 percent in the United States, and more than 36 percent in Japan.[27]

Unfortunately, diseases of affluence are spreading from industrial nations to the Third World as television, films, and other communication media transmit images of prosperous lifestyles from country to country. These diseases of affluence have been called "the new communicable diseases."[28]

Calories and Protein: How Much Do We Need from Plants and Animals?

Calories from food are the source of energy to power muscle activity, maintain the body's functions, support growth, and repair damage from injury or illness. Protein in food provides amino acids necessary to maintain the body's structure and its many biochemical processes.[29]

For calories, the U.S. National Research Council's recommended dietary allowances for adult men and women are 2,700 and 2,000, respectively. The 1988 Surgeon General's *Report on Nutrition and Health* estimated that American men and women received, respectively, 94 and 82 percent of the recommended caloric allowances, while young children received 100 percent of the recommended level.[30]

Proteins are essential components of all living cells. They consist of complex combinations of amino acids, and are found in both plant and animal products. Protein consumption can be expressed either by weight or as a percent of total calories consumed. In terms of weight, the FAO recommends a daily protein intake from plant and animal sources of 41 grams per person. Various studies show that Americans consume roughly twice this amount, and perhaps twice as much protein as their bodies can utilize.[31] And many studies have linked excessive protein intake to osteoporosis and kidney disease, among other health problems.[32]

Various organizations have set minimum or recommended daily protein requirements expressed as a percentage of total calories consumed; these range from 2.5 percent to about 8 percent. The National Research Council's (NRC) recommended daily requirement of just over 8 percent includes a substantial safety margin and is designed to be more than adequate for 98 percent of Americans.

The 1988 *Surgeon General's Report* estimated that protein accounted for about 16 percent of the calories consumed by Americans, nearly twice the NRC recommendation.[33]

In her pioneering book, *Diet for a Small Planet*, Frances Moore Lappe designed a hypothetical plant food diet without meat, eggs, or dairy products to meet the daily allowances for calories and protein set by NRC; more than 11 percent of the calories in the diet are derived from protein.[34] Lappe's book also refutes the popular myths that high-protein diets are healthful, and that meat contains the "best" protein. Her research shows that:

- Beans, some nuts, cheese, and fish contain as much protein as meat.

- All essential nutrients can be obtained from non-meat sources.

- The quality or usability of the protein in some grains and legumes is equivalent to that of meat.

- The caloric content of most plant foods by weight is similar to or lower than that of meat.

- Most Americans could completely eliminate meat, fish, and poultry from their diets and still get adequate protein from other foods in the typical American diet.[35]

If all animal products including eggs and dairy products are excluded from the diet, a deficiency of vitamin B12 can result, which can cause a serious disease called pernicious anemia. Such a deficiency can be avoided by taking inexpensive B12 supplements that are readily available from plant sources.[36]

During World War I, Americans obtained nearly 40 percent of their protein from grains and cereal products; now we get less than 20 percent from these sources. And animal products, which then accounted for about half of our dietary protein, now supply two-thirds of our protein. U.S. consumption of beef and poultry doubled between 1950 and 1975.[37]

Grains, cereals, legumes, and other plants are much more efficient sources of food for human consumption than animals.

An average Third World resident receives only about two-thirds to three-quarters as many calories as an average person in the industrial world, only about half as much protein, and only one-fifth as much animal protein. Because it requires 3 to 16 pounds of grains and legumes to produce

a pound of animal protein, the average industrial-nation citizen with an animal-centered diet accounts for several times as much agricultural produce as a Third World resident with a mostly-vegetarian diet.[38] Of the various forms of animal protein, grain-fed beef is the most inefficient to produce, requiring about 16 pounds of grain and soybeans to produce a pound of beef. To produce a pound of pork in the United States consumes on average about 6 pounds of grain and soy, while a pound of poultry or eggs requires 3 to 4 pounds of grain and soy.[39] (See Figure 5.4 on the following page.)

In the United States, livestock consume 10 times more grain than that eaten by the country's people. U.S. livestock are fed enough grain and soybeans to support more than five times the country's human population. Of the 145 million tons of grain and soy fed to beef cattle, fowl, and hogs in 1979, only 21 million tons, or 14 percent by weight, of meat, poultry, and eggs were produced.[40]

According to a report from the Feinstein World Hunger Program at Brown University, if the world's food supply were distributed equally among all people so that everyone received the United Nations-recommended intake of 2,350 calories per day, primarily from grains, there would be enough food for 6 billion people. But if 10 percent of the calories came from animal sources, as in the average South American diet, only 4 billion people could be sustained. And if 30 percent of the calories were from animal sources, only 2.5 billion people could be fed.[41] In short, the greater the human consumption of animal products, the fewer people can be fed.

In summary, many people in the United States and other industrial nations eat much more protein than they need. Excessive protein consumption of animal protein causes harmful health effects and is an inefficient use of agricultural resources. In addition, livestock production is a major cause of environmental degradation, an issue discussed later in this chapter.

Land, Food, and People: The Growing Pressure

In 1983, the FAO published a study of land resources for future populations in the Third World. The report showed that the potential of land for food production in 117 developing countries is limited, and that this potential varies greatly between and within countries. The study suggests that populations of some nations are already greater than their land can support, and that there are many areas where land resources are insufficient to feed projected populations. According to the report, Africa represents the most serious challenge; it has the lowest levels of farming inputs, the fewest sources of income to buy food imports, and the highest population growth rate.[42]

Figure 5.4
Pounds of Grain and Soy Fed to Produce One Pound of Meat, Poultry, Eggs, or Milk

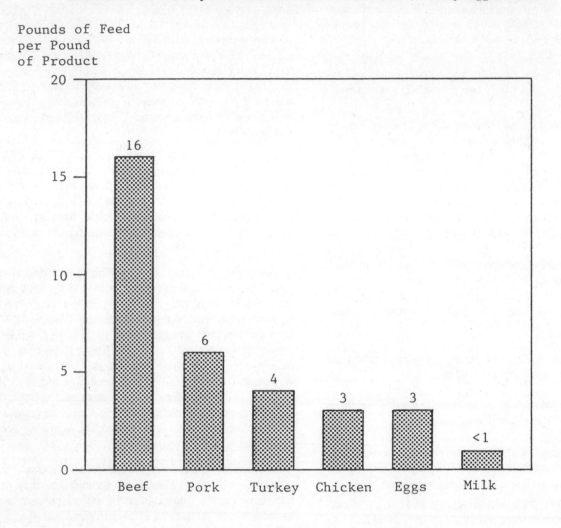

Pounds of Feed
per Pound
of Product

Source: U.S. Department of Agriculture, Economic Research Service, as cited in Frances Moore Lappé, *Diet for a Small Planet*, Revised Edition (New York: Ballantine, 1982), p. 70.

Details of the Study

The study made projections to the year 2000 for three levels of agricultural inputs: *low-level inputs* (no fertilizers or other chemicals, no long-term conservation measures, and traditional plant varieties and crop mixtures); *intermediate-level inputs* (basic fertilizers, improved plant varieties, simple conservation measures, and most productive crop mixtures); and *high-level inputs* (full use of fertilizers and chemicals, full conservation measures, high-yielding plant varieties and crop mixtures). Most of Africa is currently at the low level, while most other developing regions are between the low and intermediate levels.

If *all* potential cropland—more than twice the present cultivated area—were used to grow *only* food crops (not crops to feed livestock), the study indicated that developing countries *as a group* would be able to feed their expected populations in the year 2000, even with low inputs.

Arable land, however, is very unevenly distributed, and the extent to which potentially cultivable land is actually under cultivation varies greatly. About 60 percent of the total potential arable land lies in just 29 countries, home to only 15 percent of the world's population. In the industrial countries, about 70 percent of cultivable land is in use. In developing countries, 36 percent is being cultivated overall, but this ranges from 92 percent in Southeast Asia to only 15 percent in South America.[43]

Sixty-five of the 177 countries in the FAO study were classified as "critical," that is, unable to support their populations in the year 2000 using low inputs on their own land. Of these, 30 were in Africa; 15 in Southwest Asia (which includes the Middle East); 14 in Central America and the Caribbean; and 6 in East and South Asia. If the level of farming inputs is raised to the intermediate level, 29 of these 65 countries would cease to be critical, and 17 would become self-sufficient if inputs were raised to the highest level. But 19 countries would be unable to feed their projected populations even using the most advanced farming methods.[44]

Future Food-Deficit Areas

Taking into account the *likely* level of inputs each region might reach by the year 2000, the FAO study found that most of Asia and South America would be able to feed their projected populations in the year 2000 from their own cultivated lands. This conclusion assumes that *all* cultivable land is brought into production, the land is used only to grow food crops, and the food is distributed equally among social groups. If the Middle East reaches the intermediate level, it could support about 60 percent of its population, and if Africa remains at the low-input level, only 55 percent of its population could be supported in the year 2000.

Looking beyond the year 2000, for many poorer countries population projections for the year 2025 are well above what local land resources could support even with high inputs, and projections indicate that populations will eventually stabilize at levels 70 to 150 percent higher than in the year 2000. Unfortunately, many poorer countries lack any obvious long-term source of income to pay for food imports and for agricultural inputs such as fertilizers and improved seed varieties.[45]

Limits to Increasing Food Supply and Relieving Hunger

As demand for food increases with growing populations and rising living standards, numerous factors limit efforts to expand food production and alleviate hunger and malnutrition. Expansion of agricultural land is limited by many factors, including soil fertility, topography, climate, and prevalence of agricultural pests and human disease.[46]

According to the FAO, only about 11 percent of the Earth's land area is suitable for agriculture. On the rest, the soil is either frozen, too wet, too shallow, too dry, or chemically unsuited for crops.[47] (See Table 5.3.)

After land is prepared for agriculture and crops are planted, many factors can limit the amount of food available for human consumption. Insects, pests, plant disease and weeds are important causes of pre-harvest agricultural loss, and may cause the loss of between 5 percent and 40 percent of the world's crops each year. Rodents destroy enough food

Table 5.3
World Land Area Suitable for Agriculture

Limitations	Percent of Land Area
No Limitations	11
Too wet	10
Too shallow	22
Chemical problems	23
Too dry	28

Source: Essam El-Hinnawi and Mansur Hashmi, *The State of the Environment* (London: Butterworths, 1987), p. 36.

every year to feed nearly 200 million people. Such pests can be controlled, but improved food storage methods may be more effective in cutting losses.[48]

Post-harvest agricultural losses can occur during harvesting, threshing, cleaning, drying, storage, transport, processing, packaging, and distribution of food. Estimates of these losses in developing countries are about 10 percent of durable crops (cereals and legumes), and about 20 percent of perishable crops (vegetables, fruits, root crops). Such losses may cost as much as $10 billion a year.[49]

In the United States, food losses during harvest, transport, and storage due to rodents, insects, and microbes have been estimated at 14 percent of the food produced, with an annual value of about $25 billion. And an estimated 15 percent of food purchased by U.S. households and restaurants is discarded; this wasted food is probably worth at least $50 billion a year.[50]

Inefficient global food distribution systems are another constraint on efforts to alleviate hunger and malnutrition. Food surpluses —"mountains of wheat"— in industrial nations may be thrown away or left to rot, while food shortages exist in the Third World.

In many countries that use high levels of fossil-fuel-based inputs such as fertilizers, pesticides, herbicides, irrigation, and heavy machinery, agriculture has advanced to the point where only marginal gains in productivity are achieved by increasing these inputs. In many cases, extensive use of such inputs has made agrosystems unstable, vulnerable, and unsustainable. A growing portion of the potential gain from additional inputs is negated by land degradation caused by these energy-intensive farming methods.[51]

A number of factors that affect or limit efforts to increase food supplies and relieve hunger are discussed in detail below; these include soil erosion and loss of soil fertility, desertification, energy supply, irrigation and water supply, livestock production, the use of chemical fertilizers and pesticides, cropland conversion, loss of genetic diversity, socioeconomic influences, the role of women, and government policies. Table 5.4 summarizes some of the environmental impacts of various agricultural practices.

Soil Erosion

Serious soil erosion is occurring in most of the world's important agricultural regions, and erosion is increasing as more marginal land is farmed. Soil erosion, especially in developing countries, is seriously damaging agricultural productivity, shortening the life of dams and irrigation projects, filling in canals and harbors, and harming productive wetlands and coral reefs.[52]

In many regions, rates of soil loss exceed rates of soil formation by at least tenfold.[53] Much of the world's cropland is believed to be losing topsoil at rates that are reducing productive capacity. In 1984, Worldwatch Institute estimated that about 25 billion tons of topsoil were being lost from the world's croplands each year in excess of new soil formation.[54]

A study published in 1982 estimated that global soil loss from all land areas due to water erosion totaled nearly 77 billion tons a year.[55] For the United States, estimates of soil loss from cropland range from 1.7 to more than 3 billion tons a year, and the direct and indirect effects of soil erosion and

Table 5.4
Selected Environmental Effects of Agriculture

AGRICULTURAL PRACTICES	SOIL	GROUND WATER	SURFACE WATER	FLORA	FAUNA	OTHERS: Air, noise, landscape, agricultural products
Land development: land consolidation programmes	Inadequate management leading to soil degradation	Other water management influencing ground water table		Loss of species		Loss of ecosystem, loss of ecological diversity. Land degradation if activity not suited to site
Irrigation, drainage	Excess salts, water logging	Loss of quality (more salts), drinking water supply affected	Soil degradation, siltation, water pollution with soil particles	Drying out of natural elements, affecting river ecosystems		
Tillage	Wind erosion, water erosion					
Mechanisation: large or heavy equipment	Soil compaction, soil erosion					Combustion gases, noise
Fertilizer use — Nitrogen		Nitrate leaching affecting water				
— Phosphate	Accumulation of heavy metals (Cd)		Run-off, leaching or direct discharge leading to eutrophication	Effect on soil microflora		
— Manure, slurry	Excess: accumulation of phosphates copper (pig slurry)	Nitrate, phosphate (by use of excess slurry)		to excess algae and water-plants	Eutophication leads: to oxygen depletion affecting fish	Stench, ammonia
— Sewage sludge, compost	Accumulation of heavy metals, contaminants					Residues
Applying pesticides	Accumulation of pesticides and degradation products	Leaching of mobile pesticide residues and degradation products		Affects soil microflora; resistance of some weed	Poisonning; resistance	Evaporation; spray drift, residues

Source: Organization for Economic Cooperation and Development (OECD), *The State of the Environment 1985* (Paris: OECD, 1985), p. 189.

associated water runoff are estimated to cost $44 billion annually.[56]

Soil losses due to cultivation of steep marginal lands, reduced forest and vegetative cover, and improper irrigation are expected to accelerate, especially in North and Central Africa, the humid and high-altitude areas of Latin America, and much of South Asia. It will be difficult to improve erosion control without major changes in agricultural practices throughout the world.

In the United States, federal soil and crop conservation agencies have concluded that U.S. crop production cannot be indefinitely sustained at present levels unless soil losses from wind and water erosion are cut in half. As part of a major soil conservation reserve program, 12 million hectares of highly-erodible cropland have been planted with grass or trees since early 1986 to prevent further erosion.[57]

Fuelwood Shortages and Declining Soil Fertility

Wood is the primary energy source for nearly half the world's population, and in the early 1980s, almost 1.3 billion people could only meet their fuelwood needs by depleting wood reserves.[58]

The loss of fertile land is frequently associated with the shortage of fuelwood. Many rural poor live in areas where forests, formerly a source of firewood, have been cleared for farmland. Important soil builders, such as animal dung and crop residues, are now used for cooking and heating fuel, thereby depriving the soil of essential organic matter and vegetative cover. Such soils lose both fertility and ability to hold water. The worldwide impact of burning dung for fuel is to reduce grain production by as much as 20 million tons annually—enough to feed tens of millions of people.[59]

Desertification and Climate Change

Desertification—involving the spread of desert-like conditions in arid and semi-arid regions—is a critical problem in many parts of the world. The United Nations Environment Program (UNEP) estimates that nearly a third of the Earth's land surface is affected by desertification to varying degrees. (See Figure 5.5 on the following page.) Desertification threatens the livelihood of at least 850 million people, and has a severe impact on nearly 200 million. Each year, desertification degrades some 21 million hectares, an area the size of Kansas, to a condition of near or complete uselessness. Of this total, some 6 million hectares are agricultural lands that become unsuitable for food production.[60]

Much of the land at high risk of desertification is in Africa and Asia, but substantial portions are in the American West and in northern Mexico. As much as 20 percent of the contiguous U.S. land area has experienced or is threatened by desertification.

Desertification is caused almost entirely by human misuse and overuse of land, especially in fragile areas with low or erratic rainfall. Overgrazing by livestock and deforestation for fuelwood are the major causes, but over-cultivation of marginal lands and salinization from poorly-managed irrigation are also important factors.[61]

Soils expert Harold Dregne has estimated that an effective program to curb desertification of rangelands and agricultural land would cost a total of $141 billion, an enormous amount, but a sum nearly equal to the estimated loss of agricultural production due to desertification over a five-year period. Dregne calculated that about 90 percent of the $10 billion allocated to desertification-related projects between 1977 and 1984 went to development activities that did not combat desertification, and that in some cases may have worsened it.[62]

Energy

The "green revolution" in agriculture, with its high-yielding plant varieties, requires careful technical management and extensive use of water and agricultural chemicals. Most of the measures that have increased agricultural productivity since World War II have required high energy inputs, including mechanization, irrigation, and the use of chemical fertilizers, pesticides, and herbicides. Food production, excluding processing, transportation, and storage, accounts for about 4 percent of the world's total commercial energy consumption. The FAO estimates that energy use in Third World agriculture will rise nearly 50 percent by the year 2000, with fertilizer accounting for about 60 percent of the increase.[63] Because much of the world's agriculture depends heavily on petroleum, the expected decline in world oil production will limit efforts to expand food production.

In the United States, agriculture is highly energy-intensive. U.S. energy use per acre has risen almost fourfold since 1945. On average, each acre of crop production requires energy inputs equivalent to about 150 gallons of oil. Food processing and packaging account for about 6 percent of U.S. energy consumption.[64]

Water and Irrigation

Worldwide, the irrigation of croplands accounts for about 70 percent of total water use. Each year, some 3 trillion cubic meters of water are used for irrigation, but only about 1.3 trillion cubic meters actually reach crops; the remaining 57 percent is lost in storage and transport.[65]

Between 1950 and 1985, the total world area under irrigation nearly tripled. By 1988, FAO estimated that the world's irrigated croplands totaled 271 million hectares, about 15 percent of all cropland, and equivalent to an area

Figure 5.5
Deserts and Areas at Risk of Desertification

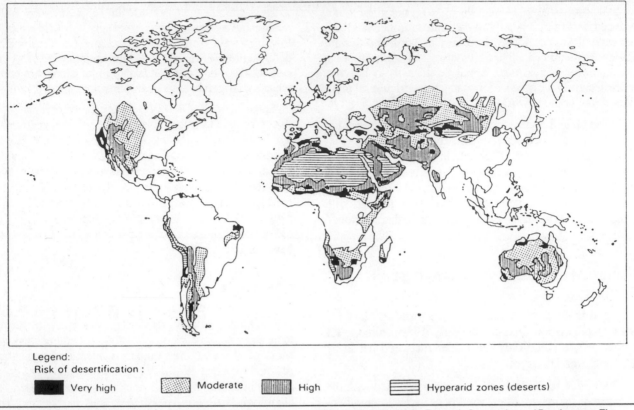

Legend:
Risk of desertification :

[■] Very high   [⦂] Moderate   [|||] High   [≡] Hyperarid zones (deserts)

Source: Adapted from United Nations Map of World Desertification, 1977, as published in Organization for Economic Cooperation and Development, *The State of the Environment 1985* (Paris, 1985), p. 96.

the size of Argentina. In India, China, and Pakistan, irrigated land accounts for between 30 and 65 percent of all cultivated land, and provides between 55 and 80 percent of the food supply.[66]

In the United States, agriculture accounts for 81 percent of total water use. In the Southwest, most land is arid and must be irrigated to grow food. To produce a pound of corn in California requires about 170 gallons of water; to grow a pound of rice uses about 2,000 gallons. And to irrigate enough grain to produce a pound of grain-fed meat takes from 500 to 1,000 gallons of water.[67]

Both industrial and developing countries are experiencing cropland degradation from excessive or inappropriate use of irrigation. Overuse of irrigation wastes large quantities of water, washes nutrients out of the soil, and can cause land to become waterlogged and excessively salty or alkaline. Salinization and alkalinization have damaged millions of hectares of productive cropland, and may be making as much land infertile as is now being brought under irrigation.[68]

Salinization occurs when water evaporates from soggy, irrigated soils, leaving behind salt deposits from the water and from fertilizers. Salinity now seriously reduces the productivity of at least 20 million hectares, or about 7

percent, of the world's cropland—an area the size of South Dakota. At least 1 million hectares become salted each year. Salinization is a persistent problem in North Africa, the Middle East, Central Asia, India, Pakistan, Australia, Argentina, and Mexico. In the United States, salinization has lowered the productivity of a quarter of irrigated cropland. In California, salinity has reduced yields on nearly 25 percent of the San Joaquin Valley's highly productive cropland.[69]

Waterlogging occurs when poorly drained soils receive more water than they can absorb; this damages the soil's structure and promotes chemical reactions in the soil that hinder plant growth. In India and Pakistan, some 22 million hectares of cropland have lost fertility because they are waterlogged.[70]

Intensive use of irrigation is depleting surface and subsurface water supplies, and the quality of irrigation often declines as supply sources become polluted by agricultural runoff and, in some cases, by industrial and municipal wastes. In some areas, fertilizers and pesticides form a large part of the chemical pollutants in runoff. In California's San Joaquin Valley, irrigation of 42,000 acres was curtailed because runoff waters became so contaminated with salts and toxic chemicals that they harmed waterfowl and threatened underground water supplies in the San Francisco Bay Area.

As demand for water increases, aquifer depletion is joining soil erosion and energy costs as limits to growth in world food production. Falling water tables and growing competition for water suggest that future increases in irrigation will be limited. Major food-producing areas that now face water constraints include the United States Great Plains, the Soviet Central Asian Republics, and the North China Plain.[71] (See Fresh Water chapter.)

Livestock

About half the Earth's land area consists of rangeland that can support grazing animals. Herds of wild animals, along with periodic natural fires, maintained the productivity of grasslands before humans domesticated animals. Today, more than 200 million people worldwide engage in some form of livestock production, and 30 to 40 million depend on animal herds for their livelihood. Well-managed grazing by domestic livestock can help maintain the productivity and biological diversity of rangelands.[72]

Yet as human and livestock populations grow, many rangelands are being overgrazed. As vegetation becomes sparse and topsoil is disturbed, grazing land becomes vulnerable to water erosion in wetter areas, and to wind erosion in drier regions. As livestock herds continue to expand and grazing land is degraded, forests are cleared and woodlands are used for grazing. In areas where cattle are fed grain, cropland must be used to produce cattle feed rather than food for direct human consumption. In the United States, more than half the harvested acreage is used to feed livestock.[73]

In addition to 5.2 billion people, the Earth's farms and rangelands must now provide food for more than 4 billion livestock, including nearly 1.3 billion cattle, and for more than 9 billion chickens and other domestic fowl.[74] It has been estimated that the world's cattle now consume an amount of food equivalent to the caloric needs of nearly 9 billion people.[75]

Three-fourths of the world's nearly 3 billion cattle, sheep, and goats are fed crop residues and crops such as hay and alfalfa. Range grazing has declined in importance as intensive livestock production has expanded.[76]

Deforestation　In some areas, the clearing of land for livestock production and related agriculture is a major cause of forest loss. Cattle ranching is causing much of the rapid loss of tropical forest in Brazil, which now has 10 percent of the world's cattle population.[77] In Central America, between 1950 and 1980 the area of pastureland and the number of beef cattle more than doubled, while more than 40 percent of the forest cover was lost. Most of the forest has been converted to pastureland.[78] (See Tropical Forest chapter.) In the United States, the area covered by forests and woodlands declined nearly 10 percent between 1965 and 1985; much of this decline was livestock-related. For each acre of U.S. forest cleared for development, about 7 acres are converted to land for grazing livestock or growing livestock feed.[79]

Desertification and Soil Erosion　Overgrazing by livestock is an important cause of the expansion of desert-like conditions into arid and semi-arid lands discussed earlier. When livestock consume protective vegetation and compact or disturb the soil, the regrowth of vegetation is impaired. The risk of livestock-induced desertification is especially high in sub-Saharan and Southern Africa, parts of Central Asia, Australia, Argentina, northeastern Brazil, Mexico, and the southwestern United States.[80]

Partly because of overgrazing, about 10 percent of the contiguous United States is classified as severely desertified. Such land can be highly prone to soil loss by wind erosion. Sub-Saharan Africa experiences some of the most severe wind erosion on Earth.[81]

In areas with moderate rainfall, overgrazing and the production of livestock feed can lead to serious loss of topsoil from water erosion. In Central America and southern Africa, erosion is widespread in deforested areas that have been used for cattle ranching. In Botswana, the World Bank supported a major cattle-raising project that resulted in extensive overgrazing, desertification, and loss of wild animal population.[82] For the United States, some estimates of annual topsoil loss are as high as 7 billion tons, and one analyst has attributed nearly 6 billion tons of that total to the combination of overgrazed rangeland and the production of cattle feed.[83]

Water and Energy Consumption　The production of grain-fed livestock consumes large quantities of water and energy for irrigation, as well as energy for fertilizers, pesticides, herbicides, and transportation. More than half the water consumed in the United States goes to grow feed for livestock. In California, the cost of water subsidies for livestock production has been estimated at between $2 billion and $4 billion a year. To produce a pound of meat requires up to 100 times more water than to grow a pound of wheat. And to produce a pound of feedlot beef takes between 20 and 40 times the energy required to grow grain or beans with equivalent caloric value.[84]

Water Pollution　In the United States, livestock produce nearly 2 billion tons of waste annually. Much of the nitrogen in the waste is converted into nitrates, which are a significant source of contamination of groundwater and surface waters. Food geographer Georg Borgstrom has estimated that U.S. livestock account for 5 times more harmful organic water pollutants than do people, and twice the amount of pollutants that comes from industry.[85]

Fertilizer Use

Fertilizers have contributed greatly to increased food production in the past two decades. For example, since World War II, China has raised agricultural productivity by more than 60 percent, largely through increased fertilizer use and other high-level inputs such as irrigation.[86] Between 1950 and 1986, world fertilizer use increased more than ninefold, from 14 million to 131 million tons, although the annual rate of growth in use declined from 6 percent in the 1970s to 3 percent in the 1980s.[87]

In many industrial nations, fertilizer use has reached the level where additional inputs do not raise yields. And about half of all fertilizers applied are lost through leaching, runoff, and evaporation. Fertilizers contain phosphates and nitrogen, which can cause eutrophication of surface water and create excessive nitrogen content in water that is hazardous to health.[88] Chemicals from fertilizers have destroyed the natural balance of soils and polluted water sources. In addition, nitrous oxide from fertilizers is an atmospheric pollutant and contributes to the greenhouse effect. (See Chapter 12.)

Pesticide Use

Worldwide, pesticide use has increased rapidly since 1950, and its growth is expected to continue. In 1987, world pesticide sales reached an estimated $16.8 billion. In many parts of the world, chemical pesticides have contributed to raising agricultural productivity and reducing pre- and post-harvest food losses.[89]

In India, pesticide use increased 40-fold between the mid-1950s and the mid-1980s, from about 2,000 tons a year to more than 80,000 tons. Nearly half of India's cropland is now treated with chemical pesticides, compared with less than 5 percent in the 1960s.[90]

In the United States, pesticide use in agriculture rose from less than 1 million pounds per year during the 1950s to more than 300 million pounds by 1965. Between 1965 and 1982, agricultural use increased 2.6-fold to nearly 900 million pounds, but has since declined about 7 percent. (See Figure 5.6.) In 1987, 815 million pounds were used, at a cost of $4.4 billion.[91]

About 70 percent of U.S. cropland (excluding land used to raise animal fodder) receives some pesticides, including 95 percent of land planted in corn, soybeans, and cotton. Despite this heavy pesticide use, each year about 37 percent of U.S. crops are lost to pests. This rate of crop loss to insects has nearly doubled since 1945, in spite of a tenfold increase in pesticide use.[92]

While pesticides have helped increase food production, they have also created serious problems. Many chemical pesticides are not only toxic to target pests, but also to people and wildlife. They have contaminated some food

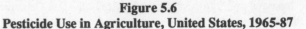

Figure 5.6
Pesticide Use in Agriculture, United States, 1965-87

*The 1983 drop was largely due to reduction in planted acreage under the Payment-in-Kind Program; see Sandra Postel, *Defusing the Toxics Threat*, Worldwatch Paper 79 (Washington, DC: Worldwatch Institute, September 1987), p. 9.

Source: U.S. Environmental Protection Agency, *Pesticide Industry Sales and Usage: 1987 Market Estimates* (Washington, DC:, 1988), Table 8.

products on which they are used, and have seriously polluted drinking water supplies in many areas. Pesticides have been linked to a wide range of serious health effects. In the United States, the indirect environmental and public health costs of pesticide use have been estimated at between $1 and $2 billion each year.[93] (See Chapter 13.)

Many pesticides that industrial nations have outlawed or restricted are still widely used in developing nations, and pesticides that are exported to developing countries are often used by farmers who are unable to read the label directions and precautionary warnings. The number of unintentional acute pesticide poisonings around the world may be as many as 2 million each year, with as many as 40,000 fatalities. In the United States, about 45,000 accidental poisonings occur each year, some 3,000 cases are hospitalized, and about 50 prove fatal. In addition, pesticide exposure has been linked to cancer and other diseases; the National Academy of Sciences estimates that pesticides may cause 20,000 cancer cases each year.[94]

The amount of pesticide actually reaching target pests is a very small percentage of the amount applied. Often less that 0.1 percent of the pesticide used on crops reaches its target, and much of the rest may contaminate soil and water supplies.[95]

When pesticides are first used in an area, crop yields may increase sharply for several seasons, but then level off or even decline. This occurs because most pesticides kill both a target pest and its natural enemies. As enemies are eliminated, the target pest population often recovers. Pesticide use can also cause previously unimportant insect species to become major pests.

Continuous use of a pesticide to control a target pest often leads, by natural selection, to evolution of individuals resistant to the pesticide. Although resistance to pesticides has been known since the early 1900s, it has greatly accelerated since widespread pesticide use began in the 1950s. Many species have developed resistance to an increasing number of chemical pesticides. By 1980, more than 400 arthropods (insects, ticks, and mites) had developed resistance, along with more than 100 plant pathogens (bacteria and viruses), and several rodents and parasitic worms. In addition, a number of weed species have developed resistance to herbicides.[96] (See Figure 5.7.)

Figure 5.7
Species Resistant to Pesticides, 1940-80

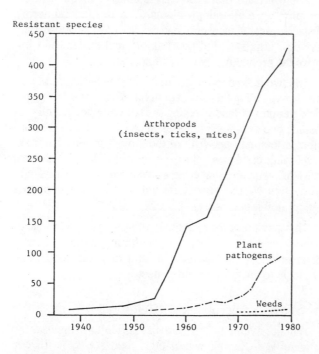

Resistant species

Source: G.P. Georghiou and R.B. Mellon, "Pesticide Resistance in Time and Space," in G.P. Georghiou and Tetsuo Saito (eds.), *Pest Resistance to Pesticides* (New York: Plenum Press, 1983), pp. 1-46.

Cropland Conversion

The conversion of fertile cropland to non-agricultural use is a serious problem in most countries. Cities, industries, and highways are often located on or near a nation's best agricultural land—rich, well-watered soils in gently sloping valleys.

Between 1957 and 1977, 30 percent of China's arable land was lost to urban and rural construction projects and to natural disasters.[97] In countries belonging to the Organization for Economic Cooperation and Development (OECD), during the 1970s between 1 and 3 percent of farmland—much of it highly productive—was converted to urban uses.[98]

In the United States, between 1945 and 1975 an area of farmland about the size of Nebraska was covered with asphalt; and each year about 1 million hectares of fertile cropland—an area twice the size of Delaware—are converted to highways, urban developments, and other non-agricultural uses. This loss is partially offset by additions of new cropland, which in the 1970s amounted to about 0.5 million hectares a year.[99] Continued conversion of cropland for development purposes will contribute to future increases in food prices.

Loss of Genetic Diversity

As the availability of fertile cropland continues to decline under pressure from expanding populations, humanity depends increasingly on genetic manipulation of plants and animals to provide enough food and meet changing environmental conditions. The genetic raw material needed to improve crops and livestock, whether through new gene-splicing technologies or more traditional breeding techniques, is found in the natural and bred genetic diversity of local varieties and the wild relatives of crops and livestock. These valuable genetic resources are rapidly disappearing because of habitat destruction—especially in the tropics—and the worldwide reliance on a few modern, highly-inbred crops and livestock. (See Chapter 6.)

Efforts are being made to diversify the world's food base by utilizing a wide range of wild crops and improving existing varieties. One objective of genetic engineering is to move away from crops that depend on environmentally-harmful chemicals. Unfortunately, this goal may be thwarted by efforts to develop "herbicide-tolerant crops" that will allow continued use of harmful chemicals.[100]

Socioeconomic Influences

Social and economic factors have major impacts on the world's food production and distribution systems. In many regions, population pressures are already diminishing the land's carrying capacity.[101] This is particularly true in rural regions with large populations of subsistence farmers.

In Africa, rural population densities are increasing rapidly in many areas. In some rural regions, the number of people supported per hectare of cropland has almost doubled since the 1950s. Factors such as low income or no-wage income, inequitable land distribution, lack of resources to develop productive capabilities, and sex discrimination that limits the role of women all undercut efforts to increase agricultural productivity.

In developing countries, other factors add to population pressures on agricultural land. Large families result from several factors, including traditional values, improved public health, limited access to family planning, and economic necessity. Poverty is the main reason why many Third World residents lack adequate access to food.

In some developing countries, the use of intensive and efficient methods is increasing agricultural productivity despite growing population pressures. But in many areas, social and economic factors combine to prevent progress in agriculture. For example, in Nepal, migration of skilled farmers to the cities has caused a drop in food productivity as remaining unskilled farmers have tried to use farming methods inappropriate to the region's mountainous terrain.[102]

The Role of Women

The importance of the role of women in agriculture has often been underestimated and misunderstood. Women's labor may account for as much as 85 percent of the food production in Africa, 50 to 60 percent in Asia, 45 percent in the Caribbean, and about 30 percent in North Africa, the Middle East, and Latin America. Women comprise the majority of all subsistence farmers (those who produce food primarily for family needs), and thus women contribute a substantial share of the world's total agricultural production. In a study of six developing nations, the percentage of total agricultural output attributable to women ranged from 20 percent in Colombia to 98 percent in Nepal.[103]

Women's efforts to provide food for their families are often undermined by conflicting agendas. In many regions, the production of cash crops has replaced the production of food for domestic use as a priority, a policy reinforced by many governments and international agencies, which tends to emphasize the role of men in the production process. Although women are often the primary food producers, extension programs and services are frequently directed toward men. Economic and technical assistance is often denied to women due to lack of credit, land tenure, and other sex-biased limitations.[104] Today, however, the central role of women in food production is gradually being recognized, and governments and international agencies are beginning to redirect their efforts toward women as they recognize their role as primary food producers.

Government Policies

Land distribution patterns and agricultural policies can limit food output. When ownership of land is highly concentrated—as in Latin America, where 7 percent of the population controls 93 percent of the land—domestic food production may be inefficient and limited in favor of cash crops for export.[105]

Trends in food production are affected by policies such as price controls and tax incentives. For example, by setting low market prices for staple foods and emphasizing the production of cash crops for export, a government can discourage the production of basic food crops for local consumption. Also, local government and international development agencies often favor high-technology and large-scale agricultural and industrial projects, rather than projects that meet local food and employment needs and are compatible with local conditions. To meet obligations of increasing debt and rising interest rates, many developing countries are compelled to continue or expand production of cash crops for export. But increased production of export crops has depressed prices for many food commodities; export earnings of developing nations for 11 of these commodities fell 17 percent between 1980 and 1985.[106]

Priorities for Improving Food Supply and Achieving Sustainable Agriculture

There are no quick or easy solutions to the world's complex agricultural problems. Yet despite numerous obstacles, many measures can be taken to increase food supplies, improve food distribution, and sustain and improve the productive capacity of arable land.

The two main approaches to increasing agricultural output are expanding the area of cultivated land, and raising the yield per unit of land by using the land more efficiently. In the last two decades, increases in cropland area have contributed less than one-fifth of the growth in food output in developing countries, and even less in industrial nations. The FAO estimates that at least 60 percent of increased food production by the year 2000 will come from improved agricultural inputs and land management.[107]

This section of the chapter summarizes a wide range of priority measures to make agricultural production more efficient and more sustainable, and to make food more available to those who need it most.

Improve Agricultural Efficiency

A major overall priority is to use agricultural inputs more efficiently, including water, high-yield seeds, fertilizers, pest control measures, mechanized equipment, and technology. Special emphasis should be given to the use of low-input, regenerative, and organic farming methods that maintain or improve soil structure and fertility through the addition of organic materials (such as crop residues and animal wastes) and the selective and controlled application of inorganic fertilizers.[108]

- *Improve Irrigation and Drainage Systems.* As irrigated cropland expands and demands for fresh water increase, improved irrigation efficiency becomes essential to conserve water supplies and prevent cropland damage from excessive irrigation. Micro-irrigation techniques, such as drip or trickle systems that deliver water through perforated tubes directly to crop roots, can greatly increase irrigation efficiency. The use of such methods grew nearly eightfold between 1974 and 1982, but still account for less than 1 percent of all irrigated cropland. Less expensive improvements, such as leveling cropland and recycling excess water using pumps and ponds, can also reduce water use. Water and energy use can be reduced as much as 40 percent by releasing irrigation water at intervals instead of continuously, allowing periodic wettings to seal the soil.[109]

- *Promote Intensified Food Production on Fertile Land, and Relieve Pressures on Marginal Lands.* Nations should increase the agricultural capacity of fertile, less fragile lands to help relieve pressure on marginal lands and other more fragile and previously unused, usually lower-potential areas. Research is needed in more fragile agricultural areas to find less-destructive farming methods. Infrastructure should be strengthened in the more fragile areas to relieve environmental stresses resulting from hunger in years of poor production, and from lack of nonfarm jobs.[110]

- *Improve Yields of Indigenous Crops.* Farmers should make greater use of appropriate and environmentally-sound technologies to increase the yield of local food crops, especially in the Third World, in order to lessen dependence on grain imports.[111] It is also important to use indigenous crop varieties and animal breeds that are well-suited to local conditions.

- *Improve the Efficiency of Farming Systems.* Farming systems should be made more efficient and sustainable by avoiding overcropping (repeated plantings of the same crop without a fallow period between plantings), and by the use of techniques such as intercropping (growing more than one crop on the same plot), sequential cropping, crop rotation, and intensive yet careful use of small plots that maximizes crop diversity. Intercropping can help control weeds, improve soil fertility, and make crops less vulnerable to pests.[112] In developing countries, some low-input farming systems rely on a combination of methods, many of them traditional, such as crop rotation, shifting cultivation, the use of fallow periods, and slash-and-burn clearing. Experiments show that these methods generally preserve the diversity of agroecosystems and contribute to sustainable food production.[113]

- *Promote Agroforestry.* Agroforestry techniques that combine the planting of productive, multipurpose trees with food crops should be implemented. Trees can provide timber, fuelwood, food for people and animals, and other products in addition to water retention, erosion protection, soil enrichment and conditioning, and shade to enhance other food crops. Tree planting can also relieve pressures to cut virgin forests.[114]

- *Integrate Livestock and Food Crops.* Mixed cropping and animal husbandry systems should be employed in regions with marginal land areas. The addition of animals to a farm provides more protein as well as wastes that can be used to improve soil fertility.[115]

Reduce Soil Erosion and Desertification

The FAO has estimated that one-fifth of all potential cropland could become useless or marginal unless long-term conservation measures are taken. A high priority should be given to the use of farming methods that minimize the loss of topsoil from cropland, including contour plowing, strip cropping, cover cropping, terracing, tree planting, and the use of minimum tillage methods that limit soil disturbance. Agricultural programs should be designed to:

- Maintain vegetative cover on cropland.

- Increase the organic content of soils to improve water retention.

- Review commodity and water pricing and other agricultural policies to ensure their consistency with soil conservation goals.

- Limit livestock grazing on arid and semiarid lands.

- Intensify underused grazing systems in less arid areas, especially in Latin America, to reduce conversion of forests and cropland to rangeland.

- Use methods in tropical areas that protect the soil from sun and rain, such as mulching and interplanting trees and food crops.[116]

The World Bank recently reported on four years of field trials with vetiver grass as a means of reducing soil erosion and increasing soil moisture. The grass will grow on practically any soil, is resistant to pests, and forms a barrier to erosion without encroaching into cropland. The grass is being introduced in a number of World Bank programs in Asia, Africa, and Latin America.[117]

In arid and semi-arid areas prone to desertification, efforts should be made to limit its major causes, including overgrazing by livestock, deforestation, overcultivation of marginal land, and salinization from poorly-managed irrigation. When overgrazing causes serious land degradation, reduction in herd size can increase available food per head of cattle and actually increase livestock productivity.[118]

In Africa, Somalia has begun a major anti-desertification program that includes a ban on cutting trees for fuel. And in northern China, where the desert claims some 155,000 hectares a year, a green barrier of trees, shrubs, and grasses stretching more than 4,000 miles is being planted to stem the desert's advance.[119]

Decrease Dependence on Chemical Fertilizers, Pesticides, and Herbicides

An important priority for achieving sustainable agriculture is to increase the use of organic fertilizers, including crop, animal, and human wastes, to maintain and improve soil fertility and water retention. Many Asian cities recycle human wastes onto nearby croplands, reducing the need for chemical fertilizers.[120] Since fertilizers probably account for about one-third of the energy consumed in agriculture, reduction of chemical fertilizer use is an important way to conserve energy.[121] Because about half of the chemical fertilizer applied to cropland never reaches the crops, more efficient ways to use fertilizers are being developed. For example, a process known as "fertigation" applies measured amounts of water and nutrients around the roots of individual plants, improving the efficiency of water and fertilizer use.[122]

Another major priority is to increase the use of integrated pest management (IPM) methods, which rely on biological controls (such as natural predators of pests), changes in cultural practices (such as planting patterns), genetic modifications (such as pest-resistant crop varieties), and selective use of chemicals to maintain food production while limiting health and environmental risks. IPM seeks not to eliminate pests, but to prevent them from causing serious economic losses.[123]

Perhaps the most successful use of IPM and biological controls has been in China, where extensive pest-forecasting and data-collection systems have helped farmers identify and control agricultural pests. There are also successful programs in a number of regions, including Brazil, Central America, Africa, and the United States. (See Table 5.5 on the following page.)

In Africa, by the early 1980s mealybugs and spider mites had spread over a large part of the 34-nation cassava-growing region, and cut cassava production—on which some 200 million people depend—by up to 60 percent. Several natural enemies of the pests were located in Latin America, where cassava originated, and introduced in Africa. One of these, a parasitic wasp, now effectively controls mealybug damage in 13 countries of the cassava region.[124]

In the United States, the U.S. Department of Agriculture has monitored IPM programs for some 40 crops, and found them to be economically beneficial. In a survey of nine commodities from 15 states, farmers using IPM realized profits $579 million greater than they would have shown without IPM. U.S. IPM programs have also shown impressive reductions of pesticide use on cotton, grain sorghum, and peanut crops.[125]

A recent study showed the feasibility of decreasing pesticide use in the United States by one-half, using a variety of currently-available pest management strategies. The program would cost an estimated $830 million a year, and would add less than 0.2 percent to the consumer's cost of food. In Denmark and Sweden, similar plans are being implemented, with the goal of cutting pesticide use about 50 percent over the next few years.[126]

Non-chemical methods can also be used to control weeds. Intercropping—for example, growing a legume between rows of wheat—can suppress weeds and add nitrogen to the soil. The use of cover crops that inhibit weed growth by releasing natural toxins can also be effective. Mulches, including the residues of certain crops, can temporarily inhibit weed growth and minimize the use of chemical herbicides. Reduced use of chemical fertilizers, pesticides, and herbicides can lessen groundwater pollution (see Fresh Water chapter) and conserve energy.[127]

Reduce Fossil Fuel Use in Agriculture

In addition to decreasing the use of energy-intensive chemical inputs, there are other ways to curb the use of fossil fuels in food production and thereby help slow disruption of agriculture from carbon dioxide-induced climate change. One way is to use solar and wind energy, especially in the Third World, to dry crops and pump water. Another method is to price energy-intensive agricultural inputs to reflect their full costs in terms of long-term supply and demand factors, and thus encourage their conservation.[128]

A recent study showed that the fossil fuel energy required to produce an acre of corn in the United States could be reduced 50 percent by rotating crops to eliminate pesticide use, using mechanical cultivation to eliminate herbicide use, substituting livestock manure for most of the commercial fertilizer, and curbing soil erosion.[129]

Expand the Use of Low-Input Agriculture

Low-input agriculture seeks to reduce dependence on material inputs characteristic of high-input farming, including fertilizers, pesticides, herbicides, and fossil fuels; and to increase reliance on inputs of information and ecological knowledge about agricultural systems.[130] According to Garth Youngberg, Executive Director of the Institute for Alternative Agriculture, low-input farming systems are designed to optimize the use of internal or on-farm inputs in order to produce acceptable, sustainable, and profitable yields. Low-input methods stress practices such as crop rotation, conservation tillage, and recycling of animal manures to control soil erosion and nutrient loss, and to

Table 5.5
Selected Successful Applications of Integrated Pest Management (IPM) and Biological Control (BC)

Country or Region	Crop	Strategy	Effect
Brazil	Soybean	IPM	Pesticide use decreased 80–90 percent over seven years.
Jiangsu Pr., China	Cotton	IPM	Pesticide use decreased 90 percent; pest control costs decreased 84 percent; yields increased.
Orissa, India	Rice	IPM	Insecticide use cut by third to half.
Southern Texas, United States	Cotton	IPM	Insecticide use decreased 88 percent; average net return to farmers increased $77/hectare.
Nicaragua	Cotton	IPM	Early to mid-seventies effort cut insecticide use by a third while yields increased.
Equatorial Africa	Cassava	BC	Parasitic wasp controlling mealybug pest on some 65 million hectares.
Arkansas, United States	Rice/ Soybean	BC	Commercially marketed, fungus-based "bioherbicide" controlling noxious weed.
Guangdong Pr., China	Sugarcane	BC	Parasitic wasp controlling stemborers at one-third the cost of chemical control.
Jilin Pr., China	Corn	BC	Fungus and parasitic wasp providing 80–90 percent control of major corn pest.
Costa Rica	Banana	BC	Pesticide use was stopped; natural enemies reinvaded to control banana pests.
Sri Lanka	Coconut	BC	Parasite found and shipped for $32,250 in early seventies prevents pest damage valued at $11.3 million annually.

Source: Sandra Postel, *Defusing the Toxics Threat: Controlling Pesticides and Industrial Waste*, Worldwatch Paper 79 (Washington, DC: Worldwatch Institute, September1987), p. 27.

maintain or improve soil productivity. Low-input systems seek to lower production costs, avoid water pollution, reduce pesticide residues in food, and increase farm profitability.[131] The U.S. Department of Agriculture recently established a research and education program on low-input farming systems; early results of the program are described later in this chapter. Figure 5.8 on the following page summarizes some of the differences between high-input and low-input agriculture.

Improve Utilization of Existing Agricultural Production

A number of strategies can help make better use of currently available productive capacity.

- *Reduce Pre-Harvest Food Losses.* Food loss can be decreased by developing more disease- and pest-resistant crops, and improving control of plant diseases, insects, and other pests, especially through integrated pest management and biological controls.[132]

- *Reduce Post-Harvest Food Losses.* Losses occurring between the time the food is harvested and delivery to the consumer can be decreased by improving facilities for food processing, handling, storage, and transportation, especially in the Third World.[133]

- *Improve Utilization of Agricultural Residues.* Crop wastes and other residues are usually organic and biodegradable. When left on the soil, they can prevent

Figure 5.8
The Concept of Sustainable Agriculture

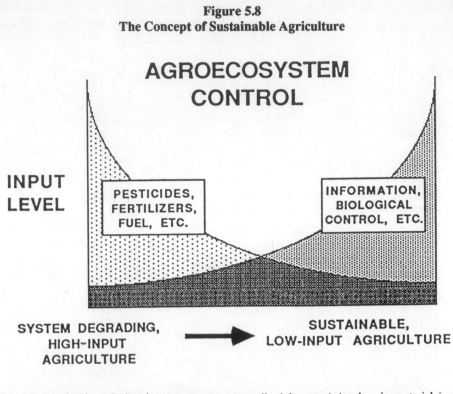

AGROECOSYSTEM CONTROL

INPUT LEVEL

PESTICIDES, FERTILIZERS, FUEL, ETC.

INFORMATION, BIOLOGICAL CONTROL, ETC.

SYSTEM DEGRADING, HIGH-INPUT AGRICULTURE

SUSTAINABLE, LOW-INPUT AGRICULTURE

Figure 1. Sustainable agricultural systems are conceptualized here as being low in material input (pesticides, inorganic fertilizers, etc.) and high in information input (applied ecological knowledge of the system). High chemical input practices conceal and depreciate the importance of ecological processes occurring in agricultural systems. However, as pesticides, fertilizers, etc. are reduced, greater knowledge of the interactions occurring in agroecosystems is required for success. Furthermore, this knowledge must be applied in a practical manner to maintain agroecosystem productivity.

Source: Benjamin R. Stinner and Garfield J. House, "Role of Ecology in Lower-Input, Sustainable Agriculture: An Introduction," *American Journal of Alternative Agriculture*, Vol. II, No. 4, Fall 1987, p. 146.

erosion. And when promptly worked into the soil, crop residues can provide nitrogen, an essential agricultural nutrient. Whenever possible, crop wastes should be used to protect the soil and enhance fertility.[134]

- *Produce Less Grain-Fed Livestock.* As discussed earlier, a reduction in the consumption of grain-fed livestock would save water, energy, soil, and other agricultural resources and make more grain available for direct human consumption. Reduction of excessive meat consumption also has important health benefits.[135]

- *Improve Food Distribution.* The distribution of available food can be improved in a number of ways, including:

  » Establishing national, regional, and international food reserves, where the surplus from good harvests can be stored until needed in years of poor harvests.

  » Establishing emergency food distribution systems that include early warning systems to enable govern-

ments and international agencies to act before a crisis exists.

» Recognizing the seasonal nature of hunger.

» Enabling the poor to buy food below market prices.

» Creating jobs and augmenting income for the poor so they can afford an adequate diet.[136]

- *Improve Public Education About Nutrition.* Available food supplies can be better utilized through improved public understanding of basic principles of health and nutrition, including knowledge of daily requirements for calories, protein, and other nutrients, and knowledge of the nutritional content of food.

Use Biotechnology to Improve Food Production

Biotechnology—the integrated use of biochemistry, microbiology, and chemical engineering—seeks to accelerate plant and animal breeding techniques, thus increasing food output within a shorter time period.[137] For example, genetically altered microbes are injected into corn seedlings containing toxins lethal to the corn borer—an insect that destroys corn.[138] Parasitic worms are now available that kill crop-damaging insects without harming humans, flora, or fauna.[139]

Biotechnology can be used more systematically to ensure future agricultural productivity. Biotechnology's potential is especially needed in Africa and Latin America, where food production per person is declining.[140] Among the existing and emerging technologies that can improve productivity are conventional plant breeding, genetic engineering of plants, biological fixation of nitrogen, and increased photosynthetic efficiency. Modern biotechnology has the potential to provide new crop varieties that mature early, are more pest-resistant, and can tolerate very acid soils, excess salt in water or soil, drought, and temperature extremes.[141] To provide genetic material for developing new crop varieties, germ plasm banks must be maintained to preserve the germ plasm of a wide variety of traditional crop plants and plant varieties with newly-discovered agricultural potential.

While biotechnology has great potential to improve agriculture in developing countries, it could also cause problems. For example, within a few years cocoa production, important in some developing countries, will be possible in the laboratory without cocoa trees. Laboratory production of coffee and tea is also under investigation.

There is also the potential for creating new pests through the release of genetically engineered organisms. For example, in the United States, more than 120 species of intentionally introduced crop plants have become weed pests.[142]

In addition to genetic modification of plant varieties, genetic improvement programs to develop breeds of livestock (cattle, sheep, goats) and draft animals (oxen, water buffalo) adapted to conditions in Third World countries can lead to increased animal protein and energy supplies for regions currently short of both.

Utilize Newer Methods of Food Production

Newer methods include hydroponics, the cultivation of plants in a medium of fertilized water rather than in soil; and controlled-environment agriculture, including the use of greenhouses, cold frames, and desert "tunnel" farms. Greenhouse experiments have shown that a doubling of carbon dioxide in the air could increase biomass yield by about 40 percent.[143]

Expand Aquaculture

Worldwide, direct consumption of fish provides between 2 and 3 percent of protein eaten by humans. Fish meal accounts for about 24 percent of protein fed to animals. Fish are highly nutritious and contain between 15 and 20 percent protein. Because the world's ocean fish catch is approaching the level that the FAO considers to be the maximum sustainable yield, expansion of aquaculture—the cultivation of fresh or salt water fish in controlled settings—is an important priority for meeting future food needs. Aquaculture now accounts for about 10 percent of the world's fish production, and has great potential to add to the world's food supply.[144] (See Ocean and Coastal Resources chapter.)

Strengthen Agricultural Research and Education

Efforts are needed to develop closer links between international agricultural research centers and national research institutions in developing nations, and to improve the scientific and technological capabilities of national institutions. High priority should be given to using biological means to increase productivity in fertile areas, and to developing productive mixed-cropping and animal-husbandry systems for marginal areas. Because the lack of appropriate, area-specific technology impedes efforts to improve food production and alleviate poverty, research should be focused on specific geographic areas.[145]

There is a need to improve data on current land use, the potential of agricultural land, and the extent of land degradation to provide a basis for planning of land use and agricultural investment.

There is also a need to improve the dissemination of research findings to farmers through agricultural extension programs that make research results available to individual farmers.[146]

Perhaps most important is the need to develop interactive and participative programs for agricultural research and education.

In the United States, a major achievement of the Department of Agriculture's Low-Input/Sustainable Agriculture (LISA) program (described later in this chapter) is the involvement of farmers in planning research goals, and the on-farm nature of much of the research itself. According to the Institute for Alternative Agriculture, this approach can speed the process of technology transfer, ensure that research addresses the most pressing farm needs, and improve the skills of farmers. Richard Harwood of Winrock International has observed that in Asia, such approaches are much further developed than in the United States.[147]

Implement Political and Administrative Measures to Improve Agriculture

There is a wide range of policy measures that can promote agricultural development in the areas of land use, land reform, rural development, and local participation.

• *Strengthen Land-Use Control to Protect Fertile Cropland*. Governments should take legal, fiscal, and administrative measures to strengthen land-use control and limit the conversion of fertile land to nonagricultural use. For example, Japan's comprehensive zoning plan designates all land for one of three uses: agricultural, industrial, or other. The system has been effective in preserving fertile land for food production.[148]

• *Implement Land Reform*. Where ownership of fertile cropland is concentrated among a few landholders, governments should redistribute land and promote smallholder agriculture rather than new plantation agriculture or large land-settlement projects. Land-tenure patterns should be altered so farmers can gain legal rights to the land as an incentive to practice soil-conservation methods.[149] World Bank studies show that farmers with small holdings generally produce more food per hectare than those with large holdings. Industrial nations and international organizations should provide consistent economic and political support for governments dedicated to agrarian reform.[150]

• *Promote Political and Administrative Decentralization*. The planning and administration of agricultural programs should be decentralized as much as possible to assure full participation of local people and organizations. Priority should be given to agricultural development in rural areas, and to strengthening rural infrastructures, especially for storage, marketing, and transportation of agricultural products.[151]

Implement Economic Measures to Improve Agriculture

A number of economic policy measures can promote agricultural development and help alleviate hunger and malnutrition. These include setting agricultural prices at levels that encourage investment in agriculture; creating jobs to provide adequate income for the poor so they can afford to buy food they cannot raise themselves; providing incentives and resources for small food producers, including credits and loans, especially in rural areas; improving food marketing systems; and stimulating production of labor-intensive food commodities for export, such as fruit, vegetables, and certain livestock commodities.[152]

Enhance the Role of Women in Agricultural Development

In many Third World countries, women are the primary food producers, and women often have the best knowledge of how to manage the land and protect its productivity. It is especially important to include women in all stages of planning and implementation of agricultural projects, and to ensure their access to land ownership, loans, credit, and information about new technology.[153]

Limit Population Growth

As the FAO study of land resources in the Third World makes clear, without progress in slowing population growth, many nations will be unable to grow enough food to feed their people. A high priority must be given to expanding family planning services and improving basic health services, education, and the status of women.[154]

In summary, it seems clear that world food needs can only be met through a variety of strategies to increase agricultural output, implement sustainable agricultural methods, reduce food losses, improve food distribution, limit food consumption and waste in industrial countries, and lower population growth rates, especially in food-deficit areas.

Progress Toward Sustainable Agriculture

There are many examples of progress toward sustainable agriculture from around the world; they range from small plots in tropical forest regions to larger farms in temperate regions to regional and country-wide programs. Each example has distinct characteristics shaped by local geography and climate, and by local economic, social and cultural factors. Yet there are some features that many cases of sustainable agriculture have in common.

Tropical Agriculture

In their study of sustainable agriculture in the tropics, Gradwohl and Greenberg concluded that, to be sustainable, an agricultural system must maintain crop production over time, supply adequate nutrients for crops, provide some protection against pests and weeds, and be able to survive changes in environmental conditions. In ten case studies of sustainable farming in tropical regions, the authors found a variety of types of agriculture, including subsistence and market-oriented farming, a number of ways of producing animal protein, and several types of indigenous agriculture, some traditional and some relatively new. They noted that the methods used in each case were shaped by that region's growing conditions, its economic situation, and its social and cultural conditions. In spite of their diversity, however, some of the cases shared several characteristics of sustainable farming, including:

» intensive and careful use of small plots;

» high diversity of crops;

» careful attention to local topography, soil, and other conditions;

» crop sequences that imitate natural forest succession (for example, from root crops and annuals to perennial shrubs to trees);

» integration of trees, shrubs, and non-woody plant crops;

» nutrient recycling of crop residues and plant, animal, and human wastes;

» integration of domestic animals; and

» combination of subsistence and cash crops.

The case studies were drawn from Mexico, Costa Rica, Panama, Brazil, Peru, Indonesia, and New Guinea.[155]

Niger

In their recent review of sustainable development projects around the world, Reid, Barnes, and Blackwelder included several projects that emphasized sustainable agriculture. In the sub-Saharan African country of Niger, CARE initiated a project in 1975, with support from the U.S. Agency for International Development, to plant double rows of trees as windbreaks on 3,000 hectares of cropland. The land had been losing nearly 20 tons of topsoil per hectare each year to wind erosion. The trees slowed wind velocity, protecting the soil and improving its moisture retention. As a result, yields of food crops and fuelwood increased significantly. The benefits to farmers were so great that many of them have started their own tree planting projects.[156]

Dominican Republic

In 1979, the Plan Sierra project was initiated in the Dominican Republic with support from several governments and foundations, and has focused on restoring productivity to a region that was 80 percent deforested and badly eroded as a result of timber harvesting beginning in the 1950s. The project combines reforestation, sustainable agriculture, improved health care, education, and transportation. Plan Sierra has trained more than 3,000 farmers in soil conservation, reforestation, composting, the use of legumes as mulch and green manure, agroforestry techniques, and the conservation of genetic diversity. While the project's efforts to develop sustainable solutions for the small farmer have been only partly successful, the resiliency and adaptability of Plan Sierra offer cause for optimism about future progress.[157]

Honduras

In the Guinope region of Honduras, by the late 1970s declining soil fertility, caused by soil erosion and continuous monocropping of maize, had caused most farmers to leave the area or resort to shifting cultivation. In 1981, an agriculture development program was initiated by World Neighbors, with public and private support from Honduras, in an effort to reverse the region's declining productivity. The Guinope program introduced soil conservation methods, such as contour and drainage ditches, contour grass barriers, and rock walls. The program also trained farmers in soil improvement measures, including the use of chicken manure, green manure (intercropping of leguminous plants), and limited amounts of chemical fertilizers.

In the first year of the Honduran program, the yields of farmers who adopted the measures tripled or quadrupled. During the next five years, additional villagers were trained. Soil conservation methods have reclaimed much of the land, and outmigration has largely been reversed as people have been attracted by the growing agricultural potential and new employment opportunities. The project has also promoted the use of small-scale experimental plots to test new methods of boosting farm productivity. Over the first six years of the program, 1,200 families adopted conservation methods, and 60 villagers were trained as agricultural extensionists.[158]

Chile

Due to historical settlement patterns, many peasant farmers in Chile have not inherited knowledge of sustainable farming methods from their ancestors. To fill that gap, in 1981 several Chileans founded a private group to give poor farmers information about sustainable agriculture alternatives. The group established an experimental farm near Santiago to demonstrate intensive gardening methods appropriate both for small-scale subsistence farming and for larger farms. The group helped other Chilean organizations and university personnel to develop a research program focusing on sustainable agricultural techniques for the small farmer. This effort now includes some 75 research programs and university courses in alternative agriculture.

Three demonstration farms have been established; each year they receive more than 10,000 visitors. And each year, between 600 and 700 farmers, extension agents, and community leaders live or work briefly at the farms. Methods taught at the facilities include cropping patterns and management techniques. Crop rotation methods are used to improve soil fertility and control insect pests. The farms grow a diversity of crops, including fruits, vegetables, grains, and farm animals; and utilize nitrogen-fixing plants, animal manure, and composting as nutrient sources. The Chilean program stresses appropriate technologies, self-help efforts, and social organization. Its success comes from its enhancement of traditional practices, its emphasis on the use of multiple techniques rather than a single "best" method, and its grassroots training activities.[159]

People's Republic of China

Before the revolution in 1949, several million people in China starved each year. Today, the country has more than twice as many people, and as a result of an ambitious national agricultural program, almost all are adequately fed. The government has given food production a high priority, spread new technologies, raised the prices farmers receive for food, and let them sell more of their crops privately, thus giving them incentives to produce more. China grows more than one-third of the world's rice and is a leader in "ecological agriculture," recycling nearly half of its crop residues and much of its human and animal wastes.

China has achieved its impressive growth in agricultural output by using labor-intensive rather than capital-intensive farming methods. About 60 percent of the population is involved in agriculture, compared with less than 3 percent in the United States. Human labor is utilized so that several crops can be grown in one field in alternate rows, and labor-intensive pest control methods make crop spraying necessary only against specific infestations. As noted earlier, China has a nationwide program for monitoring agricultural pests and controlling them through integrated pest management and biological methods.[160]

While there are many successful aspects of Chinese agriculture, intensive farming practices in China have also created problems, including extensive soil erosion, loss of soil organic matter, water pollution from increasing use of nitrogen fertilizers, and pathogenic organisms from recycling of organic wastes.[161]

United States

The U.S. Department of Agriculture (USDA) has recently begun a program known as "Low-Input/Sustainable Agriculture," or LISA. It is designed to help farmers across the nation substitute management, scientific information, and on-farm resources for some of the purchased agricultural inputs, such as chemical pesticides and fertilizers, that they rely on today. The program has congressional support and is partly a result of public concern about groundwater contamination and human health risks associated with the use of pesticides and nitrate fertilizers. The LISA program is also a response to farmers who are concerned about the growing costs of high-input agriculture, and who want to increase their net returns while reducing risks and making production and environmental goals more compatible. Initial federal funding for the program was $3.9 million and began at the end of 1987.[162]

The LISA program is interdisciplinary, and involves private research and education organizations, government and public agencies, universities, and farmers. In 1988, 49 projects were funded, and another 77 projects would have been funded if money had been available. For fiscal year 1989, Congress appropriated $4.45 million, a 14 percent increase from 1988.

Results of the program have been encouraging. For example, during the 1987 drought, studies at North Carolina State University showed that experimental plots of corn grown using conventional methods averaged a net loss of $422 per hectare, compared with a net loss of only $235 per hectare for plots using low-input agriculture. And in 1986, soybeans grown using conventional techniques yielded a net profit of $64 per hectare, compared with a profit of $116 per hectare under low-input management. In a South Dakota study during the 1988 drought, a conventional farming system with chemical inputs incurred a net loss of $25,000, while a comparable low-input system earned about $5,000—a difference between the two systems of $30,000.[163]

Because low-input farming reduces the cost of agricultural inputs, farmers in the LISA program have seen their profits rise, even in cases where their crop yields declined. At the same time, they have improved the long-term productivity of their land by using it sustainably, and they have avoided potential health risks. Not all of the low-input methods being tested are effective or profitable, but in general the outlook is optimistic. A large and growing number of farmers, extension workers, and scientists and educators at universities are forming project teams to carry out studies that will develop and test low-input farming methods. With the help of USDA, these profitable and sustainable methods are likely to become an important part of U.S. agriculture.[164]

Eight U.S. states have established formal sustainable agriculture programs, according the Renew America's 1989 *State of the States* report. The states are California, Iowa, Minnesota, Montana, Nebraska, Ohio, Texas, and Wisconsin. Seven states now allocate funds for integrated pest management—California, Connecticut, Delaware, Massachusetts, New Jersey, New York, and Texas. Three states—Minnesota, Texas, and Washington—have set up programs to certify farms as organic.[165]

In September 1989, the prestigious National Research Council released a report entitled *Alternative Agriculture*, recommending that the U.S. government revise its farm policy to encourage sustainable farming methods that preserve the soil and reduce the use of chemicals. The report supports growing numbers of farmers and consumers who advocate a fundamental change in agriculture away from reliance on harmful chemicals and toward more sustainable practices.[166]

What You Can Do

Inform Yourself

Get more information on the issues surrounding local and global hunger so you can speak and write about the problem with authority.

- Contact organizations (such as those listed under Further Information) that work to improve agriculture, conserve soil and water, and educate people on world hunger and agricultural issues. Request publication listings and subscribe to newsletters and action alerts to keep up-to-date on food issues and related legislation.

- Find out about the food you eat. Where does it come from? Is it organically grown or are pesticides, fungicides, and chemical fertilizers used? What crops are native to your area? How do climate changes affect local farmers and the price of agricultural goods?

- Try to locate a farm in your area that uses low-input, regenerative, sustainable agricultural methods, and arrange a visit. Find out what methods they use, where they market their produce, and how their costs of operation compare to conventional farms. Examples of such farms include the Rodale Research Center near Emmaus, Pennsylvania, and the Thompson farm near Boone, Iowa. For further information, contact the Institute for Alternative Agriculture or the Low-Input/Sustainable Agriculture (LISA) program at the U.S. Department of Agriculture, listed under Further Information.

Join With Others

Contact others who are concerned about world hunger and agriculture issues, and work together to educate your community. Join existing local and national groups or start your own grass-roots education effort. Volunteer your services to hunger organizations.

- Join a local Farmer's Alliance or 4-H Club in your area. Find out how you can support these groups and work on agricultural-awareness projects on the local level.

- Help create cooperative marketing networks to bring the products of low-input, regenerative farms to a wider cross section of consumers.

Review Your Habits and Lifestyle

Consider what actions you can take to affect the issues surrounding hunger and food production in your everyday life.

- Consider taking the "Noble Pledge" formulated by GTC board member and former president Jan Hartke, to eat fewer animals. Hartke notes that apart from limiting family size, eating fewer animals probably does more to reduce environmental degradation than any other personal action.[167] By lowering your consumption of animal products, you can lead a healthier life and help conserve topsoil and other valuable natural resources.

- Find out about recommended daily requirements for calories and protein for yourself and members of your family. Learn about protein complementarity—how to combine grains, legumes, and other foods to produce high-quality protein. Plan diets with adequate but not excessive protein content. For guidance, refer to Frances Moore Lappé's *Diet for a Small Planet*.[168]

- Buy organically-grown fruits and vegetables that have not been grown with chemical fertilizers or treated with highly toxic insecticides or fungicides. Encourage your local supermarkets to carry organic produce or produce grown using biological pest controls or integrated pest management techniques.

- If you have a pet, consider replacing part of the canned meat protein in its diet with vegetable protein. Pets consume enough animal protein annually to feed millions of hungry people.

Work With Your Elected Officials

Contact your local, state, and national legislators, and find out their positions on issues such as low-input, sustainable agriculture, foreign food aid, and international agricultural development projects. In brief, clear letters or visits, urge your representatives to support or initiate legislation to promote sustainable agriculture and alleviate world hunger. More specifically, urge them to:

- Support better protection and management of U.S. farmlands, including:

  » local, state, and federal programs to preserve agricultural land and conserve soil and water;

  » federal financial incentives to help preserve farmland;

  » growth management policies that discourage farmland conversions; and

  » increased support for the U.S. Department of Agriculture's Low-Input/Sustainable Agriculture (LISA) program.

- Work to ensure that the development policies of the U.S. Agency for International Development, the World Bank, and the regional development banks emphasize the following objectives:

  » Promotion of general economic development in the poorer countries through programs that promote labor-intensive employment in agriculture and industry and help increase incomes of the poorest people.

  » Adoption of sustainable agriculture management techniques, including greater use of crop residues, animal wastes, and other organic materials to improve soil fertility; more efficient use of commercial fertilizers; biological fixation of nitrogen; and integrated pest management.

>> Expansion of regional food reserve systems.

>> Preservation of crop germ plasm and prevention of agricultural genetic losses.

>> Wider involvement of women in the planning and management of agricultural projects.

>> Support for provision of family-planning information and assistance worldwide.

>> Protection of tropical forests and preservation of biological diversity.

● Support increased U.S. contributions to United Nations agencies that promote sustainable agricultural development, including the Food and Agriculture Organization and the U.N. Development Program that supply technical agricultural assistance, the U.N. Environment Program that works to reverse environmental degradation, the World Food Program that supplies food as partial payment in labor-intensive projects promoting social and economic development, and the International Fund for Agricultural Development that works to increase Third World food production by the rural poor for their direct benefit.

Publicize Your Views

Educate community leaders, policymakers, and the general public on food issues as they affect the United States and other nations. Using print and electronic media, create public service announcements, write letters to the editors of local newspapers, and offer to write stories on hunger and agricultural issues for local publications. Ideas you may want to incorporate include:

>> the links between food production, agricultural land, and population growth;

>> the severity of hunger and malnutrition in the Third World and the need for internal reforms and external economic and technical assistance to alleviate it;

>> the serious economic and environmental effects the export of cash crops has on developing countries;

>> inherent weaknesses in U.S. agriculture such as loss of topsoil and soil fertility, depletion of water supplies, concentration of farm ownership, and farm debt; and

>> the possible effects of the projected global climate changes on the world's ability to produce enough food.

Raise Awareness Through Education

Find out whether schools, colleges, and libraries in your area have courses, programs and adequate resources related to world hunger and agricultural development. Organize community activities to educate people about problems such as world hunger, the loss of fertile topsoil, and water pollution from farm chemicals and livestock.

● Help supply teaching materials on hunger to elementary and high schools in your area. Offer to lead school discussion groups and field trips to local farms, food processing and distribution facilities, and soup kitchens for the poor.

● Set up exhibits on agriculture and hunger in museums, schools, libraries, shopping malls, department stores, community centers, and other sites where they will be seen. The most effective displays are usually multimedia and could include a background film, presentations by local organic farmers, children's artwork, poetry, games, music from different parts of the world, maps, and participatory games. For further suggestions, contact the National Outreach Division, Save the Children, 54 Wilton Road, Westport, CT 06880.

● Organize a group of several dozen people for a day-long "hunger experience" in which a randomly-selected part of the group is offered gourmet-style meals in an elegant First-World manner, while the rest of the group sits on the floor and is given only rice and tea. Combine this experience with background sessions on the scope and dimensions of world hunger, and group discussions on personal reactions to the experience.

● Organize an activity that stresses how much the United States relies on other nations for many imported foods such as fruits, vegetables, meat, and fish; and how it depends on other countries to buy U.S. food exports such as wheat and other grains. Such an activity could be part of a celebration each year of World Food Day on October 16th (the founding date of the Food and Agriculture Organization). The National Committee for World Food Day, 1001 22nd Street, NW, Washington, DC 20437, can provide information on activities and resources of its sponsoring organizations.

Consider the International Connections

If you travel to another country, especially in the Third World, find out what agricultural products it exports to and imports from the United States. Learn what role (if any) the

United States plays in the nation's economic and technological development.

- Before you travel, find out what U.S. and international agencies (such as the U.S. Agency for International Development, the World Bank and regional development banks, and U.N. development agencies) and what private nongovernmental organizations (see those listed under Further Information) have agricultural development projects in the country you plan to visit. When you arrive, try to visit and observe such projects, and learn how they contribute to the country's agricultural potential and to its overall economic and social development.

- Consider the agricultural and other products you consume in your own home, and try to find out how many of them are imported from countries with widespread hunger.

Further Information

A number of organizations can provide information about their specific programs and resources related to food and agriculture.[*]

Books

American Farmland Trust. *Future Policy Directions for American Agriculture*. Washington, DC: 1984. 106 pp. Paperback, $5.00.

Berg, Norman, and Robert Gray. *Soil Conservation in America: What Do We Have to Lose?*. Washington, DC: American Farmland Trust, 1984. 133 pp. Paperback, $8.25.

Center for Science in the Public Interest. *Organic Agriculture: What the States Are Doing*. Washington, DC, 1989. 39 pp. Paperback, $3.00.

Center for Study of Responsive Law. *Eating Clean: A Consumer's Guidebook to Overcoming Food Hazards*. Washington, DC: undated. 156 pp. Paperback, $8.00. A diversified selection of readings to educate the consumer on the dangers of pesticides in foods and what can be done about them.

Concern, Inc. *Farmland: A Community Issue*. Washington, DC: 1987. 23 pp. Paperback, $3.00.

Concern, Inc. *Pesticides: A Community Action Guide*. Washington, DC: 1985. 23 pp. Paperback, $3.00.

Conservation Foundation, The. *Eroding Soil*. Washington, DC: 1985. 252 pp. Paperback, $15.00.

Doyle, Jack. *Altered Harvest: Agriculture, Genetics, and the Fate of the World's Food Supply*. New York: Viking Penguin, 1985. 502 pp. Hardback, $25.00. Examines the environmental and biological costs of the American food production system and raises questions about the genetic agricultural revolution that is changing the way we grow, process, and harvest the food we eat.

Dover, Michael and Lee Talbot. *To Feed the Earth: Agro-Ecology for Sustainable Development*. Washington, DC: World Resources Institute, 1987. 100 pp. Paperback, $10.00.

Edwards, Clive et al. (eds.). *Sustainable Agricultural Systems*. Ankeny, IA: Soil and Water Conservation Society, in press. Discusses various aspects of sustainable agriculture, including maintenance of crop yields, lower input costs, higher farm profits, and improved environmental quality.

Food and Agriculture Organization. *Land, Food and People*. Rome: 1984. 96 pp. Paperback.

Francis, Charles, and Richard Harwood. *Enough Food: Achieving Food Security Through Regenerative Agriculture*. Emmaus, PA: Rodale Institute, 1985. 20 pp. Paperback.

The Hunger Project. *Ending Hunger: An Idea Whose Time Has Come*. New York: Praeger, 1985. 430 pp. An extraordinary development education tool with a positive, constructive approach to the issue of world hunger. Examines differing views of food, population, foreign aid, and national security issues.

Lappé, Frances Moore. *Diet for a Small Planet*. 10th Anniversary Revised Edition. New York: Ballantine Books, 1982. 469 pp. Paperback, $11.95. Offers a new philosophy for changing yourself and the world by

---

[*] These include: American Farm Foundation, CARE, Center for Science in the Public Interest, Concern, Inc., Conservation Foundation, Environmental Policy Institute, Food and Agriculture Organization of the United Nations (FAO), The Hunger Project, Institute for Alternative Agriculture, Institute for Food and Development Policy, Inter-American Development Bank, International Food Policy Research Institute (IFPRI), International Fund for Agricultural Development (IFAD), National Association of State Departments of Agriculture, National Committee for World Food Day, National Wildlife Federation, Natural Resources Defense Council, New Forests Project, Renew America, Resources for the Future's National Center for Food and Agricultural Policy, The Rodale Institute, Rural America, Sierra Club, U.N. Development Program, U.N. Environment Program, U.S. Agency for International Development, U.S. Department of Agriculture, Winrock International Institute for Agricultural Development, World Bank, World Resources Institute, Worldwatch Institute. Addresses for these organizations are listed in the appendix.

changing the way you eat. Features simple rules for a healthy diet, meal preparation guide, recipes, and a listing of food and hunger education resource organizations.

Lappé, Frances Moore and Joseph Collins. *World Hunger: Twelve Myths.* A Food First Book. New York: Grove Press, 1986. 208 pp. Paperback, $7.95. Maintains that hunger is the ultimate symbol of powerlessness, and that the root cause of hunger is a scarcity of democracy, not food or land. Available from the Institute for Food and Development Policy.

LeMay, Brian (ed.). *Science, Ethics, and Food.* Washington, DC: Smithsonian Institution Press, 1988. 144 pp. Paperback, $10.95. Proceedings of an interdisciplinary colloquium on developing long-term remedies for the human dilemmas associated with the world's food supply.

Mott, Lawrie, and Karen Snyder. *Pesticide Alert: A Guide to Pesticides in Fruits and Vegetables.* A publication of the Natural Resources Defense Council. San Francisco: Sierra Club Books, 1987. 179 pp. Paperback, $6.95. A survey of the most commonly found pesticides, their known health effects, their environmental impacts, and ways to reduce their danger. Reviews the strengths and weaknesses of government pesticide regulations.

National Research Council. *Alternative Agriculture.* Washington, DC: National Academy Press, 1989. 417 pp. Paperback, $19.95. Recommends that the U.S. government revise its farm policy to encourage alternative farming methods that preserve the soil and reduce the use of chemicals.

Paulino, Leonardo. *Food in the Third World: Past Trends and Projections to 2000.* Washington, DC: International Food Policy Research Institute, June, 1986. 76 pp. Free.

Pimentel, David and Carl W. Hall (eds.) *Food and Natural Resources.* San Diego, CA: Academic Press, 1989. 512 pp. $98.00. Provides the information needed to ensure the effective management of resources for food production.

Reid, Walter V., James N. Barnes, and Brent Blackwelder. *Bankrolling Successes: A Portfolio of Sustainable Development Projects.* Washington, DC: Environmental Policy Institute and National Wildlife Federation, 1988. 48 pp. Paperback, $6.00. A review of successful development efforts, including sustainable agricultural projects in Latin America, Africa, and Asia.

Robbins, John. *Diet for a New America.* Walpole, NH: Stillpoint Publishing, 1987. 423 pp. Paperback, $12.95. Examines our dependence on animals for food and the inhumane conditions under which they are currently raised; reveals how individual food choices affect our environment, health, happiness, and the future of life on Earth.

U.S. Department of Health and Human Services. *The Surgeon General's Report on Nutrition and Health—Summary and Recommendations.* Washington, DC: 1988. 78 pp. Paperback, $2.75. Presents highlights of the Surgeon General's findings on the relationship between nutrition and health, including policy implications and illustrative tables.

Volunteers in Technical Assistance (VITA). *Environmentally Sound Small Scale Agricultural Projects.* Revised Edition, 1988. Paperback, $9.75. Available from VITA Publications, P.O. Box 12028, Arlington, VA 22209.

Wennergren, E. Boyd, et al. *Solving World Hunger: The U.S. Stake.* Published for the Consortium for International Cooperation in Higher Education. Cabin John, MD: Seven Locks Press, 1986. 99 pp. Paperback. Presents an overview of development issues including the world food problem, the history and nature of U.S. foreign assistance; contains an excellent development education resource section.

Withers, Leslie and Tom Peterson (eds.). *Hunger Action Handbook: What You Can Do and How To Do It.* Decatur, GA: Seeds Magazine, 1987. 153 pp. Paperback, $7.95.

Wittwer, Sylvan et al. *Feeding a Billion—Frontiers of Chinese Agriculture.* East Lansing, MI: Michigan University Press, 1987. 462 pp. $30. Documents agricultural changes that have enabled China, with 7 percent of the Earth's arable land, to essentially eliminate hunger for 22 percent of the world's people.

World Hunger Education Service. *Who's Involved with Hunger: An Organization Guide for Education and Advocacy.* Washington, DC: 1985. Paperback, $8.00. Describes the purposes and nature of the work of about 450 organizations from which information is available on the technical factors relating to hunger and poverty in the United States and the Third World.

For an extensive annotated listing of publications on food and agriculture, obtain a current catalog from the Agribookstore, Winrock International, 1611 North Kent Street, Arlington, VA 22209, telephone 703/525-9455.

Articles

Bradley, P.N. and S.E. Carter. "Food Production and Distribution—and Hunger." In R.J. Johnston and P.J. Taylor (eds.), *A World in Crisis?.* Second Edition. Cambridge, MA: Basil Blackwell, 1989.

Brown, Lester R. *The Changing World Food Prospect: The Nineties and Beyond* Worldwatch Paper 85. Washington, DC: Worldwatch Institute, October, 1988. 58 pp. $4.00.

Conservation Foundation, The. "Agriculture and the Environment in a Changing World Economy," *State of the Environment: A View Toward the Nineties.* Washington, DC, 1987, pp. 341-406. Paperback, $19.95.

Crosson, Pierre R. and Norman J. Rosenberg. "Strategies for Agriculture." *Scientific American*, September 1989, pp. 128-35. To feed the world's growing population, social and economic changes are needed to persuade farmers to adopt methods that will boost food production without further degrading the environment.

El-Hinnawi, Essam and Manzur H. Hashmi. "Land, Water and Food Production," *The State of the Environment.* London: Butterworth Scientific, 1987, pp. 35-69. Hardback.

Gradwohl, Judith and Russell Greenberg. "Sustainable Agriculture," *Saving the Tropical Forests.* Washington, DC: Island Press, 1988, pp. 102-37. Hardback, $24.95.

Hendry, Peter. "Food and Population: Beyond Five Billion." *Population Bulletin*, Vol. 43, No. 2, April 1988. 40 pp. Published by Population Reference Bureau.

Hrabovszky, Janos P. "Agriculture: The Land Base." In Robert Repetto (ed.), *The Global Possible: Resources, Development, and the New Century.* A World Resources Institute Book. New Haven: Yale University Press, 1985, pp. 211-254.

InterAction. "Hope for the World's Hungry?" American Council for Voluntary International Action. Undated. 6 pp. Reviews successful agricultural projects in Bangladesh, Botswana, Ecuador, Niger, and the Philippines.

Mellor, John W. "Agricultural Development: Opportunities for the 1990s," *International Food Policy Research Institute 1988 Report.* Washington, DC, 1988, pp. 8-14.

Myers, Norman (ed.). "Land," *Gaia: An Atlas of Planet Management.* Garden City, NY: Anchor Books, 1984, pp. 24-67. Paperback, $18.95.

Organization for Economic Cooperation and Development (OECD). "Agriculture," *The State of the Environment 1985.* Paris, 1985, pp. 187-99. Paperback, $30.00.

Pimentel, David et al. "World Agriculture and Soil Erosion." *BioScience*, Vol. 37, No. 4, April 1987, pp. 277-83.

Ridley, Scott. "Reducing Pesticide Contamination," *The State of the States 1988.* Fund for Renewable Energy and the Environment. Washington, DC: Renew America, 1988, pp. 12-17. Paperback, $10.00.

Ridley, Scott. "Soil Conservation," *The State of the States 1987.* Fund for Renewable Energy and the Environment. Washington, DC: Renew America, 1987, pp. 10-13. Paperback, $10.00.

Wolf, Edward C. "Raising Agricultural Productivity." In Lester R. Brown et al., *State of the World 1987.* New York: Norton, 1987, pp. 139-56. Paperback, $9.95.

World Resources Institute (WRI) and International Institute for Environment and Development (IIED). "Food and Agriculture," *World Resources 1986*, pp. 43-60; *World Resources 1987*, pp. 39-55; *World Resources 1988-89*, pp. 51-67. New York: Basic Books, 1986, 1987, 1988. Paperback, $19.95.

Periodicals

Alternative Agriculture News. Published monthly by the Institute for Alternative Agriculture (IAA), a non-profit research and education organization. The newsletter is free to members, and $12 a year for non-members, from IAA, 9200 Edmonston Road, Suite 117, Greenbelt, MD 20770.

Food First News and *Food First Action Alert.* Published quarterly by Food First, 1885 Mission Street, San Francisco, CA 94103; sent upon payment of $25 annual membership fee.

Hunger Action Forum. A monthly report on poverty and hunger in America. Published by The Hunger Project, 1 Madison Avenue, New York, NY 10010. Free.

Hunger Notes. Published six times a year by the World Hunger Education Service, 3018 4th Street, N.E., Washington, DC 20017. Each issue covers a specific topic in depth, with a guide to further sources of information and program materials. Annual subscription $15 for individuals, $40 for institutions.

International Ag-Sieve. A monthly newsletter about regenerative agriculture, published by Rodale International, 222 Main Street, Emmaus, PA 18098.

Nutrition Action Health Letter. Published ten times a year by the Center for Science in the Public Interest, 1501 16th Street, N.W., Washington, DC 20036. Annual subscription $20.

Organic Gardening. Published ten times a year by Rodale Press, 33 East Minor Street, Emmaus, PA 18098. Annual subscription $15.

Regeneration. Published bimonthly by Rodale Press, 33 E. Minor Street, Emmaus, PA, 18098. Annual subscription $15.

Films and Other Audiovisual Material

The Desert Doesn't Bloom Here Anymore. A NOVA film examining water and irrigation policies and their effect on soil quality. 58 min. 16 mm film ($310), Video ($220), Rental ($125). Coronet/MTI Film & Video.

The End of Eden. Documents the severe environmental damage that cattle ranching projects financed by multilateral development banks have caused in Botswana and other areas in sub-Saharan Africa. Filmed by Rick Lomba, a longtime African resident. For information, contact Turner Broadcasting System.

Fragile Harvest. 49 min. 16 mm film ($750), video ($595), Rental $95. Grade 9 to Adult. Explores the growing crisis in global agriculture and the threats to the genetic diversity of the world's food plants. Bullfrog Films.

Land Use Today for Tomorrow. 36 min. 16 mm color. Narrated by Eddie Albert. Explores solutions to soil erosion and competition for land use. Conservation Film Service.

The Miracle of Guinope. 16 min. VHS video, loan, free; purchase, $12 including postage. Depicting the transition of Guinope, Honduras, from a dying village to a thriving community, the film offers a model for ending hunger and slash-and-burn agriculture. World Neighbors.

Nutrition: Eating Well. 25 min. Color film ($384), video ($235). Grades 4-9. A group of students learns that there is more than one path to good nutrition. Offers lively information on the importance of protein, minerals, carbohydrates, fat, fiber, and water in eating well. National Geographic.

On American Soil. Produced by the Conservation Foundation. Discusses soil erosion in the United States. 28 min. Grade 7-adult. 16mm film ($550), Video ($250), Rental ($50). Bullfrog Films or Conservation Film Service.

Will the World Starve? Produced by the BBC-TV and WGBH-TV for public television. Examines the relationship between farmers, government economic policies, land degradation, poverty, and soil erosion in Nepal, China, and Ethiopia. 58 min. 1/2 inch VHS. Conservation Film Service or Coronet/MTI Film & Video.

Teaching Aids

Tackling Poverty in Rural Mexico. Students meet farmers in a poor village and see the changes that take place when they build an irrigation system and learn how to use new farming methods. Includes student materials, sound filmstrip, and teaching guide. $60. World Bank Publications.

For a listing of curricula, slide shows, research reports, and other educational materials on food and development issues, obtain the *Catalog of Resources* from the Institute for Food and Development Policy (Food First), 1885 Mission Street, San Francisco, CA 94103.

Biological Diversity

Without knowing it, we utilize hundreds of products each day that owe their origin to wild animals and plants. Indeed our welfare is intimately tied up with the welfare of wildlife. Well may conservationists proclaim that by saving the lives of wild species, we may be saving our own.

—Norman Myers[1]

Nearly half the world's species of plants, animals, and microorganisms will be destroyed or severely threatened over the next quarter century.

—Peter H. Raven[2]

Artwork by Carol Connet

Major Points

- By early in the next century, we could lose a million or more species of plants, animals, and other organisms—more than all the mass extinctions in geologic history, including loss of the dinosaurs.

- Wild plants and animals are vital to human survival. They produce foods, medicines, and essential raw materials, and are important for future improvement of crops and livestock, and for development of new medicines and industrial products. Plants and animals also provide services such as pest and flood control, maintenance of soil productivity, and degradation of waste.

- Habitat destruction by human activities, especially in the tropics, is the primary cause of extinction and the worldwide loss of biological diversity.

- An adequate global system of parks and reserves, guided by an overall conservation strategy and coupled with well-conceived programs for sustainable economic development, limiting population growth, controlling exploitation of species, and restoring damaged ecosystems, could preserve a substantial part of our planet's biological diversity.

The Issue: Extinction of Plants and Animals

The preservation of biological diversity is an issue of unprecedented urgency. Science is discovering that the genetic variety contained in wild species can relieve human suffering and improve the quality of life, yet, the activities of exploding human populations are degrading the environment at an accelerating rate, and diversity is being irreversibly diminished through extinction as natural habitats are destroyed.[3] In the words of the Club of Earth, the extinction crisis is "a threat to civilization second only to the threat of thermonuclear war."[4]

Conservation of biological diversity is vital to human survival and well-being, in part because wild species of plants, animals, and other organisms provide people with important products, including food, medicine, and industrial raw materials. The invaluable services they provide include pest control, flood control, and the natural recycling of waste.[5] Yet hundreds of thousands of the Earth's species will become extinct in the next 20 years because we are destroying their natural habitats and excessively depleting their populations. Tropical forests and coral reefs, which contain the greatest variety of known and undiscovered species, will suffer the most. Tropical regions contain more than two-thirds of all the world's species. For example,

- a single river in Brazil harbors more species of fish than all the rivers in the United States;

- a square mile in lowland Peru is home to more than twice as many butterflies as the entire United States and Canada—more than 1,500.[6]

- one Peruvian wildlife preserve contains more bird species than can be found in the entire United States; and

- 10 one-hectare plots in Borneo contain 700 species of trees, as many as grow in all of North America.[7]

At least two-thirds of all land-dwelling species are found in the tropics and, of those, roughly two-thirds—representing about half of the Earth's biological diversity—live in forested areas that will be cut or heavily damaged in the next two or three decades. This projected forest destruction in the tropics poses a major threat to all forms of life, including plants, insects, fish, birds, reptiles, amphibians, and mammals. Once species become extinct, they are not renewable. The loss of large numbers of plant and animal species could greatly limit the options of future generations.[8]

In addition to the species that will be lost outright, many thousands will be reduced to populations teetering on the edge of extinction. When a species becomes rare, it loses genetic diversity and becomes vulnerable to many factors that threaten its long-term survival—habitat destruction, changes in climate, disease outbreaks, human exploitation, and political upheavals.[9]

Species Diversity

About 1.4 million living species have been named and described by scientists. Approximately 750,000 are insects, 41,000 are vertebrates, and 265,000 are plants; the remainder includes invertebrates, fungi, algae, and microorganisms.[10]

No one knows the true number of species on Earth; current estimates range from 5 million to 30 million or more. Recent studies in tropical forests suggest that there might even be 30 million species of insects alone.[11] Table 6.1, on the following page, compiled by the Worldwatch Institute and based on a tally by E. O. Wilson, summarizes current knowledge about the Earth's diversity of species.[12]

Species Extinction

In 1984, based on then-current estimates of 5 million to 10 million species overall, Norman Myers estimated that the world might be losing one species a day, or about 400 plants and animals each year, and that the annual rate of species loss might reach some 10,000 species by 1990 and approach 50,000 by the year 2000.[13]

Table 6.1
Known and Estimated Diversity of Life on Earth

Form of Life	Known Species	Estimated Total Species
Insects and other arthropods	874,161	30 million insect species, extrapolated from surveys in forest canopy in Panama; most believed unique to tropical forests.
Higher Plants	248,400	Estimates of total plant species range from 275,000 to as many as 400,000; at least 10-15 percent of all plants are believed undiscovered.
Invertebrates[1]	116,873	True invertebrate species may number in the millions; nematodes, eelworms, and roundworms each may comprise more than 1 million species.
Lower Plants[2]	73,900	Not available.
Microorganisms	36,600	Not available.
Fish	19,056	21,000, assuming that 10 percent of fish remain undiscovered; the Amazon and Orinoco rivers alone may account for 2,000 additional species.
Birds	9,040	Known species probably account for 98 percent of all birds.
Reptiles and Amphibians	8,962	Known species of reptiles, amphibians, and mammals probably comprise over 95 percent of total diversity.
Mammals	4,000	
Total	1,390,992	10 million species considered a conservative estimate; if insect estimates are accurate the total exceeds 30 million.

[1]Excludes arthropods, includes 1,273 miscellaneous chordates.
[2]Fungi and algae.

Source: Edward C. Wolf, *On the Brink of Extinction: Conserving the Diversity of Life*, Worldwatch Paper 78 (Washington, DC: Worldwatch Institute, June 1987) p. 7

Based on more recent estimates of up to 30 million species total, E.O. Wilson has calculated that we may already be losing as many as 17,500 species each year. Many are plants and animals we have never seen, and of whose existence we are unaware. Some scientists forecast that losses could reach one million by the year 2000.[14] Peter Raven estimates that we could lose 10 percent of the world's species by the end of the century and more than 25 percent within the next couple of decades.[15] This would amount to the greatest biological debacle since the dinosaurs disappeared 65 million years ago.[16]

Daniel Simberloff's recent projection of plant extinctions in Latin American forests, based on current rates of population growth and forest clearing, indicates that as these forests shrink to half their original size by the year 2000, 15 percent of the forest plant species, or about 13,600 kinds of plants, could be lost from an original total of over 92,000 species.[17]

Of the 11,500 plant species originally found in Europe, 20 species are already extinct, and 21 percent (2,420) are endangered. Of some 20,000 plant species originally found in the United States, 90 species are extinct and 11 percent (2,040) are endangered or threatened.[18] In Germany, the Netherlands, Denmark, and Spain, more than 20 percent of bird species, and over 40 percent of mammalian species are classified as threatened.[19] And in the Western Hemisphere, populations of many forest birds and migratory shorebirds are declining—as much as 60 to 80 percent in some cases.[20]

Causes of Species Loss

The most serious loss of biological diversity is occurring in the tropics, due to explosive growth of human populations, widespread poverty, growing demand for fuelwood, and failure to use sustainable methods in agriculture and forestry.[21]

Figure 6.1
Estimated Annual Rate of Species Loss, 1700-2000

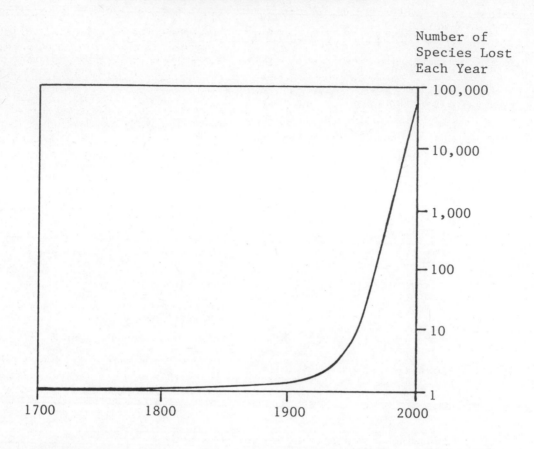

Number of
Species Lost
Each Year

Source: Based on estimates in Norman Myers (ed.), *Gaia: An Atlas of Planet Management* (Garden City, NY: Anchor Books, 1984), p. 155.

Historically, competition between species, overexploitation of species, and habitat destruction have all been important causes of species loss, but habitat destruction is now the primary direct cause of extinction.[22] Human population is expected to increase by about 40 percent in the next 20 years, primarily in the tropics; as the number of people grows, more areas of forest, savannah, and desert will be converted to cropland. The need for fuelwood will cause additional forests to be cut. At least 40 percent of the Earth's tropical forests have already been destroyed,[23] and growing demand in the United States, Europe, and Japan for hardwoods, beef, lumber, and paper pulp will put additional pressures on tropical forests.

Wetlands are among the Earth's most important and biologically diverse ecosystems. Between 25 and 50 percent of the world's swamps and marshes have been lost, and the United States had lost 54 percent of its original wetlands by the mid-1970s.[24] The habitats within shallow coastal areas and coral reefs are also threatened. These estuaries and marine areas, which serve as nurseries for commercially important fin and shellfish, are being damaged by siltation, dredging, and pollution from petroleum and other chemicals.

Studies have shown that as habitats become smaller, the number of species they support decreases.[25] Preliminary data from the World Wildlife Fund's Minimum Critical Size of Ecosystems Project in Central Amazonia shows that when forest areas are isolated, the number of bird and mammal species living inside test areas invariably declines, particularly in smaller areas.[26] However, other studies show that species richness increases with the degree of habitat subdivision; these studies suggest that to maintain biological diversity, it is better to set aside several smaller parks than one large one.[27]

Competition from introduced species is another cause of extinction. When humans bring exotic (nonnative) animals or plants into an environment, the native species may not have evolved appropriate defenses, and the introduced species edge out the native ones. This is a particular problem on islands, where many native species have evolved in complete isolation from certain predators. In Hawaii, for example, 40 percent of the native flora is considered extinct or endangered due to habitat loss combined with predation by exotic species, including cattle, goats, and rats.[28] A third of the Hawaiian islands' native bird species has become extinct in the two centuries since Europeans arrived, and most of those remaining are threatened or endangered.[29]

A third cause of extinction is excessive harvesting: rhinos are killed for their horns, and elephants for their tusks; blue whales are sought for their oil; sea turtles are hunted for their eggs, leather, shells, and meat; and cacti are collected for their decorative shapes. In some cases, one species is depleted during the harvesting of another. For instance, porpoises, sea turtles, and sea birds are often snared in commercial fishing nets.[30]

In Africa, the elephant population, estimated at around 10 million in the 1930s, was reduced to about 1.3 million by 1979, and is now believed to number less than 750,000. Some 80,000 elephants are killed each year. The number of African rhinos declined from about 15,000 in 1980 to less than half that number in 1985; the southern white rhino has dropped from almost 1,000 to less than 15. The world's population of blue whales, originally about 200,000, has been depleted to about 15,000; the humpback whale population has dropped from some 50,000 to about 8,000.[31]

The Importance of Biological Diversity

The diversity of biological species is the Earth's most important natural resource. Humans depend on the wide variety of species in healthy ecosystems for air to breathe, water to drink, and productive soil for farming. Green leaves absorb carbon dioxide and release oxygen during photosynthesis. The root systems of plants—along with other species such as worms, insects, fungi, and soil bacteria—regulate stream flows and groundwater levels, cleanse pollutants from surface waters, and help recycle soil nutrients. Insects are important as pollinators; 90 of the most important U.S. crops, valued at more than $4 billion, are pollinated exclusively by insects,[32] and nine others are more productive when pollinated by insects.

Bats, which represent almost a fourth of all mammal species, are among the most important pollinators and seed-dispersers in tropical regions.[33] Wild birds, bats, and parasitic insects prey on insect pests. Although we lack the scientific knowledge to determine which and how many species can be eliminated before a given system deteriorates significantly, we do know that if the current rate of extinction continues, we will lose these free services and our lives will be altered.[34]

Agriculture

Although laboratory synthesis has freed us from total dependence on wild plants and animals for organic chemicals, these species still provide us with important products, including food, medicine, and industrial raw materials, as well as luxury goods. Just 20 plant species provide more than 80 percent of the world's food; three of them—corn, wheat, and rice—constitute 65 percent of the food supply.

Plant breeders strive constantly to improve these crops genetically in order to make them resistant to a continually evolving array of pests. The most important sources of such genetic material are the wild or locally cultivated relatives of these crop species, which are found where they were originally domesticated. Most of the remaining populations of wild and local varieties exist in Third World tropical countries.[35] Even with modern genetic technology, genes to improve crop plants must come from existing wild varieties.[36]

As with agricultural crops, livestock production tends to rely on only a few of the available species. By preserving the diversity of wild species suitable for game ranching, more efficient use can be made of land, water, and other resources without decreasing nutritional returns. For example, certain African wild animals require little water and are much more disease-resistant than introduced cattle.[37]

Medicine

Today, over 40 percent of the prescription drugs sold in the United States contain chemicals originally derived from wild species: about 25 percent of these drugs come from plants; another 12 percent are derived from fungi and bacteria; and 6 percent come from animals. The value of medicinal products derived from such sources approaches $40 billion a year.[38]

Among the currently used drugs derived from wild species are digitoxin and digoxin, originally extracted from the foxglove or digitalis plant, used to treat heart disease; and vincristine and vinblastine, from the rosy periwinkle, used to treat Hodgkin's disease, leukemia, and other cancers. Endod, from an Ethiopian plant, shows promise in controlling the spread of bilharzia, a disease affecting 300 million people throughout the tropics. Marine organisms are considered by scientists to be a still untapped source of new chemicals for the study and treatment of disease.[39]

Industry

Many essential industrial products or raw materials are derived from wild plants, and a smaller number from wild animals. Still others come from semidomesticated plants that are highly dependent on their wild relatives for genetic improvement to enhance their economic productivity or usefulness. Timber and other wood products, including lumber, paper, and wood-based chemicals such as rayon, are the most economically important category of industrial products derived from living resources.

Rubber, another major industrial product, is derived from trees. While a synthetic substitute has been increasingly used since its invention during World War II, natural rubber constitutes about one third of current world use because of its superior qualities. Over 70 percent is used for tires—mainly heavy-duty tires for airplanes, trucks, buses, and off-road farm and construction equipment, as well as radial tires for cars.[40]

The meadowfoam plant contains a unique oil that can lubricate high-speed and high-temperature machinery that generate extreme pressures; the oil can also be used for precision instruments in medicine and space technology.[41]

Psychological and Philosophical Values

In addition to the many practical reasons for ensuring the survival of a diversity of species, psychological and philosophical reasons can be cited. Many people seem to have a psychological need to observe, admire, photograph, collect, or be surrounded by diverse living things. Many feel that it is morally wrong to allow or force a species to become extinct. They hold that to do so not only unjustly deprives future generations of their right to enjoy the possible benefits of that species' existence, it also violates that species' right to exist. The ethical and legal debates over the rights of nonhuman life are complex, but a reverence for all life is fundamental to many religions and moral systems. Even in today's predominantly secular society, the uniqueness and inherent value of life are deeply felt by many people.[42]

Species Conservation

Given the importance of the Earth's plant and animal species to human welfare and economic productivity, and the seriousness of the threat to their survival, a major worldwide effort to conserve biological diversity is needed. Michael Robinson, Director of the National Zoological Park, has called for an effort to preserve the future of life on Earth equivalent to the Manhattan Project, which developed the first atomic bomb.[43]

As noted earlier, the greatest variety of species exists in tropical developing nations. Although these countries are rich in genetic resources, they have low per capita incomes and need technical and financial assistance from wealthy nations to protect and manage these biological resources. There are several approaches to conserving species and biological diversity:

- Protecting species of recognized value or those known to be in danger of extinction, through provisions such as the U.S. Endangered Species Act, the Convention on International Trade in Endangered Species of Wild Fauna and Flora, and the International Whaling Convention.

- The so-called Noah's Ark strategy, in which part of an organism, such as its seed or semen, is stored *off-site* in a gene bank; or in which whole organisms are kept off-site in a zoo, aquarium, botanical garden, or plantation.

- The establishment of biological reserves that preserve genetic diversity *on-site* by protecting entire ecosystems.[44] This approach conserves not only those species that elicit public concern, but also the less conspicuous plants, animals, and microorganisms on which those popular species depend. Protected areas can provide benefits such as control of soil erosion and maintenance of air and water quality. At the same time, they serve as scientific repositories for study and as natural sources of the germ plasm that contains each species' heredity.

Protected Areas

To safeguard the world's wild plants and animals, a global network of parks and reserves has been developed. The number of protected areas worldwide has grown rapidly, from about 600 areas covering less than 100 million hectares in 1950 to some 3,500 areas containing about 425 million hectares receiving varying degrees of protection.[45] In spite of this growth, these protected areas amount to less than 3 percent of the world's total ice-free land area. The protected areas include substantial tracts of tundra, warm desert, and tropical dry forest, but are dangerously short of such important habitats as tropical moist forests and grasslands, Mediterranean-type zones, and islands and coral reefs. More protected areas are urgently needed; our present network meets less than one third of estimated needs.[46]

A biosphere reserve is a unique kind of protected area designed to combine both conservation and sustainable use of natural resources. To carry out these activities, biosphere reserves include three interrelated zones: a *core area* containing a minimally disturbed ecosystem characteristic of a major type of natural environment, a *buffer zone* where uses and activities are managed in ways that help protect the core, and a *transition area* combining conservation and sustainable activities such as forestry, agriculture, and recreation. (See Figure 6.2 on the following page.)

"Biosphere reserve" is an international designation made by the United Nations Educational, Scientific, and Cultural Organization (UNESCO) as part of its Man and the Biosphere (MAB) Program. As of early 1987, of more than 110 countries participating in the MAB Program, 70 countries had established some 260 biosphere reserves.[47]

If protected areas are to survive, they must be seen as meeting the real needs of people—not simply fulfilling the esoteric interests of nature enthusiasts. This is especially true in the Third World, where the pressure on protected areas for agricultural use is greatest.

When a critical habitat is made into a national park or a nature preserve, laws must be enforced and local people educated so they can understand the preserve's importance and help support its existence.[48]

**Figure 6.2
Schematic Plan of a Biosphere Reserve**

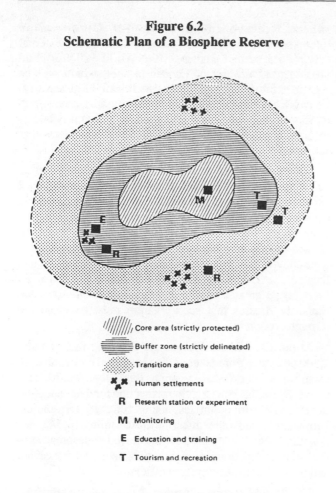

/////// Core area (strictly protected)

Buffer zone (strictly delineated)

Transition area

xⁿx Human settlements

R Research station or experiment

M Monitoring

E Education and training

T Tourism and recreation

Source: Michel Batisse, "Developing and Focusing the Biosphere Reserve Concept," *Nature and Resource*, Vol. 22, No. 3 (Paris: UNESCO, Man and the Biosphere, 1986), p. 3.

In addition to on-site preservation of genetic diversity in parks, reserves, and other protected areas, the off-site preservation and breeding of animals and plants in zoos, aquariums, and botanical gardens is an important part of the overall strategy for preserving biological diversity.[49]

Restoration of Ecosystems

Designating parks and reserves free from human interference has long been considered the best way to conserve plants and animals. But creating parks and reserves is no longer sufficient to prevent a global extinction crisis. A strategy of preservation accompanied by rehabilitation of degraded land and restoration of ecosystems is needed to avoid a mass extinction of plants and animals.[50]

One promising approach to curbing the loss of habitats and species is restoration ecology, based on the study of natural ecosystem recovery, which can promote the recovery of damaged environments and reestablish native communities of plants and animals.[51] Restoration projects now under way include programs to restore degraded pastureland in Brazil, to regenerate dry tropical forest in northwestern Costa Rica, and to restore a prairie ecosystem in Illinois. Proposed projects include an ambitious tropical restoration program in the Caribbean and a plan to test prairie restoration and management in Oklahoma.[52]

Cultural Keys to Biological Diversity

As we become increasingly separated from natural environments and their resources, we become more dependent on the knowledge of those who live in forests and other natural surroundings. Many of our foods and pharmaceuticals were first examined because of their use by indigenous cultures. Tropical forest peoples are the key to understanding, utilizing, and protecting tropical biological diversity.[53] We should see these "human keys" to plants and animals as an integral part of the ecosystems we seek to conserve. By respecting and preserving human diversity, we can draw on the knowledge of traditional cultures and tap the hidden wealth in conserved areas.[54]

Many forest peoples today face a crisis of survival. These cultures, along with their oral traditions, developed over centuries but are disappearing within decades. As ethnobotanist Mark Plotkin of World Wildlife Fund puts it, "Each time a medicine man dies, it is as if a library has burned down."[55] Many of these cultures will have disappeared by the end of the century.[56]

Native cultures in every tropical region include traditions of forest conservation and regeneration. These traditions show how food and other useful products can be obtained without disrupting forest ecosystems, and they suggest ways that forest preservation can support economic development.[57] Once a forest culture is gone, its wealth of information is lost. Much more must be done to protect these cultures and preserve their priceless knowledge.

The Need for Scientific Research

Tropical conservation is hampered by a lack of knowledge and the scarcity of ongoing research programs. There are probably no more than 1,500 professional scientists in the world competent to classify the millions of species found in humid tropical forests, and their number may be dropping because of declining research funding and professional opportunities.[58] The number of publications in tropical ecology dropped by over 50 percent from 1979 to 1983.[59] According to Peter Raven of the Missouri Botanic Garden, in the early 1980s fewer than 25 scientists worldwide were competent to supervise large-scale studies of tropical ecosystems.[60] And Michael Robinson notes that, if the world's spending on tropical biology were increased by a hundredfold, it would still be a tiny fraction of the world expenditure on defense.[61]

Linking Conservation and Development

E.O. Wilson maintains that many major human problems—such as overpopulation, hunger, and habitat destruction—are primarily biological in origin, and can be solved in part by making biological diversity a source of economic wealth. He suggests that wild species are both one of our most important resources and the least utilized, noting that we depend completely on a pool of living species for our survival, but that that pool represents only about 1 percent of the species in existence.[62]

In 1982, the Paris-based International Union of Biological Sciences launched a research program called Decade of the Tropics, in part to seek ways to use the abundance of tropical plants and animals to benefit local populations without destabilizing tropical ecosystems.[63]

The World Wildlife Fund has initiated a Wildlands and Human Needs program with support from the U.S. Agency for International Development. The program will emphasize small-scale, sustainable rural development projects based on ecosystems that supply food and fuelwood to local communities.[64] Projects must link improvement in the quality of life for local populations with maintenance of an important ecosystem, and must involve local people in planning and execution.

In response to growing recognition that traditional development projects have often been ecologically unsound and environmentally destructive, in 1987 the World Bank created an environment department to help design and monitor development projects in order to minimize environmental damage.[65]

The World Bank has also adopted a wildlands policy specifying that new development projects should be sited on converted or degraded land rather than virgin land. Projects requiring clearance of substantial virgin land must include protection of an equivalent area of natural habitat. The Bank's wildlands policy will begin to compensate for the biological losses resulting from conventional development projects.[66]

In 1987, in response to congressional legislation supported by a broad coalition of nongovernmental groups, the U.S. Agency for International Development earmarked $2.5 million to preserve biological diversity in AID-funded development projects. For 1988, AID was committed to spend $4.5 million as part of an effort to integrate biological diversity and other environmental concerns into all its development programs.[67]

In 1984, Thomas Lovejoy recommended that debtor nations with biologically valuable habitats be given discounts or credits for conservation programs.[68] In 1987, an innovative agreement linking conservation to partial debt relief was reached between Bolivia and Conservation International, a U.S.-based environmental group. Bolivia created a 1.6-million hectare protected zone around its Beni Bio-

sphere Reserve in exchange for Conservation International's purchase of $650,000 of Bolivia's external debt.[69] In a second exchange, the World Wildlife Fund purchased $1 million of Ecuador's debt, which will be converted into funds to maintain parks and wildlife reserves. Costa Rica has recently announced a program to convert up to $5.4 million of its external debt. Additional debt-for-nature exchanges are being considered in at least eight other countries.[70]

Such economic initiatives are essential to environmental conservation, writes Jeffrey McNeely. He notes that past experience shows the free market alone does too little to conserve biological diversity. Effective government regulations on the local, national, and international levels are needed to help this effort. People's attitudes toward the environment usually represent only their self-interests, and those are defined today primarily in economic terms. To encourage a greater popular concern for the environment, McNeely stresses that "conservation must be promoted through economic incentives."[71]

At the 1987 World Wilderness Congress in Colorado, 1,600 delegates from 60 countries approved an initiative to establish a World Conservation Bank that would raise money by trading conservation protection for debt reduction in Third World countries, using blocked currencies, cofinancing, and other measures. According to Michael Sweatman, the Bank concept has received widespread support from international commercial bankers, businessmen, and international development officials.[72]

The World Resources Institute is now investigating this and other innovative conservation financing concepts in an effort to solve two pressing global problems at once.

Costa Rica has developed one of the most successful wildlands conservation programs in the developing world. A quarter of its territory and over 80 percent of its remaining wildlands are now protected. The success of these programs may depend on their clear links to sustainable economic development. Most of the planned irrigation, hydropower, and potable water supply projects depend on protection of high-rainfall, steep-sloped wildlands to ensure their viability. Growing recognition of the success of Costa Rica's conservation and development programs is stimulating tourism and an influx of foreign exchange.[73]

In spite of a number of successful efforts to preserve habitats and protect biological diversity, however, much of the world's diversity is in debt-ridden developing countries where a large portion of national income has come from unsustainable cutting of tropical forests for timber and to create agricultural land. Those countries will have difficulty preserving biological diversity if their economies continue to rely on unsustainable activities and situations. With over 75 percent of the world's people and only about 15 percent of its monetary resources, developing countries need much more than debt exchanges in seven figures to relieve their trillion-dollar external debt burden and enable them to protect their biologically-rich natural resources.[74]

Laws and Treaties

There are several ways to create a legal basis for wildlife conservation. One is to establish laws to protect individual species, such as the vicuna, or a group of species, such as marine mammals. The Endangered Species Act, for example, mandates the listing of endangered species and provides for their protection and revival. The act was passed in 1973, then expired in 1985 but was kept alive by annual appropriations until 1988, when the Congress voted to extend and strengthen the act.

A second approach is to develop a regional treaty, such as the Convention on Conservation of European Wildlife and Natural Habitats. A third way is to establish a global treaty, such as the Convention on Wetlands of International Importance. Of these three, the first is easiest to establish, while the third may have the greatest impact.[75]

International treaties and conventions were first used to protect wildlife less than a century ago. The first treaties were largely concerned with economically important species. Since then, there have been a number of treaties and conventions, the most important of which is CITES—the Convention on International Trade in Endangered Species—now adopted by nearly 100 nations. This convention is one of four negotiated in the 1970s that are limited neither to a few species nor to geographic regions. A second is the World Heritage Convention. Two separate conventions protect migratory animals and preserve wetlands, especially those important as waterfowl habitats.[76]

In spite of this progress, existing treaties are quite limited, and wildlife continues to disappear faster than ever. A global treaty or "species convention" to protect biological diversity is needed.[77] Under such a treaty, each nation could accept responsibility for species within its borders. In return, a nation could apply for support from the community of nations to assist it in protecting plant and animal species.[78]

A Global Strategy

The World Conservation Strategy was prepared in 1980 by the International Union for Conservation of Nature and Natural Resources (IUCN) with the support of several United Nations agencies, the World Wildlife Fund, and many other government agencies and conservation organizations. (See Figure 6.3 on the following page.) The Strategy embodies three important propositions:

- Plant and animal populations must be helped to retain their capacity for self-renewal.

- The Earth's basic life-support systems, including climate, the water cycle, and soils, must be conserved intact if life is to continue.

- Genetic diversity, a major key to our future, must be maintained.[79]

Following the pattern of the World Conservation Strategy, some 40 countries have begun or completed national strategies to identify conservation priorities and integrate sustainable resource management into their development plans.[80] The list includes the biologically rich countries of Venezuela, Indonesia, and Malaysia, but unfortunately does not include Brazil and the West African countries containing Africa's most threatened tropical forests.[81]

Recently, the Environmental Law Center of IUCN began work on an international convention to protect biological diversity by preserving critical wildlife habitats.[82] At its May 1989 meeting, the United Nations Environment Program's Governing Council approved a plan to begin drafting a binding legal convention to protect the full range of biological diversity. The convention would operate alongside and supplement existing conservation measures. The new convention will probably focus on habitat protection, and may incorporate IUCN's work on protecting habitats.[83]

What You Can Do

Inform Yourself

Get more information about biological diversity by consulting sources and contacting organizations listed under "Further Information." Learn about issues on the local and international level so you can write and speak on them with authority and express your ideas to others.

- Learn about endangered species living in or near your area, and work to ensure their rehabilitation. You can start by contacting local environmental organizations, university biology departments, state fish and game departments, or natural heritage programs.

- Visit and support zoos, botanical gardens, and aquariums with captive propagation programs. Get to know zoo, garden, or aquarium staff, and volunteer your time as a guide. Attend special presentations on endangered species and captive propagation programs.

- Visit National Wildlife Refuges in your state. Acquaint yourself with the important habitats in your area. Get to know refuge managers, show your support, and ask how you can help. Consult the *Guide to the National Wildlife Refuges* listed in the section on "Further Information."

- Familiarize yourself with fish and wildlife and game management programs or agriculture departments in your state or county that protect endangered or

Figure 6.3
World Conservation Strategy

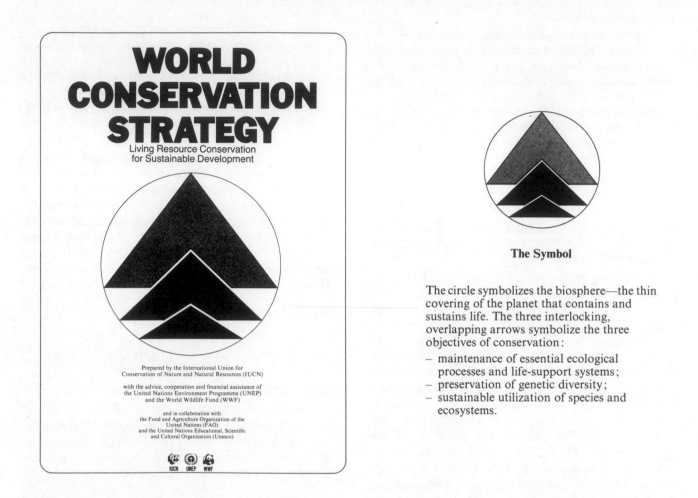

WORLD CONSERVATION STRATEGY

Living Resource Conservation
for Sustainable Development

Prepared by the International Union for
Conservation of Nature and Natural Resources (IUCN)

with the advice, cooperation and financial assistance of
the United Nations Environment Programme (UNEP)
and the World Wildlife Fund (WWF)

and in collaboration with
the Food and Agriculture Organization of the
United Nations (FAO)
and the United Nations Educational, Scientific
and Cultural Organization (Unesco)

IUCN   UNEP   WWF

The Symbol

The circle symbolizes the biosphere—the thin
covering of the planet that contains and
sustains life. The three interlocking,
overlapping arrows symbolize the three
objectives of conservation:

– maintenance of essential ecological
  processes and life-support systems;
– preservation of genetic diversity;
– sustainable utilization of species and
  ecosystems.

Source: IUCN/UNEP/WWF, *World Conservation Strategy: Living Resource Conservation for Sustainable Development* (Gland, Switzerland: International Union for Conservation of Nature and Natural Resources, 1980).

threatened species. Ask how you can help as a concerned citizen.

● Learn about how your community or state could use a real estate transfer tax to establish a "land bank" to purchase land and protect open space that would otherwise be lost to development. In what has become a model program, Nantucket, Massachusetts, passed legislation to establish a land bank, and has used real estate transfer tax funds to purchase nearly 800 acres of wetlands, woodlands, and ocean front property.[84]

Join With Others

Join a local group or a national organization with a regional chapter working to preserve biological diversity;

learn about current priorities and pending legislation on the issue, and volunteer your time to work on the issue.

● Support nongovernmental organizations that work for improved conservation at home and abroad. These groups pursue a variety of approaches, including establishment of biological reserves, biological research, training of foreign scientists and wildlife managers, legislation, and lawsuits. A number of organizations active in the protection of biological diversity are listed under "Further Information."

● Support the Nature Conservancy and other organizations that acquire ecologically-important land and then transfer it to other public or private jurisdiction for preservation or protection.

Review Your Habits and Lifestyle

Consider what role biological diversity plays in your own life, and how your can help protect it as a student, citizen, voter, and consumer.

- Become aware of plant and animal species you consume. Learn to appreciate the many ways species diversity enhances the quality of your life.

- Promote biological diversity in your home and on your property. Keep plants in your home; learn where they came from; propagate them. Avoid strict monoculture of grass on your property. Use a variety of trees, shrubs, and ground covers. Maintain shrubby and weedy areas for the benefit of birds and insect-eating mammals. Plant hedgerows—rows of shrubs or trees—where fences are needed.

- Don't buy endangered plants, animals, or products made from overexploited species unless you are sure they were obtained or propagated by legitimate means rather than taken from the wild. Inform retailers of your concern that they might be selling endangered or threatened species. Products to be concerned about include: cactus plants, orchids, wild furs, sea turtle and reptile skin products, ivory, tropical birds and fish and other exotic pets, and exotic tropical wood products such as mahogany and teak. For further information, contact the World Wildlife Fund and request information on WWF's Traffic (U.S.A.) Wildlife Trade Education Kit, listed below under teaching aids.

- Support and participate in non-consumptive uses of wildlife such as birding, whale watching, nature photography, and scuba diving. Use these activities as a means for increasing your own awareness and getting to know others interested in the issues.

Work With Your Elected Officials

Contact your local, state, and national legislators, find out their positions on the issues of wildlife preservation, and urge them to take constructive action.

- Promote conservation of local biologically rich and diverse habitats and endangered and threatened species. Persuade local government agencies to give consideration to natural resource conservation.

- Urge Congress and the administration to:

  » support programs that help conserve species and ecosystems, including the Endangered Species Act;

  » direct the National Science Foundation to support research in conservation biology;

  » establish a National Endowment for Biological Diversity that would provide grants to groups engaged in conservation research;[85]

  » increase funding for protection of tropical forests inside the United States—in Hawaii, Puerto Rico, and the Virgin Islands; and

  » increase funding—through the U.S. Agency for International Development, National Park Service, and Fish and Wildlife Service—to train Third World students in conservation biology and wildlife management; to help pay for national parks, species inventories, and other projects to conserve biological diversity[86]; and to promote land reform and provide employment and sources of food for people who might otherwise endanger critical ecosystems and wildlife reserves.

Publicize Your Views

Express your views and concerns about preservation to leaders in your community and to the general public through use of the local media.

- Determine regions of ecological diversity in your area that are threatened by development (wetlands, marshes, woodlands, etc.), and express the value of these regions and the importance of their preservation in letters to the editor, speeches at town meetings, or public service announcements on local television or radio stations.

- Encourage local garden clubs to discuss biological diversity. Volunteer yourself or suggest someone else as a speaker. Try to involve garden clubs in promoting local community floral diversity. Designate rare and spectacular trees for special protection. Discuss gardening practices which promote a diversity of wildlife.

- If you are a farmer, grow some of your own food, or live in a farming community, organize a meeting to discuss genetic diversity in food crops. Promote awareness programs at local 4H groups and other agricultural education organizations.

Raise Awareness Through Education

Find out whether biology courses in your local schools and colleges stress the importance of biological diversity.

- Encourage biology and science teachers at local schools to include biological diversity in their curricula. Assist in contacts between teachers and institutions such as zoos, research centers, and environmental education organizations. Organize training workshops for teachers.

- Encourage and support environmental education in your area. If you have children, enroll them in environmental education programs. Make books about animals, plants, and ecology available in your home. Teach your children about the endangered species problem. Take them on walks in natural areas near your home or vacation area. If there is a center for nature education in your area, visit it, and encourage local schools and organizations to use it. If your community lacks a nature center, discuss with community leaders the possibility of organizing one. A national directory of natural science centers is listed in the "Further Information" section below.

- Consider participating in the Student Conservation Association's (SCA) resource assistant program for adults or its student work group program (see the Appendix for SCA's address). SCA places about a thousand adults and high school and college students each year in voluntary conservation projects throughout the United States and its territories.

Consider the International Connections

Consider the challenges of preserving wildlife both as a local priority and as a global issue affecting all people.

- Learn more about tropical regions of the world. Try to visit a tropical rain forest. Hawaii, Puerto Rico, and the Virgin Islands all contain tropical forests. For information, contact the Tropical Research Institute, Smithsonian Institution, Washington, DC 20560. For information on nature tours of tropical areas, contact companies specializing in natural history tours advertised in magazines such as *Audubon*, *Natural History*, and *Smithsonian*.

- If you travel to another country, learn about its endangered species and its programs to protect them. Learn about U.S. organizations that have links with wildlife conservation organizations in other countries and work to strengthen those links.

Further Information

A number of organizations can provide information about their specific programs and resources related to biological diversity.*

For a comprehensive listing of U.S. conservation organizations and agencies, see the National Wildlife Federation's *Conservation Directory* listed under "Further Information." For other general information about the issue, contact Faith T. Campbell, Natural Resources Defense Council, 1350 New York Avenue, N.W., Suite 300, Washington, DC 20005. Telephone: (202) 783-7800.

Books

Berger, John J. *Restoring the Earth: How Americans Are Working to Renew Our Damaged Environment.* New York: Anchor Press, 1987. 241 pp. Paperback, $9.95. Presents case studies of ecological restoration—the repair of damaged resources and the re-creation of ecosystems.

Directory of Natural Science Centers. 1984. Lists over 1100 centers in the United States and Canada. Available from Natural Science for Youth Foundation, 130 Azalea Drive, Roswell, GA 30075. $15.00 plus $2.50 postage.

Durrell, Lee. *State of the Ark.* Garden City, NY: Doubleday, 1986. 224 pp. Paperback, $22.95. A lavishly illustrated atlas of conservation in action documenting what is being done to save life on Earth.

Eckholm, Erik P. *Down to Earth: Environment and Human Needs.* New York: W.W. Norton, 1982. 238 pp. $14.95. Analyzes the web of global environmental problems and describes how the well-being of all humanity is intimately linked to the environment.

Ehrlich, Paul R. and Anne H. Ehrlich. *Extinction: The Causes and Consequences of the Disappearance of Species.* New York: Ballantine, 1983. 384 pp. Paperback, $4.50. One of the best non-technical overviews of the issue available.

Martin, Vance (ed.). *For the Conservation of Earth.* Golden, CO: International Wilderness Leadership Foundation, 1988. 418 pp. Paperback, $15.95. Contains the proceed-

---

* These include: African Wildlife Foundation, American Cetacean Society (aquatic mammals), American Committee for International Conservation, Bat Conservation International, Center for Conservation Biology, Center for Environmental Education (marine mammals), The Conservation Foundation, Conservation International, Defenders of Wildlife, Environmental Defense Fund, Friends of the Earth, International Union for Conservation of Nature and Natural Resources, Izaak Walton League, Missouri Botanical Garden, National Audubon Society, National Parks and Conservation Association, National Wildlife Federation, Natural Resources Defense Council, The Nature Conservancy, New York Botanical Garden, New York Zoological Society, North Carolina Botanical Gardens, Sierra Club, Student Conservation Association, Wilderness Society, Wildlife Conservation International, World Resources Institute, Worldwatch Institute, and World Wildlife Fund. Addresses for these organizations are listed in the appendix.

ings of the 4th World Wilderness Congress held in Colorado in September, 1987.

Myers, Norman. *The Sinking Ark: A New Look at the Problem of Disappearing Species.* Oxford: Pergamon Press, 1979. 307 pp. $22.50. Contains Myers' famous estimate that by the year 2000, species will become extinct at a rate of one per second.

Myers, Norman. *A Wealth of Wild Species: Storehouse for Human Welfare.* Boulder, CO: Westview Press, 1983. 272 pp. Paperback, $13.95. Reviews the contributions that a vast range of wild species make, and can make, to our daily lives. Emphasizes the dollar values of their benefits to humans. Stresses that the loss of wild species is a loss of potential to deal with present as well as yet-unknown future problems.

Myers, Norman (ed.). *Gaia: An Atlas of Planet Management.* Garden City, NY: Anchor Press, 1984. 272 pp. Paperback, $18.95. The section on evolution graphically portrays the loss of habitats and species as well as progress made in conservation.

National Wildlife Federation. *Conservation Directory.* Published annually by the Federation in Washington, DC. The 1989 edition contains 313 pp.; the price is $15.00 plus $3.25 shipping charge. A comprehensive listing of U.S. organizations, agencies, and officials concerned with natural resource use and management. Also includes major Canadian agencies and citizens' groups. Lists U.S. National Forests, Parks, Seashores, and Wildlife Refuges.

Norton, Bryan (ed.). *The Preservation of Species: The Value of Biological Diversity.* Princeton, NJ: Princeton University Press, 1986. 305 pp. Paperback, $16.95.

Office of Technology Assessment, U.S. Congress. *Grassroots Conservation of Biological Diversity in the United States,* Background Paper No. 1, OTA-BP-F-38. Washington, DC: U.S. Government Printing Office, February 1986. 67 pp. Paperback, $3.50.

Office of Technology Assessment, U.S. Congress. *Technologies to Maintain Biological Diversity.* Washington, DC: U.S. Government Printing Office, March 1987. 340 pp. Paperback, $15.00.

Wilson, E.O. (ed.). *Biodiversity.* Washington, DC: National Academy Press, 1988. 521 pp. Paperback, $19.50. A comprehensive survey based on the 1986 National Forum on Biodiversity sponsored by the National Academy of Sciences and the Smithsonian Institution. Stresses the rapidly accelerating loss of plant and animal species due to increasing human population pressure and the demands of economic development.

World Conservation Strategy: Living Resource Conservation for Sustainable Development. Gland, Switzerland: International Union for Conservation of Nature and Natural Resources (in cooperation with the United Nations Environment Program and the World Wildlife Fund), 1980. 68 pp. Available from UNIPUB, 4611-F Assembly Drive, Lanham, MD 20706-4391, $7.50 including postage. Represents a major step toward a unified approach to protecting the global environment. The strategy stresses that sustainable development depends on a healthy environment and conservation of the Earth's living resources.

Articles, Pamphlets, and Brochures

"The Quiet Apocalypse: Biologists Warn that a Mass Extinction is Happening Now." *Time,* October 13, 1986, p. 80.

U.S. Fish and Wildlife Service, Office of Endangered Species. Has many free publications. Write to the above, c/o Department of the Interior, Washington, DC 20240.

Wilson, Edward O. "Threats to Biodiversity." *Scientific American,* September 1989, pp. 108-16. The elimination of species-rich tropical forests and other habitats is driving many plant and animal species to extinction. The accelerating loss of diversity is a moral, scientific, and economic tragedy.

Wolf, Edward C. "Avoiding a Mass Extinction of Species." In Lester R. Brown et al., *State of the World 1988.* New York: Norton, 1988, pp. 101-17. An expanded version of this article is available as Worldwatch Paper 78, *On the Brink of Extinction: Conserving the Diversity of Life.* Washington, DC: Worldwatch Institute, June 1987. 54 pp. $4.00.

World Resources Institute (WRI) and International Institute for Environment and Development (IIED). *World Resources 1986.* New York: Basic Books, 1986, pp. 85-101. Covers species, threatened species, vulnerable groups, protection of threatened species, ecosystems, and protected areas.

WRI and IIED. *World Resources 1987.* New York: Basic Books, 1987, pp. 77-92. Focuses on the economic value of wildlife and regional trends in management.

WRI and IIED. *World Resources 1988-89.* New York: Basic Books, 1988, pp. 89-107. Covers loss of biological diversity, indicators of wildlife's condition, island habitats, migratory species, and illegal trade.

World Wildlife Fund. *Future in the Wild: A Conservation Handbook.* Washington, DC. No date. 24 pp.

Periodicals

Many magazines and journals regularly cover topics related to biological diversity. The better known ones such as *Audubon*, *National Geographic*, and *Natural History* are a good place to start. More specialized periodicals are listed below.

Bioscience, published monthly by the American Institute of Biological Sciences, 730 11th Street, N.W., Washington, DC 20001.

Center for Environmental Education, 1725 DeSales Street, N.W., Suite 500, Washington, DC 20036, publishes newsletters and bulletins on the protection of marine mammals.

Diversity: A News Journal for the Plant Genetic Resources Community, published quarterly by Genetic Resources Communications Systems, 727 8th Street, S.E., Washington, DC 20003.

New Scientist, published in London by IPC Magazines, Ltd. Gives excellent coverage of the field.

Films and Other Audiovisual Materials

Biodiversity: The Videotape. Adapted from the teleconference comprising the concluding session of the 1986 National Forum on Biodiversity. 45 min. Video ($24.50). National Academy Press.

Diversity Endangered. 1987. A concise introduction to the meaning and importance of biological diversity; focuses on the remarkable variety of the world's land and aquatic environments. 10 min. 3/4" Video ($25.00). Smithsonian Institution Traveling Exhibition Service, Washington, DC 20560.

Garden of Eden. 1983. Documents the necessity of protecting natural diversity; shows that corporate and conservation interests can be pursued simultaneously. 28 min. 16 mm color film, 3/4" Video (VHS, BETA). Direct Cinema.

Teaching Aids

Global Tomorrow Coalition. *Biological Diversity Education Packet*. Includes background information, lesson plans, and activities emphasizing interactive learning aimed at upper elementary and intermediate students; can be upgraded for high school. $7.00 including postage. GTC Education Services, 1325 G Street, N.W., Suite 915, Washington, DC 20005-3104.

Nature Conservancy. *Rare and Endangered Species: Understanding Our Disappearing Plants and Animals*. An Activities Guide. Four-page pamphlet and poster. 1987. American Gas Association, 1515 Wilson Boulevard, Arlington, VA 22209.

Smithsonian Institution Traveling Exhibition Service (SITES). *Diversity Endangered: A Resource and Program Guide*. 1987. 22 pp. Available from SITES, Education Department, Smithsonian Institution, Washington, DC 20560.

World Wildlife Fund. *Wildlife Trade Education Kit*. Includes 80-slide show and script, educator's guide, fact sheets, and color poster. $40.00. Order from: Traffic (U.S.A.), WWF-U.S., 1250 24th Street, N.W., Washington, DC 20037.

Tropical Forests

Tropical forests form part of the global heritage. We all find our daily lives enhanced by virtue of their existence. We shall all lose if they disappear....They "belong" to us all.

—Norman Myers[1]

The rainforests are being destroyed not out of ignorance or stupidity but largely because of poverty and greed.

— Michael H. Robinson Director, National Zoological Park[2]

Major Points

- Tropical rainforests cover only seven percent of the Earth's land surface, yet they contain at least two-thirds of all plant and animal species. These forests are our richest source of raw materials for agriculture and medicines, and yield many products of industrial and commercial value.

- Tropical forests are rapidly being cleared for farming, logging, cattle ranching, and large-scale development projects. About half of all forests are already gone, and each year an area the size of Pennsylvania is eliminated or degraded. Moreover, pressures on tropical forests are increasing.

- Loss of tropical forests already affects hundreds of millions of people through increased flooding, soil erosion and silting of waterways, drought, shortages of fuelwood and timber, and displacement of societies and cultures. Forest loss causes the extinction of numerous plant and animal species, and the loss of many valuable forest products; deforestation may alter regional and even global climate.

- Deforestation can be slowed by establishing forest reserves, improving management of exploited forests for sustainable use, preventing unsustainable development projects, increasing reforestation, and reducing pressures for forest clearing through land reform, improved agriculture, and population planning.

- Concerned U.S. citizens and their local groups can help save tropical forests by promoting greater community awareness, supporting private organizations engaged in tropical conservation, and influencing the U.S. government to increase support for foreign assistance programs that promote protection and sustainable use of tropical forests.

The Issue: Tropical Deforestation

Near the equator in Latin America, Africa, and Asia are some of the most complex, important, and beautiful ecosystems on Earth: tropical rainforests. Three large blocks of forest remain: the Amazon Basin in South America, the Congo Basin in West Central Africa, and the Malay Archipelago between Southeast Asia and Australia. Tropical forests once covered some 1.6 billion hectares (4 billion acres), twice the area of the United States. But, in recent

Figure 7.1
Extent of Tropical Deforestation

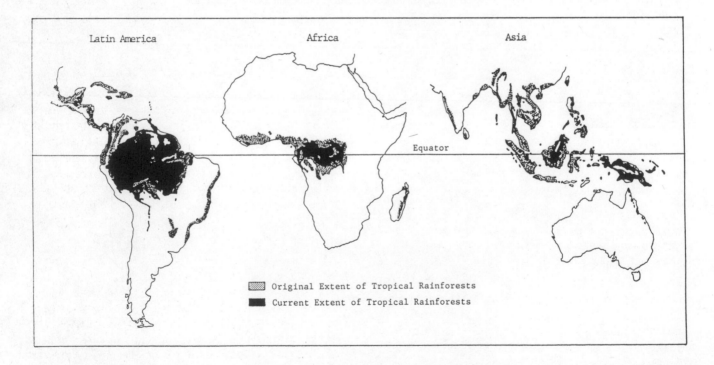

Source: Smithsonian Institution Traveling Exhibition Service, *Tropical Rainforests: A Disappearing Treasure* (Washington, DC: Smithsonian Institution, 1988).

decades, forests have been cleared for farming, cattle ranching, timber, and fuelwood, and today only about half of them remain. Of those surviving, many will disappear in the next few decades.[3]

Why should we save the tropical forests? They support the richest diversity of life on Earth, including millions of species of plants and animals and hundreds of thousands of people. Tropical forests provide products of immense value to agriculture, medicine, and industry, and the value of their diversity and beauty cannot be measured in monetary terms alone.[4]

What Are Tropical Forests?

Tropical forests can be separated into four different types. *Evergreen forests*, also known as rainforests, are drenched with at least 400 centimeters (150 inches) of rain annually. Rainforest trees form a closed canopy that allows little light to reach the ground. *Moist forests* also form a closed canopy, but receive less rain than evergreen forests and experience a short dry period. *Deciduous forests* lose their leaves during the four to six months when they get almost no rain. Finally, *open dry woodlands* receive less than 100 centimeters (38 inches) of rain per year and sometimes experience prolonged droughts.[5] Evergreen and moist forests are generally referred to as moist tropical forests and are the focus of this chapter. They are complex ecosystems, with predators and prey, hosts and parasites, all involved in an intricate web of interdependence.

Despite the abundance of lush green vegetation, tropical forests often have surprisingly poor soil. Most forest nutrients are stored in plants, animals, and micro-organisms, not in the soil. When leaves or trees fall or animals die, they are quickly decomposed by various organisms, and their nutrients support other forest life. In tropical forests, this process occurs so quickly that there is little time for nutrients to be absorbed into the soil; the few nutrients that do reach the soil are washed away by heavy tropical rains.

Although they cover only about seven percent of the world's landmass, these forests support at least two-thirds of all the Earth's plant and animal species—which could number as many as 30 million.[6] The Malay Peninsula in Southeast Asia, for example, supports 7,900 different flowering plant species and 2,500 native tree species, while Great Britain, with twice the area, supports only 1,430 flowering plants and 35 native trees.[7] Peru contains some 30,000 species of plants.[8] In Colombia, a country slightly larger than Texas and New Mexico combined, bird species number more than 1550, well over twice the number found in North America from the Mexican border to the Arctic Circle.[9]

The Importance of Tropical Forests

In addition to sustaining native populations and providing a habitat for millions of plant and animal species, tropical forests supply many products that we take for granted. Timber, fruits, vegetables, spices, nuts, and medicines, as well as oils, waxes, rubber, and other industrial goods are all derived from tropical forests. Further, tropical species can supply genetic engineers with the information they need to create new products for industry, medicine, and other commercial uses.[10]

Recent studies in Brazil and Peru have shown that "the most immediate and profitable way" of combining the goals of development and conservation is to exploit the forests for products such as foods, oils, rubber, and medicines. The studies indicated that net revenues from long-term harvesting of such non-timber forest products are two to three times greater than from commercial logging or clearing the forest for cattle ranching.[11]

Timber

For many developing nations, exports of tropical forest products are a major source of income; annual revenues from these exports averaged about $7 billion between 1975 and 1984.[12] Highly prized woods including teak, mahogany, and rosewood grow in tropical forests. Demand for hardwood in industrial nations has risen dramatically since 1950, and many tropical countries have hardwood timber as well as low labor costs. Consequently, much of this demand has been met by developing nations. Tropical nations also use large quantities of wood for fuel.

Agricultural Products

The extraordinary diversity of edible plants in tropical forests provides indigenous populations and people around the world with a wide variety of foods. Staple grains such as rice, wheat, corn, and millet have wild relatives found only in the tropics. Yams, bananas, pineapples, coffee, and sugarcane are only a few of the many fruits, vegetables and nuts that originated in tropical forests. Many other tropical plants await discovery and use as new food sources.[13] In many parts of the tropics, forest animals are an important source of protein.

Disease and Pest Control

Many of our food crops originated in tropical regions, and we must turn to the tropics to find useful relatives of those species. Such relatives have genetic materials that can be used to modify current crops and produce varieties resistant to pests and diseases, that have higher yields and better nutritional quality, and that can adapt to difficult soils and climates. For example, wild strains have been used to protect commercial barley in California from a lethal virus, sugarcane in the southeastern United States from disease, and Asian rice from four major diseases. A perennial corn recently discovered in Mexico is valuable for its immunity or resistance to the seven major diseases affecting domesticated corn.[14]

Pests destroy at least 40 percent of the world's crops each year, but natural pest controls exist in tropical forests that could greatly reduce this loss. Natural pesticides including rotenone, pyrethrum, and methyl carbamate can be derived from tropical plants.[15] Tropical insects also control pests, by acting as predators and parasites that attack eggs, larvae, and adult insects. In Florida, citrus growers now save at least $35 million a year after a one-time expenditure of $35,000 to import three kinds of tropical parasites.[16]

Medicines

A quarter of all medically active substances came originally from tropical plants. Some 1,400 plants in tropical forests may contain effective anti-cancer agents; a drug made from the rosy periwinkle plant is now used to treat Hodgkin's disease and childhood leukemia.[17]

Industrial Products

Many of the products used in industry come from tropical forests, including gums, resins, latexes, waxes, rubber, fibers, dyes, tanning agents, turpentine, lubricants, rattan, and bamboo.[18] Tropical plants also yield numerous by-products used in industry and commerce. Examples are the cauassu plant, whose leaves contain a high quality wax; the Buriti palm, a source of edible oil, starch, and industrial fiber; the Seje palm, which yields valuable oil; and the Babassu palm, a rich source of edible oil, waxes, and steroids. Rubber trees in Southeast Asia are a major cash crop — natural rubber is the fourth largest agricultural export from the developing world, valued at more than $3 billion annually.[19]

Environmental Services

Tropical forests provide essential services, not only for their immediate surroundings, but also for the Earth as a whole. Forests moderate air temperature, maintain the hydrologic cycle by absorbing rainfall and releasing moisture to the atmosphere, and take in carbon dioxide and generate oxygen through photosynthesis. They recycle nutrients and wastes, control soil erosion and sedimentation of waterways, and regulate stream and river flows, helping to moderate floods and droughts. Tropical forests also prevent or limit landslides and rockfalls during rainstorms and earthquakes, and moderate damage from tropical cyclones. When tropical forests are degraded, these essential functions are lost or jeopardized.[20]

Indigenous Societies and Cultures

Tropical forests are home to hundreds of peoples whose unique cultures contain valuable knowledge, developed over millennia, about tropical plants and animals, and about the sustainable use of forest resources. Forest destruction is displacing these peoples and causing their knowledge to be lost before it can be recorded.[21]

Aesthetic Values

The rich diversity and visual beauty of tropical forests have generally been available to most of us indirectly through books, films, zoos, and museums

Recently, a growing number of conservation-oriented organizations have focused on the importance of rainforests; some groups are now producing high-quality audiovisual materials that convey more of the visual richness and sense of wonder that characterize tropical forests. These efforts are helping to increase controlled tourism in tropical areas, making the aesthetic values of rainforests directly accessible to travellers, and also demonstrating that, if preserved in their wild form, the forests can be a source of much-needed foreign exchange.

Rates of Deforestation

Tropical moist forests may have once covered over 1.5 billion hectares (3.7 billion acres), an area nearly twice the size of the continental United States. Today, less than 900 million hectares remain, and the number is shrinking dramatically.[22] This figure includes evergreen tropical forests but excludes deciduous or semi-deciduous moist tropical forests. Estimates from the early 1980s of the worldwide rate of forest clearance range from 10 million to 20 million hectares each year, depending on which types of tropical forest are included, and what level of disturbance is defined as clearance. Based on the Smithsonian Institution's conservative estimate of about 11 million hectares, an area of moist forest the size of Pennsylvania is eliminated or degraded each year. That is equivalent to ten city blocks every minute, and an area the size of Philadelphia every day.[23]

Recent satellite data indicated that in 1987, some 8 million hectares of forested land may have been burned in Brazil alone, which suggests that the 11 million hectare estimate may be as much as 50 percent too low.[24]

The rate of deforestation varies among the tropical regions. Africa as a whole has lost 23 percent of its moist tropical forest in a little over 30 years.[25] The Ivory Coast has cut 66 percent of its forests and woodlands in the last 25 years, and Madagascar has lost 93 percent of its forest cover.[26] In the early 1980s, more than 5 percent of the moist coastal forests of West Africa were being cleared each year.[27] In Latin America, 99 percent of Brazil's moist Atlantic Coast forest has been cut and 98 percent of the dry tropical forest that once covered the Pacific coast of Mexico and Central America has been eradicated.[28] Haiti, which originally was heavily forested, today has tropical forest over only 2 percent of its land.[29]

Causes of Deforestation

The primary direct causes of tropical deforestation are agriculture, logging, cattle ranching, and large-scale development projects. Underlying these factors are a number of indirect causes, including human population growth, rising world demand for forest products, unequal distribution of land and wealth in tropical nations, poverty, growing Third World debt, and pursuit of military objectives.[30] Norman Myers, a leading tropical ecologist, notes that tropical forests are often seen as obstacles to expanding civilization rather than as permanent resources. Forestry departments in tropical countries may be inclined to emphasize quick exploitation of timber for cash rather than sustainable use and realization of broad future benefits.[31] Many Third World governments provide economic incentives that encourage overuse of forest resources and the rapid conversion of forests to cropland, cattle ranches, and other unsustainable uses.[32]

Agriculture

Each year some 8 million hectares of tropical forest are cleared for cultivation, an area the size of South Carolina.[33] Much of this land is used for slash-and-burn agriculture, also known as shifting cultivation. This practice is an ancient and productive means of growing crops and can be sustainable if used properly in an area with low population density.[34] A farmer traditionally clears a small plot of forest by cutting the vegetation and then burning it, allowing the ashes to fertilize the poor soil. Crops are grown on that plot until its nutrients are depleted several seasons later. The farmer then moves on to another patch of forest, clearing and cultivating it, while the previous one lies fallow, reverting to a wild state and gradually regaining its fertility. The plot lies fallow for up to thirty years before a farmer returns to it. Today, however, higher population density makes forest clearing more frequent and widespread, and old fields are soon recultivated rather than allowed to lie fallow long enough to regain their fertility. Once cleared, the land is vulnerable to soil and nutrient loss from rainfall, and can become virtually useless, leaving behind an agricultural wasteland. Such misuse of land is responsible for 35 percent of the tropical deforestation in Latin America, nearly 50 percent in Asia, and 70 percent in Africa.[35]

Figure 7.2
Loss of Tropical Forests between 1950 and 1982

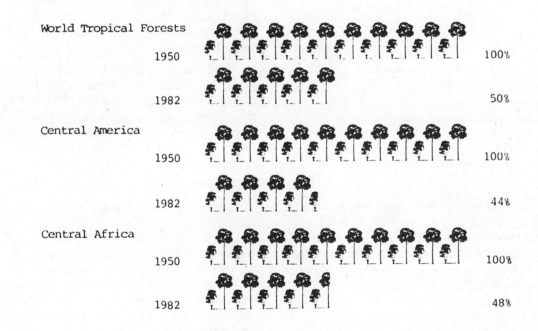

Source: Michael H. Robinson, "The Fate of the Tropics and the Fate of Man," *ZooGoer* (Washington, DC: Friends of the National Zoo), May–June 1986, p. 6; graphic prepared by the Office of Graphics and Exhibits, National Zoological Park.

In some countries, large-scale agricultural plantations are rapidly replacing tropical forests. These large farms produce commercial crops (cash crops), grown primarily for export. In a given region, one or two cash crops are usually dominant. In Central America, for example, bananas and coffee are the main crops. In Southeast Asia, rubber, oil palm, and cacao are common.[36]

Figure 7.3
Tropical Forest Depletion and Demand for Agricultural Land, 1950-2025

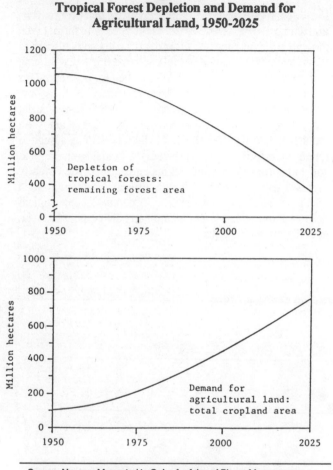

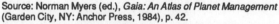

Source: Norman Myers (ed.), *Gaia: An Atlas of Planet Management* (Garden City, NY: Anchor Press, 1984), p. 42.

In the Peruvian Amazon, coca farms are believed to account directly or indirectly for the destruction of some 700,000 hectares of tropical forest, or roughly 10 percent of Peru's rainforest loss in this century. In addition, the production of cocaine from coca has seriously polluted more than 150 streams and rivers in Peru with vast quantities of toxic wastes, including kerosene, sulfuric acid, acetone, and toluene.[37] Coca, along with marijuana, has also become a major crop in deforested areas of Bolivia, Colombia, Ecuador, and Brazil because of the lucrative, rapidly-growing drug trade, and because coca thrives in direct sunlight and on steep, eroded soils where few other tropical plants can survive.[38]

Logging for Timber and Industry

Each year at least 5 million hectares of tropical forests are cut for timber, paper pulp, and other wood products, an area the size of Vermont and New Hampshire.[39] Forest loss from logging is greatest in Asia and West Africa, and though still limited, is accelerating in the Amazon. With many developing countries facing staggering foreign debts, governments are eager to export timber to obtain foreign currency. Indonesia, for example, owns about 10 percent of the world's tropical forests and is felling up to one million hectares a year. In 1986, its timber exports earned $1.4 billion.[40] Roughly half of all tropical hardwood timber cut is exported. Japan is the largest importer, consuming 40 percent of the world's tropical hardwoods, followed by the United States.[41] In Amazonia, much of the logging is carried out by large timber corporations from North America, Western Europe, and Japan, mainly to meet market demand in developed countries.

Because of rapid exploitation of their tropical forests for timber exports, many nations may eventually find themselves without enough wood for domestic needs. Ivory Coast, with the world's highest deforestation rate (5.2 percent per year), may have lost more than half of its forest cover in two decades.[42] And a consortium of research groups has warned that the Philippines, traditionally a major timber exporter, may be unable to produce enough timber for its own consumption by the year 2000.[43] Careless logging methods can destroy tropical forests. For example, in 1983, negligent logging practices in Borneo contributed to a fire that burned for four months, devastating an area larger than Massachusetts and Connecticut and destroying timber worth $6 billion.[44]

Wood is an industrial fuel in many tropical regions, and industrial development depletes the resource. In the Brazilian state of Para, 20 pig iron plants will soon begin to exploit the world's largest iron ore deposit, using local charcoal. To produce that charcoal, the project could require the cutting of 230,000 hectares of forest every year.[45]

Fuelwood

Wood is the principal source of energy for cooking and heating for almost half the world's population, and for 80 percent of the people in the developing world.[46] Fuelwood is already acutely scarce for over a billion people in developing countries. Fuelwood harvesting is a major cause of tropical deforestation in some regions, especially in drier areas. Two of the world's largest fuelwood-producing countries, Brazil and Indonesia, contain the largest remaining areas of closed tropical forest.[47]

Cattle Ranching

In Latin America, land clearing for cattle production eliminates some 2 million hectares of forest each year, an area the size of New Jersey.[48] That accounts for at least a quarter of annual tropical forest loss worldwide. Cattle ranching is often encouraged by government policies. Until recently, Brazil offered a large tax reduction to cattle ranching investors in Amazonia.[49] As a result, over the last two decades, some 10 million hectares of Amazonia has been cleared for pasture — an area the size of Austria.[50] In Central America, between 1950 and 1980 about two-thirds of the primary forests were cleared, largely for cattle ranching. During this period, the area of pasture land and the number of beef cattle more than doubled, and beef exports increased almost fivefold. Most of the beef went to the United States, often to fast food chains and for pet foods. Today, most beef imported by the United States comes from Australia and New Zealand; most Latin American beef is consumed in Latin America and Europe.[51] It has been calculated that for each hamburger exported from Central America, five square meters of forest, or a patch the size of a kitchen, must be cleared.[52]

Once cleared for cattle ranching, pastureland is productive for only a limited time. After a few years, the land becomes useless because of erosion, compaction by vehicles and cattle hooves, and burning to control weeds that are toxic to cattle.[53] Useless land is abandoned; the rancher clears more forest, and the cycle repeats. Ranchers could greatly increase their beef production from existing pastures by using more efficient methods. But government incentives, often supported by international aid agencies, encourage extensive ranching to produce export commodities rather than more intensive methods.[54]

Large-Scale Development Projects

In many tropical countries, large-scale development projects such as roads, hydroelectric dams, and programs to resettle landless peasants are designed to create jobs and raise living standards. However, many of these projects achieve only limited success, and often cause major damage to tropical forests and injury to tribal people.[55]

Beginning in the 1880s, Brazil encouraged colonization of the forest-rich Bragantina zone of the Amazon by building a railroad into the region. Nearly 3 million hectares were converted into agricultural land, but the soil quickly gave out and was no longer able to support intensive cultivation. Today, Bragantina is a vast, unproductive expanse of scrubland. Other Brazilian programs to encourage colonization of the Amazon are estimated to have caused over 25 percent of the Amazonian deforestation between 1966 and 1975.[56]

Expansion of large-scale deforestation in many parts of Southern Amazonia is continuing.[57] Each year, millions of hectares of forest and scrub are burned by plantation owners, cattle ranchers, and small landholders who do not take the time or have the money to cut down the trees. In 1988, as many as 20 million hectares of land may have been burned,

which would amount to 4 percent of the entire Amazon region, equivalent to an area the size of Nebraska. The fires released an estimated 600 million tons of carbon into the atmosphere, and the smoke spread over millions of square kilometers, causing health problems and hampering air and surface travel.[58] Scientists have estimated that these fires may account for one-tenth of the carbon dioxide being added to the atmosphere.[59] Between 23 and 43 percent of the total increase in global atmospheric carbon dioxide may come from the burning of forests in developing countries.[60]

Following an international outcry in 1988 over the rapid destruction of Amazonian forests, the Brazilian government indicated that it would restrict subsidies for agriculture in the region, curb the export of logs, and place environmental controls on agricultural and industrial projects.[61]

Table 7.1
Causes of Tropical Deforestation in Selected Regions

Region	Principal Causes
Colombia, Ecuador	Commercial logging, shifting cultivation
Brazil (Rondonia Province)	Government policy encouraging colonization
Ivory Coast, Ghana, Sierra Leone	Shifting cultivation, fuelwood needs, commercial logging
Madagascar	Shifting cultivation, cattle grazing, charcoal production
Southwest India	Shifting agriculture, fuelwood needs
Malaysia	Agricultural development, large-scale plantations, commercial logging
Philippines	Shifting agriculture, logging

Source: American Forestry Association, "Tropical Deforestation: Why Is It Happening?," *American Forests*, November-December 1988.

Population Growth and Unequal Land Distribution

The world's population grows by more than eighty million each year, and most of this increase occurs in tropical Africa, Asia, and Latin America. In these regions, the most fertile land is often controlled by a few wealthy landowners, so that many landless poor are forced to clear forests to obtain cropland. In Latin America, for example, only 7 percent of the population controls 93 percent of the agricultural land. Often facing hunger, the poor lack opportunities to use land and forest resources in sustainable ways.[62]

Tropical countries with the highest population growth rates usually have the highest deforestation rates. In western Africa, for instance, the five countries with the highest deforestation rates have an average population growth rate of 2.9 percent, significantly above the Third World average of 2.1 percent. In Latin America, the eight countries with the highest deforestation rates experience a 3.0 percent population gain, also well above the average. Those five African countries lose 3.4 percent of their forests per year, and the eight Latin American countries lose 3.1 percent, far above the average of 0.6 percent per year for tropical countries.[63]

In Brazil, 70 percent of thé population is either landless or lacks secure access to land. Crowded coastal cities and large numbers of landless poor have led to government policies to encourage resettlement of hundreds of thousands of people in inland forest areas. In the territory of Rondonia near the Bolivian border, the population has doubled during the 1980s, and over 20 percent of the rain forest has been cleared.[64]

In Indonesia, a majority of the people lack access to land. High population density on the island of Java has led the Indonesian government to move millions of people to less populated islands, where they are clearing the forests for farming. If completed, this project will destroy some 3.3 million hectares of tropical forest in areas where there is little evidence that the deforested land will support sustainable agriculture.[65]

The Debt Crisis

In the late 1970s and early 1980s, many tropical countries borrowed heavily to compensate for sharp increases in world oil prices and to maintain adequate economic growth for their growing populations. By 1988, Third World debt totaled more than $900 billion, and loan payments now consume a large portion of export earnings. Brazil, for example, must spend nearly 40 percent of its export earnings simply to pay interest on its loans, which now exceed $100 billion. To meet loan payments, many tropical nations rely on rapid and often unsustainable extraction of timber, minerals, and agricultural products from their forests.[66]

Military Activity

During the Vietnam War, the United States used bulldozers, explosives, napalm, and 19 million gallons of Agent Orange and other herbicides to defoliate and destroy tropical forests in Southeast Asia in an effort to deny cover and food to the opposition forces. According to a 1985 study, Vietnam's forests have not recovered, wildlife has not returned, cropland and fisheries productivity remains depressed, and a third of the country is still considered to be wasteland.[67]

Military activities in many parts of the world threaten forestlands. In Latin America, Brazil's National Security Council plans to establish military frontier posts in Amazonia with airstrips, highways, rural electric projects, and housing in order to strengthen security and promote development in border areas. A number of Amazon nations, including Brazil, Ecuador, Guyana, Peru, and Colombia, are seeking to assert sovereignty over new sectors of Amazonia by establishing settlements in the forest areas and filling the "demographic void."[68] The Indonesian government's Transmigration Program has moved people from Java to Borneo, Sumatra, and other sparsely populated areas. In a number of tropical areas, forestland on the outer edges of national territory is being settled in order to claim the land for strategic purposes.[69]

Consequences of Deforestation

When a tropical forest is eradicated, much more than just trees are lost; an entire ecosystem is destroyed. Countless animal and plant species must either adapt to their dramatic new environments or perish. Human populations are displaced, and their cultures and sources of livelihood are disrupted or lost. Fragile tropical soils become infertile and prone to erosion, which can cause widespread damage. Regulated stream flows are replaced by alternating floods and droughts. When deforestation is extensive, local climate may become drier. Valuable sources of timber, food crops, medicine, and industrial goods are lost. These losses, in turn, can contribute to a range of social, economic, and political problems.[70]

Extinction of Plants and Animals

Tropical forest plants and animals are highly interdependent. For example, most trees and plants in a rainforest rely on insects, birds, bats, and other creatures for pollination, since there is little air movement within the dense forest canopy to carry the pollen. The disappearance of a particular pollinating animal can cause the extinction of a plant that it pollinates. The extinction of that plant can then adversely affect animals that feed on it, and so on. Continued degradation and elimination of these interdependent and vulnerable tropical forests could eliminate several hundred thousand to over a million species by the year 2000.[71] Since less than 5 percent of all tropical forests receive any protection, mass extinctions of species appear certain.[72]

When a tropical forest is razed, it is not only the year-round inhabitants that are endangered, but seasonal visitors as well. The tropical forests of Latin America provide essential winter habitat for many migratory birds that breed in the United States. Migratory birds often concentrate in small areas during the winter, and so even the clearing of a small patch of forest can cause a large number of birds to lose their winter habitat. Once tropical winter habitats for these species are lost, the species will also be lost. Breeding bird censuses in the United States over the past 40 years have shown dramatic declines in many species that winter in

Latin America. At the same time, the population of many non-migrating birds has remained relatively stable. For example, census data from Rock Creek Park in Washington, D.C. show that the population of migrants such as vireos, warblers, flycatchers, and thrushes declined from 85 percent of the total breeding population in 1948 to only 34 percent in 1985, and that some species have disappeared altogether. While some of the decline of forest songbirds in the eastern United States is probably due to fragmentation of eastern forests, a recent analysis of 20 years of bird surveys from across North America indicated that tropical deforestation may be the primary cause of the decline in populations of many species of migratory birds.[73]

Displacement of Indigenous Cultures

Traditional forest-dwelling peoples have evolved ways of life over thousands of years that permit them to exist in a state of equilibrium with their environment. Where food is easy to produce, as on fertile floodplains along rivers with abundant fish, many live in permanent villages. In upland areas, where soil fertility is not renewed by seasonal flooding, tribal people practice shifting cultivation supplemented by hunting and gathering.[74]

Indigenous peoples' knowledge of forest species and ecology is critical to their own survival and valuable to the outside world. For example, one study of traditional plant use by indigenous forest dwellers of the Amazon led to the identification of over a thousand plants with potential scientific and commercial value. These include at least six sources of contraceptives, a cure for fungal skin infection, and a high-protein fruit useful for cattle-feed.[75]

In even the most remote regions, virtually all forest people have experienced some form of contact with the outside world. Such contact often leads to exploitation and destruction of the forests and disruption of the indigenous people's way of life. With deforestation, a whole range of goods that these people rely on vanishes and their economies collapse. Further, when they first come in contact with outsiders, many forest people succumb to diseases to which they have no immunity. Those who survive this initial hazard are often forced into reservations or submerged into the dominant culture as landless peasants or low-paid laborers. The forest people who refuse to abandon their traditional lands and accept a new and incompatible lifestyle may simply be killed.

As their homelands are destroyed, the eventual loss of their cultures and their intimate knowledge of forest ecology seems inevitable. In Brazil, the Indian population has dropped from more than 6 million in the 16th century to about 200,000 today. Many little-known South American forest cultures will not exist by the year 2000.[76]

Degradation of Forest Soils

In spite of the biological richness of tropical forests, they are often based on infertile and unproductive soils.[77] Most nutrients are stored in decaying vegetation on the forest floor rather than in the soil, and when a forest is cleared or burned and the vegetation disappears, the nutrient cycle is broken, and the land quickly becomes infertile.

Dense forest vegetation also absorbs a substantial portion of the rainfall it receives, and when the forest cover is removed or reduced, flooding and soil erosion increase.[78] The tropical sun bakes exposed soil into a hard, brittle surface, easily eroded by rainfall. Estimates of current maximum erosion rates in deforested areas of the Dominican Republic, El Salvador, Madagascar, and Burma range from 140 to 350 tons per hectare every year, compared with 18 tons per year for U.S. cropland.[79]

Siltation of Waterways

Soil erosion is also harmful to streams and rivers. Fish are the primary source of protein for many people living in forested regions. Soil erosion reduces the productivity of their fisheries, by raising water temperatures and sediment content. Downstream estuaries, mangrove swamps, and near-shore coral reefs can also be damaged by increased siltation, and navigation can be hampered.

Hydropower dams have been built in many tropical areas, at great financial, social, and environmental cost, to create reservoirs for flood control, power generation, and irrigation. Deforestation has greatly accelerated the sedimentation of these reservoirs and reduced their usefulness, especially in parts of Latin America and Southeast Asia.[80] In Costa Rica, soil erosion from a deforested watershed has greatly shortened the useful life of a major hydroelectric dam.[81]

Much of the soil eroded from deforested land is carried by rivers to the sea. Worldwide, rivers discharge some 13 billion tons of sediment into oceans each year, with over half of this total coming from rivers in tropical South America and South and Southeast Asia.[82] Experts calculate that 40 percent of the man-made reservoir that feeds the Panama Canal will be silted by the year 2000, which could jeopardize the canal's capacity to handle large freighters.[83]

Disruption of Water Flow

Many tropical regions have distinct wet and dry seasons. In these regions, the forest functions as a natural water reservoir, absorbing heavy rainfall and releasing the water slowly to the forest soil and into waterways. By slowing the movement of rainwater into streams, forests minimize flooding during wet seasons and help mitigate drought by maintaining water flow during dry seasons.

In tropical developing countries, an estimated 160 million hectares of upland watershed have been grossly degraded over the last three decades.[84] In Southeast Asia, extensive cutting of tropical forests has caused flooding that becomes more intense and widespread every year. In Malaysia, deforestation has contributed to drought and resulted in reduced rice harvests and insufficient water flow to hydroelectric plants. In Panama, where much of the Panama Canal's watershed has been deforested, on one occasion a drought forced the closing of the Canal to large vessels. Low river flow during the dry seasons reduces the amount of water available to irrigation and diminishes agricultural productivity. It also threatens both the quantity and quality of urban water supplies, and makes them subject to contamination.[85]

Regional Climate Change

Tropical forests moderate local climates by heating and cooling the air and water, helping maintain humidity, and offering protection from wind. More than half the precipitation that falls on land is "breathed out" by plants, and contributes to the rainfall received by downwind regions.[86] It is estimated that half of the Amazon region's rainfall is generated by the forest itself. If half of the Amazon forest were cleared, the remaining forest could become considerably drier and unable to sustain many of its species.[87] Forest destruction in both Haiti and Senegal is believed to be linked to significant decreases in rainfall over deforested areas.[88]

Alteration of Global Climate

Tropical forests may play a critical role in global climate patterns, and continuing deforestation could alter those patterns. Deforested lands reflect sunshine back into space, affecting convection patterns, wind currents, and rainfall in areas far from the tropics.[89] In 1980, the clearing of tropical forests added an estimated 1.7 billion tons of carbon to the atmosphere each year as carbon dioxide.[90] That CO_2 contributes to the "greenhouse effect," which is believed to be causing global warming, rising ocean levels, and increased climate variability. (See Chapter 12.)

Loss of Valuable Products

The many agricultural, industrial, and medical products that come from tropical forests are becoming scarce in some regions with rapid deforestation. Further, deforestation is limiting the potential for creating new products. The genetic engineers who create innovative agricultural, industrial, and medical products rely on the variety of rainforest species to supply genetic blueprints, which can be modified for commercial application. Products derived from rainforest species promise to bolster economies and improve health and living standards, and the destruction of rainforests is proceeding at a time when genetic engineers are just learning to take advantage of previously unexamined organisms.[91]

Rapid deforestation and lack of sustainable forest management have resulted in sharp declines in exports of forest products. If current trends continue, by the year 2000 the group of 33 developing nations that are now net exporters of forest products will shrink to fewer than ten, and the annual value of tropical timber exports from developing nations is projected to drop from the 1984 level of over $7 billion to less than $2 billion.[92] Nigeria's annual tropical timber exports declined from 773,000 cubic meters in 1964 to less than 100,000 meters in recent years, while Ghana's timber exports fell by two thirds between 1974 and 1985.[93] As deforestation reduces fuelwood supplies, many Third World families are forced to spend many hours collecting wood, and to burn animal dung and crop residues instead of using them to fertilize their cropland. An estimated 400 million tons of dung are burned each year which, if used as manure, could raise grain production by 20 million tons.[94]

Social and Economic Consequences

Where deforestation has eliminated plants and animals and has degraded water supplies and soil fertility, families can no longer support themselves. Major deforestation can cause the displacement of whole communities, and contribute to political instability or even civil war. Such disruption can force people to flee and seek livelihoods elsewhere. Several million people have left their home countries in Central America and the Caribbean to escape poverty, conflict, and environmental deterioration related to deforestation. In Central America, forest cover has dropped from about 60 percent in 1960 to no more than 33 percent today, and in Haiti, where over 100,000 people have emigrated, the once abundant forests now cover less than 2 percent of the land. In Indonesia, over a million people have abandoned deforested and eroded areas of Java and migrated to Borneo and other islands.[95]

Slowing Deforestation

Vigorous efforts are needed to slow and reverse current rates of deforestation

These include establishing tropical forest reserves to protect forested land, and improving management of unprotected forests to ensure their sustainable use. Development projects involving unsustainable land use practices must be halted. Reforestation projects and other programs to make deforested land productive must be expanded. Land reform and improvements in agricultural productivity are also important. Population growth must be slowed, and the demand for forest products curbed.

Establishing Forest Reserves

Protecting substantial areas of remaining tropical forest will slow deforestation, maintain critical watersheds, and preserve the Earth's diversity of plant and animal life.

However, it is impractical and probably unnecessary to seek total protection in most remaining tropical forest areas. Successful forest reserves can allow limited and balanced economic use of forests, and healthy forests should be a part of rural life, rather than isolated from the people. When indigenous people take part in forest maintenance and in sustainable productive activities in and around forests, the survival potential of forests is increased.[96]

Worldwide, less than 5 percent of the remaining tropical forests are protected as parks or reserves. However, Brazil has established a system of forest parks and conservation areas covering nearly 15 million hectares, while Costa Rica has protected 80 percent of its remaining wildlands through parks, wildlife refuges, and Indian reserves.[97] Unfortunately, such programs are beyond the ability of many governments that lack the personnel, expertise, and money to set aside and administer protected lands, and guard them against illegal settlers, poachers, and other exploiters.

A recent review of a dozen promising tropical forest reserves in Latin America, Asia, and Africa has identified qualities that contribute to establishing successful reserves. In such reserves, local people have participated directly in planning and management, and all plans have been designed to benefit people as well as natural resources and wildlife. Such reserves had strictly protected core areas; buffer zones for low-density habitation by indigenous people and low-impact activities such as scientific research, education, and nature tourism; and transition zones for sustainable agroforestry to provide food and income to park residents. Local benefits from the reserves have included income from tourism, environmental education, community services, and watershed protection.[98]

Improving Forest Management

There are many ways to improve the management of tropical forests. For instance, damage caused by harvesting timber can be restricted. Instead of clear-cutting large sections of forest to facilitate the removal of timber by truck, light machinery, cables, or helicopters can be used to remove the logs, minimizing soil compaction and leaving more of the forest intact. Selective removal of trees can be part of an extended cutting cycle that allows the forest to regenerate between cuttings. And the leaves, branches, and bark of trees can be left to decompose and return nutrients to the soil. Careful extraction techniques like these are currently being used with success in Latin America.[99]

Some tropical countries have taken more direct action to curb the loss of their forests. For example, Panama recently passed a decree making it illegal to cut any tree older than five years. The Ivory Coast, which lost two-thirds of its forests in 25 years, has announced a ban on timber exports to protect its remaining 400,000 hectares.[100] Indonesia has also moved to end timber exports, and Thailand has announced a ban on logging operations.[101]

Norman Myers has suggested that if tropical forest nations were to develop facilities to process their own hardwoods and produce finished furniture for export, their profit margin could be five to twenty times greater, and they would have a greater incentive to manage their forests in sustainable ways. Myers believes that we have the knowledge and experience necessary to meet our current needs for wood, food crops, and industrial products by reusing and making more efficient use of forestlands that have already been exploited.[102]

Tropical forests provide sustainable supplies of many non-wood crops, including fruits, nuts, animals, fibers, rattan, bamboo, latex, gums, resins, oils, and medicinal plants. Exploitation of non-wood products should be maintained, but also controlled to prevent over-harvesting and destruction of the trees and plants that provide these products. Sustainable harvesting of non-wood products is currently being carried out throughout the tropics for subsistence and commercial purposes.[103] Eighty thousand rubber tappers, organized by the assassinated Chico Mendes in Brazil, are working to prevent the conversion of their forest into farms and cattle ranches, so that some 1.5 million Brazilians can continue to harvest latex and other products from the forest.[104]

Great success has been achieved in areas where large trees were removed to allow seedlings to grow for later harvest. Malaysia has profited from that technique, and similar methods have been successfully applied in Indonesia, Burma, Gabon, and Guyana.[105]

In mountainous or hilly areas, tropical forests protect watersheds that provide water and hydropower. Careful management of watershed forests is needed to protect and maintain water supplies. The Rio Nima watershed project in Colombia and the Dumoga Bone Forest Reserve in Indonesia exemplify sound forest management to preserve critical water resources, although Dumoga Bone has caused relocation of local people.[106]

Preventing Unsustainable Development Projects

Soils need to be carefully tested before development activities are approved for tropical forest regions, to determine whether conditions are appropriate. Unsustainable use of the forest, such as conversion to pasture that can support grazing for only a short time, should be discouraged or prevented. Other destructive activities, such as road construction through prime forest areas and hydroelectric projects which flood large areas of forest, should not be supported by international development assistance and should be avoided unless the long-term benefits clearly outweigh the long-term costs. Past projects, including the recent settlement of the state of Rondonia and construction of the Tucurui Dam in Brazil, have been carried out despite their apparent unsustainability. In an encouraging move, the

World Bank recently adopted new development guidelines designed to prevent loans for unsustainable projects.[107]

Restoring Deforested Land

According to the Food and Agriculture Organization's most recent assessment of tropical forests, 11.3 million hectares of closed and open forests were being cleared annually, while only 1.1 million hectares were being replanted. In Asia, only five hectares are being cleared for each hectare replanted, but in Africa, twenty-nine are cleared for every one replanted.[108]

Although a natural tropical forest, once destroyed, can never be fully replaced, the restoration of cleared or damaged land in the tropics is receiving increasing attention as a means to recover some of the productivity and ecological services lost through deforestation.

Successful restoration projects are planned or under way in a number of tropical countries.[109] Most such projects utilize a few species of non-native trees that have qualities such as rapid growth, drought resistance, and nitrogen-fixing roots, and that stabilize the soil and provide useful products such as timber, fuelwood, food, and animal fodder. When planted in rows, the trees can provide protection, soil enrichment, and moisture retention for food crops planted between the rows. Such methods are known as agroforestry, and they have proven highly successful in a number of countries. Many observers feel that agroforestry should be expanded.[110]

United Nations agencies and the World Bank are shifting their emphasis from large-scale, industrial timber programs to small-scale, community-based, people-oriented agroforestry programs. Many countries are developing training programs for such projects.

On a larger scale, China is one of the few countries with tropical or semitropical terrain that has had success in reforesting major areas of its land. Between 1979 and 1983, in spite of planting more than 4 million hectares each year, China could not keep pace with growing demand for timber. However, in 1985, through enormous effort, China was able to reforest 8 million hectares and achieve a net gain in forest cover. In all, China has replanted some 30 million hectares with trees — an area more than double that devoted to tree plantations and fuelwood throughout the Third World.[111]

Getting rural communities to participate in growing badly needed trees is difficult. The experience of countries such as China that have implemented participatory forestry programs on a wide scale shows that effective reforestation programs require changes in attitudes, redirection of activities of governments and aid agencies, and restructuring of village land use and social organization. Tree planting cannot be imposed from above or carried out among hostile people. New methods of land use affect the daily lives of everyone. Without active local support, saplings may disap-

pear overnight, and uncontrolled grazing animals can ruin reforestation efforts. In most cases, the cost of planting and maintaining trees would be prohibitive without voluntary local help. Community involvement and consideration of local needs are a necessity if rural forestry needs are to be met.

Improving Agriculture

Instead of clearing primary forest, farmers should be encouraged to use alternative sites for cropland such as fertile patches of grassland, floodplain, and secondary or degraded primary forest. Qualities that contribute to sustainability and efficiency include small plot size, high crop diversity, inclusion of tree crops (which require less light and nutrients than root crops), inclusion of domestic animals, recycling of plant and animal wastes, and the sale of surplus products.[112] At present, too little is known about how the rainforest's mineral-poor, easily eroded land can be used in sustainable ways. Research is urgently needed on techniques such as agroforestry so that products can be extracted from tropical forests while still preserving their vital environmental functions.

By combining several levels of vegetation and many kinds of crops, multi-species gardens can produce an astonishing variety of useful products in a small area. For example, in Thailand, the Lua Tribe grows over a hundred crops in such gardens, including plants that provide food, medicine, and materials for weaving and dyes.[113]

Another successful strategy is to mimic the forest's natural response to a tree fall. First, a space is cleared and root crops are grown for a few years, followed by shrub crops. Then tree crops are harvested as the forest cover returns.[114]

In tropical regions where cleared forest is used for livestock production, there is much room for improving management of established pasturelands. Through more efficient methods, cattle ranchers could double their beef production from existing pastureland.[115]

Redistributing Land

In rural areas of the Third World, at least 800 million people are either landless or nearly landless, without work, and without significant cash income. Of this number, the majority lives in countries with tropical forests.[116] As noted earlier in this chapter, in many of these countries most good cropland is owned or controlled by a small percentage of the population. If land were distributed more evenly, fewer people would be forced to clear forests to support themselves, and forest land could be used in more sustainable ways. Land reform and measures to achieve more secure access to farmland are important priorities for the protection of tropical forests.

Slowing Population Growth

Excessive population growth is a basic contributing factor in tropical deforestation, as well as in much other environmental degradation. Tropical countries should give the highest priority to integrated health and family planning programs in order to stabilize their populations at levels that can be supported by sustainable use of forest and other resources.

Curbing Demand for Tropical Products

A number of actions can help reduce demands for timber and other forest products. In the energy sector, more efficient use of fuelwood and other energy sources is needed, along with development of alternative energy sources such as non-wood vegetation, waterpower, wind power, and sunlight. Increased recycling of paper and more efficient use of lumber in construction would also help conserve tropical forests. Higher prices for tropical forest products could limit demands for these products or help cover the costs of developing methods for sustainable forest use. Wider public education is needed to reduce the demand in industrial nations for tropical goods produced in unsustainable ways. When pressures for such products diminish, more attention can be devoted to meeting the needs of local people. A disproportionate amount of land in the tropics is used to produce cash crops for export, including beef, coffee, tea, rubber, coca, sugar, and palm oil. A shift to greater reliance on basic staple crops for local consumption would relieve pressures to cut forests.

Of course, economic pressures to grow cash crops for export are enormous, since they are a major source of income for struggling tropical nations. One innovative approach to this dilemma is the forgiveness of foreign debts in exchange for the creation of domestic forest reserves, in so-called "debt for nature" programs. As an example, Bolivia has agreed to create a 3.7 million acre rainforest reserve in exchange for forgiveness of $650,000 of external debt. With a seminal foundation grant, the debt was purchased at a discounted price of $100,000 from banks that had little other hope of recovering the debt.[117] A number of other "debt-for-nature" exchanges are in progress or planned (see Biodiversity chapter).

Global Action Plans

In 1984, Norman Myers reviewed priorities for slowing tropical deforestation and making better use of tropical resources. Drawing on cost estimates by the U.N. Food and Agriculture Organization (FAO) and Dr. Ira Rubinoff, Director of the Smithsonian's Tropical Research Institute in Panama, Myers proposed a ten-year global program of funding to cover four areas of action: (1) establishing a network of parks and reserves; (2) expanding tree plantations; (3) developing facilities for processing hardwoods in tropical forest countries; and (4) improving research, training, and watershed management. The annual budget for the program was estimated to be $6.4 billion, of which $5.4 billion would come from private sources such as timber corporations, and from tropical Third World governments. The remaining $1 billion would be supplied by developing nations and oil-exporting (OPEC) countries through a progressive tax based on their per capita GNP, and a value-added tax on tropical hardwood imports.[118] (See Table 7.2.)

Table 7. 2
A Ten-Year Program for Curbing Deforestation

(Annual Expenditures in Millions of U.S. Dollars)

Priority	Developed and OPEC Nations	Private Sources and Tropical Forest Nations	Total
Parks and Reserves (Rubinoff)	$300	–	$300
Tree Plantations (FAO)	500	$2,500	3,000
Processing of Tropical Hardwoods in Tropical Forest Nations (FAO)	100	2,900	3,000
Research, Training, Watershed Management (Myers)	100	–	100
Total	$1,000	$5,400	$6,400

Source: Adapted from estimates by Norman Myers, *The Primary Source* (New York: Norton, 1984), pp. 319-28.

In 1985, a worldwide Tropical Forestry Action Plan was developed jointly by FAO, development agencies, the World Resources Institute and other nongovernmental organizations, and representatives from over 60 nations. The plan focuses on needed action in five areas:

- integration of forestry and agriculture to promote sustainable land use and increase agricultural productivity;

- development of sustainable forest-based industries;

- conservation of fuelwood and development of new fuelwood resources;

- conservation of tropical forest ecosystems;

- strengthening of institutions for planning, research, education, and training.[119]

The Action Plan proposed an initial budget of $8 billion over a five-year period, with half the total to be raised by development assistance agencies and international lending institutions, and half coming from private sources and national governments. Roughly a third of the total investment would be agriculture-related and designed to provide landless people in tropical forest areas with alternatives to clearing forests. This budget represents only a "down payment" on what will have to be a greatly expanded, long-term effort.[120]

Critics of the Action Plan say that it overemphasizes "top-down" approaches and plantation forests. The plan's designers admit that it is not a comprehensive solution to the problem of deforestation, and they stress the importance of greater protection of ecologically valuable forests and stronger grassroots participation in their protection and management.

As part of a proposed budget for achieving sustainable development worldwide by the year 2000, Worldwatch Institute has recommended increased funding for global reforestation efforts, beginning at $2 billion per year in 1990 and rising to $7 billion by 1998.[121]

Signs of Progress

There is considerable evidence of progress in efforts to curb deforestation. Within the framework of the Tropical Forest Action Plan, development agencies are coordinating their loans and grants in forestry, resulting in increased funding for forestry and conservation projects. By the end of 1988, some 55 countries were in various stages of planning or implementing national action plans. Forestry sector reviews were under way in 38 countries, and had been completed in 12 countries. In addition, five more countries had requested assistance in developing national plans.[122]

On the donor side, France, West Germany, and The Netherlands are committed to doubling their bilateral development assistance in forestry, and at the initial meeting of the International Tropical Timber Organization in 1987, Japan pledged $2 million for research on sustainable forest management.[123] Throughout the tropics, a growing number of international and local nongovernmental groups are undertaking projects designed to promote the preservation and sustainable use of forest lands.[124]

In Brazil, where there are some 500 conservation organizations, a coalition of legislators, environmentalists, and leaders from industry and the media has been organized to protect the country's remaining coastal forests.[125]

During the past decade, there has been a significant increase in grass roots efforts by nongovernmental organizations in developing countries. Worldwide, there are now thousands of nongovernmental forestry and conservation organizations. Many of these groups are working to protect critical forest ecosystems while meeting local needs for forest products.[126]

What You Can Do

Tropical forests are disappearing with every word you read on this page. While the problems may seem overwhelming and solutions hopelessly beyond reach, there are many ways that you can help arrest destruction of the rainforests. Action on all levels is needed. Even a small effort can make a difference.

Inform Yourself

To help save tropical forests, you must start with an understanding of the issues involved.

- Find out more about tropical deforestation and what you can do to stop it; contact some of the organizations and sources listed in the Further Information section.

- The issue of tropical deforestation is now receiving broad media coverage. Read current newspaper and magazines articles about deforestation in Central America, the Amazon Basin, the Congo Basin, and Southeast Asia.

- Subscribe to one or more periodicals that carry articles about tropical forests; several are listed under Further Information.

- Visit a zoo or botanical garden; this may be as close as you get to seeing actual plants and animals of tropical forests. An afternoon spent strolling through a park or zoo with tropical wildlife can spur your enthusiasm for helping to preserve tropical forests.

- Arrange to see the Smithsonian Institution's excellent traveling exhibition, "Tropical Rainforests: A Disappearing Treasure." The exhibition will be shown at the following locations (dates are approximate and should be confirmed):

> July-September 1989
> Discovery Center, Charlotte, NC
>
> November 1989-January 1990
> Indiana State Museum, Indianapolis, IN
>
> March 1990-May 1990
> Los Angeles County Museum of Natural History, Los Angeles, CA
>
> July 1990-September 1990
> Missouri Botanical Garden, St. Louis, MO
>
> November 1990-January 1991
> Boston Science Center, Boston, MA
>
> February 1991-May 1991
> American Museum of Natural History, New York City, NY
>
> September 1991-December 1991
> Denver Museum of Natural History, Denver, CO

January 1992-April 1992
Louisiana Science Center, New Orleans, LA

May 1992-August 1992
Houston Museum of Natural History,
Houston, TX

September 1992-November 1992
Fernbank Science Center, Atlanta, GA

January 1993-March 1993
Chicago Botanic Garden, Chicago, IL

May 1993-August 1993
Science Museum of Minnesota, St. Paul, MN

October 1993-January 1993
Cleveland Museum of Natural History,
Cleveland, OH

- After you have a good grasp of the deforestation issue, pass along your knowledge and insights to your family, friends, and professional contacts.

Join With Others

Contact a local group or a national organization that is working to save tropical forests, and become a member. Contribute some time or money to their efforts. Even a small contribution can help spread awareness of the issue and aid conservation efforts. Working with a group on a project can be a valuable learning experience, as well as a good way to meet new people who share common interests.

- If you can donate money to help save tropical forests, be sure to verify whether your employer has a matching fund program.

- Join in the Periwinkle Project of the Rainforest Alliance to urge that doctors and pharmacists display an attractive poster highlighting dependence on tropical rainforests for medicines and other pharmaceuticals.

Review Your Habits and Lifestyle

Much of the damage being done to the forests is a direct result of our demands for tropical hardwoods and other products from the tropics.

- Avoid buying furniture or other timber products which may use tropical hardwoods, such as teak, mahogany, and plywood containing mahogany.

- Ask retailers where their tropical products come from and how they are obtained. If the retailer is not sure, do not buy their merchandise and inform them about your concern for deforestation.

- Encourage retailers not to carry any products that were produced in natural tropical forests in unsustainable ways.

- If you must have a tropical plant or animal, buy one that was raised in the United States.

- Use conservation and recycling programs in your community to help bring attention to the issue of tropical deforestation as well as other environmental issues.

Work With Your Elected Officials

Contact your representatives in the U.S. Congress, find out whether they support efforts to slow deforestation, and tell them where you and the organizations you represent stand on the issue. Newsletters of several national conservation organizations listed under Further Information regularly report opportunities for contacting public officials on legislation that affects tropical forests.

- Every letter you write to a politician will be read. Often just a few hundred letters can strongly influence an elected official. Members of Congress and other public officials should receive many letters, telephone calls, and visits supporting environmentally-sound foreign assistance programs when legislation affecting these programs is being considered. Urge public officials to oppose development projects that would flood or clear tropical forests such as large-scale dams and cattle ranches. Ask them to support greater federal assistance for tropical forest conservation.

- Urge your congressional representatives to increase support for the U.S. Agency for International Development and the Peace Corps, including their programs to slow tropical deforestation. In particular, the United States should increase support for improving agroforestry techniques and sustainable forest management methods, and create incentives for tropical nations to set aside protected areas.

- In addition, write to the Director of the U.S. Agency for International Development, the Director of the Peace Corps, the Presidents of the World Bank and the Inter-American Development Bank, the Director of the International Monetary Fund, and the Secretary-General of the United Nations (addresses are given below), since all these organizations have programs that affect the future of tropical forests.

  » Director, U.S. Agency for International Development, 320 21stSt. NW, Washington, DC 20532

  » Director, Peace Corps, 1990 K St. NW, Washington, DC 20526

» President, The World Bank, 1818 H St. NW, Washington, DC 20433

» President, Inter-American Development Bank, 1300 New York Ave. NW, Washington, DC 20577

» Director, International Monetary Fund, 700 19th St., NW, Washington, DC 31

» Secretary General, United Nations, New York, NY 10017

Publicize Your Views

The more people hear about the problem of deforestation, the more they will be willing to help protect the forests.

• Encourage the local media to devote more attention to the destruction of tropical forests. Suggest that environmental writers examine local connections to tropical deforestation. Offer to provide the needed information and even act as a "local resident expert." Volunteer to write articles about tropical forests for local publications.

• Consider organizing a festive "rainforest day" to generate community interest and involvement through speeches, music, games, and theater, all invoking the theme of tropical forests. Any profits from the event can support work to save the forests.

• Consider organizing a rainforest town meeting in your community. Invite congressional representatives, civic leaders, and representatives of conservation groups to discuss U.S. actions and policies on global concerns such as tropical forests and biological diversity, economics and sustainable development, international cooperation and foreign assistance, and ethics and values in a global community. Devote much of the program to questions and comments from the audience. Following the town meeting, hold a reception with representatives of local organizations active on environmental, resource, population, and development issues. Contact the Global Tomorrow Coalition for further information on organizing a global town meeting.

Raise Awareness Through Education

Encourage your local schools, colleges, and libraries to obtain and present adequate information on deforestation and how it threatens the rich diversity of life in tropical forests.

• If they lack these resources, volunteer to help gather needed materials. Encourage teachers to include this information in their courses, and to stress the links between foods, medicines, and other materials on which we all depend and the tropical sources of these products. The

Global Tomorrow Coalition and other organizations have curriculum materials for use in upper elementary and secondary classes (see Further Information section).

• Offer to help a local library put together an exhibit on tropical forests, and provide the library with a list of recent books and articles taken from those listed in the Further Information section.

Consider the International Connections

Tropical deforestation is truly a global issue that affects all parts of the Earth.

• If you travel to tropical countries, see the wonders of a tropical rainforest yourself, with a group of friends, or with an organized expedition. Hawaii, The Virgin Islands, Puerto Rico, southern Mexico, and Central and South America all have tropical forests. For information on tropical nature tours, contact companies specializing in natural history tours that advertise in magazines such as *Audubon*, *Natural History*, and *Smithsonian*. Before you go, find out whether any of the nongovernmental organizations (NGOs) listed under Further Information have programs in the country you plan to visit, or links with local NGOs there. If so, plan to contact local representatives and learn about their activities.

• Once you are there, find out what local organizations are doing to save the forests, and learn about their programs and their needs for money and other resources. Try to visit a tropical forest reserve maintained by a local organization. Learn how the reserve is used to protect a critical habitat and to educate the public about forest conservation and sustainable forest use. Let the local officials know your impressions of the reserve, and of the area's forests in general.

• When you return home, organize a presentation about your experience and write about it for your local media. Consider what you can do to establish a "twinning program" or link between a U.S. conservation group and an NGO in the country you visited, or to strengthen a link that already exists. Possible ways to help a tropical NGO include raising money for the group through car washes, by selling posters or T-shirts from the group, or by adding contributions for the group to tour packages for traveling in the country. In addition, you can solicit donations of goods such as books, binoculars, typewriters, and computers to help support the group's programs.

• Write to the Chief Executive Officer of a multinational corporation with operations in tropical forest countries, and suggest that the corporation cooperate with local programs to establish a sustainable yield forestry policy in those countries.

Further Information

A number of organizations can provide information about their specific programs and resources related to tropical forests.[*]

Table 7.3 (on the following page) lists resources and services offered by some of these organizations.

Books

Caufield, Catherine. *In the Rainforest*. Chicago: University of Chicago Press, 1984. 304pp. Paperback, $11.95.

Caufield, Catherine. *Tropical Moist Forests*. London: Earthscan, 1982. Available from International Institute for Environment and Development. 67pp. Paperback, $5.50.

Denslow, Julie S. and Christine Padoch. *People of the Tropical Rainforest*. Berkeley, CA: University of California Press, 1988. 231 pp. Paperback.

Gradwohl, Judith. *Tropical Rainforests: A Disappearing Treasure*. Washington, DC: Smithsonian Institution Traveling Exhibition Service, 1988. 32 pp. Paperback.

Gradwohl, Judith and Russell Greenberg. *Saving Tropical Forests*. London: Earthscan, 1988. 234 pp. Paperback, $12.95; and Washington, DC: Island Press, 1988. 207 pp. Hardback, $24.95.

Lanly, Jean-Paul. *Tropical Forest Resources*. Forestry Paper No. 30. Rome: Food and Agriculture Organization, 1982. Available from UNIPUB. 106pp.

Myers, Norman. *The Primary Source: Tropical Forests and Our Future*. New York: Norton, 1984. 399pp. $17.95.

Repetto, Robert. *The Forest for the Trees?: Government Policies and the Misuse of Resources*. Washington DC: World Resources Institute, May, 1988. 105pp. Paperback, $10.00.

U.S. Forest Service. *Global Neighbors Growing Together: A Tropical Forestry Program*. Washington, DC: U.S. Department of Agriculture, 1989. 16 pp. Paperback.

World Resources Institute (WRI), World Bank, and United Nations Development Program. *Tropical Forests: A Call for Action*. Washington, DC: WRI, 1985. 3 booklets, encased. 144 pp. $14.50.

Articles, Pamphlets and Brochures

Lovejoy, Thomas E. "Hope for a Beleaguered Paradise." *Garden: Special Amazon Issue*, January-February, 1982, pp. 32-36. Available from the Garden Society, New York Botanical Garden, Bronx, NY 10458.

Myers, Norman. "Tropical Rain Forests and Their Species: Going, Going...? " In E.O. Wilson (ed.), *Biodiversity*. Washington, DC: National Academy Press, 1988, pp. 28-35.

Myers, Norman. (ed.). "Land." In Norman Myers (ed.), *Gaia: An Atlas of Planet Management*. Garden City, NY: Anchor Books, 1984. pp. 26-32, 42-46.

National Geographic Society. "Tropical Rainforests." *National Geographic Magazine*, January 1983, pp. 2-65.

Postel, Sandra and Lori Heise. "Reforesting the Earth." In Lester R. Brown et al., *State of the Earth 1988*. New York: Norton, 1988, pp. 83-100. For an expanded version, see Worldwatch Paper 83.

Raven, Peter H. "The Cause and Impact of Deforestation." In National Geographic Society, *Earth '88: Changing Geographic Perspectives*. Washington, DC, 1988, pp. 212-29.

Raven, Peter H. "Our Diminishing Tropical Forests." In E.O. Wilson (ed.) *Biodiversity*. Washington, DC: National Academy Press, 1988, pp. 119-21.

Wilson, E.O. "Diversity at Risk: Tropical Forests." In E.O. Wilson, (ed.), *Biodiversity*. Washington, DC: National Academy Press, 1988, pp. 119-154.

World Resources Institute (WRI) and International Institute for Environment and Development (IIED). "Forests and Rangelands." *World Resources 1986*, pp. 61-83; *World Resources 1987*, pp. 57-76; *World Resource 1988-89*, pp. 69-88. New York: Basic Books, 1986, 1987, 1988.

Periodicals

A number of periodicals carry articles about tropical forests, including the following (publishing organizations are in parentheses): *Audubon* (National Audubon Society), *The Canopy* (Rainforest Alliance), *Focus* (World Wildlife Fund), *International Wildlife* (National Wildlife Federation), *National Geographic*, *Natural History* (American

---

[*] These include: American Forestry Association, Conservation International, Cultural Survival, Environmental Defense Fund, Environmental Policy Institute, Friends of the Earth, Global Tomorrow Coalition, International Union for the Conservation of Nature and Natural Resources, Missouri Botanical Garden, National Audubon Society, National Wildlife Federation, Natural Resources Defense Council, Nature Conservancy International, New Forests Fund, New York Botanical Garden, Rainforest Action Network, Rainforest Alliance, Sierra Club, National Museum of Natural History and National Zoological Park (both divisions of the Smithsonian Institution), Smithsonian Tropical Research Institute, Survival International U.S.A., Wildlife Conservation International (a division of New York Zoological Society), World Resources Institute, Worldwatch Institute, and World Wildlife Fund. Addresses for some of these organizations are given in Table 3; all the addresses are listed in the Appendix.

Table 7.3
Selected Organizations Involved in Saving Tropical Forests

Organization	Classroom Materials	Issue/Legislative Updates	Publications	Slides/Slide Shows	Speakers	Technical Reports	Tourism/Trips to Tropical Countries
Conservation International 1015 18th Street, NW, Suite 1002 Washington, DC 20036		•		•			•
Cultural Survival 11 Divinity Avenue Cambridge, MA 02138			•	•	•		
Environmental Defense Fund 1616 P Street, NW, Suite 150 Washington, DC 20036		•	•		•	•	
Friends of the Earth/U.K. 2628 Underwood Street London N17JU UNITED KINGDOM	•		•				
Global Tomorrow Coalition 1325 G Street, NW, Suite 915 Washington, DC 20005	•	•	•				
International Institute for Environmental Development 1717 Massachusetts Avenue, NW Washington, DC 20036				•		•	•
International Union for the Conservation of Nature and Natural Resources Avenue Mont Blanc 1196 Gland SWITZERLAND	•		•		•		
Missouri Botanical Garden P.O. Box 299 St. Louis, MO 63166	•		•		•	•	•
National Audubon Society 645 Pennsylvania Avenue, SE Washington, DC 20003	•	•	•	•	•	•	
National Wildlife Federation 1412 16th Street, NW Washington, DC 20036	•	•		•	•		
Natural Resources Defense Council 1350 New York Avenue, NW Washington, DC 20005		•					
The Nature Conservancy 1800 North Kent Street Arlington, VA 22209		•					•
New York Botanical Garden Bronx, NY 10458			•	•	•	•	•
Rainforest Action Network 300 Broadway, Suite 28 San Francisco, CA 94133	•	•	•	•	•		•
Rainforest Alliance 295 Madison Avenue, Suite 1804 New York, NY 10017		•	•	•	•		
Sierra Club 730 Polk Street San Francisco, CA 94009				•			•
National Museum of Natural History Smithsonian Institution Washington, DC 20560					•	•	
National Zoological Park Smithsonian Institution Washington, DC 20008	•				•	•	
Smithsonian Tropical Research Institute APO Miami 34002-0011	•					•	
Wildlife Conservation International New York Zoological Society Bronx, NY 10460	•		•				•
World Resources Institute 1735 New York Avenue, NW Washington, DC 20006			•			•	
World Wildlife Fund/U.K. Panda House Godalming Surrey GU7 IXR UNITED KINGDOM	•						
World Wildlife Fund/Conservation Foundation 1250 24th Street, NW Washington, DC 20037	•		•	•	•	•	•

This partial list does not include the many fine organizations located in tropical countries. These can be reached through the organizations listed above.

Source: Smithsonian Institution Traveling Exhibition Service, *Tropical Rainforests: A Disappearing Treasure* (Washington, DC: Smithsonian Institution, 1988).

Museum of Natural History), *Not Man Apart* (Friends of the Earth), *Smithsonian* (Smithsonian Institution), and *World Rainforest Report* and *Action Alert* (Rainforest Action Network).

American Forests. Special issue on tropical deforestation, November-December 1988. Includes 18 articles on various aspects of tropical forests. Published by The American Forestry Association.

Films and Other Audiovisual Materials

Cry of the Muriqui. 1982. Relates the story of endangered monkeys in the largely deforested mountains of southeastern Brazil, north of Rio de Janeiro. 28 min. 16mm color film, 3/4" or 1/2" Video. State University of New York at Stony Brook.

Rain Forest. 1983. 60 minutes. 16mm color film, Video (Beta and VHS). ($26.95). National Geographic Society.

Amazonia: A Burning Question. 1987. Surveys efforts to understand and conserve the Amazonia region of South America. Dr. Thomas Lovejoy walks us through the forest, explains its importance and illustrates the role of scientific research in its preservation. 58 min. 16mm color film, 3/4" or 1/2" Video. State University of New York at Stony Brook.

Amazonia: A Celebration of Life. 1984. Set in Peru, this World Wildlife Fund production presents a kaleidoscope of tropical species, from giant otters to tiny tree frogs. It serves as an introduction to the diversity of life in a tropical forest. 20 min. 16mm film, 3/4" or 1/2" Video. State University of New York at Stony Brook.

Conservation of the Southern Rainforest. 1988. In the southeastern region of Peru, a group of scientists are working with local conservationists and native healers to study and protect the environment. The film documents their work and discusses alternative solutions that address both local and global needs, including debt-swaps for conservation and the creation of national parks. 60 min. 16mm film, 3/4" or 1/2" Video. Distributed by Biosphere Films.

Creatures of the Mangrove. 1986. The coastal island of Siarau and its tidal forest of mangroves are the offspring of Borneo's steamy interior rainforest, built by silt washed down from the mountains and deposited in the sea. Portrays the diversity of life, from trees that grow in water and breathe through roots to archer fish that shoot down insects from overhanging branches. 59 min. 16mm film, 3/4" or 1/2" Video. National Geographic.

Korup. An African Rainforest. 1984. Describes the interdependence between plants and animals, and the intricate relationships which sustains one of the most fragile and finely tuned ecosystems on earth. Filmed in Cameroon, it allows the beauty and complexity of one forest to speak for the survival of many. 55 min. 3/4" or 1/2" Video. Partridge Films.

The Last Forest. 1989. Portrays one teenager's awakening to the importance of tropical forests and how her life will be different if their destruction is not stopped. Includes five lesson plans and background information for the teacher. Grades 6-12. 20 min. 1/2" video with curriculum packet. CARE Film Unit.

The Living Planet. Part 4, The Jungle. 1984. Sir David Attenborough examines the workings of the earth's rainforest ecosystem, as he guides the viewer down a kapok tree from the crown above the jungle canopy to the floor below. 55 min. 16mm film, 3/4" or 1/2" Video. Pennsylvania State University.

Our Threatened Heritage. 1987. Presents the environmental consequences and political aspects of tropical deforestation. Through interviews with ecologists and policymakers, it explains the changes in climate and loss of biological diversity that may arise from continued forest destruction. Covers cattle ranching, road projects and other impacts on the forest as well as the global connections between multinational banks and deforestation. 50 min. 3/4" or 1/2" Video. National Wildlife Federation.

Teaching Aids

CARE. *The Last Forest*. See description under Films and Other Audiovisual Materials above.

Global Tomorrow Coalition. *Tropical Forest Education Packet*. Includes background information, lesson plans, and activities emphasizing interactive learning aimed at upper elementary and intermediate students: can be upgraded for high school. $7.00 including postage. GTC Education Services, 1325 G Street, N.W., Washington, DC 20005-3104.

World Wildlife Fund. *The Vanishing Rain Forest Education Kit*. Designed for grades 2-6. Includes the following: a 28 page booklet with color photographs that introduces the rain forest: its ecology, its importance to our world, and the threats to this endangered ecosystem; a Teacher's Manual; two posters; and a lively, six-minute video describing the beauty of the rain forest, especially its interesting animals and plants. The cost is $20 (or $15 for the video alone); prices include postage for mailing only in the United States. Publications Department OB, World Wildlife Fund.

Ocean and Coastal Resources

The coastal zone may be the single most important portion of our planet. The loss of its biodiversity may have repercussions far beyond our worst fears.

—G. Carleton Ray, University of Virginia[1]

There's no mystery to marine pollution. The worst problem today is the huge quantity of raw sewage and industrial effluent spewed into the sea, with no thought to consequences, from coastal cities all over the world.

—Stjepan Keckes, United Nations Environment Program[2]

Major Points

- Oceans cover more than two-thirds of the Earth's surface and contain animal life that rivals tropical forests in its diversity of species.

- Coastal waters within 200 miles of land contain more than half the ocean's total biological productivity and supply nearly all the world's fish catch.

- Oceans and coastal zones are delicately balanced ecosystems that are threatened by overharvesting, by construction and development, and by pollution from oil, municipal and industrial wastes, and other land- and sea-based sources.

- Yields from a number of the world's major ocean fisheries have leveled off or are declining, and some have collapsed due to overfishing. Excessive harvesting of whales has depleted some species to near extinction.

- The world fish catch is approaching the annual total—100 million metric tons—that the U.N. Food and Agriculture Organization (FAO) believes is the most that can be taken sustainably over the long term.

- Recent progress in protecting marine resources includes improved management of ocean fisheries and a number of international and regional initiatives to control pollution and protect marine environments.

The Issue: Degradation of Marine Resources

Protecting our ocean and coastal resources is of fundamental importance to all of us, and to future life on Earth. The Earth is largely a water planet; oceans cover more than 70 percent of the planet's surface,[3] and contain some of the earth's most complex and diverse ecosystems. Life on Earth began in the ocean, and recent research suggests that the oceans contain animal life that rivals tropical forests in its diversity of species. Of the world's 71 phyla (groups of life forms), 43 include marine species, while only 28 contain land species.[4] Marine species play a central role in the biological, chemical, and physical cycles on which all life depends.

In addition to serving as the habitat for a vast array of plants and animals, the oceans also supply people with food. Over half the population of developing nations obtains 40 percent or more of their total animal protein from fish.[5] In the United States alone, seafood represents a $10-billion industry. Fish products are used for animal feed and fertilizer, and for soaps, pharmaceuticals, and other commercial products. The oceans are also a storehouse of mineral wealth and a potential source of energy. In addition, they

influence weather patterns and stabilize climate around the world.[6]

The present harvest of many marine species is dangerously high, given their limited ability to reproduce and the environmental stresses we are placing on these resources. These stresses include serious pollution from industrial, municipal, and agricultural sources, oil spills, the dumping of toxic wastes, and the destruction of rich coastal ecosystems by the development of seashore regions.

Marine Ecosystems

The most productive parts of the ocean are the coastal areas within 200 miles of land. These areas account for more than half the ocean's biological productivity, and supply nearly all the world's catch of fish.[7] Along the coasts, nutrients are washed down from the land, surface winds and ocean currents dredge up nutrient-rich sediments from the sea bottom, and sunlight promotes plant growth on the shallow seafloor. Coastal and island regions contain many kinds of ecosystems that are vital to marine life and humankind; four of the most productive are salt marshes, mangroves, estuaries, and coral reefs.

Salt Marshes and Mangroves

Salt marshes are tidal wetlands found in temperate zones, and mangroves are their tropical counterpart; both are associated with the growth of offshore seagrasses. In salt marshes, these grasses provide winter food for ducks and geese; mangroves contain food sources for sea turtles and aquatic mammals.

Mangroves are found on over half the world's tropical shores; they are probably the most productive coastal ecosystem, supporting large quantities of finfish and shellfish. Seagrasses also trap nutrients for shellfish and finfish, while filtering out pollution and preventing coastal erosion.[8]

Estuaries

Coastal estuaries are regions where freshwater rivers carrying fertile silt meet ocean tides. Containing both land and ocean nutrients, estuaries support a long and elaborate chain of life, from protozoa to fish-eating mammals, and they provide essential habitats and breeding grounds for a wide range of wildlife.[9]

In one of the world's richest fishing areas, the continental shelf of the eastern United States, at least three quarters of commercially valuable fish spend part of their life in estuaries.[10] Because of their structure, estuaries trap and retain pollutants rather than carrying them out to sea, thus exposing their marine life to ever-increasing levels of concentrated contaminants.[11]

Coral Reefs

Coral reefs are highly diverse tropical ecosystems, containing more plant and animal groups than any other ecosystem on Earth, while supporting a third of all fish species.[12] Coral reefs need clear water, bright light, constant high salinity, and a water temperature of over 21 degrees Celsius (70 degrees Fahrenheit).

Because the reefs are complex and delicate, they are easily disturbed by changes in their environment. Coral reefs depend on a constant flow of water for nourishment. Excessive soil erosion or dumping of dredged waste can make the water opaque and block sunlight from the reef, or simply bury the reef.[13] Coral mining, blast fishing, overfishing, and collection of coral are additional pressures on these fragile and slow-growing ecosystems.

Productivity

Compared with the open ocean, coastal ecosystems have much greater biological productivity. Estuaries and coral reefs are 14 to 16 times more productive than the open ocean; mangroves are over 20 times more productive.[14] (See Table 8.1.)

Table 8.1
Biological Productivity of Marine Areas

Ecosystem Type	Mean Net Primary Productivity*
Open Ocean	57
Continental Shelf Regions	162
Upwelling Areas	225
Saltmarshes	300
Estuaries	810
Coral Reefs and Seaweed Beds	900
Mangroves	1,215

* in grams of carbon per cubic meter per year

Source: Norman Myers (ed.) "Ocean," *Gaia: An Atlas of Planet Management* (Garden City, NY: Anchor Press, 1984) p. 74.

Damage to Marine Resources

Human activity is responsible for a wide range of threats to the rich diversity of marine life. Causes of the damage to coastal and marine environments are varied and complex, but all are related to the high concentration of people in coastal regions. In the United States, for example, over half the population lives within 50 miles of the sea.[15] Roughly two-thirds of the world's people live along coastlines and rivers draining into coastal waters.[16] This high concentration of people is responsible for most of the harm done to marine and coastal resources.

The main sources of damage to marine resources are land-based coastal pollution from municipal sewage, industrial waste, urban and agricultural runoff, and inland deforestation; construction and development in coastal areas; ocean dumping of dredged material, sewage sludge, and hazardous wastes; discharge and spills of petroleum; discarded plastics; and overfishing and the use of wasteful fishing methods.[17]

Pollution from a variety of sources has already had a serious effect on the ability of marine areas to provide us with food. For example, one third of U.S. shellfish beds have already been closed due to water pollution. Eagle Harbor in Washington's Puget Sound is so polluted that much of its rapidly declining population of English sole cannot reproduce. New York State officials have warned residents to limit consumption of fish from the Hudson River or coastal waters to one serving per week. And illness from eating contaminated seafood—including cholera, hepatitis, and gastroenteritis—is increasing around the United States.[18]

Degradation of Marine Ecosystems

Salt marshes, mangroves, and coral reefs are threatened by human activities in many parts of the world. For example, in the Northern Persian Gulf, land reclamation for urban development has destroyed several types of ecosystems, including seagrasses, mangroves and coral reefs. Mangrove forests on the East African coast have been depleted for firewood and building materials. Along East Asian coasts, extensive conversion of mangrove forest to rice fields has eliminated natural barriers to flooding from storms. Indonesia has already converted a quarter of its coastal mangroves to rice fields and shrimp ponds. In Central and South America, mangroves are being cleared for shrimp farming; one third of Ecuador's mangroves have been converted to shrimp ponds.

Coral reefs face a variety of threats and are being damaged in many tropical regions. In addition to destroying one of the Earth's richest habitats, damage to coral reefs increases the chances of storm damage and beach erosion. For example, along the coast of Sri Lanka some 75,000 tons of coral reefs are mined each year for construction and industry; loss of the reefs is causing serious coastal erosion.[19] Reefs along East Africa's coast are being disrupted by fishing using dynamite explosives, and in the Philippines, most of the coral reefs have been damaged by cyanide used to stun fish. Silt deposits resulting from deforestation and poor land management inland threaten coral reefs in East Asia, the South Pacific, and the Caribbean.[20]

Land-based Coastal Pollution

Over 80 percent of all ocean pollution comes from land-based activities.[21] This pollution reaches the oceans either from "point" or "nonpoint" sources. Point sources include pipes, ditches, canals, or similar channels that regularly release pollution in a specific area. Sewage and industrial waste are commonly introduced into waters from point sources. Nonpoint pollution covers all types of unregulated runoff from the land, including runoff from urban and agricultural areas.[22]

Municipal Sewage   The discharge of sewage and other municipal and industrial effluents from coastal cities into the sea is probably increasing, especially in the Third World. Even in relatively developed areas, sewage and industrial wastes are often pumped into the sea without treatment. The most polluted seas are off the densely populated coasts of India, Pakistan, and Bangladesh, and near the coastal cities of Thailand, Malaysia, Indonesia, and the Philippines.[23]

In the United States, publicly-owned sewage treatment plants are among the most important point sources of coastal pollution. Because much of the U.S. coast is densely populated, about 35 percent of all U.S. sewage ends up in marine waters. The National Oceanic and Atmospheric Administration (NOAA) reported that treatment plants discharged 3.3 trillion gallons of sewage into marine waters in 1980. NOAA projects that the volume will rise to 5.4 trillion gallons by the year 2000.[24]

Most of America's best-known bays and harbors are badly polluted. Inadequate sewage treatment and industrial waste have turned Boston Harbor into one of the most polluted areas in the nation; bottom-feeding flounder in the harbor are afflicted with tumors and fin rot. Much of the New York City area's 8-million-ton annual output of sewage sludge is dumped offshore, but during heavy rains the city's sewage—amounting to 1.7 billion gallons a day—goes directly into New York Harbor, which already contains high concentrations of municipal and industrial contaminants.[25]

Industrial Waste   In some areas, industrial facilities are important point sources of coastal pollution. Some industrial wastes are discharged through pipelines directly into rivers or bays. In the United States, about 5 trillion gallons of industrial waste water go directly into coastal waters each year, contaminating marine life in important fishing areas, including California, Florida, and the Great Lakes.[26] In the eastern United States, 32 rivers and streams carry industrial and other wastes into the Atlantic Ocean.[27] In Europe, discharge of polychlorinated biphenyls (PCBs) into the Rhine River is harming the seal population in Holland's Wadden Sea.[28]

Most PCBs enter the ocean by deposition from the atmosphere. Marine mammals in many areas are accumulating elevated levels of PCBs which impair reproduction. Scientists are concerned that such animals may be facing extinction.[29]

Urban Runoff   Domestic and industrial effluents such as food wastes, solid wastes, pesticides from lawns, toxic chemicals, heavy metals, and sediments from construction sites are commonly washed into storm sewers. Rain and snowmelt carry these pollutants from lawns, streets, and waste dumps through the storm sewers into waterways and eventually into coastal waters. Nearly half the oil that pollutes marine waters comes from urban runoff.[30]

Many cities combine their sewage systems with their storm sewers and carry both types of waste to treatment plants. With heavy rainfall, the volume of sewage and runoff that reaches these treatment plants can exceed their capacity. The resulting overflow of waste cannot be handled and is discharged without treatment.[31]

Agricultural Runoff   Fertilizers, herbicides, and pesticides run from farmland into streams and rivers, eventually making their way into coastal waters. Each year over 2.5 billion pounds of pesticides are used in the United States, and because they are often not easily biodegradable, these pollutants are persistent contaminators of coastal waters, steadily moving up the food chain and increasing in concentration with each step.[32]

Pollution from urban and agricultural runoff has damaged marine resources in many U.S. coastal areas. Municipal and agricultural effluents from a three-state area are a major cause of declining fish and oyster harvests in the Chesapeake Bay, North America's largest estuary; annual oyster harvests fell from 3.5 million bushels in the 1960s to less than 1 million in 1986.[33]

Nitrates from acid rain are a major pollutant in the Chesapeake Bay; they promote growth of algae that depletes oxygen supply and blocks sunlight. The main sources of these nitrates are motor vehicle exhausts and power plant emissions.[34]

Algae Growth   Both sewage and agricultural runoff introduce large quantities of nitrogen and phosphorous into coastal waters. Coming from such sources as detergents, fertilizers, and human waste, these compounds nourish algae and can cause their explosive growth. Excessive algae growth can deplete the water of oxygen and suffocate other species. Oxygen-depleted waters are known as "dead zones"; a 3,000-square-mile dead zone has been found in the Gulf of Mexico, near the mouth of the Mississippi River. In 1980, an outbreak of algae in the Adriatic Sea killed fish along a 1,000-mile-stretch of the Italian coast. Algae clusters can block sunlight and stunt the growth of other marine life.[35] Some types of algae can also be toxic to marine life. In 1987, over a dozen whales washed ashore on

Cape Cod; their death was attributed to toxins from algae. The toxins made their way up the food chain, finally tainting mackerel that were eaten by the whales. In that same year, thousands of mullet and virtually the entire scallop population of the Carolinas were wiped out due to toxins from an algae bloom. A study by the National Oceanic and Atmospheric Administration concluded that between 1987 and 1989, as many as 3,000 dolphins may have died along the Atlantic Coast after eating fish contaminated with toxic algae.[36]

These carpets of algae are often red or brown, hence the name red or brown tide. Japan's Inland Sea is affected by some 200 red tides each year. In 1987, one such tide killed more than a million fish with a potential market value of $15 million. And millions of salmon and sea trout suffocated off Scandinavia after algae clung to their gills and formed a slimy film.[37]

Fallout of Atmospheric Pollutants   A 1988 study by the Environmental Defense Fund reveals that nitrates found in acid rain are a major source of the nitrogen that encourages the explosion of algae. The study found that 25 percent of the nitrogen in the Chesapeake Bay came from acid rain, making this a larger contributor to the Bay's nitrogen content than sewage and nearly as large as fertilizer runoff. The nitrogen in acid rain comes largely from nitrogen oxides emitted by motor vehicles and power plants. Preliminary studies indicate that acid rain makes a similar nitrogen contribution to Long Island Sound, North Carolina's Pamlico Sound and the New York Bight.[38]

Construction and Development

Much of the damage to marine resources has resulted from the growth of coastal communities. Rapid population growth and development pressures threaten many highly productive coastal ecosystems in the tropics, and in the United States between 1940 and 1980 a doubling of the population living within 50 miles of the coast greatly increased coastal development.[39]

Dredging, filling, and building at the water margin associated with coastal development upset the balance of marine life. Such construction can change the water's salinity or stir up enough sediment to alter its clarity. If the sediment disturbed by dredging or construction contains toxic pollutants, these can harm marine plants and animals.[40]

The most ecologically productive marine areas are usually found in calm, protected waters, which are also ideal for seaports, marinas, and other kinds of coastal development. Biologically rich coastal areas, which filter sediment runoff from the land and provide breeding habitats for many species, are among the most threatened marine ecosystems.

Ocean Dumping

A number of industrial nations still use ocean dumping to dispose of wastes, despite laws and treaties that ban or limit dumping, such as the U.S. Marine Protection, Research, and Sanctuaries Act of 1972 (also known as the "Ocean Dumping Act") and the London Dumping Convention. Wastes that continue to be dumped include dredged material, sewage sludge, incinerated hazardous waste, and industrial waste, although the dumping of industrial waste off U.S. coasts has stopped.[41] There is evidence that ocean dumping of acidic wastes off the New Jersey coast has harmed crustacean and phytoplankton populations in the dumping area.[42]

Dredging   Material dug from rivers, harbors, and channels to keep them clear for ships is the single largest source of ocean dumping, far exceeding sewage sludge and industrial and radioactive wastes combined.[43] Dredged material contains large amounts of sediment and may contain oil, grease, and heavy metals along with other bottom contaminants such as polychlorinated biphenyls (PCBs), pesticides, and pathogens, which can harm not only marine species but humans as well if the contaminants enter the food chain in high concentration. Even dredged material free of such pollutants can damage marine organisms by burying them and increasing suspended sediment, which blocks out light.[44]

Sewage Sludge   Only the United States and the United Kingdom currently dump substantial quantities of treated sewage sludge at sea.[45] Each year New York City and neighboring sewage authorities dump 8 million tons of sludge at a location 106 miles out to sea. In 1986, 60,000 tons of sewage sludge from the Hyperion outfall in the Southern California Bight were dumped only seven miles from shore, and each year about 22,000 tons of sludge are emptied annually into Boston Harbor. Both of these dumping sites are to be phased out by 1990.[46]

Depending on its content and concentration, sewage sludge may be relatively harmless, but it can also contain PCBs, pesticides, and toxic heavy metals such as lead, mercury, and cadmium, as well as disease-causing microorganisms and pathogens. These noxious substances can fatally poison marine species, or they can impede their growth and damage their sensory and reproductive functions.[47] Sewage can also introduce excessive amounts of nutrients into coastal waters, leading to explosive growth of algae and depleting the water of oxygen essential to marine organisms.

Industrial Waste   With the increasing financial costs of treating and disposing of industrial waste on land, a number of industrial and Third World nations have turned to the oceans to get rid of a wide range of organic and inorganic residues, sludges, and debris. Only in the Southwest Atlantic and the Polar Oceans is industrial pollution not yet a problem. Chemical pollutants have reduced oxygen levels and endangered marine life in coastal waters in the North Adriatic Sea, the Persian Gulf, the Black Sea, and the Arabian Sea near Karachi, Pakistan. In 1981, the United States dumped about 3 million tons of industrial waste into marine waters.[48]

The Dutch government recently reported that the North Sea is among the most polluted seas in the world. In 1984 alone, about 5.6 million tons of industrial waste, as well as 5,100 tons of sewage sludge and 97 million tons of dredged material, were dumped directly into its waters. Along the eastern coast of the North Sea, PCB pollution is believed responsible for massive fish poisonings and high rates of infertility and miscarriages among local seal populations, and there is evidence that pollution has caused the death of thousands of seals by weakening their immune systems. The U.K. government disputes these claims, asserting that the pollution is restricted to marginal areas of this body of water and claiming that the North Sea is generally healthy.[49]

Figure 8.1
Dumping of Industrial Waste, Sewage, Sludge,
and Dredged Material at Sea, 1981

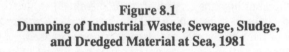

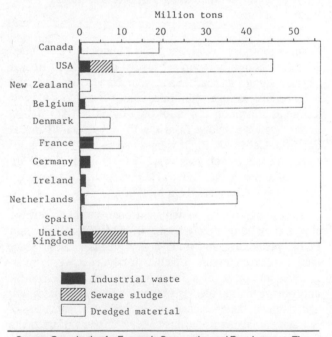

Source: Organization for Economic Cooperation and Development, *The State of the Environment 1985* (Paris, 1985), p. 79.

Radioactive Waste   In 1983, the London Dumping Convention (LDC) adopted a resolution banning the dumping of low-level radioactive waste, which is generated in medical research and industrial activities, in oceans beyond

the limits of national jurisdiction. Previously, several European countries had been dumping radioactive wastes off the coast of Spain. In 1985, despite heavy opposition and the threat of a walkout by the British, the moratorium was extended indefinitely pending further research into the effects of such dumping. Even though the United States had not dumped radioactive material since 1970, it vetoed the moratorium because high-ranking Energy and Defense Department officials wanted to retain the ocean dumping option.[50]

A number of plants that reprocess nuclear fuels release radioactive wastes into coastal areas in ways not covered by the LDC. Discharges from reprocessing plants in the United Kingdom and France have been linked to illnesses and deaths of local people.[51]

Incinerated Hazardous Waste   One way to dispose of certain types of liquid hazardous waste is to burn them at sea. Such wastes account for only 8 percent of the 250 million metric tons of hazardous waste generated each year in the United States. Yet incineration of this relatively small amount can cause significant damage. Although incineration destroys much of the waste, some remains intact and enters the ocean. In addition, spills and leaks can occur as the waste is carried to a docked incinerating boat. Hazardous waste can be toxic, carcinogenic, or lethal to marine life. People may also be at risk if they consume contaminated fish.[52] Currently, the North Sea is the only area where maritime incineration of hazardous waste takes place.

Oil Pollution

Though not the problem it was in the 1970s, oil pollution is still a major concern. Experts estimate that each year between 3 million and 6 million metric tons of oil are discharged into the oceans from land- and sea-based sources, but they disagree as to the relative importance of the two sources. The Organization for Economic Cooperation and Development estimates that marine oil pollution is almost evenly divided between land- and sea-based sources, while many experts assert that up to 90 percent of the oil originates from land-based sources.[53]

The main source of sea-based oil pollution is the shipping industry, which discharges roughly 1.5 million metric tons of oil into the oceans each year. About a ton of oil is discharged for every thousand tons transported by sea. (See Figure 8.2.) While oil spills from tanker accidents gain news headlines, they account for less than a third of all the oil released. The bulk of seaborne oil pollution results from washing tankers out with seawater and releasing oily ballast water into the ocean.[54] Most land-based marine oil pollution comes from municipal and industrial wastes and from runoff.[55]

Oil pollution can kill or seriously harm marine life, including plankton, finfish, shellfish, and marine mammals and birds. When a Swedish tanker spilled 300 tons of oil in the Baltic Sea, some 60 tons of mollusks and crustaceans perished. In 1969, the discharge of some 150 tons of oil near the Netherlands killed 40,000 seabirds.[56] And in 1989, the Exxon Valdez disaster off Alaska spilled more than 10 million gallons of oil and killed at least 23,000 migratory birds, 730 sea otters, and 50 birds of prey.[57]

Oil pollution is especially serious in regions where oil is produced, including the Persian Gulf, the Red Sea, and the Gulf of Aden. In the Kuwait region, where there are more than 800 oil wells on the seabed, spillage is probably the greatest in the world. Some beaches in these oil-producing regions are soiled beyond recovery.[58]

Discarded Plastics

In general, most plastics enter the marine environment through ocean garbage dumping and from ships that throw their trash overboard. There is no estimate of how much plastic the oceans contain, but each year more than 6 million metric tons of shipboard litter are tossed into the ocean, and over 5 million plastic containers are discarded each day.

Commercial fishermen often throw old nets and other fishing gear overboard, which account each year for about 136,000 metric tons of plastics dumped into the seas.[59] In a three-hour cleanup sweep covering 157 miles of Texas coast in 1987, volunteers collected 307 tons of litter; two thirds of it was plastics and included 31,733 bags, 30,295 bottles, and 15,631 six-pack rings.[60]

Discarded plastics pose a deadly threat to marine life. Up to 2 million seabirds and 100,000 marine mammals, including some 30,000 fur seals, perish each year after eating or becoming tangled in plastics. Sea turtles, for example, choke to death on plastic bags they mistake for jellyfish.[61] Birds eat plastic pellets floating on the surface, mistaking them for fish eggs or larvae. A recent study on Midway Island in the Pacific found that 90 percent of the shearwaters and albatross chicks examined had plastic debris in their digestive system.[62] Some marine mammals become trapped and starve while retrieving fish from old discarded nets, while young seals become entangled while playing with fishing nets.[63] With the production of plastic doubling about every 12 years, the amount of this litter in the oceans is bound to increase.[64]

Figure 8.2
Location of Visible Oil Slicks, 1980s

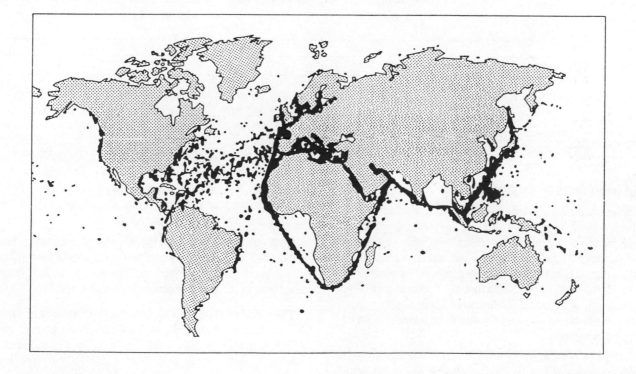

Source: International Oceanographic Commission (UNESCO), as published in Organization for Economic Cooperation and Development (OECD), *The State of the Environment* (Paris: OECD, 1985), p. 76.

Table 8.2
Major Pollutants Affecting U.S. Coastal Waters

Pollutant	Source	Effects
Nutrients, including nitrogen compounds	Fertilizers, sewage, acid raid from motor vehicles and power plants	Creates algae blooms, destroys marine life
Chlorinated hydrocarbons: pesticides, DDT, PCBs	Agricultural runoff, industrial waste	Contaminates and harms fish and shellfish
Petroleum hydrocarbons	Oil spills, industrial discharge, urban runoff	Kills or harms marine life, damages ecosystems
Heavy metals: arsenic, cadmium, copper, lead, zinc	Industrial waste, mining	Contaminates and harms fish
Soil and other particulate matter	Soil erosion from construction and farming; dredging, dying algae	Smothers shellfish beds, blocks light needed by marine plant life
Plastics	Ship dumping, household waste, litter	Strangles, mutilates wildlife, damages natural habitats

Source: Adapted from *Newsweek*, August 1, 1988, p.45.

Overharvesting of Marine Resources

Because of excessive exploitation, populations of both ocean fish and marine mammals have been seriously depleted in some regions.

Depletion of Fish Stocks

Of the nearly 20,000 known species of fish, about 9,000 are currently harvested, but only 22 species are regularly caught in significant quantities. Just six groups— herrings, cods, jacks, redfishes, mackerels, and tunas—account for nearly two thirds of the total annual catch.[65] Between 1950 and 1970, the world's fish catch rose by 7 percent annually, more than tripling from 21 million to 66 million metric tons. This steady increase was largely due to technical advances in fishing equipment.

In 1972, however, the world's catch suddenly dropped over 6 percent, due largely to the collapse of the Peruvian anchovy fishery. Over the next decade, the catch only increased by about 2 percent a year, reaching 77 million tons in 1983. Some analysts suggest that the slower growth in the 1970s can be attributed to such factors as sharply higher fuel

costs, economic recession, and the depressed Peruvian anchovy industry.[66] Since 1983, the catch has grown by about 4 percent annually, reaching 91 million metric tons in 1986. Of this, 7 million tons were produced by aquaculture. The remaining total of 84 million tons is approaching what FAO believes may be the maximum sustainable yield—100 million tons. This level of catch could be surpassed in 1991 if the rate of increase continues along its current path.[67] (See Figure 8.3 on the following page.)

Although the world fish catch grew substantially between 1983 and 1986, much of the gain was in low-value species used for animal feed and conversion to fish meal rather than in the more desirable high-value species.[68]

The world totals do not include the estimated 24 million metric tons caught annually by local fishers for private use or for sale in their communities.[69] Of the total marine catch, roughly a third is used to feed animals and fertilize croplands.[70]

In spite of recent growth in the world fish catch, the yields of some major fisheries, especially in the North Atlantic, have leveled off or are even declining. A 1983 FAO report

Figure 8.3
World and Regional Fish Catches, 1950-87

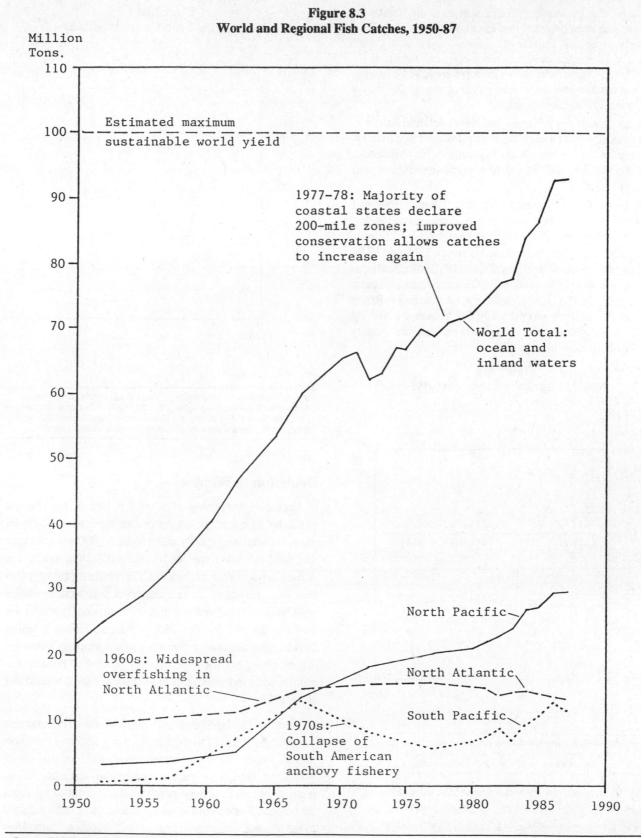

Million
Tons.

Estimated maximum sustainable world yield

1977-78: Majority of coastal states declare 200-mile zones; improved conservation allows catches to increase again

World Total: ocean and inland waters

North Pacific

North Atlantic

South Pacific

1960s: Widespread overfishing in North Atlantic

1970s: Collapse of South American anchovy fishery

Source: *FAO Yearbook of Fishery Statistics* (Rome: Food and Agriculture Organization, various years); Paul Harrison and John Rowley, *Human Numbers, Human Needs* (London: International Planned Parenthood Federation, 1984), p. 35.

concluded that the stocks of four fisheries in the Northwest Atlantic had been depleted and nine others were described as "fully exploited" (which often means overfished), with their yield reduced below their biological maximum.[71] Since 1982, North Atlantic yields have averaged less than 14 million tons a year, below the peaks of around 16 million tons in the 1970s. (See Figure 8.3.)

In addition to the Peruvian and North Atlantic fisheries, other fisheries that have collapsed or declined due to over-exploitation include the California sardine, North Sea cod, Alaskan pollack, and the South African pilchard and anchovy fisheries.[72]

When overharvesting of a particular fishery is followed by its collapse, the chances for eventual recovery to earlier yields may be slim. In the case of the Peruvian anchovy fishery, the catch more than tripled between 1969 and 1970, peaking at more than 13 million tons in 1970 before collaps-ing to less than 2 million tons by 1973 and remaining below 2 million until 1986.[73] (See Figure 8.4.) According to Roger Revelle, the collapse was probably due to both overfishing and climatic disturbances that followed an abnormally warm El Nino current in 1972.[74]

Figure 8.4
Peruvian Anchovy Catch, 1960-1987

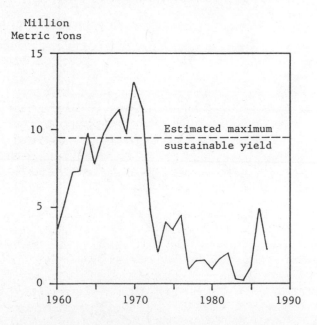
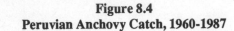

Million Metric Tons

Source: Lester R. Brown, "Maintaining World Fisheries," in Brown, et al., *State of the World, 1985* (New York: Norton, 1985), p. 79; *FAO Yearbook of Fishery Statistics* (Rome: FAO, various years).

More recently, the harvest of Alaskan king crabs grew rapidly during the 1970s and peaked in 1980 at nearly 84,000 tons before falling to less than 7,000 tons in 1985. By 1987, the catch had only recovered to about 13,000 tons.[75] (See Figure 8.5.)

Figure 8.5
Alaska King Crab Catch, 1978-1987

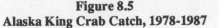

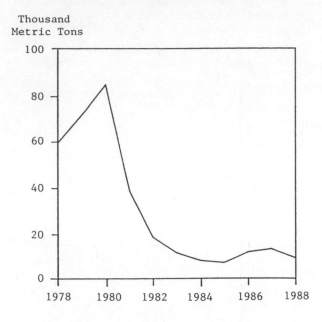

Thousand Metric Tons

Source: Conservation Foundation, *State of the Environment: A View Toward the Nineties* (Washington, DC, 1987), p. 311; National Marine Fisheries Service, *Fisheries of the United States* (Washington, DC: National Oceanic and Atmospheric Administration, various years).

Depletion of Whales

Excessive harvesting of whales is one of the clearest examples of overexploitation of marine resources. At its peak, the whaling industry killed some 66,000 whales a year and depleted many species to near extinction. Today the Antarctic population of the giant blue whale is probably less than one percent of its original level. Humpback whales have been depleted to about three percent, and fin whales are down to 20 percent. In 1985, the International Whaling Commission imposed a five-year moratorium on commer-cial whaling, but the IWC has no powers to enforce the moratorium, and more than 11,000 whales have been killed since 1985.

Japan and Iceland continue to hunt whales. The Japanese have recently turned to the smaller and relatively common minke whale. Southern waters may still contain about 400,000 minkes, but in the areas where the Japanese whalers are most active the minke population has dropped by about 50 percent. For centuries, whale stocks have been reduced to the point of "commercial extinction" because demand has exceeded the whales' ability to reproduce. Now only the minke whale remains in reasonable numbers.[76]

Protecting the Marine Environment

Relatively few successes in the management and protection of marine resources can be cited, but there are hopeful signs. Fishery management is improving, and new international and regional initiatives have been launched to control pollution and protect marine environments. A number of international treaties and agreements are being implemented, and the concept of a common marine heritage, though weaker than initially conceived, is still part of the Law of the Sea treaty.

Pollution Control Treaties

Several international and regional initiatives are designed to counter the flow of pollution into the world's oceans. The Law of the Sea, MARPOL, and the London Dumping Convention all aim to limit the disposal of various wastes in international waters, while the International Maritime Organization works to control both navigation and pollution in congested waterways. The United Nations Environment Program (UNEP) created the Regional Seas Program to control pollution in specific regions, most notably the Mediterranean Sea. Various other treaties are designed to control ocean pollution in specific regions.

Law of the Sea Convention   In the 1960s, numerous commentators proposed that the oceans be recognized as a "common heritage of mankind," managed by an international body such as the United Nations. This spirit eventually led to the Third United Nations Conference on the Law of the Sea, which first convened in 1973 and which completed the Law of the Sea (LOS) Convention in 1982. The Convention's goal was to develop an all-encompassing treaty covering marine issues, ranging from fishing and mining to navigation and pollution. More than 150 nations participated, hoping to establish a global order that would regulate use of the oceans. But they failed to agree on such a comprehensive treaty. The concept of a marine "common heritage" was confined to deep seabeds, where a potential wealth of minerals lies.[77]

The Treaty states that all nations should benefit from the deep seabed resources, mined under the supervision of an International Seabed Authority. However, this provision split the conference; industrial nations resisted the idea of sharing their ocean mining technology with developing nations.[78] They favor the concept that the deep seabeds are common to all nations, but maintain that the resources should be available on a first-come, first-served basis, with a minimum of regulation.[79] The United States, the United Kingdom, West Germany, and several other major seabed-mining nations have not accepted international development of seabed resources, and instead have enacted unilateral legislation to permit national mining of seabeds. Except for the seabed-mining provision, the United States endorses the Law of the Sea Treaty.[80]

By 1989, the LOS Convention had been signed by 159 nations and ratified by 39. But 60 nations must ratify the treaty to bring it into force; it appears unlikely that this will happen before 1990 at the earliest.[81]

In addition to addressing the issue of seabed mining, the LOS Convention also established Exclusive Economic Zones (EEZ). This provision recognizes that a nation has complete economic sovereignty over the oceans within 200 miles of its coast; foreign ships must have permission to enter an EEZ for economic purposes. Even though the United States has not signed the Convention, it declared its own EEZ in 1983.[82] These coastal zones are particularly important to the preservation of fish stocks, since they now account for 40 percent of the world's oceans, and are controlled by the coastal states. Such uniform control should make the supervision and management of fisheries easier.[83]

The LOS Convention upholds the traditional notion of "freedom of the seas" for the remaining 60 percent of the ocean, although of this area, 42 percent is deep seabed, which under the treaty would be "common heritage of mankind," to be controlled by an International Seabed Authority.[84] A final provision of the LOS is that nations are obliged to protect and preserve their marine resources, recognizing such protection to be for the common good of humanity.[85]

MARPOL Convention   In 1973, the International Convention for the Prevention of Pollution from Ships, known as MARPOL, was formulated to control all forms of pollution from ships. For example, MARPOL established minimum distances from land for dumping sewage, garbage, and toxic waste.[86] A separate annex to the Convention specifically prohibits ocean disposal of plastics, limits the disposal of other garbage, and requires ports to provide facilities for receiving trash from incoming ships.

Because the annex lacked enough ratifying votes, it languished until 1987, when the U.S. Congress, under pressure from numerous governmental agencies and environmental groups, finally ratified the annex, allowing it to take effect. Beginning in 1989, it is illegal for ships of a signatory party state and all ships within the waters of a signatory party to discard plastics into the ocean. However, much ocean dumping of plastics is attributed to military vessels. Only by 1994 will U.S. military ships be required to comply with the annex.[87]

Through strict limits or outright bans, MARPOL has successfully reduced the amount of oil discharged from ships. New technology has made it possible to identify ships that discharge more than the allotted amount of oil, and offenders can be fined. As a result, less oil is being discharged even though more and bigger oil tankers are at sea.[88]

London Dumping Convention   This convention, forbidding the dumping of certain kinds of waste from ships and aircraft, was implemented in 1975. Signed originally by 33 nations, the treaty covers all the world's oceans beyond national jurisdictions, and as of 1988 had nearly 60 signatory parties. The treaty specifically bans the dumping of "black-listed" substances such as heavy metals, petroleum products, and carcinogens. These substances may be dumped only in "trace" quantities. In addition, a "gray list" of substances including lead, cyanide, and pesticides may be dumped only with specific permits. As noted earlier, since 1983 there has also been a moratorium on dumping low-level radio-active wastes under the LDC.[89]

International Maritime Organization   This organization has developed various treaties to improve marine safety and control pollution. Among the issues it covers are liability and compensation from oil spills, the charting and removal of offshore oil rigs, the designation of marine areas for special protection, and plastic debris in the water.[90]

Regional Seas Program   Begun in 1974 under UNEP, the Regional Seas Program focuses on cleaning up specific marine areas. Starting with the successful Mediterranean Action Plan, the Program has created similar initiatives in nine other regions. Participants in the Program include more than 120 nations, 14 UN agencies, and a dozen other international organizations. Partly as a result of the region's Action Plan, Mediterranean waters and beaches are now cleaner and shellfish are less contaminated than a decade ago. Mediterranean nations have adopted a series of common measures to limit discharge of mercury and other industrial pollutants.[91]

Other Regional Treaties   As one of the earliest comprehensive marine protection agreements, the Helsinki Convention of 1974 served as a model for the UNEP's Regional Seas Program. The convention was the first to focus on both sea- and land-based pollution in the Baltic Sea. Earlier initiatives had been limited to seaborne pollution, such as the Bonn Agreement of 1969, which covers oil pollution in the North Sea.[92] The Oslo Convention was signed in 1972 by the major North Sea nations; it is designed to limit waste disposal in the North Sea and the Northeast Atlantic Ocean.[93]

In addition to international agreements and regional programs to protect marine resources, there are signs of progress from country programs. For example, Singapore now has modern sanitation facilities for 97 percent of its residents and recycles its waste water for industrial use. As a result, bacteria levels in its coastal waters have dropped sharply since 1980.[94]

U.S. Marine Protection Measures

In the United States, a number of laws have been implemented as awareness of marine issues has risen in the last 30 years. U.S. efforts to protect marine resources include the Coastal Zone Management Act, the Clean Water Act, the Marine Mammal Protection Act, the Ocean Dumping Act, and the National Marine Sanctuary Program.

Coastal Zone Management Act   This legislation, adopted by Congress in 1972, is designed to curb ecological deterioration of coastal areas due to population pressure and development. To improve the management of coastal zones, state governments set up cooperative efforts, with support from the National Oceanic and Atmospheric Administration. Of the 35 eligible states, 28 had developed federally approved plans by 1985. Implementation of the act has been controversial, however, due to inconsistencies between federal and state regulation of activities such as offshore oil and gas leasing and because of conflicting proposals for funding coastal zone management projects.[95]

Clean Water Act   Passed in 1977, this legislation is an amended version of the 1972 Federal Water Pollution Control Act. It includes provisions to control the discharge of pollutants in inland and coastal waterways, and is administered by the Environmental Protection Agency. One of its functions is to provide grants for cities to improve their sewage treatment facilities to meet established standards. This program has been hampered, however, by ill-planned urban growth and inadequate sewage treatment facilities. The act itself was flawed as well; for example, it did not require pollution control for storm sewers. In addition to collecting rainwater, storm sewers also gather oils, fertilizers, and pesticides, funneling these pollutants into city waterways and eventually into bays and estuaries.[96] The reauthorization of the Clean Water Act passed the Congress unanimously, but was repeatedly vetoed by President Reagan; Congress overrode the veto in 1987. (See Chapter 9.)

Ocean Dumping Ban Act   In 1988, the United States enacted legislation mandating a phaseout of ocean dumping of industrial waste and sewage sludge. The act is designed to end dumping through implementation of alternative waste management methods and requires current dumpers either to end ocean dumping by 1991 or pay escalating permit fees as long as dumping continues.[97]

Marine Mammal Protection Act   Devised in 1972, this act originally was passed as an attempt to reduce the killing of porpoises in tuna nets. Often swimming in schools of yellowfin tuna, porpoises become tangled or maimed in the nets and either perish in the water or are hauled onto boats, where they die. The Department of Commerce issued regulations requiring stricter measures to reduce the number of porpoises caught in tuna nets, and placed federal observers aboard tuna boats. Since 1972, these regulatory attempts have been challenged and violated, reversed and reinstated.

In addition to porpoises, each year hundreds of thousands of other sea creatures, including birds, seals, sea turtles, and whales are killed accidentally by fishing nets. Many also perish in discarded nets and other fishing gear that float freely in the oceans. An estimated 145,000 trawl net fragments are discarded annually in the waters off Alaska alone. During reauthorization hearings for the act in 1984, sentiment on this issue was so high in Congress that both houses passed bills directing the National Marine Fisheries Service to investigate the problem of entanglement and seek possible solutions. These might include economic incentives to deter fishers from discarding old nets, and the development of biodegradable nets.[98]

National Marine Sanctuary Program　Established by Congress in 1972 as a part of the Marine Protection, Research and Sanctuaries Act, this program creates marine sanctuaries for ecological, recreational, scientific, or aesthetic values.[99] Seven sanctuaries had been created by 1985, including the 342-hectare Channel Island National Marine Sanctuary off the Southern California coast. In recent years, oil drilling and fishing interests have slowed the process of creating marine sanctuaries; none has been established since 1985.[100]

Regional and State Initiatives　Some of the most important initiatives to protect U.S. marine resources are being taken at the regional and state level. In response to steadily deteriorating conditions in the Chesapeake Bay, the states of Maryland, Pennsylvania, and Virginia, along with the District of Columbia, recently signed an agreement to reverse the Chesapeake's decline by reducing runoff and the discharge of effluents into the Bay.[101]

The governor of New Jersey has proposed an ambitious plan to limit ocean pollution, involving the cooperation of municipal, state and federal governments and projecting expenditures of $200 million over five years. More than half the money would be spent to curb runoff from communities and farmlands.[102]

Washington is one of the few U.S. states with a comprehensive marine pollution cleanup program, begun in 1985. The Puget Sound water quality authority and other state agencies monitor discharge of industrial waste, work with businesses to reduce effluents, and are developing ways to limit urban and agricultural runoff. The state legislature increased cigarette taxes by 8 cents a pack to help fund the cleanup; in 1988, the tax contributed more than $20 million to the program.[103]

In the Delaware River estuary, pollution from heavy industry and the region's several million residents had become so severe by the 1940s that fish populations disappeared and airplane pilots could smell the river at 5,000 feet. A major cleanup effort was begun by the Federal Government and the four states in the region, and now shad and more than 30 other fish species have returned to the river.[104]

Managing Fisheries

A number of measures are needed to protect ocean fisheries from overharvesting and to ensure a sustainable harvest in the future. These include setting fishing limits, altering fishing methods, reducing waste, and expanding aquaculture.

Setting Fishing Limits　To improve long-term prospects for a sustainable marine harvest, nations need to agree on realistic and strictly enforced fish catch quotas and impose some moratoriums before fish stocks collapse. Seasons can be established in which certain fish can only be caught during a specific period. Instead of simply concentrating on the maximum sustainable yield of individual fish species, nations can adopt multi-species management strategies based on ecological interactions within ocean communities.[105]

Altering Fishing Methods　To prevent the capture of fish too small for commercial use, minimum net mesh sizes can be set, allowing immature fish to live and grow. The introduction of biodegradable nets and other fishing gear would also reduce the accidental trapping and death of marine mammals.[106]

Reducing Waste　Significant improvements can be made in reducing fish wastage by working to improve storage and sanitation conditions on fishing vessels, by locating fish processing plants closer to fishing areas, and by encouraging the salvaging of less desirable, lower-value fish. For example, in the pursuit of commercially valuable fish such as shrimp, tons of lower-value fish are inadvertently caught in the trawls. These less valuable fish are simply discarded, often dying from shock, injury, or suffocation. It is estimated that more than four million metric tons of lower valued species caught along with the desired fish are discarded annually from shrimp boats alone. Especially in hotter climates, valuable fish catches are lost to spoilage and to insects before the catch can be brought to market.[107]

To develop a more sustainable fishing industry, some of the conflicts of interest associated with marine fishing need to be addressed. Tensions arise when local traditional fishing communities feel their way of life threatened by "fishing factory" ships that exploit large areas of the ocean. A related source of conflict exists between people who depend on fish for part of their daily diet and those who process fish into oils, fish meal, and other fish by-products.[108]

Expanding Aquaculture   The farming of fish and other marine organisms is an ancient occupation, but only recently has its potential in the modern world been realized. Aquaculture has increased steadily during the past 20 years; in 1980, 8.7 million metric tons of seafood and edible seaweed were produced, nearly one-eighth of the total freshwater and marine-captured harvest that year. Advances in salmon production have been especially rapid; aquaculture now produces a third of a million tons of salmon yearly, more than half the oceanic catch.[109]

In developing countries, aquaculture can provide valuable protein, and for high-valued exports such as shrimp, can be a source of foreign exchange. Third World nations are indeed leading the way in this field; in 1980, they produced 74 percent of the aquaculture harvest. Aquaculture is usually labor intensive and therefore well suited to densely populated, low-income nations. It can be practiced in ponds, estuaries, and bays, or in open ocean close to shore. However, to begin an aquaculture program requires considerable capital and scientific knowledge, which are in short supply in developing nations. Both financial and technical challenges must be met to fulfill aquaculture's potential as a major food source in the twenty-first century.[110]

What You Can Do

Inform Yourself

To find out more about ocean and coastal issues, consult sources and contact organizations listed under Further Information. If you live or vacation in a coastal area, find out what agencies in the local government have responsibility for coastal issues, and how their current programs affect coastal resources. Learn how the Office of Coastal Zone Management in the U.S. Commerce Department handles your state's coast.

Join With Others

If you are in a coastal area, become active in your local civic association on coastal issues. Join a local group or a national organization with a program on ocean and coastal issues; learn about current priorities, legislation, and projects related to coastal issues. Persuade other organizations to include marine issues among their priorities.

- Support organizations, policies, and programs that seek to ensure the preservation, protection, and responsible use of ocean and coastal resources of the United States and other countries.

- Work with local fishermen and boaters on the problem of dumping plastics and other materials into the ocean.

- Support efforts to protect remaining undeveloped coastline from future development by ordinance or

through purchase.[111] Encourage plans to set up marine reserves.

- Work to halt the addition of industrial pollutants to sewage-treatment systems; urge reduced reliance on water for carrying and assimilating human wastes, and development of productive uses for these wastes.[112]

- Attend meetings that address coastal issues such as waste disposal in the ocean; industrial and power plant siting; wetlands protection; coastal access by the public; development of oil, gas, and mineral reserves; and the prevention and clean-up of oil spills and discharges. Follow activities in the U.S. Congress that relate to these concerns.

- With the assistance of a national marine environmental organization (see list in Further Information), locate a "sister city" in a developing country that is concerned with preserving a bay, estuary, coast, or marine animal species. Exchange speakers, educational materials, and funds to promote clean-up initiatives and protect endangered wildlife.

Review Your Habits and Lifestyle

If you live near a coastal region, there are many things you can do in your home and community to help protect marine resources and raise your awareness of the marine environment.

- Take up water sports like snorkeling and scuba diving. Work with local diving clubs to promote reef conservation. Go on whale watches and oceanic birding trips.

- Boycott buying rare shells and coral (especially black coral).

- Visit aquariums and attend their presentations. Consider holding special functions such as receptions, parties, or school dances in an aquarium, using "The Aquatic World" as your theme.

- In your home, avoid using or disposing of substances that could damage the marine environment. For example:

  » use phosphate-free detergents;

  » dispose of unwanted hazardous household chemicals such as pesticides in sanitary landfills, or take advantage of approved community collection points;

  » compost your food wastes or put them in the trash instead of the garbage disposal.

- Avoid unnecessary use of water in your home and on your property. (See the What You Can Do section of the Fresh Water chapter for specific suggestions.)

- On your property, take measures to protect marine resources. For example:

  » if you have a septic waste system, be sure it is emptied periodically and properly maintained;

  » never discard used motor oil or antifreeze in a storm sewer; instead, take them to a gas station with an oil recycling program;

  » landscape your garden to minimize rainwater runoff and soil erosion; select plants, trees, and shrubs suitable for the soil in your yard; don't choose plants that need lots of watering;

  » avoid using liquid fertilizers or weed-killers on your lawn;

  » use non-toxic methods to control weeds and insects in your garden and lawn; for example, rotate crops, hand-pick insect pests or dislodge them with a water spray, introduce natural bacteria and insect parasites that will kill pests without harming other organisms;

  » if you have a swimming pool, drain it only when necessary, and then only onto a large expanse of lawn to allow the chlorine to dissipate and the water to filter slowly through the soil.

- In your community, consider how you can help protect marine resources. For example:

  » encourage your local public works department to use porous asphalt and modular paving materials to reduce rainfall runoff from paved surfaces;

  » learn about your town's waste treatment facilities: how effective are they now, and how adequate will they be in the future?

- If you own or operate a boat in coastal areas,

  » protect shores by observing posted marine speed limits and not producing wakes within 500 feet of the shore;

  » dispose of boat sewage in on-shore sanitary facilities or into waters deeper than 20 feet;

  » don't throw trash overboard.

  » avoid spillage when using cleansers, paint, and antifouling compounds on your boat.

These and other suggestions are contained in the *Baybook*, available from the Alliance for the Chesapeake Bay (see Further Information section).

Work With Your Elected Officials

Contact your local, state, and national legislators, find out their positions on coastal and ocean issues, and tell them where you and the organizations you represent stand.

- Work for enforcement and strengthening of existing regulations governing ocean dumping of sewage and other wastes.[113]

- Support action to improve the water quality of the coastal zone by improving the identification and regulation of land-based pollutants entering coastal areas, and by using the leverage of federal aid to require sound conservation practices and progress in curbing marine pollution.[114]

- Urge the National Oceanic and Atmospheric Administration to complete its inventory of U.S. coastal resources, and to accelerate designation of national marine sanctuaries.

- Support efforts to achieve a sustainable fishing industry in the United States and work for expanded U.S. support to improve fisheries management in developing countries.

- Work for international compliance with the International Whaling Commission's moratorium on all commercial whaling, especially by Iceland and Japan.

- Work for United States ratification of the United Nations Law of the Sea Convention and increased support for the United Nations Environment Program.

Publicize Your Views

Express your views on ocean and coastal issues to opinion leaders in your community, and to the print and electronic media through letters to the editor and public service announcements. Create displays and presentations, and organize community activities to raise public awareness about the threats to marine resources and what can be done to protect them.

- Hold a "Plague of Plastic" beach clean-up to draw attention to plastic pollution in the marine environment.

Raise Awareness Through Education

Find out whether schools, colleges, and libraries in your area have courses, programs, and adequate resources on marine issues; if not, offer to help develop or supply them.

Encourage educators and conservation groups to initiate classes, seminars, lectures, field trips, and other programs that increase awareness of facts and policies concerning ocean and coastal resources.

Consider the International Connections

If you travel to another coastal country, take the opportunity to learn firsthand about marine issues there. Go on diving trips to the Caribbean or Mexico, for example. Visit areas with mangroves and coral reefs and observe their diversity of marine life. Learn what effects coastal development is having on these resources, and what efforts are under way to protect them. Try to contact local organizations with programs to protect marine resources and express your support for their efforts. When you return home, contact U.S. conservation groups and discuss possible links with groups in the country you visited.

Further Information

A number of organizations can provide information about their specific programs and resources related to ocean and coastal issues.*

Books

Baybook: A Guide to Reducing Water Pollution at Home. 1986. 32 pp. Paperback. Available from Alliance for the Chesapeake Bay, 6600 York Road, Baltimore, MD 21212. Single copies free, multiple copies $1 each. Explains how to prevent damage to inland and coastal resources from soil erosion, sewage, pesticides, and household chemicals.

Bullock, David K. *The Wasted Ocean: The Ominous Crisis of Marine Pollution and How to Stop It.* An American Littoral Society publication. New York: Lyons and Burford, 1989. 150 pp. Paperback, $9.95.

Carson, Rachel. *The Edge of the Sea.* Boston: Houghton-Mifflin, 1955.

Center for Marine Conservation (CMC). *A Citizens Guide to Plastics in the Ocean: More Than a Litter Problem.* Washington, DC, 1988. 131 pp. Available from CMC for $1.50 for postage and handling.

Clark, John. *Coastal Ecosystem Management.* Washington, DC: The Conservation Foundation, 1977. 188 pp. Paperback, $7.50.

Cousteau, Jacques-Yves. *The Cousteau Almanac: An Inventory of Life on Our Water Planet.* Garden City, NY: Doubleday, 1981. 838 pp. Paperback, $19.95.

Culliny, John L. *The Forest of the Sea.* San Francisco: Sierra Club Books, 1976.

Hansen, Nancy R. and others. *Controlling Nonpoint-Source Water Pollution.* A Citizen's Handbook. Washington, DC and New York: The Conservation Foundation and National Audubon Society, 1988. 170 pp. Paperback.

Millemann, Beth. *And Two If By Sea: Fighting the Attack on America's Coasts.* A Citizen's Guide to the Coastal Zone Management Act and Other Coastal Laws. Washington, DC: Coast Alliance, 1987. 110 pp. $5.00.

Myers, Jennie. *America's Coasts in the '80s: Policies and Issues.* Washington, DC: Coast Alliance, 1981.

Natural Resources Defense Council. *Ebb Tide for Pollution: Actions for Cleaning Up Coastal Waters.* New York, 1989. 43 pp. Paperback, $7.00 plus $1.50 postage and handling. Discusses the importance of coastal resources, threats to U.S. coastliines, and ways to stop coastal pollution.

Pontecorvo, Giulio (ed.). *The New Order of the Oceans: The Advent of a Managed Environment.* New York: Columbia University Press, 1986. 277 pp. $30.00.

Simon, Anne W. *Neptune's Revenge.* New York: Franklin Watts, 1984. 224 pp. $15.95.

Teal, John, and Mildred Teal. *Life and Death of a Salt Marsh.* New York: Little, Brown, 1969.

U.S. Environmental Protection Agency. *Marine and Estuarine Protection: Programs and Activities.* Washington, DC, February 1989. 42 pp.

Weber, Michael and Richard Tinney. *A Nation of Oceans.* Washington, DC: Center for Environmental Education, 1986. 95 pp. Paperback, $8.95.

Weber, Michael et al. *The 1985 Citizen's Guide to the Ocean.* Washington, DC: Center for Marine Conservation, 1985. 180 pp. Paperback, $4.95.

---

* These include: American Cetacean Society (marine wildlife), American Littoral Society, Center for Marine Conservation (marine wildlife, sanctuaries, debris), Chesapeake Bay Foundation, Coast Alliance (coastal issues), Council on Ocean Law, Environmental Defense Fund, Environmental Policy Institute, Friends of the Earth, Global Tomorrow Coalition, Greenpeace USA (ocean issues), International Marinelife Alliance USA, Izaak Walton League, National Audubon Society, National Coalition for Marine Conservation, National Parks and Conservation Association, National Wildlife Federation, Natural Resources Defense Council, Oceanic Society, Sierra Club, World Resources Institute, Worldwatch Institute, and World Wildlife Fund. Addresses for these organizations are listed in the appendix.

Articles, Pamphlets, and Brochures

Bailey, Richard. "Third World Fisheries: Prospects and Problems." *World Development*, Vol. 16, No. 6, June 1988, pp. 751-57.

Concern, Inc. "The Oceans;" "Wetlands: Where Life Begins;" "Barrier Islands." Washington, DC. Each item 30 cents per copy.

Catena, John G. *Environmental Policy Considerations for Ocean Mining and the Law of the Sea*. Washington, DC: Oceanic Society, March 1988.

Cummins, Joseph E. "The PCB Threat to Marine Mammals." *The Ecologist*, Vol. 18, No. 6, 1988, pp. 193-95.

Curtis, Clifton E. *Congressional Testimony Concerning Environmental Conditions and Trends in Marine and Near-Coastal Waters*. Washington, DC: Oceanic Society, April 1988.

"The Dirty Seas." *Time*, August 1, 1988. pp. 44-50.

"Don't Go Near the Water." *Newsweek*, August 1, 1988. pp. 42-48.

Lenssen, Nicholas. "The Ocean Blues." *World Watch*, Vol. 2, No. 4, July-August 1989. Oceans are the source of food, livelihood, and meteorological balance, but they are reeling from human assault.

Myers, Norman (ed.) "Ocean." In *Gaia: An Atlas of Planet Management*. Garden City, NY: Anchor Books, 1984, pp. 68-99.

Nierenberg, William A. "The Oceans." In *Earth '88: Changing Geographic Perspectives*. Washington, DC: National Geographic Society, 1988, pp. 284-299.

Organization for Economic Cooperation and Development (OECD). "Marine Environment." In OECD, *The State of the Environment 1985*. Paris: OECD, 1985, pp. 69-91.

Ratiner, Leigh. "The Law of the Sea: A Crossroads for American Foreign Policy." *Foreign Affairs*, Vol. 60, No. 5, Summer, 1982.

Ray, G. Carlton. "Ecological Diversity in Coastal Zones and Oceans." In E.O. Wilson (ed.), *Biodiversity*. Washington, DC: National Academy Press, 1988, pp. 36-50.

Revelle, Roger. "Present and Future State of Living Marine and Freshwater Resources." In Robert Repetto (ed.), *The Global Possible: Resources, Development, and the New Century*. New Haven: Yale University Press, 1985, pp. 431-455.

United Nations Environment Program. "The State of the Marine Environment." *UNEP News*, April, 1988.

Weiskopf, Michael. "Plastic Reaps a Grim Harvest in the Oceans of the World." *Smithsonian*, March, 1988. pp. 58-66.

World Resources Institute and International Institute for Environment and Development. "Oceans and Coasts." In *World Resources 1986*, pp. 141-160; *World Resources 1987*. pp. 125-142; *World Resources 1988-1989*, pp.143-161; New York: Basic Books, 1986, 1987, and 1988, respectively. All volumes paperback, $16.95.

Periodicals

Ocean Watch. A monthly newsletter published by the Oceanic Society. Subscription $15 a year.

Films and Other Audiovisual Materials

Are You Swimming in a Sewer? NOVA examines the effects of waste mismanagement on marine life, the costs of cleaning up damaged urban waters, and alternatives for waste management. 58 min. Video ($250), Rental ($125). Coronet MTI Film & Video.

Drowning Bay. 1970. 10 min. 16mm color film. Michigan Media.

Estuary. Uses underwater microphotography to portray the biologically important wetlands where the fresh water of a river meets the salt water of the ocean. 12 min. 16mm film ($250), Video ($65), Rental ($25). Bullfrog Films.

The Intertidal Zone. Explores the ecology of the area covered by the highest tides and exposed during the lowest; explains how pollution may affect intertidal food chains. 17 min. Grade 5-adult. 16mm film ($350), Video ($150), Rental ($35). Bullfrog Films.

The Living Ocean. 1988. Describes how the oceans formed, how they maintain salinity and affect weather and climate, and how humans affect the ocean. 25 min. 16mm film ($384), Video ($235). National Geographic Films.

The Salt Marsh. 1975. 22 min. 16mm color film. Encyclopedia Brittanica Educational Corp.

Wellsprings. 1976. 58 min. 16mm color film. Rental ($11.95). Michigan Media.

Wetlands: Our Natural Partners in Wastewater Management. 1980. 38 min. 16mm color film. Rental ($10.20). Michigan Media.

Where the Bay Becomes the Sea. Depicts the abundance and diversity of life where the Bay of Fundy meets the Atlantic Ocean. 30 min. Grade 7-adult. 16mm film ($550), Video ($150), Rental ($50). Bullfrog Films.

Teaching Aids

Global Tomorrow Coalition. *Marine Resources Education Packet*. Includes background information, lesson plans, and activities emphasizing interactive learning aimed at upper elementary and intermediate students; can be upgraded for high school. $7.00 including postage.

World Wildlife Fund. *Coral Reef Teacher's Kit*. Provides junior high teachers with background information on coral reefs, their importance as an ecosystem, and the wildlife they support. Contains teacher's guide, coloring book, poster, and color slides. 1986. $15.00.

Fresh Water

Of all environmental ills, contaminated water is the most devastating in consequences. Each year 10 million deaths are directly attributable to waterborne intestinal diseases. One-third of humanity labors in a perpetual state of illness or debility as a result of impure water; another third is threatened by the release into water of chemical substances whose long-term effects are unknown.

—Philip Quigg, Water: The Essential Resource[1]

Our planet is shrouded in water, and yet 8 million children under the age of five will die this year from lack of safe water. The same irony will see 800 million people at risk from drought.... Two-thirds of the world's rural poor have no access to safe drinking water, and while millions are made homeless from floods, hundreds of millions are coping with drought.

—United Nations Environment Program[2]

Rigorous conservation methods are necessary to protect both our groundwater reserves and our surface waters...To achieve this, considerable changes are required in the operation of our industrial system.

—"Water Fit to Drink?," *The Earth Report*[3]

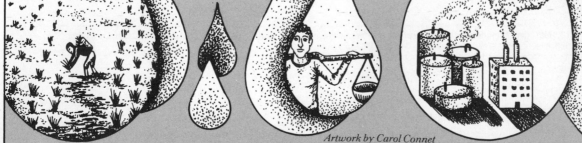

Artwork by Carol Connet

Major Points

- Adequate quantities of water are available to satisfy projected worldwide demands through the year 2000 if water is used wisely and efficiently. However, poor management, lack of adequate conservation, pollution, and rapid local increases in demand will create shortages in some areas.

- In developing nations, only about half of the people have access to safe drinking water. Worldwide, some 10 million deaths each year result from waterborne intestinal diseases.

- In industrial nations, surface and underground water supplies are being polluted by industrial and municipal wastes, and by surface runoff from urban and agricultural areas.

- Heavy demands for water by agriculture, industry, and municipalities are rapidly depleting groundwater supplies in China, India, the United States, and many other countries.

- To ensure adequate water supplies, major priorities include proper protection and management of watershed areas; creation of incentives for conservation, such as water prices that reflect the real costs of supply; and legislation that encourages water recycling.

- To ensure acceptable water quality, the highest priority should go to programs to reduce the generation of solid, liquid, and airborne wastes—especially toxic wastes—from industrial plants, mining and smelting operations, electric power production, cities and towns, and agriculture. In addition, there must be adequate containment and treatment of remaining wastes that can not be eliminated.

The Issue: Growing Water Demand, Declining Water Quality

Unlike fossil fuels or soil, fresh water is a renewable resource. If properly used and carefully conserved, the global hydrological cycle can meet current and anticipated fresh water needs on a sustainable basis.[4] However, problems of fresh water supply and water quality are of immediate and fundamental importance to all people. Population growth and rising requirements for energy and

Figure 9.1
World Water Use, Total and Per Capita, 1940-80

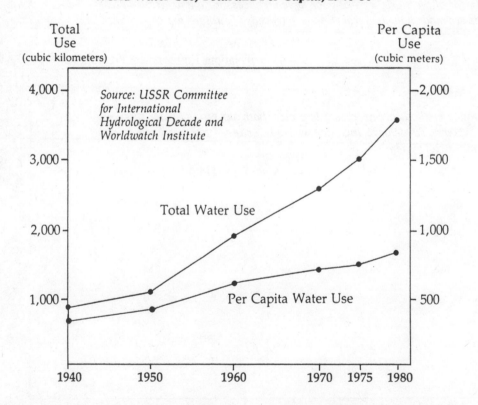

Source: Sandra Postel, *Water: Rethinking Management in an Age of Scarcity*, Worldwatch Paper 62 (Washington, DC: Worldwatch Institute, December 1984), p. 12.

food are placing greater demands on both the quantity and quality of fresh water supplies. Total world water use more than tripled between 1950 and 1980.[5] (See Figure 9.1 above.) In developing countries, problems of water pollution and depletion are dominant.

This chapter covers four aspects of the water issue:

- global supply and demand,

- water supply and sanitation in the Third World,

- water pollution in developed countries, and

- water resource management.

Global Supply and Demand

Water covers three-quarters of the earth's surface, but more than 97 percent of the Earth's water is saltwater in the oceans, and less than 3 percent is fresh water. Of the latter, 77 percent is frozen in polar ice caps and glaciers, 22 percent is groundwater, and the remaining small fraction is in lakes, rivers, plants and animals.[6] (See Figure 9.2.)

Through history, the world's lakes, streams, and rivers have provided important resources and services, including water for drinking, washing, agriculture, energy production, transportation, recreation, and waste disposal.[7]

Fresh water is very unevenly distributed around the world. (See Figure 9.3 on the following page.) Much of the Middle East, most of Africa, parts of Central America, and the western United States are already short of water.[8] Shortages can result from several factors, including limited supplies, heavy demands, and inefficient use. By the year 2000, many countries will have about half as much water per capita as they had in 1975—and many will experience much greater demands on water from agriculture and industry. Future water shortages are likely to limit growth in agriculture and industry, and could jeopardize health, nutrition, and economic development.

Water has been treated as an unlimited resource that is provided as cheaply as possible and in any quantity desired. If continued, this attitude will lead to critical deficiencies in the quantity and quality of available water. Population growth and rising requirements for energy and food are placing increased demands on fresh water supplies. To prevent shortages, nations must practice more efficient

Figure 9.2
Distribution of the World's Water

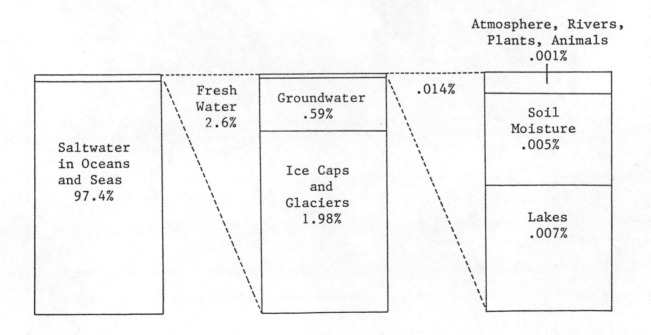

Source: David H. Speidel and Allen F. Agnew, "The World Water Budget," in Speidel et al., (eds.), *Perspectives on Water Uses and Abuses* (New York: Oxford University Press, 1988), Table 3.1, p. 28.

water management, introduce recycling, prevent pollution, and promote water conservation.[9]

Water is as important as food for human life. A person needs one or two liters of water daily to survive. The basic problem is not lack of drinking water—few people die of thirst. Instead, the problem is to obtain a sufficient supply of safe drinking water and adequate sanitation services. Recent studies show that the economic burden of disease and ill health, which in large part results from lack of water supply and sanitation facilities, is very great, particularly in the Third World. In a UNICEF survey of water supplies in developing countries, only 51 percent of the people had access to safe water.[10] Even in the United States, 15 million citizens drink from potentially unsafe water supplies, and another 30 million lack decent sanitation.

Where technology has allowed us to tap the vast underground water reservoirs of the world, we drain too many of them at rates that far exceed any hope for natural replenishment. In other cases, we drain and divert rivers for short-term, local benefits, oblivious to long-term ecological needs. As evidence of the effects of growing use, heavy demands for water are lowering underground water tables in China, India, and the United States, and shrinking the Caspian and Aral Seas in the Soviet Union.[11] Between 1950 and 1980, the amount of water drawn from U.S. lakes, steams, reservoirs, and underground aquifers increased by 150 percent, while the population increased by only half.[12]

Perhaps worst of all, the world's lakes and rivers receive enormous quantities of municipal sewage, industrial discharges, and surface runoff from urban and agricultural areas.[13] These wastes and chemicals are poisoning the natural groundwater and surface waters on which future generations depend for survival.

Much of the growing population in developing countries is moving out of rural areas. By the century's end, almost half of humanity will live in cities.[14] Providing water for such enormous populations will require greatly improved management of scarce water resources. Furthermore, with that urbanization and with greatly increasing industrialization, efforts to control water pollution must be greatly expanded if we are to protect water quality.

Figure 9.3
Global Water Surplus and Deficiency

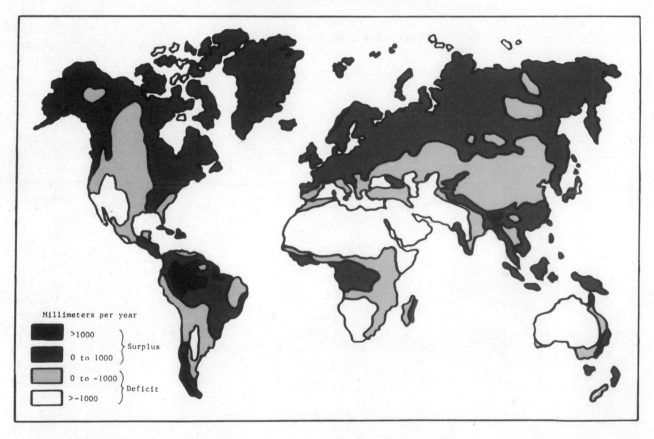

Note: Deficit areas receive less precipitation than that needed by well-established vegetation; surplus areas receive more.

Source: Malin Falkenmark, "Water and Mankind — A Complex System of Mutual Interaction," *Ambio*, Vol. 6, No. 1, 1977, p. 5.

Causes of Water Shortages

Growing populations, increasing use of water in agriculture and industry, and environmental degradation all contribute to shortages of fresh water. Worldwide, irrigation for agriculture accounts for about 73 percent of water use; 21 percent goes to industry, and the other 6 percent to domestic use.[15] In the United States, agriculture accounts for more than 80 percent of total water use.[16] Most cropland irrigation is highly inefficient, since much of the water is not absorbed by the crops and is lost by evaporation or seepage into the ground. Worldwide, irrigation systems, on average are only about 37 percent efficient.[17]

In 1980, total and per capita U.S. water withdrawals from surface and groundwater supplies were substantially greater than withdrawals for nine other nations included in a survey of water use compiled by Worldwatch Institute. Daily U.S. withdrawals were 7200 liters per person, compared with 4800 liters for Canada, 3600 for the Soviet Union, and only 1400 for the United Kingdom.[18] (See Table 9.1.)

Table 9.1
Estimated Water Withdrawals in Selected Countries, Total and Per Capita, Circa 1980

Country	Total (billion liters per day)	Per Capita (thousand liters per day)
United States	1,683	7.2
Canada	120	4.8
Soviet Union	967	3.6
Japan	306	2.6
Mexico	149	2.0
India	1,058	1.5
United Kingdom	78	1.4
Poland	46	1.3
China	1,260	1.2
Indonesia	115	0.7

Source: Sandra Postel, *Water: Rethinking Management in an Age of Scarcity*, Worldwatch Paper 62 (Washington, DC: Worldwatch Institute, December 1984), p. 16.

In the Southwestern Soviet Union, water withdrawals for irrigation have been largely responsible for a 40 percent decrease in area of the Aral Sea, formerly the world's fourth largest lake in area.[19]

In the United States, one-fifth of the water pumped out of the ground each year is nonrenewable.[20] In the West, rapid development and increasing use of irrigation have caused extensive water withdrawals from rivers and streams in the Lower Colorado and Rio Grande basins, and in much of California. To meet the demand, some states must bring in water from other areas or pump more from groundwater reservoirs. In many parts of the Great Plains and the Southwest, people are withdrawing groundwater more rapidly than it is being replenished. The Ogallala aquifer, which supplies water to one-fifth of all U.S. irrigated cropland, is already half depleted.[21] The annual overdraft of this aquifer is nearly equal to the flow of the Colorado River.[22] In the Southwest, water tables around the city of Tucson, Arizona, which depends exclusively on groundwater, have fallen about 150 feet in the last two decades.

Damage to a drawn-down aquifer can be irreversible. Groundwater overdraft can cause "subsidence," or irreversible settling of the land as the water is removed. In parts of California's San Joaquin Valley, where groundwater withdrawal for irrigation is heavy, the land has settled 27 feet.[23] A collapsed aquifer can never be replenished.[24] In Florida, groundwater overdrafts in some areas have caused the land to collapse and created numerous sinkholes.

In the United States, irrigation to grow food for livestock, including hay, corn, sorghum, and grass, accounts for about 50 percent of water consumption. Other farm uses, mainly irrigation of food crops, consumes another 35 percent, making all water use for agriculture account for more than 80 percent of total national water use.[25]

An enormous amount of water used for irrigation in western states is wasted, mainly because the cost of water is heavily subsidized and is essentially cost-free to farmers.[26] Waste of water is encouraged by the U.S. Bureau of Reclamation's policy which sells water at astoundingly low subsidized rates.[27] In one $500 million irrigation project near Pueblo, Colorado, irrigation water has been used to grow corn, sorghum, and alfalfa for cattle feed. According to a 1981 Government Accounting Office report, the full cost of water delivered was $54 per acre-foot, but farmers were charged only 7 cents per acre-foot.[28] The Bureau of Reclamation's own studies have shown that realistic water pricing can boost irrigation efficiency by 20 percent.[29]

In California, agriculture accounts for 85 percent of all water use. Billions of federal dollars have been spent since the 1940s to build the water storage systems that irrigate the state's crops, which are now worth $15 billion a year. Diversion of water for agriculture and other uses has dried up more than 90 percent of the seasonal wetlands that once supported migratory waterfowl, and salmon have disappeared from the San Joaquin River.[30]

The single biggest use of water in California is the irrigation of pasture for livestock. Cattle contribute only $94 million to that state's $500-billion economy—one five-thousandth of the total. Yet the livestock industry uses one-seventh of the state's water.[31] If farmers paid the full price for water, they would use it more carefully, and could not afford to grow low-value crops for livestock feed.[32]

Although the most severe water shortages in the United States generally occur in the Southwest, periodic droughts cause local and regional shortages in nearly every part of the country. For example, parts of the East Coast experience persistent groundwater deficits; in 1980, drought forced many mid-Atlantic communities to curtail water use. Seventeen western states have been declared "water-short."[33]

Energy production and many industrial processes require large quantities of water for steam production and cooling.[34] The production of oil, natural gas, and electricity can disrupt water supplies through destruction or contamination of groundwater aquifers, mixing of saline water with fresh water, and sulfur and fly ash fallout from power plants.

In the United States, water withdrawals increased 37 percent in the 1960s, and another 22 percent in the 1970s.[35] However, the trend of increasing water use has recently been reversed: in the first five years of the 1980s, water use was down 11 percent.[36] (See Table 9.2.)

Table 9.2
U.S. Water Withdrawals Per Day, 1940-1985

Year	Billion Gallons
1940	140
1950	180
1960	270
1970	370
1980	450
1985	400

Source: U.S. Bureau of the Census, *Statistical Abstract of the United States: 1988* (Washington, DC, 1987), p. 191.

In many Third World countries, deforestation of watersheds for commercial timber or fuelwood use, and grazing and cultivation practices in steep, high-rainfall zones reduce the soil's capacity to absorb rainfall, which increases flooding downstream and reduces the amount of water available during dry seasons.[37] Soil eroded from uplands prematurely causes siltation of reservoirs used for water storage and power generation.

In Africa, the sheer number of people seeking to survive on arid, marginal land is lessening the soil's ability to hold water, and may be literally drying out parts of the continent through deforestation, overcultivation, and overgrazing.[38]

Water Resource Conflicts

As water shortages increase, international rivers and lakes will be the focus of growing tension. Of the 200 largest river systems worldwide, 150 are shared by two nations, and more than 50 by 3 to 10 nations. These major arteries support 40 percent of the world's population. In the Middle East, water is currently a major factor in political confrontation as Israel, Syria, Lebanon, and Jordan compete for access to water from the Jordan River. There is similar competition between India and Pakistan over the Indus River headwaters, and between India and Bangladesh over the Ganges River. Water-supply problems have already increased tension in the Mekong River basis shared by Laos, Thailand, Kampuchea, and Vietnam, and in the Middle East where the Tigris and Euphrates Rivers flow through Syria and Iraq.[39] (See Table 9.3 on the following page.)

A coordinated water resource policy is needed for each of these potentially explosive regions. Senior officials from these countries took an important step towards this goal when, in the spring of 1988, they met for a seminar on the management of water resources. The group focused on the growing necessity of negotiated trade-offs and cooperation in avoiding future conflicts.[40]

Adverse Effects of Water Development Projects

Water development projects, such as dams, canals, and stream channelization programs, can seriously degrade water quality, spread waterborne diseases, destroy farmland, ruin wetlands and downstream fisheries, and contribute to species extinction through habitat destruction. Large dams built in developing countries have contributed to major outbreaks of debilitating diseases such as schistosomiasis, which is caused by a parasite that thrives in reservoirs and irrigation systems. In some villages on Ghana's Lake Volta, the largest artificial lake in the world, at one time all the children were infected with schistosomiasis.[41] Some large-scale irrigation projects, such as Egypt's Aswan Dam, have brought benefits, but at great cost. Lake Nasser is filling up with over 100 million tons of silt, clay, and sand that once fertilized downstream fields during periods of flooding. These fields now require increased chemical fertilization.[42]

The projected useful life of major dams has been consistently overestimated because silting often proves to be twice as rapid and severe as anticipated. The gradual loss of storage capacity reduces power-generating ability, while soil below the dam is deprived of its annual enrichment by silt-bearing floodwaters.[43] Also, where seasonal flooding once flushed salts from the soil, the controlled irrigation made possible by dams often leads to damaging buildups of salts. Fields that no longer have a chance to dry out become waterlogged. Moreover, the expectations that these reservoirs can be made into fisheries are often proven wrong; the deep waters of high dams tend to be sterile, with little nourishment for fish. Finally, in addition, large-scale water projects often displace many people from their ancestral homes and destroy wildlife habitats.[44]

Damaging or disasterous projects continue throughout the developing world. The Narmada Valley Project for example, a series of large dams in India, is funded by the World Bank and currently under construction despite grave

Table 9.3
Unresolved International Water Issues, Mid-Eighties

Rivers	Countries Involved in Dispute	Subject of Dispute
Nile	Egypt, Ethiopia, Sudan	Siltation, flooding, water flow/diversion
Euphrates, Tigris	Iraq, Syria, Turkey	Reduced water flow, salinization (constraints on irrigation & hydropower)
Jordan, Yarmuk, Litani, West Bank aquifer	Israel, Jordan, Syria, Lebanon	Water flow/diversion
Indus, Sutlei	India, Pakistan	Irrigation
Brahmaputra, Ganges	Bangladesh, India	Siltation, flooding, water flow
Salween/Nu Jiang	Burma, China	Siltation, flooding
Mekong	Kampuchea, Laos, Thailand, Vietnam	Water flow, flooding
Paraná	Argentina, Brazil	Dam, land inundation
Lauca	Bolivia, Chile	Dam, salinization
Rio Grande, Colorado	Mexico, United States	Salinization, water flow, agrochemical pollution
Great Lakes	Canada, United States	Water diversion
Rhine	France, Netherlands, Switzerland, West Germany	Industrial pollution
Maas, Schelde	Belgium, Netherlands	Salinization, industrial pollution
Elbe	Czechoslovakia, East and West Germany	Industrial pollution
Werra/Weser	East Germany, West Germany	Industrial pollution
Szamos	Hungary, Romania	Industrial pollution

Source: Michael Renner, *National Security: The Economic and Environmental Dimensions*, Worldwatch Paper 89 (Washington, DC: Worldwatch Institute, May 1989), p. 32.

concerns expressed by environmentalists throughout India and abroad.[45] Some large-scale projects, such as the Soviet Union's recently abandoned proposal to change the direction of major river systems, could alter regional or even global climate.[46]

Water Supply and Sanitation in the Third World

About 1.2 billion people, or a quarter of the world's population, lack access to safe drinking water. And about 1.4 billion people have no facilities for sanitary waste disposal.[47]

According to a survey conducted by the World Health Organization in 1985 and 1986, 97 percent of people in industrial countries have access to safe water, compared to only 53 percent of people in developing countries. For Asia and Latin America, the figures are 74 and 72 percent, respectively; while in Africa, only 36 percent have access to safe drinking water.[48]

Some 80 percent of all human disease is linked to unsafe water, poor sanitation, and lack of basic knowledge of hygiene and disease mechanisms. Waterborne diseases claim at least 25 million lives each year in the Third World, and the cost of polluted water to human health is enormous. There is a strong association between lack of access to clean water and high rates of infant mortality. (See Figure 9.4.)

Figure 9.4

Infant Mortality and Percent of Population with Access to Clean Drinking Water, Selected Countries, 1982

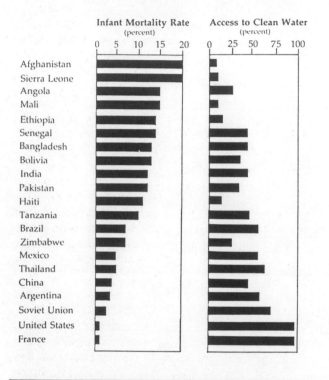

Source: Derived from UNICEF, *State of the World's Children*, 1984, as published in William U. Chandler, *Investing in Children*, Worldwatch Paper 64 (Washington, DC: Worldwatch Institute, 1985), p. 32.

Contaminated water is responsible for trachoma blindness (500 million sufferers) and elephantiasis (250 million); diarrheal diseases kill over a thousand children every hour. Elimination of diarrhea would end much of the malnutrition that afflicts the world. Contaminated water also transmits cholera, typhoid, infectious hepatitis, poliomyelitis, and intestinal worms.[49]

Schistosomiasis is estimated to affect at least 200 million people and endanger another 600 million in Asia, Africa, the Caribbean, and Latin America. Although the disease does not always kill, it frequently cripples and debilitates its victims, leaving them hopelessly immobile. The rapid spread of the disease is directly linked to the increased number of artificial lakes. Since the construction of the Aswan Dam in Egypt, the incidence of schistosomiasis has risen from 1 in 30 to 3 out of 4 persons.[50]

The disease can be controlled by interrupting the life cycle of the parasite. This can be done by preventing human excreta from contaminating water supplies, controlling the snail that acts as an intermediate host, reducing human contact with infected water by providing clean water, and, finally, by treating infected people. Environmentally sound water development projects, sanitation, and adequate water supplies are essential to control of the disease. Health education, medication, and surveillance of population movements are also crucial. However, success can only be achieved through an integrated approach that takes local factors into account.[51]

From a worldwide perspective, the most hazardous substance to human health is not toxic or radioactive wastes—it is disease-carrying feces. Currently, human excreta is the most serious source of water pollution.

Social stigmas give sanitation a low priority. Ignorance and, in many places, cultural taboos prevent the introduction of simple sanitary latrines for the safe disposal of household wastes. Many people's personal and food hygiene measures are inadequate to limit the spread of disease.[52]

Economic and topographic factors also inhibit installation of hygienic facilities. Through the loss of human productivity and income, sickness can spell disaster to any subsistence community, especially during the rainy season—a peak period for both agricultural activities and water-related diseases. As those unable to work lose income, levels of family nutrition drop, both stunting human growth and increasing susceptibility to disease.

The wide prevalence of water-related disease can hurt entire economies of developing countries, where farmers, herders, village craftspeople, forestry workers, and fishers often contribute the major share of the gross national product. In India, for example, waterborne diseases alone claim 73 million workdays every year. The cost, in terms of medical treatment and lost production, has been estimated at almost $1 billion per year.[53]

The human costs of inadequate sanitation and polluted water supplies are borne disproportionally by women and children. Women in developing countries are traditionally responsible for collecting and carrying water. This backbreaking labor can use up to one-quarter of a woman's daily caloric intake. In urban areas, families living in slums frequently use 10 percent of their earnings to buy water.[54]

Third World women should be directly involved in rural water development programs. More convenient access to safe water will allow women to engage in other productive tasks. Not only can women benefit from rural water programs, they can help assure the programs' success.

Infants and children under five are most vulnerable to water-related disease. Each year 3 million children die of dehydration caused by diarrheal diseases.[55] Improvements in sanitation can help reduce infant and child mortality;

water supply and sanitation programs thus can be an important component of family planning efforts.

Major hygiene education efforts are needed to reduce the incidence of water-related disease. Experience has shown that drilling wells and constructing latrines will not have a significant health impact unless supported by comprehensive education.[56] However, attitudes of personal cleanliness are closely connected to local customs and culture, and any attempt to change these attitudes must be handled with care and sensitivity.

In 1980, the United Nations initiated the International Drinking Water Supply and Sanitation Decade, designed to improve health through an integrated approach to water management and sanitation. The goal is to provide drinking water and sanitation to two-thirds of the world's population by 1990, and by 2000 to provide the other one third with potable water systems.[57]

By 1983, 65 developing countries had set full or partial targets for the Decade, and had completed or were preparing actual plans.[58] The original objective of the United Nations program was to provide clean water and adequate sanitation to all by 1990. However, a recent assessment shows that by 1990 only about 79 percent of urban inhabitants and 41 percent of rural people will have access to clean water. As for sanitation, only 62 percent of urban and 18 percent of rural people will have sanitary facilities.[59]

Malawi, one of the poorest nations, has already made progress toward its target of providing clean water for all its people by the year 2000. Its success is based on the use of gravity-piped water systems instead of expensive, high-maintenance bore wells, and on village-based self-help arrangements in which communities that benefit provide their own labor and local organization.[60]

Water Pollution in Industrial Countries

Along with scarcities caused by growing demand, in industrial countries the issue of fresh water pollution is a major concern.

Water is polluted when human activities have made it unsuitable for a particular use; the nature and extent of pollution can be defined by the intended use of the water. Once groundwater is contaminated, it is extremely difficult and prohibitively expensive to correct the problem.[61]

Sources of Water Pollutants

Industrial and mining operations are major sources of toxic water pollutants in developed countries. Most industrial processes produce potential water pollutants, including the production of petroleum, petrochemicals, and other commercial chemicals; pesticides and herbicides; fertilizers; steel and other metals; and paper products. Major industrial pollutants include chlorinated organic compounds, minerals and oils, phenols, nitrogen, phosphorus,

mercury, lead, and cadmium. Other important sources of water pollution include waste disposal facilities, agricultural and urban runoff, acid precipitation, and radioactive waste near nuclear weapons facilities.[62]

In the United States, a recent survey by the Environmental Protection Agency showed that more than 17,000, or 10 percent, of the nation's rivers, streams, and bays are significantly polluted. The study found that 250 city sewage facilities and 627 industrial operations routinely dump toxic waste into U.S. waterways.[63]

Acid precipitation has left thousands of U.S. lakes in the industrial north biologically dead, and thousands more are dying. Over 50 pesticides contaminate groundwater in 32 American states, and at least 2,500 toxic waste sites need cleanup.[64] Damage from all types of water pollution in the United States is estimated be $20 billion annually.[65]

Industrial Sources   In the United States, the organic chemical and plastics industries are the largest sources of toxic chemical pollution, with metal finishing and the iron and steel industries next in line. Chemical plants are concentrated in the Northeast, along the Gulf Coast, and in California. The largest concentration of metal finishing plants is in the northeastern United States. The fastest growth in both industries is in the South and West. Other important industrial sources of toxic chemicals include the pulp and paper industry, metal foundries, and petroleum refineries.[66]

Chemical and oil spills cause extensive contamination that can never be completely removed. In December 1988, a fuel storage tank beside the Monongahela River burst, emptying a million-gallon, 40-foot wave of diesel fuel into the river. First, drinking water for the Pittsburgh area was contaminated, then the diesel slick spread downstream, affecting the drinking water of more than 800,000 people. One or more industries apparently took advantage of the river pollution following the Monongahela discharge to dump cancer-causing industrial solvents into the Ohio River.[67]

Although most oil and chemical accidents are less dramatic than the Monongahela spill, such toxic discharges are fairly commonplace. Nationwide, more than 20 major oil spills occur every year, and since 1981 there have been more than two dozen spills of more than a million gallons. A recent study by EPA documents nearly 7,000 accidents involving hazardous substances between 1980 and 1985.[68]

Hazardous Waste Disposal   Only a small fraction of industrial waste is recycled, detoxified, or destroyed. Two-thirds of the hazardous waste produced in the United States is disposed of through injection wells, pits, and landfills. All of these disposal methods can contaminate groundwater.[69]

In the United States, more than 77,000 disposal sites exist for industrial, municipal, agricultural, and mining wastes, and for brines from oil and gas extraction. Each day an estimated 82 billion gallons of wastewater—equivalent to about 330 gallons per American—flow into these sites. As of July 1987, the EPA had placed 951 landfills and other waste sites on its National Priority List of sites needing urgent attention. Most of these locations pose actual or potential threats to surface groundwater.[70] In West Germany, at least 35,000 landfill sites have been declared potential threats to groundwater supplies.[71]

Other toxic pollutants enter water supplies from nonpoint sources, including runoff from mining activities, farms, and urban areas. For example, urban runoff from streets and parking lots carries heavy metals, such as lead and cadmium, as well as other pollutants. Chemicals containing cyanide that are used to de-ice streets, septic-tank cleaners containing trichloroethylene, and herbicides used in agriculture are all potential sources of pollution.[72]

Acid Mine Drainage   Drainage from mining operations pollutes at least 12,000 miles of steams in the United States. Much of the drainage comes from coal mines, but some pollution results from the mining of copper, lead, zinc, iron, uranium, and other minerals. The acidity harms or kills living organisms and corrodes metal structures; iron compounds can coat the bottoms of streams, making them uninhabitable. The mine-drainage problem can be expected to grow with the nation's increasing reliance on coal as an energy source.[73]

Agricultural Sources   Worldwide, agriculture is a major source of organic and inorganic pollutants from fertilizers, pesticides, herbicides, and animal wastes. Pesticides and fertilizers have contaminated groundwater in a number of areas in the United States, and may affect more than 50 million people. Sixty pesticides—many of them carcinogens—have been found in the groundwater of 30 American states.[74]

A major threat to water supplies is rainfall runoff from areas of high waste concentration. Runoff from barnyards, feedlots, and land treated with chemical fertilizers carries potassium and nitrogen compounds into water supplies. Contamination from agricultural sources contributes to nitrates in drinking water that can kill or disable infants and livestock.[75] Of all organic wastes that pollute surface water and groundwater in the United States, livestock produce five times as much waste as humans, and twice as much organic waste as industry.[76]

Water used to irrigate crops often becomes polluted with salts, and can then contaminate groundwater or steams and rivers. In the western United States, where water from the Colorado River is withdrawn for irrigation and later returns to the river, the water leaches large quantities of salts from the irrigated land and adds them to the river. And during some dry periods, the Red River in Oklahoma and Texas is more salty than seawater.[77]

Acid Rain   Water can become polluted not only from direct discharge of materials, but also from air pollution. Projected increases in the burning of fossil fuels, especially coal, will greatly raise the production of sulfur and nitrogen oxides, the primary cause of acid rain and other acid deposits. These substances acidify lakes, rivers, and groundwater; kill or contaminate fish and other aquatic life; harm birds and mammals that feed on fish; damage forests; and pollute drinking water supplies. Hundreds of lakes in parts of Scandinavia, the northeast United States, and southeastern Canada have turned acidic. Half the 219 high-elevation lakes surveyed in the U.S. Adirondack mountains show acid damage.[78] A 1988 EPA report found that more than 4 percent of the streams surveyed in the Middle Atlantic states were acidic, and roughly half had a low capacity to neutralize acidic rain.[79] Nitrates from acid precipitation are an important pollutant in the Chesapeake Bay and other areas along the Atlantic coast.[80]

Acid rain can harm human health. It can poison reservoirs and water supply systems by dissolving toxic metals, such as aluminum, cadmium, and mercury, from soils and bedrock in watersheds. Acid water can also corrode metals, including the copper and lead used in water-supply pipes. U.S. lakes and rivers are being polluted not only by acids from distant airborne sources of sulfur and nitrogen oxides, but also by toxic clouds carrying a range of hazardous chemicals, including DDT, polychlorinated biphenyls, and dioxins, from a variety of sources.[81]

Groundwater Pollution

Groundwater, the fresh water beneath the Earth's surface, is the most precious and least-protected resource in the United States. Groundwater is the only source of drinking water for half the U.S. population. Any pollutant that comes in contact with the ground may contaminate groundwater. A quarter of the usable groundwater is already contaminated, and up to three-quarters is contaminated in some areas.[82] Figure 9.5, on the following page, shows some of the sources of groundwater contamination.

There are increasing occurrences of groundwater contamination from saltwater, microbiological contaminants, and toxic organic and inorganic chemicals, including pesticides. Irrigation practices in the West have raised groundwater salinity, as have overdrafts in coastal areas.

Disposal of toxic industrial wastes is a major source of groundwater contamination in the United States. A survey of more than 8,000 industrial waste disposal sites found that over 70 percent of the sites are unlined. About 30 percent are located in permeable materials and lie above usable water aquifers; one-third of these sites are within a mile of a water supply well.

Figure 9.5
Sources of Groundwater Contamination

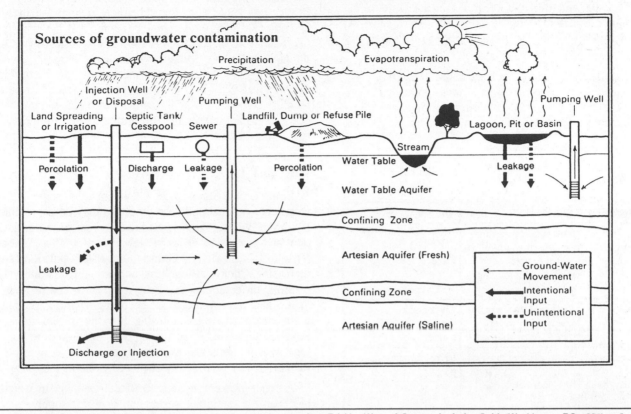

Source: U.S. Environmental Protection Agency, as published in Concern, Inc., *Drinking Water: A Community Action Guide* (Washington, DC, 1986), p. 2.

In recent years, groundwater contamination by toxic organic chemicals has forced the closing of hundreds of wells in the United States.[83]

Wetlands play an important role in replenishing groundwater and as natural water purifiers. The United States has already lost half of its wetlands to urban and agricultural development. The current annual loss is 450,000 acres, an area half the size of Rhode Island.[84]

Wetlands plants can absorb excess nutrients, and can immobilize pesticides, heavy metals, and other toxins, preventing them from moving up the food chain. Wetlands in Florida have been successfully used to treat sewage. Unfortunately, the capacity of wetlands to cleanse polluted water is limited, and many have been overwhelmed by pollution.[85]

Both groundwater pollution and groundwater depletion may be irreversible. Once contaminated, the water is likely to remain that way. Depletion can cause an aquifer to consolidate, diminishing its storage capacity. In coastal areas, groundwater withdrawals can pull saltwater into an aquifer and render it permanently unsuitable for most uses.[86]

Health Effects of Water Pollutants

Concern over waterborne viral diseases has increased in recent years. In the 1970s, contaminated groundwater reportedly caused a third of the waterborne disease outbreaks recorded by the U.S. Centers for Disease Control. And between 1971 and 1983, there were 427 reported outbreaks of waterborne disease affecting 106,000 people.[87] According to the Environmental Protection Agency, a recent study of 446 municipal water systems showed that only 60 met federal standards for bacteria content. More than half the nation's water systems are judged to be deficient in controlling bacteria.[88]

In a 1989 report to Congress assessing compliance with the 1972 Clean Water Act, the EPA found that more than two-thirds of the nation's 15,600 wastewater treatment plants have water quality or public health problems. The cost of upgrading the nation's sewage treatment facilities to meet federal standards has been revised upward to $83.5 billion—17 times more than the EPA's current budget.[89]

Toxic substances in water supplies can cause a wide variety of health and environmental effects. (See Table 9.4.) A 1988 study reported that more than 2,100 contaminants have been found in U.S. drinking water since the Safe Drinking Water Act was passed in 1974. Of the chemicals found, 97 are known carcinogens, 82 cause mutations, 28 cause acute and chronic toxicity, and 23 are tumor promoters. The remaining 1,900 contaminants have not been adequately tested for possible adverse health effects.[9]

Acute health effects—such as nausea or poisoning—occur from weak concentrations of these pollutants in surface water supplies. In groundwater supplies, toxic contaminants have reached concentrations known to cause

Table 9.4
Water Pollutants and Their Health Effects

Substance	Source	Health Risk
Chlorinated solvents	Chemical degreasing, machinery maintenance, intermediaries in making other chemicals	Cancer
Trihalo-methanes	Produced by chemical reactions in water treated with chlorine	Liver and kidney damage, possibly cancer
Polychlorinated biphenyls (PCBs)	Wastes from many outmoded manufacturing operations	Liver damage, possibly cancer
Lead	Old piping and solder in public water distribution systems, homes, and other buildings	Nerve problems, learning disabilities, birth defects, possibly cancer
Pathogenic bacteria, viruses	Leaking septic tanks, untreated sewage	Gastrointestinal illness, more serious diseases

Source: Adapted from *Time*, March 27, 1989, p. 38

acute health effects; public and private wells have been closed in several states.[91] Nitrates in drinking water can interfere with the blood's ability to carry oxygen and cause globinemia, or "blue baby" syndrome. Between 1945 and 1969, nitrates in U.S. drinking water caused 328 reported cases of infant globinemia, including 39 deaths, between 1945 and 1969.[92]

The amount of direct human exposure to most toxic chemicals may be small in surface water. The major exception is exposure to chlorinated organic compounds that are created, often in relatively high concentrations, when chlorine is used to treat industrial and municipal wastewaters and to purify drinking water. Chlorination can

create potentially toxic compounds such as chloroform, a suspected human carcinogen.

People may be exposed indirectly to high concentrations of toxic pollutants, since some chemicals accumulate in fish and other seafood. When eaten, contaminated fish can subject humans to high concentrations of toxic substances. In the Japanese town of Minamata, during the 1950s hundreds of people were killed or tormented by an incurable neurological disease. The cause was determined to be a mercury compound from industrial waste dumped into the ocean that had accumulated in seafood eaten by the victims.[93] (See Chapter 13.)

Improving Water Resource Management

In the United States, there is evidence that some of the worst conventional pollution problems may be improving. By the early 1980s, salmon had reappeared in the Connecticut and Penobscot rivers in New England. There is less algae in Lakes Erie and Ontario, and fish populations are increasing. Control of industrial and municipal pollution is restoring the ecological productivity of a few coastal areas. While such improvements are encouraging and important, toxic contamination is increasing. Fish may thrive again in Lake Michigan and Lake Huron, but—because of increasing levels of nitrates, heavy metals and organic micropollutants—restrictions have been placed on eating these fish.[94]

A major problem in the U.S. effort to manage its waters has been the traditional separation of water supply, wastewater collection and treatment, and storm and flood water management at all levels. Each unit has its own goals, constituencies, and procedures, so there is no overall strategy for managing water resources. As competition for water becomes more intense, regionalization and coordination of water management based on watershed boundaries offer inherent advantages, as long as overall federal guidelines exist to prevent regional degradation of the nation's waters.

In England and Wales, water management has been regionalized and centralized on a river-basin basis. While this effort was stimulated by a particular set of conditions, its success makes this approach worth examining for possible use elsewhere.

Maintaining Water Supply

To ensure adequate water supplies, a major priority for any region is proper watershed management, including protection of aquifer recharge areas and upland vegetation, so that rainfall can be held upstream and released gradually to meet downstream needs.

A second priority is to conserve water and avoid needless waste, in part by treating water as a valuable commodity and pricing it to reflect the real costs of supply. In many parts of the world, there is an urgent need to improve the efficiency of water use in agriculture. Irrigation for food production accounts for nearly three-quarters of world water use and, worldwide, irrigation systems on average are only about 37 percent efficient. Elimination of excessive irrigation subsidies and encouragement of the reuse of wastewater would greatly improve the efficiency of water use.

In 1986, Israel reused 35 percent of its wastewater, mostly for irrigation. By 2000, Israel plans to recycle 80 percent of its wastewater, thereby adding substantially to its renewable water supply.[95]

New micro-irrigation methods such as the trickle-drip system and precise scheduling of water deliveries can cut losses dramatically. Trickle-drip methods can be three times more efficient than conventional irrigation; so far less than half of one percent of all irrigation uses this technique.[96]

In the United States, the Natural Resources Defense Council has taken action to force the U.S. Interior Department to consider the environmental impacts and possibilities for water conservation before it renews long-term water subsidies in the Western United States, where most water is used for agriculture and where hundreds of federal water contracts are due to expire in the next decade.[97]

Industrial recycling can make a dramatic difference in a region's water use. Paper industries in Sweden, by using recirculation, have cut water consumption in half while doubling production.[98] In the United States, an Armco steel mill in Kansas City, Missouri, uses only 9 cubic meters of water per ton of steel produced, compared with as much as 100 to 200 cubic meters per ton in many other steel plants.[99]

Strong pollution standards make it economical for industry to recycle wastewater. In 1980, Israel supplied 4 percent of its total water needs through reclaimed wastewater. This share is expected to reach 16 percent by the year 2000. In contrast, the United States today meets only about 0.2 percent of its water needs with reclaimed wastewater.[100]

Household water use can be cut by one-fifth through the use of simple water-saving devices such as more efficient faucets, showerheads, toilets, and clothes washers. The use of water meters can cut domestic water consumption by as much as 45 percent.[101]

Safeguarding Water Quality

To ensure acceptable water quality for human consumption and ecosystem survival, the highest priority should go to programs that reduce the generation of solid, liquid, and airborne wastes—especially toxic wastes—from industrial plants, mining and smelting operations, electric power production, and cities and towns. In addition, there must be adequate containment and treatment of the remaining wastes from these sources that can not be eliminated.

In agriculture, contamination of water supplies must be minimized by using more efficient irrigation methods, controlling runoff from cattle feed lots, and exercising much greater caution and selectivity in the use of chemical fertilizers, pesticides, and herbicides. (See Chapter 5.)

In the treatment of municipal sewage, innovative, cost-effective approaches have often been slighted in favor of traditional capital- and energy-intensive sewage treatment plants. As developing countries begin to deal with sewage problems, they would do well to avoid the mistakes made in many industrial countries.

Some Third World countries have started to tackle their water pollution problems. For example, India has begun a $200 million project to clean up the sewage, industrial wastes, pesticides, and other pollutants in the 1500-mile long Ganges River, the source of water for a third of India's population.[102]

In Eastern Europe, Poland, which has some of the world's worst pollution from industrial sources, plans to build 188 new water purification plants to treat 5.7 billion cubic meters of domestic and industrial waste, and another 200 facilities to treat 1 billion cubic meters of toxic industrial wastewater.[103]

To protect valuable water resources and ensure survival of fragile ecosystems, all countries must restore degraded watersheds, lakes, and rivers; protect wetlands; and invest adequately in appropriate waste reduction and pollution prevention measures.

Protecting U.S. Water Quality

In the 1920s, the U.S. Public Health Service began to develop standards for contaminants in drinking water to control typhoid, cholera and other diseases carried by polluted water. However, through the next fifty years these standards changed little as the volume and number of industrial pollutants grew. Pollutants were simply diluted in the nation's waters. It was not until the early 1970s that toxic chemicals were detected in the public water supplies of major metropolitan areas. In response to public pressure, Congress passed the Safe Drinking Water Act in 1974. This required the Environmental Protection Agency (EPA) to nationalize standards for the quality and treatment of water supplies, regulate underground waste injections, and set monitoring and reporting regulations for public water systems.[104]

Drinking water originates from two sources, groundwater and surface water. Rural areas rely primarily on groundwater, while urban areas draw mainly on surface water supplies. The 1972 Clean Water Act established federal jurisdiction over all U.S. surface waters. The 1987 federal Water Quality Act gave states control over both toxic substance discharge and enforcement, and included a joint federal-state nonpoint source pollution control program.

However, groundwater protection remains largely uncoordinated among a number of agencies and a variety of federal statutes. No single authority or law exists for the protection of groundwater.[105] While the federal government sets drinking water standards for all public water systems, federal law does not require water quality standards for groundwater.[106] Instead, action has had to come from the states. According to a recent survey by Renew America, 18 states have passed legislation designed to keep agricultural chemicals out of groundwater, and six states have laws to regulate the application of agricultural chemicals through irrigation systems.[107]

Between 1974 and 1986, states assumed primary responsibility for the safety of water supplies. Standards, regulations and requirements for water safety programs are first approved by the EPA and then turned over to the state. Despite the fact that the EPA retains oversight authority, and enforcement is largely controlled by the state, EPA data showed that in 1987, a total of 36,763 public water systems, servicing 41 million residents, violated EPA standards.

Enforcement of EPA standards is weak. Both the EPA and the states lack the resources for effective enforcement. Moreover, standards and testing practices for many contaminants do not exist, making enforcement even less likely. By 1987, standards had been set for only 30 contaminants, with another 53 due to be set by 1989 as ordered by the 1986 amendments to the Safe Drinking Water Act. These amendments also require the EPA to name, every three years, 25 possible new contaminants for regulation.[108]

Successful Programs

Renew America recently evaluated the quality of U.S. state programs and policies for drinking water protection. The study rated each state on a scale of 1 to 10.

Maine, Massachusetts, and New Jersey received the highest ratings (between 9 and 10). Maine's program is considered the best in the nation for overall drinking water protection, monitoring, and enforcement. Most of the state's drinking water budget goes to train its water system operators to maintain quality control. The state reportedly allows no community water system to deviate significantly from the requirements of the Safe Drinking Water Act.[109]

The next highest rated group of states (with ratings between 7 and 8) included California, North Carolina, Virginia, Wisconsin and Iowa. Iowa has a $65 million program, funded partially by fees on pesticides and nitrogen fertilizers. The program is a comprehensive but basically nonregulatory approach using education, research, and demonstration projects to address groundwater pollution.

Seventeen states received ratings between 5 and 6, fourteen states fell in the 3 to 4 range, and eleven states were judged to have the poorest drinking water protection programs, with ratings between 1 and 2.[110] (See Table 9.5.)

Table 9.5
Ratings of U.S. State Programs and Policies for Drinking Water Protection

Rating	States
10-9	MA, ME, NJ
8-7	CA, IA, NC, VA, WI
6-5	AZ, CO, CT, GA, HI, ID, KS, MD, MI, MN, MO, NE, OK, OR, SC, VT, WV
4-3	AK, DE, FL, IL, MS, MT, ND, NH, NY, PA, RI, TX, WA, WY
2-1	AL, AR, IN, KY, LA, NM, NV, OH, SD, TN, UT

Source: Renew America, *The State of the States 1989* (Washington, DC, 1989), p. 19.

Effective clean water policies and programs provide incentives for industry to recycle water, which not only saves water, but also reduces pollution. Some industry efforts to conserve water and reduce water pollution from industrial wastes have achieved these goals and saved money as well. The Minnesota Mining and Manufacturing Company (3M) has a program called "Pollution Prevention Pays," based on the idea that waste is inefficient. In the program's first 12 years of operation, 3M claims to have cut waste generation in half and saved $300 million. Borden Chemicals, located in California, has managed to reduce organic chemicals in wastewater by 93 percent and lower sludge disposal costs by $49,000 a year. Pioneer Metal Finishing in New Jersey has cut water use by 96 percent and sludge production by 20 percent, saving $52,500 a year.[111]

There has also been progress in controlling pollution from household and municipal wastes. An example, the New Alchemy Institute has designed a solar-aquatic wastewater-treatment plant for the Sugarbush ski resort in Vermont. Set up as an experiment to compete eventually in the marketplace, the plant uses biogeochemical cycles to treat waste.

Housed in a greenhouse, the plant operates by allowing raw sewage to flow first through aeration cylinders where air, light, and bacteria digest organic components and nitrify ammonia. Next, it passes through plant-filled waterways, where algae takes up nitrate, phosphate, and other nutrients. Shrimp eat the algae, which in turn are eaten by fish. Sludge is cleaned throughout the system by both snails and shrimp, and plants are chosen according to their ability to take up toxic substances and kill pathogenic bacteria.

After five days of flowing through the greenhouse, the effluent is filtered by an artificial marsh. Although operating data for the system are still inconclusive, the treatment system promises to be both economical and effective compared with conventional methods.[112]

A Look Ahead

Future management of our water resources will require greater efforts to use water more efficiently and to protect water quality. Greater efforts to increase conservation and efficiency are needed. In the United States, only 6 states have statewide water conservation programs.[113] Because water for agriculture accounts for so much of total water use, improved irrigation efficiency is a high priority.[114]

With stricter water quality standards and increased demands for compliance also come increased costs. Regardless of the difficulties, governments must protect water quality and therefore must invest in enforcement measures. However, the polluters should contribute a major share of the resources necessary to provide safe drinking water and protect health. Investments in the future must be made today, before the costs to society become unmanageable and the contamination of our water supplies becomes a crisis.

By improving water collection and storage, by increasing water recycling and the efficiency of water use, and by protecting water quality, we should be able to stretch our global water resources almost indefinitely, provided we begin soon.[115] International cooperation in water usage, technology, and monitoring is important to the success of these efforts.[116]

What You Can Do

Inform Yourself

Find out more about the issues surrounding fresh water supply and quality so you can speak and write clearly and communicate your views to others. Learn how the problem affects your community, your state, and the country as a whole, as well as the international implications it holds.

- Find out about programs to protect drinking water in your state. Consult the ratings of state programs earlier in this chapter, and obtain a copy of Renew America's drinking water study containing those ratings (see Further Information).

- Determine the source, treatment and quality of public drinking water supplies in your area by contacting your local Department of Environmental Services or your private water supplier. Learn who may be using or polluting your water supply. Remember, safe water is your right! For assistance, consult *Groundwater: A Community Action Guide*, published by Concern, Inc. and listed under Further Information below.

- Learn about your town's waste treatment facilities: how effective are they now, and how adequate will they be in the future?

- Review results of drinking water tests in your area, attend public meetings, and keep track of other drinking water matters such as the development of new water quality standards.

- Subscribe to news and action letters issued by local and national organizations addressing fresh water quality and conservation (see the listing of organizations under Further Information). Keep up-to-date on current legislation on the issue.

Join With Others

Contact others in your area who are concerned about fresh water quality and conservation, and work together to support local and national legislation, and to educate the community. Join and support a national organization working toward similar goals, and encourage other environmental groups to stress the importance of preserving fresh water reserves.

- Work to halt the addition of industrial pollutants to sewage treatment systems; urge reduced reliance on water for carrying and assimilating human wastes; work to develop productive uses for these wastes.

- Attend meetings that address fresh water issues such as groundwater contamination; prevention and cleanup of oil spills and discharges in local streams and rivers; and industrial developments near fresh water supplies.

Review Your Habits and Lifestyle

There are many things you can do in your home and community to help conserve and preserve the quality of fresh water supplies.

- In your home, avoid using or disposing of substances that could damage surface and groundwater:

  » use phosphate-free detergents (liquid products generally contain no phosphate, while powders do);

  » dispose of unwanted hazardous household chemicals (such as pesticides, paint remover, motor oil or antifreeze) in sanitary landfills, or take advantage of approved community collection points and recycling centers;

  » compost your food wastes or put them in the trash instead of the garbage disposal.

- Avoid unnecessary use of water:

  » Check faucets and toilets for leaks and repair them if necessary; a leaky toilet can waste one gallon every six minutes.

>> Take short showers instead of baths; a bath uses 20-40 gallons; a three minute shower uses 15-20 gallons.

>> Turn off water while brushing teeth, shaving, or lathering hands, and save 1-2 gallons each time. (See Table 9.6).

>> Use low-flow showerheads and faucets; they can reduce water use by up to 50 percent.

>> Place a half-gallon plastic bottle of water in your toilet tank to reduce the amount of water used for each flush.

>> Wash cars and water lawns only when absolutely necessary; water lawns during the coolest part of the day to avoid rapid evaporation; don't overwater lawns and gardens.

● On your property, take precautionary measures to protect fresh water resources:

Table 9.6

WATER SAVING GUIDE

CONSERVATIVE USE WILL SAVE WATER		NORMAL USE WILL WASTE WATER
Wet down, soap-up, rinse off 4 gallons	SHOWER	Regular shower 25 gallons
May we suggest a shower?	TUB BATH	Full tub 36 gallons
Minimize flushing. Each use consumes 5-7 gallons	TOILET	Frequent flushing is very wasteful
Fill basin 1 gallon	WASHING HANDS	Tap running 2 gallons
Fill basin 1 gallon	SHAVING	Tap running 20 gallons
Wet brush. Rinse briefly ½ gallon	BRUSHING TEETH	Tap running 10 gallons
Take only as much as you require	ICE	Unused ice goes down the drain
Please report immediately	LEAKS	A small drip wastes 25 gallons a day
Turn off light, TV, heaters and air conditioning when not in room	ENERGY	Wasting energy also wastes water

THANK YOU FOR USING THIS COLUMN . . . . AND **NOT** THIS ONE

Source: San Francisco Convention and Visitors Bureau

>> If you have a septic waste system, be sure it is emptied periodically and properly maintained.

>> Landscape your garden to minimize rainwater runoff and soil erosion.

>> Avoid using liquid fertilizers, weed killers, and toxic pesticides on your property. Instead, consider organic methods to enhance and protect your lawn and garden. (See Chapter 5.)

● Express willingness to pay higher water rates if necessary to finance improvements in water quality.

● Consider eating fewer animal products, especially meat, remembering that about 50 percent of all U.S. water use goes to the inefficient production of feed for livestock, instead of to crops for direct human consumption.

Work With Your Elected Officials

Contact your local, state, and national legislators and find out their positions on issues concerning fresh water conservation and quality control.

● Urge your representatives on all levels to support:

>> research and action on water pollution causes and control measures, and on how to reduce water requirements for irrigation;

>> public education campaigns about water resource problems, appropriate solutions, and the need to protect and improve the quality of drinking water.

● Urge officials in Congress to support:

>> a national analysis of fresh water uses and supplies and likely future demands as a basis for future planning;

>> legislation to protect groundwater;

>> efforts to charge water users rates that encourage efficiency and reflect the costs of providing water on a sustainable basis;

>> programs for storing water underground, to stabilize year-to-year variations in supply;

>> efforts to greatly increase the efficiency of water use in irrigation by pricing water to cover the true cost of supply;

>> construction of smaller, community-based irrigation systems as alternatives to large-scale irrigation projects;

» management of fertilizers and pesticides in ways that protect surface and underground water supplies;

» control of industrial water pollutants at the source through required use of available technologies and through recycling or effective containment of hazardous wastes;[117]

» amendments to strengthen the Safe Drinking Water Act;

» legislation to reduce sulfur and nitrogen oxide emissions, and to strengthen the Clean Air Act;

» expansion of the National Wild and Scenic Rivers System;

» thorough assessment of the environmental impacts of large-scale water development projects in the United States and in developing countries, before such projects are funded;

» programs of the U.S. Agency for International Development, World Bank, and the regional development banks to improve water resource management and to develop nonstructural alternatives to traditional water resource development projects, including water conservation and water demand reduction; and

» U.S. participation in the closing years of the International Drinking Water Supply and Sanitation Decade, coordinated by the World Health Organization.

Publicize Your Views

Express your views on issues concerning fresh water resources to community leaders, and the public in general, through print and electronic media.

• Write letters to the editor of your local paper to educate the community about the wide range of water conservation and efficiency measures available to reduce municipal, industrial, and agricultural water use, and the money these measures can save the consumer.

• Create public service announcements about fresh water issues for local television and radio stations.

• If you live in an acid rain area, urge local newspapers, television and radio stations to report routinely on the acidity of rainfall.

Raise Awareness Through Education

• Find out whether schools, colleges, and libraries in your area have courses, programs, and adequate resources addressing the issues of water supply and water quality; if not, offer to help develop or provide them.

• Encourage educators and conservation groups to initiate classes, seminars, lectures and field trips that increase awareness of facts and policies concerning fresh water resources. Visit facilities where your local water supply is stored and treated.

• Meet with biology, geography, and environmental studies teachers and encourage them to examine the challenges to fresh water resources in their lessons and discussions. Suggest projects to evaluate individual lifestyles and patterns of household water use.

• Suggest an alternative to the traditional four food groups taught in school—one that emphasizes the importance of whole grains and deemphasizes the importance of meat and animal products.

Consider the International Connections

If you travel to another country, look for evidence of international competition over water resources. The United States currently has strained relations with its neighbors regarding pollution of the Rio Grande River which it shares with Mexico, and acid rain damage to lakes, trees, and wildlife in Canada. Support international groups that address fresh water issues on a global level; see the listing below.

Further Information

A number of organizations can provide information about their specific programs and resources related to fresh water.*

---

* These include: Acid Rain Foundation, American Rivers, American Water Resources Association, American Water Works Association, Association of State and Interstate Water Pollution Control Administrators, Association of State Drinking Water Administrators, Clean Water Action, Clean Water Fund, Concern, Inc., Conservation Foundation, Environmental Action, Environmental and Energy Study Institute, Environmental Defense Fund, Environmental Policy Institute, Freshwater Foundation, Friends of the Earth, International Rivers Network, League of Women Voters, National Audubon Society, National Wildlife Federation, Natural Resources Defense Council, Renew America, Sierra Club, U.S. Environmental Protection Agency, U.S. Geological Survey, Water Pollution Control Federation, World Resources Institute, and Worldwatch Institute.

Books

Alliance for the Chesapeake Bay. *Baybook: A Guide to Reducing Water Pollution at Home.* 1986. 32 pp. Paperback. Single copies free, multiple copies $1.00 each. Explains how to prevent damage to fresh and salt water resources from soil erosion, sewage, pesticides, and household chemicals.

Boyle, Robert H. and R. Alexander Boyle. *Acid Rain.* New York: Schocken Books, 1983. 146 pp. Paperback, $8.95.

Concern, Inc. *Drinking Water: A Community Action Guide.* Washington, DC: 1986. 31 pp. Paperback, $3.00.

Concern, Inc. *Groundwater: A Community Action Guide.* Washington, DC: 1984. 23 pp. Paperback, $3.00.

Conservation Foundation. *Toward Clean Water: A Guide to Citizen Action.* Washington, DC, 1976. 328 pp. $10.00.

Costner, Pat. *We All Live Downstream: A Guide to Waste Treatment That Stops Water Pollution.* Waterworks Publishing Co. (P.O. Box 548 Eureka Springs, AR 72632), 1986. 92 pp. Paperback, $6.95.

Diamant, Rolf et al. *A Citizen's Guide to River Conservation.* Washington, DC: The Conservation Foundation, 1984. 100 pp. Paperback, $7.95.

El-Ashry, Mohamed T. and Diana C. Gibbons. *Troubled Waters: New Policies for Managing Water in the American West.* Washington, DC: World Resources Institute, 1986. 104 pp. Paperback, $10.00.

Hansen, Nancy Richardson et al. *Controlling Nonpoint-Source Water Pollution.* A Citizen's Handbook. Washington, DC and New York, NY: The Conservation Foundation and National Audubon Society, 1988. 170 pp. Paperback, $7.50.

King, Jonathan. *Troubled Water.* Emmaus, PA: Rodale Press, 1985. 235 pp. Paperback, $8.95.

League of Women Voters. *Safety on Tap.* Publication No. 840. Washington, DC: 1987. 60 pp. $7.95.

Powledge, Fred. *Water: The Nature, Uses and Future of Our Most Precious and Abused Resource.* New York: Farrar Strauss Giroux, 1982. 423 pp. Paperback, $7.95. Reviews policies and practices of EPA, Interior Department, Army Corps of Engineers, industry, agriculture, and others who affect the quality of U.S. water.

Pye, Veronica et al. *Groundwater Contamination in the United States.* Philadelphia: University of Pennsylvania Press, 1983. 315 pp. Paperback, $14.95.

Reisner, Marc. *Cadillac Desert: The American West and Its Disappearing Water.* New York: Penguin Books, 1986. 582 pp. Paperback, $9.95.

Sheaffer, John, R. and Leonard A. Stevens. *Future Water: An Exciting Solution to America's Most Serious Resource Crisis.* New York: Morrow, 1983. 269 pp. $14.95.

U.S. Environmental Protection Agency. *Is Your Drinking Water Safe?* 20 pp. Free. Available from EPA, Public Information Center, PM 211-B, 820 Quincy Street, N.W., Washington, DC 20011.

Articles, Pamphlets, and Brochures

American Chemical Society. "The Water Environment," *Cleaning Our Environment: A Chemical Perspective*, pp. 188-274. Washington, DC, 1978. Paperback, $12.95.

Conservation Foundation. "Water Quantity," *State of the Environment: A View toward the Nineties.* Washington, DC: 1987, pp. 223-234.

Freshwater Foundation. "Agrichemicals and Groundwater Protection: Suggested Directions for Action." Pesticides and Groundwater: A Health Concern for the Midwest." Nitrates and Groundwater: A Public Health Concern." Sample copies free.

La Riviere, J.W. Maurits. "Threats to the World's Water." *Scientific American*, September 1989, pp. 80-94. In many regions water is in short supply and threatened by organic waste and industrial pollutants. International cooperation on integrated water resource management is needed.

Maywald, Armin, Barbara Zeschmar-Lahl and Uwe Lahl. "Water Fit to Drink?" In Edward Goldsmith and Nicholas Hildyard (eds.). *The Earth Report: The Essential Guide to Global Ecological Issues.* Los Angeles: Price Stern Sloan, 1988, pp. 79-88.

Myers, Norman (ed.). "Elements," *Gaia: An Atlas of Planet Management.* Garden City, NY: Anchor Books, 1984, pp. 108-109, 120-121, 132-136.

National Wildlife Federation, Corporate Conservation Council. "Statement of Policy and Practices for the Protection of Groundwater." Washington, DC, undated. 16 pp.

Organization for Economic Cooperation and Development (OECD). "Inland Water Resources," *The State of the Environment 1985.* Paris, 1985, pp. 49-68. Available from OECD, 2001 L St., NW, Washington, DC 20036.

Postel, Sandra. *Water: Rethinking Management in an Age of Scarcity.* Worldwatch Paper 62. Washington, DC: Worldwatch Institute, December 1984. 65 pp. $4.00.

Postel, Sandra. *Conserving Water: The Untapped Alternative.* Worldwatch Paper 67. Washington, DC: Worldwatch Institute, September 1985. 66 pp. $4.00.

Renew America. "Drinking Water," *The State of the States 1989*. Washington, DC: Renew America, 1989, pp. 18-23.

Ridley, Scott. "Groundwater Protection," *The State of the States 1987*. Fund for Renewable Energy and the Environment. Washington, DC: Renew America, 1987, pp. 22-25.

Ridley, Scott. "Surface Water Protection," *The State of the States 1988*. Fund for Renewable Energy and the Environment. Washington, DC: Renew America, 1988, pp. 6-11.

Rogers, Peter P. "Fresh Water." In Robert Repetto (ed.). *The Global Possible: Resources, Development, and the New Century*. A World Resources Institute Book. New Haven: Yale University Press, 1985, pp. 255-98.

Scudder, Thayer. "Conservation vs. Development: River Basin Projects in Africa." *Environment*, Vol. 31., No. 2, March 1989. Describes the conflict between developers who seek hydropower for electrification and industrialization to boost economic growth, and conservationists who work to preserve natural resources and habitats on which riverside communities depend. Discusses possible compromises and solutions.

White, Gilbert F. "A Century of Change in World Water Management." In National Geographic Society, *Earth '88: Changing Geographic Perspectives*. Washington, DC, 1988, pp. 248-60.

World Resources Institute and International Institute for Environment and Development. "Freshwater," *World Resources 1986*, pp. 121-140; *World Resources 1987*, pp. 111-124; *World Resources 1988-89*, pp. 127-142. New York: Basic Books, 1986, 1987, 1988.

Periodicals

Clean Water Action News, Clean Water Action, 317 Pennsylvania Avenue, S.E., Washington, DC 20003. 1 year, 4 issues, $24.

EPI's Groundwater Monitor, published occasionally by the Environmental Policy Institute, 218 D Street. S.E., Washington, DC 20003. Free.

U.S. Water News, Circulation Department, 230 Main Street, Halstead, Kansas 67056. 1 year, 12 issues, $28.

World Rivers Review, International Rivers Network, 301 Broadway, Suite B, San Francisco, CA 94133. Published bimonthly.

Films and Other Audiovisual Materials

Great Lakes: Troubled Waters. 1987. Examines the threat to one of the world's most important fresh water supplies—the sources of pollution and the failure of the United States and Canada to fully address the problem. 53 min. Video, $495. Umbrella Films.

Requiem or Recovery. Explores the transnational threat of acid rain. 29 min. 16mm color film. Available on loan without charge from the National Film Board of Canada.

The Valley Green. 1987. Examines the complex interrelationship of the natural world of Wissahickon Creek with urban Philadelphia. Looks at water quality and the threat to the Wissahickon watershed. 28 min. 16 mm. film, $475; video, $375. Umbrella Films.

Water: A Precious Resource. 1980. Explains where water comes from and how it is endlessly recycled. Demonstrates many ways in which this resource is used and abused. 23 min. Grades 6-adult. 16mm color film, $279.50; video, $69.95. National Geographic Society.

Teaching Aids

Water Watchers: A Water Conservation Curriculum. For Junior High Science and Social Studies classes. Education Development Center, Inc., 55 Chapel St., Newton, MA 02160.

Nonfuel Minerals

The more complex the society, the greater the demand for minerals and the more intricate the interrelationships between minerals, other resources, and social, political, and economic systems. Minerals are critical to national well-being and security.

—Ann Dorr, *Minerals—Foundations of Society*[1]

In an industrial era, the power, the influence, and the security of a nation depend on assured supplies of food, energy, and mineral raw materials and on possession of an industrial structure capable of converting mineral raw materials into essential manufactured goods. Despite this, in the United States there is no commitment to maintaining a strong domestic mineral base.

—Eugene Cameron, *At the Crossroads*[2]

Major Points

- Modern industrial society could not exist without adequate supplies of a wide variety of mineral materials. Worldwide demand for mineral commodities is growing due to increasing population and rising per capita consumption.[3]

- International trade in minerals is essential because no country is entirely self-sufficient. Many industrial countries, including the United States, are heavily dependent on mineral imports.

- Environmental issues—especially air and water pollution, and land use—arise at every step from the search for and production and processing of minerals to their disposal as used products.

- Recycling of metals, building materials, and other mineral products can save money, create jobs, reduce pollution, and conserve energy and water as well as mineral supplies.

- Sound economic, environmental, political, and social decision-making require an understanding of society's reliance on minerals, the uneven worldwide distribution of mineral deposits, and the finite nature of each deposit.

The Issue: Growing Demand, Finite Supply

Minerals are one of the foundations of modern society. They are essential raw materials for economic growth. The issue is how to provide adequate supplies of finite mineral resources, which are unevenly distributed on Earth, in environmentally sound, economically viable ways to meet the long-term needs of a growing, diverse world population whose per capita demands are increasing.

Minerals, the naturally-formed inorganic solids (elements and compounds) that make up the Earth, provided the names which distinguished periods in the development of human civilization: Stone Age, Bronze Age, Iron Age. Each period brought major advances in the ability to use earth materials. New patterns of life evolved.

More than a hundred different mineral commodities are now used daily, in many forms and combinations, to provide the special properties necessary for the materials and products of modern society. For example, modern agriculture and medicine could not have developed without use of minerals. Production and delivery of energy and water depend on availability of minerals, even as production of minerals and mineral products depends on adequate supplies of energy and water.

Mineral deposits (concentrations of minerals that contain particular elements or compounds with potential value) are unevenly distributed on Earth. They are the products of geologic processes that have occurred over tens of thousands to hundreds of millions of years; they are finite resources, nonrenewable in a human time frame. Since the Stone Age, there has been trade in mineral materials. The international dimension of mineral supply is critically important in today's industrial world.

Land must be disturbed in order to extract minerals from the Earth, and costs of land reclamation must be included when planning for mining operations. Often, complex and expensive technologies are required to minimize water and air pollution, especially during mineral processing to separate the desired material from the ore (for example, aluminum from bauxite) or to prepare the commodity for use.

Because minerals rarely are included in public discussions of resource and environmental issues, few people are aware of their important role in modern societies and economies. In 1988, for example, the value of materials of mineral origin produced in the United States was about $300 billion, and U.S. exports of mineral materials were worth about $35 billion.[4] (See Table 10.1.)

Table 10.1
The Importance of Nonfuel Minerals in the U.S. Economy, 1988

	Estimated Value in Billions of Dollars
Domestic production of mineral raw materials	30
Domestic production of processed materials of mineral origin	300
Imports of mineral raw materials	4
Imports of processed materials of mineral origin	40
Exports of mineral raw materials and processed materials of mineral origin	35
Nonfuel mineral trade deficit	9

Source: U.S. Bureau of the Census and U.S. Bureau of Mines

Overview

The use of nonfuel minerals is inherently different from the use of fuels. When petroleum, natural gas, and coal are used, they no longer exist as fuels, and are converted to energy and other substances. They cannot be retrieved and reused for fuel. But when minerals are used, they usually continue to exist—in altered form to be sure, but in most cases the material has not disappeared. The mineral components of many products are being recycled and used again. Mineral recycling has many benefits.

Minerals

Minerals are naturally formed inorganic solids (elements or compounds) with a limited range in chemical composition and with an orderly internal atomic arrangement that determines physical properties. The term mineral also is used in an unscientific way for ores, metals, and materials processed from ores. When people speak of minerals in seawater, they are speaking of ions dissolved from minerals (sodium, chlorine, carbonate, gold); indeed, almost all elements are present to some extent in seawater. Rocks are composed of the most abundant minerals.

Mineral Commodities

Minerals are used in many forms. For example, building stone and road paving aggregates are derived from rocks; mica (in electronics) and halite (salt) are themselves minerals; gold, silver, and others are found as "native" elements; steel and bronze are alloys; cement, concrete, and glass are made by combining mineral components. These are a few of the forms in which commodities of mineral origin are used by society.

Metallic and Nonmetallic Minerals

Metallic minerals can be *ferrous*, meaning that they contain iron or elements alloyed with iron to make steel, or *nonferrous*, those that contain metallic elements not commonly alloyed with iron.

Nonmetallic minerals are classified as *structural materials* such as sand and gravel, stone, cement; or *industrial materials* such as sulfur, salts, fertilizers, abrasives, asbestos, industrial diamonds.

Nitrogen, phosphorus, and potassium are major elements in fertilizers. And at least 24 other elements are essential to plant, animal, or human nutrition. (See Table 10.2.)

Table 10.2
Elements Essential to Plant, Animal, or Human Nutrition

Boron	Copper	Magnesium	Silicon
Calcium	Fluorine	Manganese	Sodium
Carbon	Hydrogen	Molybdenum	Sulfur
Chlorine	Iodine	Nickel	Tin
Chromium	Iron	Oxygen	Vanadium
Cobalt	Lithium	Selenium	Zinc

Source: U.S. Bureau of Mines.

Abundance of Minerals

About 46 percent of the Earth's crust (by weight) is oxygen that is combined with various other elements to make minerals. Silicon comprises a little more than 28 percent. The third and fourth most abundant elements are aluminum and iron, then come calcium, sodium, magnesium, and potassium. Together, these eight elements make up more than 99 percent of the Earth's crust. (See Table 10.3.) Other elements are relatively rare, including many that we know well: copper, which averages only about 55 parts per million (ppm) in the Earth's crust, tin (2 ppm), and gold (0.004 ppm).[5] A few of the elements occur as "native" minerals, but most occur as compounds in complex assemblages of minerals. In scattered locations worldwide, geologic processes have concentrated these sufficiently and in such form that they can be mined.

Table 10.3
Eight Most Abundant Elements in the Earth's Crust

Element	Symbol	% by Weight
Oxygen	O	46.4
Silicon	Si	28.2
Aluminum	Al	8.3
Iron	Fe	5.6
Calcium	Ca	4.2
Sodium	Na	2.4
Magnesium	Mg	2.3
Potassium	K	2.1
		99.5

Source: Ann Dorr, *Minerals—Foundations of Society* (Alexandria, VA: American Geological Institute, 1987), p. 6.

Society depends for its mineral materials (even for such abundant elements as aluminum and iron) on these concentrations which are called *mineral deposits*—naturally occurring mineral concentrations that have potential value, or *ore deposits*—mineral deposits from which specific desired substances can be extracted economically.[6]

Mineral Resources and Reserves

If minerals are indeed finite nonrenewable resources, what is our present supply situation, and what is the outlook for the future? Definitions again are essential.

Resources

The term refers to already discovered or still undiscovered concentrations of mineral materials in such form that a usable commodity can be extracted now, or can be expected to be extracted, in the future. This is a very broad term. Factors to be considered include geologic environments, present and anticipated technologies, limits imposed by depth, and by water and energy availability. It is impossible to say when, or whether, undiscovered resources may become available. Estimates of resources therefore cannot be exact.

Reserves

These are known, identified resources from which a usable mineral commodity can be extracted technologically and economically at the time of designation. Confusion often arises because reserve estimates change with time. When mines are first opened, reserves are estimated on the basis of preliminary exploration indicating that enough ore is present to make mining economically feasible. As mining proceeds, drilling and testing continue until the entire ore body is outlined, and improved estimates of the amount and quality of the ore can be made. Reserve estimates generally increase over time while exploration is under way. However, as mining continues, the amount of ore remaining decreases until it is exhausted or the costs of production exceed the value of the ore. Each ore body is finite. Varying market prices also affect reserve estimates; lower-grade deposits or those more difficult to mine may be dropped from reserves if prices fall. They may, however, still be included in the U.S. Bureau of Mines reserve base.[7]

Current Supply and Demand

Nonfuel minerals are not renewable. Many of them are abundant in the Earth's crust and will never be totally consumed, but only transformed. Yet over time, as human numbers and needs increase, minerals become scarcer in the sense that the most accessible deposits are depleted. Most materials derived from minerals continue to exist in innumerable forms and in even more numerous varieties of products. In many cases, mineral materials contained in manufactured products are readily recyclable. But often they are made into products that are very hard to recycle.

Mineral resources are very unevenly distributed in the Earth's crust. The largest countries (Australia, Brazil, Canada, China, Soviet Union, United States) with diverse geologic environments have many different kinds of mineral resources; however, of the leading industrial countries, only the Soviet Union is today largely self-sufficient. Developing countries possess more than half of the world's supply of nickel, tin, bauxite, and phosphate rock.

Many small developing countries are major world sources of certain materials. For example, Guinea and Jamaica are important producers of bauxite, the main source of aluminum; Zaire is the world's largest producer of cobalt; Chile is the world's largest producer of copper; and Morocco and the western Saharan region have by far the largest reserves of phosphate rock, a source of fertilizer.

Worldwide, consumption of major metallic minerals grew rapidly during the 1960s and 1970s. Between 1965 and 1980, consumption of aluminum, copper, lead, nickel, tin, zinc, and iron ore increased at an average annual rate of almost 4 percent. Since 1980, however, the average yearly growth in consumption of these metals has slowed, and was less than 1 percent per year for the period 1980-1986.[8]

In the United States, many major mineral deposits have been depleted. The lead deposits of southwest Missouri, which for many years provided lead for U.S. industry and export, are exhausted. Even though there are other lead deposits in Missouri, we now are net importers of lead. The bauxite deposits in Arkansas, which were so valuable during World War II, now produce only about three percent of our requirements, and no other U.S. deposits have been found. The high-grade iron deposits of northern Minnesota and Michigan have been depleted, and higher costs of using lower grades of domestic iron ore have led to greater reliance on imports and contributed to the problems in the U.S. iron and steel industry.

The Western European countries that were among the world's early industrial leaders have exhausted many of their resources. Japan has very limited mineral resources of any kind, and has relied almost entirely on imports to build its industrial capacity to world prominence.

The greatest pressure on remaining mineral supplies comes from industrial countries. Though they have only 25 percent of the world's people, they consume about 90 percent of the world's yearly mine production of the three most used metals: iron, aluminum, and copper. The industrial countries use an even higher percentage of elements for high-technology applications. In 1980, for example, while representing five percent of the world's population, the United States produced 11 percent, and consumed more than 13 percent, of the global nonfuel mineral production.[9]

Critical and Strategic Minerals

Minerals are considered *critical* when they are required for the production of essential industrial and consumer goods. This term implies that suitable substitutes are not available.

The term *strategic* is often used to refer to critical minerals for which a country depends on imports, and which are considered vulnerable to supply disruption for causes ranging from instability within a producing country, boycotts, or war.[10]

Industry depends heavily on some 80 minerals. A number of these, including aluminum and iron, are relatively abundant and widely distributed, although bauxite deposits are located chiefly in the tropics and subtropics. A few critical and strategic mineral commodities—chromium, cobalt, manganese, and platinum—are of particular concern because the largest reserves are in central and southern

Africa and the Soviet Union. As the world's largest consumer of strategic minerals, the United States is highly dependent on imports of these minerals, and is particularly vulnerable to supply interruptions. In 1988, the United States imported 75 percent of its chromium (the balance came from recycling), 84 percent of its cobalt (most of the remainder recycled), 93 percent of its platinum group metals, and 100 percent of its manganese.[11]

Japan and members of the European Economic Community (EEC) are also highly dependent on mineral imports. In contrast, the Soviet Union is much less dependent on imports, and is largely self-sufficient in most important mineral materials.[12] (See Figure 10.1.)

To illustrate the importance of strategic minerals, Table 10.4 shows the quantities of several of these minerals contained in a jet aircraft engine.[13]

Table 10.4
Strategic Mineral Requirements for a Pratt & Whitney F100 Turbofan Aircraft Engine

Titanium	5440 lb	Aluminum	670 lb
Nickel	4504 lb	Columbium	145 lb
Chromium	1485 lb	Manganese	23 lb
Cobalt	885 lb	Tantalum	3 lb

Source: U.S. Bureau of Mines

Future Demand

Because of rapid population growth and increasing per capita consumption, mineral demand is increasing sharply.

Figure 10.1
Reliance on Imports of Selected Nonfuel Mineral Materials

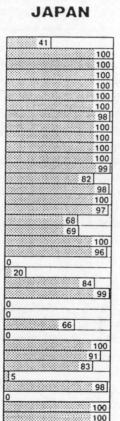

1987 NET IMPORT RELIANCE SELECTED NONFUEL MINERAL MATERIALS

1988 NET IMPORT RELIANCE SELECTED NONFUEL MINERAL MATERIALS

	E.E.C	JAPAN	U.S.A.	U.S.S.R.
ARSENIC	23	41	100	0
COLUMBIUM	100	100	100	0
MANGANESE	98	100	100	0
MICA (sheet)	100	100	100	12
STRONTIUM (celestite)	0	100	100	0
YTTRIUM	100	100	100	0
BAUXITE & ALUMINA	65	100	97	60
PT-GROUP METALS	100	98	93	0
FLUORSPAR	6	100	91	53
DIAMOND (industrial stones)	100	100	90	0
TANTALUM	6	100	89	0
COBALT	100	100	84	47
ASBESTOS	56	99	78	0
TUNGSTEN	61	82	75	43
CHROMIUM	96	98	75	0
NICKEL	94	100	75	0
TIN	90	97	73	33
BARITE	17	68	71	50
ZINC	65	69	70	10
POTASH	15	100	69	0
ANTIMONY	93	96	65	5
CADMIUM	81	0	47	12
GYPSUM	0	20	40	0
FERROSILICON	40	84	35	0
IRON ORE	92	99	22	0
CEMENT	0	0	19	0
IRON & STEEL	0	0	18	0
LEAD	75	66	14	8
SULFUR	14	0	14	13
BERYLLIUM	100	100	13	8
COPPER	99	91	13	0
SALT	0	83	11	0
NITROGEN	9	5	11	0
ALUMINUM	40	98	10	0
TITANIUM (sponge metal)	65	0	8	0
MOLYBDENUM	100	100		9
PHOSPHATES	100	100		0

Source: U.S. Bureau of Mines

In the past two generations, the world has consumed more minerals than were used in previous human history. Between 1950 and 1987, while world population doubled, production of the three most used metals increased far more: copper nearly 300 percent, steel almost 400 percent, and aluminum more than 1000 percent. Mineral resources that were formed over hundreds of millions of years are now being consumed in only decades.[14] In 1984, the global demand for most major nonfuel mineral commodities was projected to increase by 3 to 5 percent annually, almost doubling between then and the year 2000.[15] Recent estimates of future increases in demand vary widely from mineral to mineral. Largest increases in the ferrous metals are anticipated in developing countries for the building of infrastructure (highways, bridges, airports, and energy and water systems, for example). Major increases in the use of high-technology materials will occur in the highly industrialized countries.

Future Supply: Contrasting Views

The question of future availability of minerals is often seen from two different points of view: one of geologists and other physical scientists, another of economists.

Geologists consider the availability of minerals ultimately to depend on the physical nature of the Earth: first, the distribution of mineral concentrations, and second, the physical laws of nature that impose limits on our ability to locate deposits, extract desired materials, and process them. Better understanding of geologic processes makes it easier to anticipate where certain types of minerals may have been formed. Modern exploration techniques have greatly facilitated the search for surface and near-surface deposits. But there are limits to the depths at which mining is feasible, due to steadily increasing heat and pressure. As depth increases, costs of production rise rapidly, not just in dollars but also in energy, water, and other materials required for mining. At some point, economic extraction becomes impossible. These and other factors also set lower limits on the grade of ore that can be processed.

Each mineral deposit is finite, even the abundant and widespread deposits of sand, gravel, and stone. It is impossible to know when, where, or whether deposits located beneath the surface will be discovered, what their size and grade will be, and if they will be minable.

Economists, by contrast, tend to view the supply of minerals as essentially infinite because the entire solid Earth is made of minerals. Economists assume that market mechanism will take care of supply and demand, reflecting increased prices when supplies are limited, thereby encouraging exploration and production as well as development of new technologies to make possible both mining and extraction of desired commodities even from lower-grade ores. Economists point out that the cost of raw material is such a small part of the final cost of mineral products that

large increases in prices of those materials would make little difference.[16]

It is difficult to forecast the long-run adequacy of many mineral commodities. Estimates of worldwide resources of some minerals are large and could seem to imply long-term security for hundreds of years. Considering their accelerating rates of use, however, reserves of many minerals are not large and could be depleted within decades. Based on changes in annual production rates from 1973-1980 — which ranged from -5.1 percent for mercury (now expected to decrease more slowly) to +7.3 for barite (now expected to increase more slowly) — possible life-expectancies for reserve bases of mineral commodities vary from almost infinite to less than 50 years.[17] (See Table 10.5.)

Table 10.5
Possible Life Expectancy of Selected Mineral Resources

Mineral	Estimated Life Expectancy
Salt, magnesium metal	Almost infinite
Lime, silicon	Up to thousands of years
Potash, manganese ore, cobalt	Over 200 years
Iron ore, chromite, feldspar	100 - 200 years
Bauxite (aluminum ore), phosphate rock, nickel	50 - 100 years
Copper, mercury, zinc, lead	Less than 50 years

Source: Eugene N. Cameron, *At the Crossroads: The Mineral Problems of the United States.* (New York: Wiley-Interscience, 1986), pp. 158-59.

As available supplies of many of the mineral commodities on which society depends today are depleted, international competition for remaining supplies will increase. Changes in the use of these materials will be essential and almost certainly will cause changes in life-styles. We must anticipate these inevitable developments and plan for them in time to extend the supplies of critical materials and provide for a smooth transition to whatever replaces them.[18]

Many uncertainties make it difficult to forecast how long we have to devise viable changes in the use of mineral resources. These include:

- Increases in demand for mineral resources;

- Probability of price increases that could spur exploration and exploitation of lower-grade deposits;

- Rate of discovery of hidden resources;

- Development of new technologies that will make either the production of mineral materials or their use more efficient;

- New technologies that may require materials for which we do not now anticipate a need;

- Discovery or development of adequate substitutes for materials in short supply; and

- Extent to which recycling can reduce demands for primary (new) raw materials.

Environmental and Land-Use Issues

With the first large-scale use of minerals and fossil fuels around 1850, when world population was about one billion, atmospheric pollution became obvious. Pollution was especially apparent in urban areas where coal was the major fuel, and near smelters that extracted minerals from their ores. In the United States, evidence of the devastation from copper smelting in the early 1900s is still visible near Copperhill and Ducktown, Tennessee, in spite of reclamation efforts that began in the 1930s. There were no environmental controls in the early 1900s, and few people had thought seriously about the long-term effects of pollution.

Beginning in 1970 with passage of the National Environmental Policy Act, followed by the Clean Air Act of 1970 and the Clean Water Act of 1972, a series of actions was taken to regulate mining and mineral processing activities with the goal of reducing air and water pollution. Air pollution, even from surface mining, consists mostly of dust from crushing, screening, and concentrating, and has been limited by using wet processes. Stabilizing and restoring piles of tailings to environmentally-acceptable conditions and other land reclamation projects for mined-out areas are required by existing regulations.[19]

In 1986, the Environmental Protection Agency determined that regulation of mining waste as hazardous waste was not warranted, because metals and other materials for industrial use had been removed, and the remaining waste consisted largely of common rock materials.

Under the Clean Air Act, stack emissions from smelting and refining operations are closely monitored and regulated. Both capital costs and operating expenditures for pollution-control equipment are very high. Air quality requirements caused the closing of several smelters at which pollution control would have raised production costs to the point at which the plants could not compete with smelters abroad. Today, U.S. copper smelters must remove 90 percent of sulfur dioxide emissions. Japanese smelters remove 95 percent, but removal in other countries, including Canada, ranges from 0 to 35 percent.[20] (See Figure 10.2). At two Mexican smelters near the Arizona border, initially operated without pollution controls, prevailing winds carry emissions into the United States that are thought to cause acid rain in several western states.[21]

Recently, a U.S. Federal Court recognized the health hazard of emissions from lead processing when it ruled that impounded primary lead smelter wastes must be listed and treated as hazardous substances.[22]

Land use for exploration, mining, and mineral processing, particularly on Federal lands, is another issue that has attracted public attention, even though most mines are far from population centers. From 1930 through 1980, mining, including coal mining, used only 5.7 million acres of land, (2.7 million of which has been reclaimed) compared to 7 million acres for railroads and airports, and 21.5 million acres for highways.[23]

While the United States was expanding westward, mineral extraction was recognized by law to have the highest priority among land uses. In recent years, because of pressure for other uses of federal lands, access for mineral exploration and development has been greatly restricted. Much of the resistance to use of federal lands for mining stems from past abuses when there were no regulations to preserve or restore the land.

Geological conditions rather than human laws are the ultimate limitations on mining sites. In the United States, the best and largest potential sites remaining for mineral discoveries are in the West. Mines and support facilities occupy small areas, but larger areas must be explored (requiring only minimal disturbance to the land) in order to locate mineral deposits.

The basic law governing most mining activities on federal lands was enacted in 1872. This law has since been amended and made more restrictive several times, most radically by the Leasing Act of 1920 (also amended several times), which governs fuels, fertilizers, and chemical minerals.[24]

State and sometimes local governments also have roles to play in regulating mining. Restrictions are often imposed by environmental regulations. Many mineral resources, particularly stone, gravel, and sand, which are required in enormous amounts in urban areas, have been covered over by development or preempted for other uses. Zoning restrictions and citizen opposition often prevent mining and quarrying, even when the material would be of benefit to the jurisdiction.

Prospects for the Future

Two possibilities are available to help delay the depletion of mineral resources on which present industrial societies depend. We can make more efficient use of traditional resources. And, we can develop technologies to extract minerals from sources that we have been unable to tap effectively. Both possibilities will require citizen understanding and support, as well as international cooperation.

Figure 10.2
Sulfur Dioxide Control in Copper Production in Selected Countries

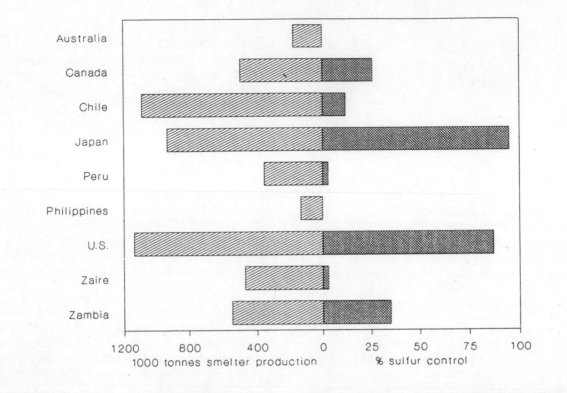

Source: Duane Chapman, "The Economic Significance of Pollution Control and Worker Safety Cost for World Copper Trade," Cornell Agricultural Economics Staff Paper (Ithaca, NY: Cornell University, 1987), as published in U.S. Congress, Office of Technology Assessment, *Copper: Technology and Competitiveness* (Washington, DC, 1988), p. 17.

Improving Efficiency

There is room for improved efficiency at all stages of mineral production and use—including mining, processing, manufacturing, product use, and recycling of used materials. For copper, Figure 10.3 on the following page illustrates the many production stages in which improved efficiency may be possible.

Mining and Processing

When a mineral deposit is mined, the practice of "high-grading"—removing high-grade portions and leaving lower-grade materials behind—should be avoided as much as possible. High-grading has been practiced for two reasons: lack of technology to extract low concentrations of a mineral, and high cost of processing low-grade materials.

Some technologies exist now and others are being developed to extract minerals that are present in very low concentrations. Success depends on what the desired element is and how it is combined with other elements. In many cases, mixing of high-grade with low-grade ores makes it possible to extract elements economically that otherwise would have to be abandoned. Deliberate high-grading to maximize present profits regardless of future costs should be prohibited.

Some elements can be obtained only as by-products of mining for other minerals. In the past, the importance of many of these elements was not recognized, and they were left behind in tailings or carried off in liquid processing wastes. Three such elements are rhenium, selenium, and tellurium, for which copper production has been the major source. When by-product materials are present in ores, they should be extracted whenever possible, even if they have to be stockpiled until a market is available.

Progress is being made continually on more efficient methods of smelting and refining to extract the desired commodity from ores.

Manufacture, Use, and Recycling

Manufacturing for efficient use involves producing goods with the longest possible useful life rather than constantly modifying products to make previous ones obsolete. Products should be designed to permit rebuilding and refurbishing. Size reduction and miniaturization also decrease

Figure 10.3
Principal Stages of the Copper Production Process

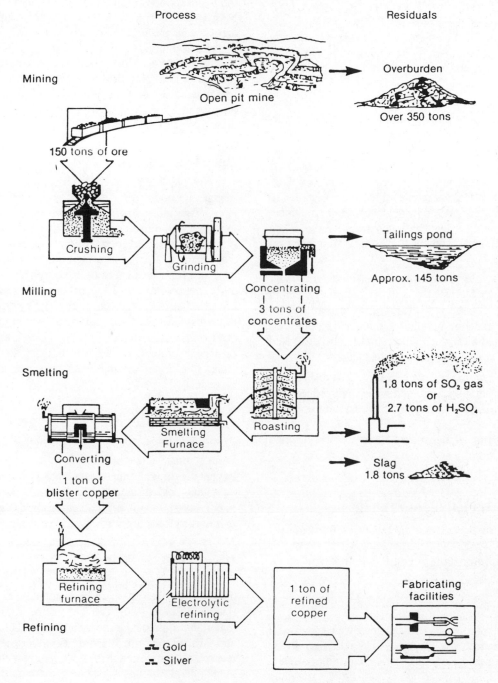

Process   Residuals

Mining — Open pit mine → Overburden — Over 350 tons

150 tons of ore

Crushing — Grinding — Concentrating

Milling

Tailings pond — Approx. 145 tons

3 tons of concentrates

Smelting

Roasting — Smelting Furnace — Converting

1.8 tons of SO₂ gas or 2.7 tons of H₂SO₄

Slag 1.8 tons

1 ton of blister copper

Refining furnace — Electrolytic refining — 1 ton of refined copper — Fabricating facilities

Refining

Gold
Silver

NOTE: Tonnage of residuals is based on experience in the Southwestern United States assuming an ore grade of 0.6 per cent copper

Source: J.F.McDivitt and G. *Manners, Minerals and Men* (Baltimore, MD: The Johns Hopkins University Press, 1974, a Resources for the Future publication), as published in U.S. Congress, Office of Technology Assessment, *Copper: Technology and Competitiveness* (Washington, DC, 1988), p. 4.

the amount of material required: the production of smaller cars and printed electronic circuits are examples.

Recycling of materials from both manufacturing and consumer waste offers a major opportunity to conserve nonfuel minerals and reduce environmental damage from their use.[25] Some mineral products are dispersed during use and cannot be recycled (for example, fertilizers, cleaning fluids, and manganese required for steel-making), but most can be reprocessed and used again if they are properly disposed of. (See Chapter 14.)

In 1988, about 36 percent of apparent consumption of aluminum metal in the United States was from new purchased scrap, and another 18 percent was from old scrap such as cans. In 1989, 25 percent of U.S. nickel consumption was from purchased scrap. Almost 50 percent of the lead used was from scrap batteries. Recycling reduced U.S. chromium imports by 25 percent and cobalt imports by 16 percent—both are critical and strategic materials for which, without recycling, the United States would have relied completely on imports, mostly from southern and central Africa.[26]

Aluminum recycling is being facilitated by reverse vending machines which dispense coins in return for empty aluminum cans. By the end of 1985, more than 12,000 reverse vending machines could be found in 15 countries. Sweden alone had a third of the machines and the United States, Norway, and France together had half.[27]

Metal recycling has many benefits. It saves capital, creates jobs, conserves energy and water, and reduces environmental impacts. For example, when aluminum is recycled instead of produced from virgin ore, the process reduces energy use by between 90 and 97 percent, and lessens air pollution by 95 percent and water pollution by 97 percent.[28] (See Chapter 14.)

Minerals from the Sea

Although extraction of mineral resources from the sea poses economic, technological, and legal problems, the oceans are an important potential source for many minerals.[29]

Seawater

Almost every element is present in seawater, but most occur in concentrations too low for extraction to be economically feasible. Sodium chloride (common salt) and magnesium sulfate (Epsom salts) can be obtained by evaporation, and some magnesium metal has been produced at very great energy costs. Bromine and iodine also are extracted from seawater.[30]

Continental Borders

The sediments of the continental shelf and slope contain various minerals eroded from land. Some of them—gold,

platinum, diamonds, titanium, phosphorite, and tin—have been mined by dredging. Sulfide minerals are present in the Gorda Ridge and Juan-de-Fuca regions off the northwest coast of United States; these minerals may be of future interest, although neither their value nor the costs of mining have as yet been fully assessed. These continental border areas are in the Exclusive Economic Zone (EEZ), which extends 200 miles beyond the shores of the United States and its territories. This zone was proclaimed by President Reagan in 1983 after the United States declined to sign the international Law of the Sea Treaty. Minerals in the EEZ are managed by a new U.S. agency, the Minerals Management Service.[31]

Deep Sea Deposits

Since the 1960s, there has been major interest in ferromanganese nodules widely scattered over the deep ocean floor. They consist of manganese oxide and iron oxide minerals with nickel, copper, cobalt and minor amounts of other elements. The composition of nodules varies widely from place to place, as do their shape, structure, and concentration on the ocean floor. Research has been conducted and experimental technologies have been developed, but so far there has been no commercial exploitation of these minerals. Costs of exploration and mining are enormous, environmental aspects are not fully understood, and provisions of the Law of the Sea Treaty do not offer sufficient protection to mining companies to encourage them to proceed.[32]

International Aspects

Since mineral resources are distributed unequally in the Earth's crust, and since some countries consume much larger quantities of minerals than others, a country's domestic supply seldom matches its mineral needs. As a result, international trade in mineral commodities is an important part of the global economy.

Industrial Countries

Most industrial nations consistently run trade deficits in minerals because of their high consumption and limited reserves. Since 1983, the U.S. trade deficit in mineral materials required by industry has varied between $9 billion and $16 billion.[33] In the case of shortages on the world market, industrial countries become competitors for available mineral supplies.

Soviet Union

The Soviet Union is largely self-sufficient and is a major exporter of strategic minerals. In the early 1980s, the U.S.S.R. accounted for about 23 percent of the world's nickel production, 26 percent of chromium, 40 percent of manganese, and 49 percent of platinum group metals.[34] In recent years, it has reduced exports of iron ore, manganese, chromium, and nickel, and has begun to import small amounts of some minerals for the first time. It is not certain to what extent these changes reflect depleted reserves or political or economic motives. Undoubtedly, large resources remain in Siberia, but mining conditions there are difficult and costs of extraction are great.

Developing Nations

The term "developing nations" includes countries of widely varying size, population, resource potential, level of economic development, and social and political characteristics. With respect to minerals, they are grouped together mainly because their per capita consumption of minerals is very low compared with industrial countries. Yet most developing nations face large mineral needs if they are to develop their infrastructure.

Developing countries differ greatly in their domestic mineral resources. Brazil and China are very large countries with a wide variety of mineral resources; others are small and have only one or two mineral resources of world value. Guinea is the world's second largest producer of bauxite; 97 percent of its export earnings are from mineral exports, almost all from bauxite. Copper and byproduct cobalt accounted for 90 percent of Zambia's export earnings in 1984, when it was the world's fifth largest producer of copper and second largest of cobalt. Cobalt is a strategic material, not subject to wide market fluctuations, but copper is very vulnerable to price changes resulting from supply-demand factors. Low copper prices have caused serious economic problems for Zambia.[35]

Some have suggested that Third World nations might cut production or limit markets in order to raise the prices of mineral commodities, but because they are so dependent on income from mineral exports, these countries probably would be reluctant to risk retaliation.

Increasingly, developing countries that mine ores are finding it advantageous to build processing plants (often financed by international institutions or companies) to upgrade their exports; for example, processing of bauxite to alumina (anhydrous aluminum oxide), or pelletizing iron to increase the grade of raw material shipped. Many of the ferroalloys required for steel-making are produced in Third World countries.

This trend toward greater processing benefits the producing countries by increasing the export value of the product and creating jobs requiring higher skills. It benefits the importing countries by reducing energy, water, and environmental costs of processing ores.[36] Yet serious questions remain. What will these developing countries, which are dependent on a few finite resources, do as their deposits are depleted? Will they be able to develop alternative sources of income to support their growing populations with aspirations for higher living standards? How can the developed countries help? Satisfactory answers to these questions are important to all countries.

In developing countries, much of the exploration to locate mineral deposits, money to finance development, and technical skills to bring mines into production have come from foreign sources. In many cases, foreign ownership has been greatly reduced in the last few decades as some countries have expropriated mines and others have insisted in majority national ownership, often by government companies.

The international minerals picture is extremely complex and constantly changing. Few countries have the resources to develop as Brazil has, but the dramatic changes there since World War II, based largely on mineral development, are worth noting. From a country with small local mines worked by hand labor, Brazil has become the world's second largest producer of iron ore and the third largest supplier of bauxite. It has become the largest producer of manganese among market-economy countries. Brazil has developed large steel and aluminum industries using some of the most advanced technologies. Brazil has not only exported both raw and processed materials to many parts of the world (it is a major supplier of raw materials for Japan), but it also has developed manufacturing industries and exports automobiles, airplanes, and electronic equipment. Brazil buys large amounts of aluminum scrap from the United States, which it converts into automotive parts for major U.S. and European automobile manufacturers. In many ways, Brazil is still a Third World country, but it has developed its infrastructure and industries in an amazingly short time. It is now beginning to tackle some of its growing environmental and population problems.

A Look Ahead

Scientists are searching for new materials—for new uses or to serve as substitutes for materials in short supply. Substitutes have been found for many materials, but often there are drawbacks: the substitutes may not be as effective as the material they replace, they may cost more, or their availability may be limited. Extensive research on new materials is being conducted, including work on advanced metals, ceramics, polymers, and composites. There have been many exciting discoveries, and much remains to be learned.

It is clear, however, that whatever materials we develop and use in the future will have to come from the Earth. In that process, land, air, water, and energy supplies will be affected.

What You Can Do

Inform Yourself

Get more information about minerals by consulting sources and contacting organizations listed under Further Information.

- Make an effort to become aware of the things that you use daily that are made of minerals or have mineral components; for example, the mineral materials required for your television set, your automobile. Where do they come from? What fertilizers do you use? What minerals do they contain and where did they come from?

- Check in your library to see what sources of information on minerals they have. If the United States Geological Survey or Bureau of Mines has an office near you, call or visit and ask what they have published that would be of use to you as a citizen.

Join With Others

There are not many organizations that focus on minerals (some are listed under Further Information), but there are many that have committees or working groups interested in environmental, natural resource, and population issues. Minerals are related to all of these. Suggest that these organizations include minerals in one of their studies.

- If your community is involved in recycling, look into the whole issue of mineral recycling—not just cans and bottles, but also automobiles, appliances, building materials, batteries, and other products. Try to visit a metal recycling operation and learn how metals are collected and reprocessed. If there are no recycling facilities in your area, consult with community leaders about the possibility of organizing a recycling program.

- If you live in an area where there are mines, quarries, mineral processing plants, or manufacturing plants that use minerals, try to visit these facilities. Ask what methods they use to limit pollution. Find out where the building and road materials used in your area come from.

Review Your Habits and Lifestyle

Do whatever you can to use and reuse products containing mineral materials more efficiently.

- By recycling cans, bottles, and paper, you can help save minerals, energy, and water used in their production. Recycle your own reusable mineral-based goods and encourage your friends and neighbors to do the same. When you dispose of a car battery, be certain that it goes to a dealer who will recycle it for its lead content.

- When you purchase an appliance or other consumer product, ask the dealer how it can be repaired, and whether replacement parts are available. Try to buy products that can be refurbished or repaired rather than those that must be replaced. Look for products that are durable and will not become obsolete or unusable in a short time.

Work With Your Elected Officials

Encourage your elected representatives to consider efficient use of minerals in planning and allocating funds for major governmental projects.

- If mining and mineral processing take place in your area, suggest that your representatives consider their importance to citizens, to other local industries and businesses, and to the economic development of your region.

- Urge your elected state and national representatives to:

  » promote the recycling of metals and other mineral materials by working to remove economic and institutional barriers to recycling;

  » exclude highly sensitive or valuable areas from minerals development;

  » establish and enforce reasonable standards for restoring mining sites and preventing offsite pollution; and

  » support the "polluter pays" principle for environmental protection in the production of minerals.[37]

Publicize Your Views

Express your view on the need for efficient use of mineral resources to opinion leaders in your community, and to the print and electronic media through letters to the editor and public service announcements.

- Volunteer to write articles for local publications about the importance of minerals. In your writing, refer to the significance of minerals in our everyday lives and to their importance in the U.S. and the world economy.

Raise Awareness Through Education

Find out whether schools, colleges, and libraries in your area have courses, programs, and adequate resources on mineral materials; if not, offer to help develop or supply them.

- If earth science courses are taught in local schools, inquire whether sections on minerals emphasize their importance to society.

- Encourage schools to arrange field trips to see mining or mineral processing operations, or to see how metallic and nonmetallic mineral materials are used or recycled in your area.

Consider the International Connections

Learn as much as you can about the international distribution and consumption of minerals and about international mineral trade. When you travel to other countries, keep your eyes open and take advantage of opportunities to learn about mineral production, use, and regulation.

Further Information

A number of organizations can provide information about their specific programs and resources related to nonfuel mineral resources.*

Books

Abelson, Philip H. and Allen L. Hammond (eds.). *Materials: Renewable and Nonrenewable Resources.* Washington, DC: American Association for the Advancement of Science, 1976. 196 pp. Paperback, $7.50. Contains 38 articles originally published in *Science* magazine.

Barrett, Thomas S. *Self-Initiation: The Hardrock Miner's Right.* The Public Resource Foundation, 1987.

Bates, Robert L. and Julia A. Jackson. *Our Modern Stone Age.* Los Altos, CA: Kaufmann, 1982. 150 pp. Hardback, $12.95. An excellent book for the lay person on the many uses of nonmetallic minerals. Available from the American Geological Institute.

Cameron, Eugene N. *At the Crossroads—The Mineral Problems of the United States.* New York: Wiley, 1986. 320 pp. Paperback, $19.95. Suitable for grade 10 to adult. Concludes that domestic policies must be modified to create an environment more favorable to the discovery and development of new mineral deposits.

Dorr, Ann. *Minerals—Foundations of Society.* Second Edition. Alexandria, VA: American Geological Institute, 1987. 96 pp. Paperback, $9.95. An overview of nonfuel minerals—their significance, their origin and distribution, the complexities of making them available, the role of minerals in U.S. development, and information on present and future U.S. and world minerals problems and challenges. Includes a bibliography.

Environmental Policy Institute et al., *Minerals and Public Lands: An Analysis of Strategic Minerals Issues and Public Lands Policy.* Washington, DC, October 1981. 105 pp. $5.00.

Johansen, Harley E. et al., (eds.). *Mineral Resource Development: Geopolitics, Economics, and Policy.* Boulder, CO: Westview Press, 1987. 369 pp. $34.00. Essays on the economic and business roots of the global minerals industry, conflicting goals of mining companies and host governments, and mineral development and environmental and community concerns.

Leshy, John D. *The Mining Law.* Washington, DC: Resources for the Future, 1987. A history of the West from the viewpoint of mineral exploration and production. Gives a detailed analysis of the 1872 mining law, its application, and continuing problems.

Mineral Information Institute, Inc. *Mineral and Energy Information Sources,* Fourth Edition. Denver, CO, 1987. 145 pp. Paperback. An extensive listing of booklets, films, slide shows, and other resources related to energy and minerals.

Strauss, Simon D. *Trouble in the Third Kingdom.* London: Mining Journal Books, 1986. 240 pp. Available from American Metal Market for $35.00 plus $1.00 postage. A worldwide view of minerals and the problems that exist in making them available.

U.S. Department of the Interior, Bureau of Mines. *Mineral Commodity Summaries.* Washington, DC, issued annually. The best source of concise, up-to-date information on mineral commodities.

U.S. Department of the Interior, Bureau of Mines. *Mineral Facts and Problems.* Washington, DC, published every five years. An overall view of all the nonfuel mineral commodities, covering industry structure, uses of material, reserves and resources, technology, supply and demand, and other aspects for each commodity.

Wolfe, John A. *Mineral Resources: A World Review.* New York: Chapman and Hall, 1984. 293 pp. Paperback.

Articles, Pamphlets, and Brochures

Abelson, Philip H. (ed.). "Advanced Technology Materials." *Science,* Vol. 208, No. 4446, May 23, 1980.

---

* These include: American Association for the Advancement of Science, American Geological Institute, American Institute of Mining, Metallurgical, and Petroleum Engineers, American Mining Congress, The Conservation Foundation, Environmental Defense Fund, Environmental Policy Institute, Mineral Information Institute, U.S. Bureau of Land Management, U.S. Bureau of Mines, U.S. Forest Service, U.S. Geological Survey, and World Resources Institute. Addresses for these organizations are listed in the appendix.

An entire issue on special material needs and new materials to meet those needs.

American Mining Congress. "The Resources War." A booklet that tells why America's minerals base is essential to the nation's security, economic well-being, and global competitiveness. Illustrated with color photographs and charts. 16 pp. For grade 9 to adult.

Chandler, William U. *Minerals Recycling: The Virtue of Necessity.* Worldwatch Paper 56. Washington, DC: Worldwatch Institute, October 1983. 52 pp. $4.00.

"Deep Ocean Mining." *Oceanus*, Vol. 25, No. 3, Fall 1982. Includes articles on minerals, technology, environment, policy, and law of the sea.

Myers, Norman (ed.). "The Elemental Potential." In Myers (ed.), *Gaia: An Atlas of Planet Management.* Garden City, NY: Anchor Books, 1984, pp. 102-11.

Vogely, William A. "Nonfuel Minerals and the World Economy." In Robert Repetto (ed.), *The Global Possible: Resources, Development, and the New Century.* New Haven: Yale University Press, 1985, pp. 457-73.

World Resources Institute and International Institute for Environment and Development. "Energy and Minerals," *World Resources 1987*, pp. 299, 306-12; "Energy, Minerals, and Waste," *World Resources 1988-89*, pp. 311-316. New York: Basic Books, 1987, 1988.

Periodicals

AMC Journal. Promotes public understanding of the central role of minerals and mining in American society. Published monthly by American Mining Congress. $36 a year.

Earth Science. Contains news for a general audience about geologic phenomena worldwide. Published quarterly by American Geological Institute. $12 a year.

Geotimes. Covers recent research, geologic events, scientific meetings, trends in government policy and education. Published monthly by American Geological Institute. $20 a year.

Films and Other Audiovisual Materials

The Minerals Challenge. Shows how the Nation's growing needs for fuels, metals, and other mineral materials are being met by technological advances. 27 min. Grade 9 to adult. 16 mm color film. U.S. Bureau of Mines.

Mining: Who Needs It? Award-winning audiovisual on mining's contribution to our way of life and our economy, and how mining strives to be sensitive to environmental considerations. Accompanying booklet allows the audience to take the message home. 15 min. Grade 9 to adult. American Mining Congress.

Resource in Crisis. 1975. A film on the mineral industry, calling on both critics and industry spokespersons to articulate viewpoints on mineral industry taxation, profits, environmental impacts, and occupational health issues. Produced by the Canadian Institute of Mining and Metallurgy. 60 min. College-adult. 16 mm color film. Noranda, Inc.

Riches from the Earth. 1982. Includes precious metals, commercially-valuable ores, and other minerals. Covers discovery and depletion of minerals, as well as technology, conservation, and recycling. 20 min. Grade 7 to adult. 16 mm color film, $230; video tape, $60. National Geographic.

There's More to Mining than Meets the Eye. Points out that construction aggregates and other mining products will be essential to provide homes, buildings, schools, and roads in the coming years. Explains mining processes and extensive reclamation efforts of the mining industry. 25 min. Grade 10 to adult. 16 mm film. American Mining Congress.

For additional audiovisual materials, obtain a copy of *Mineral and Energy Information Sources*, listed under Books above.

Teaching Aids

Christensen, John W. *Global Science: Energy, Resources, Environment.* Second Edition. A complete curriculum—including 355-page student text, laboratory manual, and teacher's guide—designed as a secondary level science program. Helps students understand interactions between people and their environment, with special emphasis on energy and mineral resources. Stresses conservation and environmental considerations. A project of the Mineral Information Institute. Thirty-day review copies available; contact Kendall/Hunt Publishing Co., 2460 Kerper Boulevard, Dubuque, IA 52001, telephone 1-800-258-5622.

Mining, Minerals, and Me. A multidisciplinary program designed to help students understand and appreciate the Earth and its mineral resources. Contains ten modules with activity-oriented programs; each takes about 20 days to teach. Students observe, draw, describe, construct, discuss, manipulate, act, and write as they learn. Grades K-6. Average cost per module, $65. Delta Education, Box M, Nashua, NH 03061-6012. Telephone (800) 258-1302.

Energy

The United States consumes one-fourth of the world's energy each year. Yet, for a given amount of energy, the United States produces less than half as much economic output as Japan and West Germany.

—*Time* Magazine[1]

The cost-effectiveness of "efficiency" as the most environmentally benign "source" of energy is well established. The energy consumption per unit of output from the most efficient processes and technologies is one-third to less than one-half that of typically available equipment.

Every effort should be made to develop the potential for renewable energy, [which] should form the foundation of the global energy structure in the twenty first century.

The generation of nuclear power is only justifiable if there are so   olutions to the presently unsolved problems to which it gives rise.

—World Commission on Environment and Development, *Our Common Future*[2]

Major Points

- Air pollution from growing use of fossil fuels is harming human health, causing acid rain damage to entire ecosystems, and increasing the buildup of atmospheric carbon dioxide and the likelihood of global warming and climate instability.

- Despite the current world surplus of petroleum, obtaining adequate and affordable energy supplies over the long term presents serious economic and environmental problems for both industrial and developing countries.

- The basic vulnerability of the world economy to oil supply disruptions is far from being eliminated. Unpredictable events could precipitate another oil price shock.

- Oil purchases are a major source of debt for most developing countries, and the world's poorest regions face increasing shortages of fuelwood, their primary energy source.

- The spread of nuclear power has expanded the potential for nuclear accidents, proliferation of nuclear weapons, and blackmail and terrorism involving nuclear materials, and has intensified the problem of nuclear waste disposal.

- In view of the uncertainties, costs, and risks associated with energy obtained from oil, gas, coal, and nuclear materials, it is essential to increase energy efficiency and conservation, and to develop a variety of renewable, solar-related energy sources as soon as possible.

- Government and private incentives to conserve energy and use energy more efficiently can slow or even reverse growth in energy consumption, and can result in important environmental, economic, and social benefits.

The Issue: Growing Energy Demand, Inefficient Use, Environmental Pollution

Energy is among the most essential of our Earth's resources. Without the heat, light, and food it provides, human civilization would not exist. Since World War II, the world's energy consumption has increased about fourfold. The use of fossil fuels has grown rapidly, and enabled many nations to achieve high standards of living. Yet most energy is used inefficiently, and widespread consumption of coal and oil produces pollution that threatens air quality, vegetation, and climate stability. To a large extent, the ability of our environment to support life depends on the kinds of energy choices we make, and especially how efficiently we use our energy supplies.

Current Conditions and Trends

Patterns in world energy use are marked by striking contrasts and inequities. Differences in wealth, in economic development, and in the priorities assigned to energy efficiency and conservation create huge disparities in the amounts of energy consumed from country to country. A knowledge of how energy is used is essential for understanding the importance of improving energy efficiency and for alleviating the wide range of energy-related environmental problems.

Oil accounts for 43 percent of the world's commercial energy production, coal and other solid fuels 31 percent, natural gas 21 percent, and electricity from nuclear and hydro power 5 percent. Over 2 billion people, mostly in the Third World, rely on fuelwood for energy, yet many developing countries face serious fuelwood shortages.[3]

The United States and the Soviet Union use far more energy than other countries, and they use it much less efficiently than many countries. With little more than one-tenth of the world's population, Soviets and Americans together use 44 percent of the world's commercial energy.[4] Western Europe and Japan maintain high living standards with more energy-efficient economies. These nations have a total population close to that of the two superpowers, but consume less than half as much energy.[5]

In the Third World, per capita use of commercial energy is small compared with that in industrial countries; an average Third World resident uses less than one-twelfth the energy consumed by an average U.S. citizen.[6] Even so, the size and rapid growth of Third World populations, along with the relatively inefficient use of energy in many developing countries contribute significantly to the rapid growth in global energy use.[7]

Figure 11.1 on the following page shows the consumption of energy from major sources for the United States and Canada, other members of the Organization for Economic Cooperation and Development (OECD), the Soviet Union, China, and other developing nations. In 1988, the United States and Canada together (with a population of 270 million) consumed 21 percent more energy than all the other 22 OECD member nations together (with a combined population of 526 million).[8]

The United States obtains about 42 percent of its energy from oil, 24 percent from coal, 23 percent from natural gas, 7 percent from nuclear power, and 4 percent from hydropower. As for consumption, U.S. industry uses about 36 percent of total energy, residential and commercial use accounts for another 36 percent, and transportation consumes 28 percent.[9]

Until recently, the traditional approach to energy policy was simply to meet increasing demands for energy—resulting from growing populations and rising living standards—by expanding energy supply. But during the past two

Figure 11.1
Primary Energy Consumption by Source, 1988

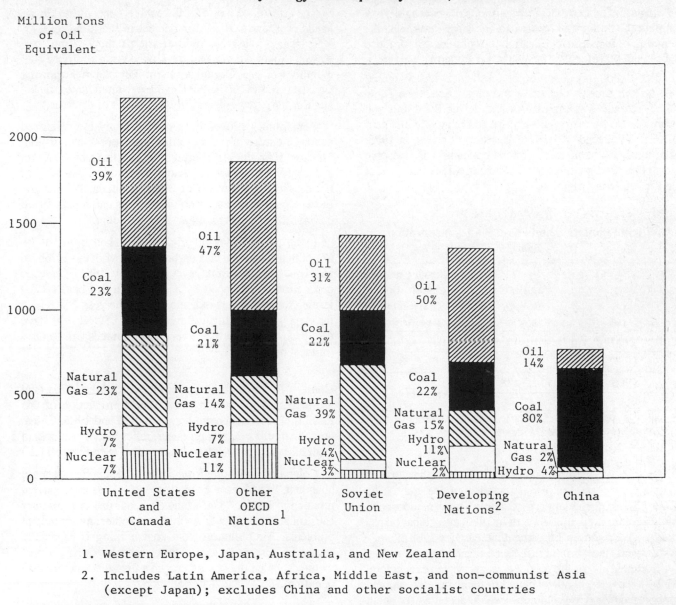

Million Tons
of Oil
Equivalent

1. Western Europe, Japan, Australia, and New Zealand
2. Includes Latin America, Africa, Middle East, and non-communist Asia (except Japan); excludes China and other socialist countries

Source: British Petroleum, *BP Statistical Review of World Energy, 1989* (London, 1989), p. 34.

decades, several factors have challenged this approach: growing realization that oil and natural gas supplies are being depleted; steep oil price increases during the 1970s and early 1980s; dimming prospects for nuclear power due to high costs, accidents, and technical problems; and growing environmental damage from fossil fuel use. In short, rising energy consumption is leading to mounting economic and environmental costs.

As a result, energy policy is shifting from the traditional emphasis on increasing supply to approaches that emphasize the end or final uses of energy and the services it supplies, and that stress how to provide these services most efficiently, at the lowest cost, and with the least environmental damage from energy-related pollution.[10]

After reviewing current trends in energy use, this chapter examines the environmental impacts of energy consumption, describes the vast differences between nations in their efficiency of energy use, discusses "end use" and "least cost" strategies for improving energy efficiency, and reviews prospects and progress in the use of renewable energy.

Fossil Fuels

Fossil fuels—oil, coal, and natural gas—are the remains of plants and animals that lived millions of years ago. They contain the biologically-stored solar energy that fueled the building of industrial civilization. Fossil fuels still produce more than 90 percent of the world's commercial energy.[11]

Since 1900, consumption of fossil fuels has increased nearly four times as fast as world population. (See Table 11.1.) Growth was especially rapid during the 1960s and early 1970s, when oil was plentiful and relatively inexpensive. In the decade following the first oil crisis, in 1973, shortages and rising prices forced consumers to conserve fossil fuels and use them more efficiently. (See Figure 11.2 on the following page.)

Table 11.1
World Population and Fossil Fuel Consumption, 1900-86

Year	Population (billions)	Fossil Fuel Consumption (billion tons of coal equivalent)
1900	1.6	1
1950	2.5	3
1986	5.0	12

Source: Lester R. Brown and Sandra Postel, "Thresholds of Change," in Brown et al., *State of the World 1987* (New York: Norton, 1987), p. 5.

On today's world market, however, supplies exceed demand and fossil fuel use is increasing once again. The United States, for example, used nearly 7 percent more coal, oil, and natural gas in 1987 than in 1983, and global oil use is expected to show a 3.6-percent increase for 1988 alone.[12] As a result, the harmful effects of fossil fuel use on the environment are a more serious threat now than ever before. Many types of pollution, including acid rain, smog, and carbon dioxide (CO_2) buildup—which accounts for about half of the greenhouse effect—can be traced to the burning of these fuels.

Oil   Petroleum is the most important energy source for industrialized countries and the most indispensable of the fossil fuels. For many of the services it provides, particularly in the transportation sector, there is no readily available fuel substitute. Modern agriculture also depends on oil, which fuels farm equipment and is used to make fertilizers and pesticides.

After rising during the early 1980s, the ratio of proven world oil reserves to the world oil production rate dropped in 1986. By the end of 1986, there were sufficient proven reserves to last somewhat more than three decades at the 1986 production rate.[13]

The bulk of the world's oil wealth is concentrated in the hands of a few. In 1987, 62 percent of world oil reserves were located in just five Middle Eastern nations—Iran, Iraq, Kuwait, Saudi Arabia, and the United Arab Emirates—all members of the Organization of Petroleum Exporting Countries (OPEC). The Soviet Union, United States, Japan, and Europe together have only 12 percent of reserves.[14]

Remaining petroleum reserves in many industrialized nations are relatively inaccessible and expensive to exploit. The cost of producing a barrel of oil in Alaska or the North Sea is 5 to 10 times greater than in the Middle East. Most of the developing countries of South America, Africa, and Eastern Asia have few or no oil reserves and must depend on imports, which are a major source of debt.[15]

In the United States, oil production peaked around 1970, as predicted two decades earlier by Dr. M. King Hubbert, and has been declining ever since. By 1985, domestic production had dropped by about 25 percent below 1970 levels. Some analyses indicate that by the year 2020, both the supply and quality of remaining U.S. oil will have declined to the point that other energy sources will be used for most purposes.[16]

Coal   Following the oil price increases of the 1970s, coal use increased in many regions. World production of this fossil fuel rose 32 percent between 1976 and 1986. China, now the world's largest producer, relies heavily on coal to meet its rapidly growing energy needs.[17] (See Figure 11.1.)

Compared with other fossil fuels, coal supplies are more abundant. World reserves total at least 200 years at current production levels.[18] Unfortunately, coal use is a primary cause of acid rain, global warming, and other environmental problems. Air pollution from coal burning is especially severe in Eastern Europe. The Czechoslovakian government has called the city of Prague a disaster zone because of extreme coal-related air pollution.[19]

New clean-coal technologies can reduce air pollution from coal use. But coal's emissions of climate-altering carbon dioxide—25 percent higher than an equivalent amount of oil, and 80 percent higher than natural gas—appear to be an intractable problem. The production of synthetic liquid and gaseous fuels from coal offers versatility but also increases CO_2 emissions per unit of energy by 50 percent or more.[20] (See Chapter 12 for a discussion of CO_2-induced climate change.)

Natural Gas   A relatively clean fuel, natural gas creates less pollution when burned than either oil or coal. It is a widely-used commercial and residential energy source. In the transportation sector, some 30,000 U.S. motor vehicles now operate on compressed natural gas.[21]

Figure 11.2
World Oil Price and Oil Consumption Rates, 1968-87

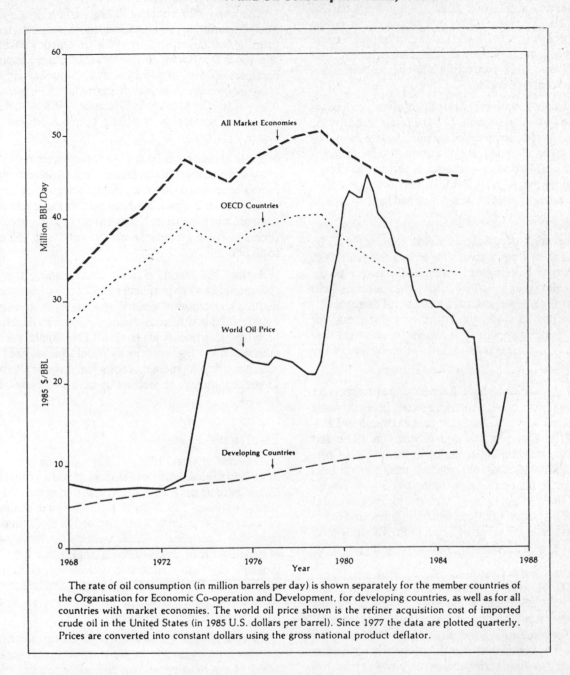

The rate of oil consumption (in million barrels per day) is shown separately for the member countries of the Organisation for Economic Co-operation and Development, for developing countries, as well as for all countries with market economies. The world oil price shown is the refiner acquisition cost of imported crude oil in the United States (in 1985 U.S. dollars per barrel). Since 1977 the data are plotted quarterly. Prices are converted into constant dollars using the gross national product deflator.

Source: José Goldemberg et al., *Energy for a Sustainable World* (Washington, DC: World Resources Institute, 1987), p. 2; oil price plot extended to 1987.

World reserves of natural gas are estimated to contain at least a 60-year supply at current rates of use. Many industrial countries have greater supplies of gas than oil. Still, the bulk of known reserves are concentrated in a few areas; more than half are in the Soviet Union and Iran. The Soviets have discovered immense natural gas fields in Siberia and are now the world's leading consumer and exporter of this fuel.

Soviet gas production more than doubled between 1976 and 1986.[22]

Nuclear Power

The nuclear industry's contribution to world energy supplies, now about 5 percent, has risen substantially in the last two decades, but less than predicted earlier. In France, nuclear power supplies close to 25 percent of total energy, compared with about 7 percent for the United States. Worldwide, the production of electricity from nuclear power doubled between 1980 and 1985, with the largest increases occurring in the Soviet Union and France. Nuclear plants supply about 60 percent of France's electricity, compared with around 18 percent for the United States, and about 15 percent worldwide.[23]

Although nuclear power plants themselves produce no greenhouse gases that threaten global climate stability, the energy used to mine and prepare uranium fuel for the plants releases substantial quantities of carbon dioxide, a major greenhouse gas. In addition, a range of serious economic, environmental, safety, and security concerns are undercutting public confidence in nuclear power and have slowed its growth.[24]

The proportion of the world's electricity supplied by nuclear plants will begin to decline before the year 2000. The expense of building and maintaining nuclear power plants has risen steadily. In the United States, recently-built nuclear plants generate power at a total cost of more than 13 cents per kilowatt hour, twice the prevailing rate for electricity. Largely because of increasing costs, every new order for a U.S. nuclear plant since 1974 has subsequently been cancelled.[25]

A number of accidents have dimmed future prospects for nuclear power, including serious incidents at Kyshtym in the Soviet Union in 1957; at Windscale in Britain in 1957; at Three Mile Island in the United States in 1979; and especially the catastrophic explosion at the USSR's Chernobyl plant in 1986. Most of these accidents were at least partially the result of mistakes made by fallible human operators. The well-publicized Chernobyl disaster caused at least 31 deaths, some 1,000 immediate injuries, and direct financial losses in the range of $10 billion. The long-term impacts on human health and productivity remains to be defined.[26]

In the Three Mile Island accident, a malfunctioning pump and valve combined with a series of operator errors to drain water from the reactor's core and cause a partial melt-down as core temperatures soared above 5000 degrees Fahrenheit. The containment system of the reactor held in an estimated 18 billion curies of radioactivity—at least 100 times the amount released in the Chernobyl accident. By 1989, the cleanup, still nearly two years from completion, had cost about $1 billion.[27]

About 430 nuclear plants are now in operation around the world. Little additional construction is expected after the completion of about 100 plants currently being built. New plants are effectively prohibited in West Germany, and in Sweden, as a result of a nationwide plebiscite, all nuclear power will be phased out by the year 2010. The disastrous Chernobyl accident has strengthened anti-nuclear political movements worldwide. Another major accident could result in the forced closing of many plants by public demand.[28]

The safety of the nuclear industry is a major concern in many countries. In 1987, U.S. nuclear power plants reported nearly 3,000 mishaps, at least 430 emergency shutdowns, and 104,000 incidents in which workers were exposed to measurable doses of radiation.[29] U.S. public support for nuclear power has gradually eroded since 1979, and according to a Harris poll taken in December 1988, 61 percent of Americans oppose the building of more nuclear power plants.

Little attention has been paid to the problem of decommissioning current nuclear power plants when they must be retired after about 30 years of operation. High radiation levels present in shutdown reactors will make this process complex and expensive. Estimates of the costs involved in decommissioning a nuclear reactor range from $50 million to $3 billion.[30]

In its 1987 report, *Our Common Future*, the World Commission on Environment and Development concluded that the generation of electricity from nuclear energy "is only justifiable if there are solid solutions to the unsolved problems to which it gives rise." The Commission concluded that the highest priority should be assigned to research and development on ecologically viable alternatives to nuclear energy, as well as to means of increasing its safety.[31]

Electricity

Between 1960 and 1984, world production of electricity more than tripled. Many developing countries cannot meet the present demand for electricity and, during the exceptionally hot summer of 1988, some U.S. cities were on the verge of blackouts as utilities were stretched to the limit of their capacities. Centralized electric systems can leave millions of people without power when a single plant fails. These problems emphasize the need to stress energy conservation, improve the efficiency of energy use, and eventually to replace large coal and nuclear plants with more decentralized sources of renewable energy.[32]

Electricity is a versatile form of energy, both in the services it provides and in the ways it can be generated. Many of the most promising renewable technologies are well suited to producing electricity. Currently, however, most power is generated in plants that burn fossil fuels or split uranium atoms. Coal now accounts for 57 percent of the energy used by U.S. electric utilities, and nuclear power contributes 18 percent.[33]

Conventional electric power production represents a very inefficient use of fossil fuels. In producing and transporting electricity, a typical U.S. power company loses about 2.5 times more energy than it delivers. Heating a home or office with electricity produced by a coal-fired plant requires much more coal than would heating the same building with an efficient coal-burning stove.

In the United States, 25 percent of all energy used in 1985 was wasted in the production and distribution of electricity. And in 1987, U.S. electricity consumption was 28 quadrillion BTUs ("quads"), but only 8 quads were sold to consumers; the remainder was used to generate, transmit, and distribute the electricity.[34]

Renewable Energy

Unlike fossil fuels, carefully managed renewable energy sources could last indefinitely. Most renewable technologies draw on the abundant flow of energy from the sun, both directly through the light and heat it provides, and indirectly from plants, winds, and falling water. Others employ heat from the earth's molten core and utilize ocean heat, tides, and currents. These energy sources are essentially free; the challenge is to harness them in ways that are both economically competitive and environmentally benign. Many promising technologies can meet these criteria and provide significant amounts of energy.

Over half the world's people use renewable energy sources for cooking, heating, and other purposes. Worldwide, renewable sources provide more than one-fifth of all primary energy. Traditional fuels, primarily wood and other biomass, are the largest contributors; these sources supply about 40 percent of energy in developing nations. Hydroelectric and geothermal sources together provide more than 21 percent of the world's electricity.[35] The present contribution of other renewable energy sources is small, but some are growing rapidly and probably will continue to expand as costs come down. In fact, electricity from many renewable electric technologies is already competitive with power from new conventional facilities. In the United States, electricity from biomass, geothermal, and hydroelectric facilities, at a cost of between 3 and 6 cents per kilowatt-hour, is already competitive with most new conventional baseload electrical supplies.[36]

Renewable energy sources accounted for nearly 9 percent of U.S. domestic energy production in 1988; this is expected to double by the year 2000, when renewables are likely to supply at least 15 to 19 percent of the country's energy needs. While hydropower now supplies about 40 percent of total U.S. renewable energy, biomass, geothermal, wind, and direct solar sources account for much of the current growth. By 2000, non-hydro sources could account for about three-quarters of the energy from renewables.[37]

This section summarizes the current status of major renewable energy sources; a later section reviews a selection of successful renewable energy projects and promising new technologies.

Hydropower   Electricity generated from falling water supplied nearly 7 percent of the world's commercial energy

in 1985. The use of hydropower in the Third World is increasing rapidly. At least 31 developing countries doubled their hydroelectric capacity between 1980 and 1985. Although most power is provided by large dams, small projects generate electricity for many people in remote areas. In China, over 86,000 small hydroelectric plants provide power to 99 percent of the counties. Many developing countries have considerable unused hydroelectric potential, but environmental, economic, and social concerns could prevent numerous available sites from being exploited.[38]

Geothermal Energy   Heat from the Earth's interior can be used to generate electricity or to heat buildings directly. The world's geothermal capacity grew by over 16 percent a year from 1978 to 1985. Some 12,000 megawatts of geothermal energy are now used for direct heating worldwide. By 1990, electricity generated from geothermal sources is expected to reach 6,400 megawatts, more than the capacity of six large nuclear plants. The United States is the leading producer of geothermal energy. Mexico, the Philippines, Japan, Italy, and New Zealand are also important producers of geothermal energy.[39]

The production of geothermal energy is not without environmental impacts. The water from some geothermal reservoirs contains salts and minerals that can be harmful pollutants, and geothermal plants can release hydrogen sulfide and other noxious gases.

Well blowouts have released steam and salt water, and the venting of steam from some installations can create a noise problem. There are still some technological obstacles to be overcome before geothermal resources can be widely utilized.[40]

Solar Thermal Energy   Solar water heating, one of the most publicized renewable technologies, is now widely used in some areas. In Israel, 65 percent of all domestic hot water is heated by solar energy and in Cyprus, 90 percent of all houses have solar water heaters. The U.S. solar water heating industry, the world's largest through the mid-1980s, has been forced to make severe cutbacks since 1985 due to lower oil and gas prices and the elimination of tax credits for renewable energy.

A number of countries are experimenting with solar ponds, which use lined cavities filled with salt and water to collect solar energy. The denser salt water on the bottom absorbs heat while the surface water traps it.

More complex solar systems use reflectors to concentrate the sun's rays and heat liquids to as high as 3,000 degrees Celsius. This heat can be used to turn a turbine and generate electricity. "Trough collector" systems, which employ U-shaped mirrors and produce temperatures up to 400 degrees Celsius, have been the most commercially successful. One corporation has built several 30-megawatt trough collector plants in California's Mojave Desert that now produce almost 200 megawatts of electricity.[41]

Photovoltaics   Solar photovoltaic (PV) cells are thin wafers made from silicon that convert sunlight directly into electricity. They require no moving parts or heat, and only enough light to displace electrons from their orbits and produce electric current. PV cells are considered to be the cleanest and most environmentally safe of any energy source. The total capacity of PV units sold in 1986 was nearly 25 megawatts, compared with only 0.5 megawatts a decade earlier. (See Figure 11.3.) Japan is now the world's leading supplier of PV cells, having surpassed the United States in 1985. The U.S. share of the world PV market dropped from 75 percent in 1980 to 32 percent in 1988; Japan's share has risen from 15 percent to 37 percent.[42]

The primary market for photovoltaic power is currently in remote areas, where electricity from power lines is not available. Rural electrification projects powered by PV units are gradually spreading in the Third World. More than 15,000 homes worldwide obtain their electricity from PV units. Collector costs, the major obstacle to more widespread use of PVs by electric utilities, dropped eightfold between 1976 and 1986 and should continue to fall. (See Figure 11.3.) PVs can now generate electricity for 25 to 35 cents per kilowatt hour (KWH), and the cost is expected to decrease to between 6 and 18 cents per KWH by 2000. The total cost of government subsidies and environmental damage from coal and nuclear power is estimated to be between 10 and 15 cents per KWH. If these costs were taken into account, photovoltaics would already be a competitive source of electricity for utilities in some areas.[43]

Wind Power   An important source of energy generations ago, wind power has experienced a renaissance in recent years. Since 1974, nearly 50,000 units have been installed, primarily in Denmark and California. Clusters of turbines have been installed in several California mountain passes. In 1987, these "wind farms" generated some 1.6 billion KWH of electricity, worth an estimated $100 million. Wind turbines can now be installed at costs that are competitive with some conventional generating technologies. In the United States, the cost of power from wind turbines has dropped from 25 cents per KWH in 1980 to about 7-9 cents per KWH today. Overseas, Greece, China, the Netherlands, and India are planning ambitious wind projects.[44]

Ocean Energy   The ocean thermal energy conversion (OTEC) process now under development uses tropical sea water as a vast solar collector. OTEC works like an air conditioner in reverse. Instead of using electricity to create a temperature difference, OTEC utilizes the contrast between the ocean's warm surface and its cool, deep waters to

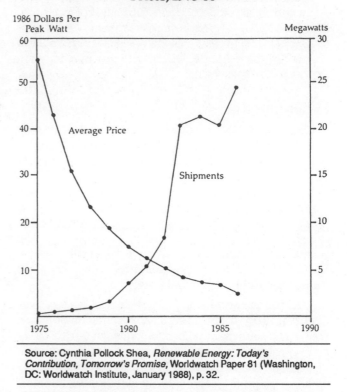

Figure 11.3
World Photovoltaic Shipments and Average Market Prices, 1975-86

Source: Cynthia Pollock Shea, *Renewable Energy: Today's Contribution, Tomorrow's Promise,* Worldwatch Paper 81 (Washington, DC: Worldwatch Institute, January 1988), p. 32.

generate power. In addition to electricity, OTEC creates fresh water as a by-product.

OTEC has only been demonstrated by small prototype plants so far, but there are plans to build several full-size, 100-megawatt generators over the next few years. A preliminary study has shown that OTEC could provide electricity at under 7 cents per kilowatt hour, less than current prices in many developing countries. In other attempts to tap the ocean's energy, Norway has two prototype wave power plants that began producing electricity in 1986. France operates a tidal power plant capable of generating 240 megawatts of electricity.[45]

Biomass   Wood and other plant or animal matter that can be burned directly or converted into fuels are important energy sources, especially in rural areas of the Third World. Biomass provides 15 percent of all energy consumed worldwide and over 90 percent in some developing countries. While wood has been used as a fuel throughout history, technologies are now available to convert a variety of biomass materials into more-efficient fuels. Plants, garbage, and animal dung produce gas for cooking and heating in many countries. In Brazil, the production of clean-burning alcohol fuels from sugarcane is now a major industry, providing about half of the country's automotive fuel. And in the United States, pulp and paper companies now obtain

55 percent of their energy needs from wood and other biomass sources.[46]

In some areas, however, exploitation of biomass fuels is reducing the land's fertility. In developing countries, excessive use of dung and crop residues for fuel instead of soil enrichment deprives the soil of essential nutrients, and jeopardizes future crops. In Nepal, for example, where fuelwood has become scarce, the diversion of biomass from the fields has caused an estimated 15-percent loss in grain yields.[47]

Fuelwood

Wood, the oldest of fuels, is still the principal energy source for the majority of people in developing countries. As Third World populations have soared, excessive cutting of trees for cooking fuel has created severe fuelwood shortages. In 1980, the U.N. Food and Agriculture Organization reported that nearly 1.2 billion people were relying on wood that was being harvested unsustainably from shrinking forests, and that this number might reach 2.4 billion by the year 2000. Well over 100 million people already face conditions of acute scarcity, unable to acquire enough wood to meet their daily cooking needs.

Buying and gathering wood is a major burden for many people. Around some Third World cities, a wide swath of land has been completely stripped of trees, and families must buy wood that is shipped long distances from the countryside. A typical family can spend 20 to 40 percent of its income on fuelwood products. In rural areas, where wood is "free," people are forced to spend increasing amounts of time gathering it. In the African Sahel and the Himalayas, women and children labor up to 300 days each year collecting fuelwood.[48]

Energy and Security

Continued reliance on oil and the spread of nuclear power have arguably made the world a less secure place to live. And growing reliance on complex, centralized, and interdependent energy systems that are vulnerable to technological failure and terrorist attack raises serious questions about the reliability and security of our energy sources. Economic security, personal safety, and peace between nations can all be affected by the kinds and amounts of energy we use, and by how that energy is supplied.

Dependence on Petroleum

Will the world experience another energy crisis in the near future? Experts disagree about the likelihood of a return to the shortages, skyrocketing prices, and world recession of the late 1970s. Most do concur, however, that the trend toward increased oil dependency makes a future crisis more likely.

OPEC members, attempting to regain the market share they once enjoyed, have increased oil production since the mid 1980s. Prices have fallen substantially, from $27 per barrel in 1985 to around $13 in late 1988. As a result, the United States, the world's largest oil-consuming nation, is importing more petroleum and producing less. U.S. oil imports rose from 27 percent of total consumption in 1985 to 42 percent at the end of 1988. Most studies project that this figure will exceed 50 percent in the 1990s.[49]

In 1988, imported oil accounted for 18 percent of total U.S. energy consumption, while domestic oil supplied 24 percent. Domestic coal and natural gas together supplied 46 percent, and nuclear and hydro provided 7 percent and 4 percent, respectively. (See Figure 11.4.)

Figure 11.4
Sources of U.S. Energy, 1988

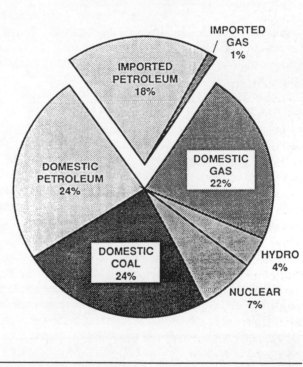

Source: U.S. Energy Information Administration

Relatively expensive domestic oil is being priced out of the market. The U.S. industry is shrinking and may not be able to expand again quickly in response to a future crisis. Domestic oil production has dropped from 9 million barrels per day in 1985 to less than 8 million barrels in December 1988, and is projected to fall to between 6 million and 7 million barrels in the 1990s.[50]

Currently, a surplus in world petroleum production capacity insulates the energy market from price shocks. Oil-rich nations, primarily in the Middle East, have the capacity to pump about 10 million barrels per day more than they currently produce.[51] Increased demand for petroleum by large consumer countries like the United States, which can no longer expand their own production, could erode this surplus. The resulting tighter market would be much more vulnerable to the wars, revolutions, and other potential disruptions common in the Middle East.

In the 1970s, OPEC demonstrated its willingness to raise oil prices when market conditions are right. And by spending $47 billion on military escort operations in the Persian Gulf in 1987, the United States has shown what oil-consuming nations will do to secure their supplies.[52] Petroleum that costs $47 billion to protect can hardly be called a bargain. A stronger commitment to fuel efficiency and alternative energy sources could help end "gun-boat commerce" and make future price shocks and oil wars less likely.

Nuclear Power and Weapons Proliferation

The technological "secrets" necessary to build nuclear weapons are now widely known. The major challenge that would-be bomb builders face is to obtain the "missing ingredient," plutonium. One of the most toxic substances known, plutonium is not found in nature but is produced from spent nuclear reactor fuel. Recent developments could make plutonium much more accessible to countries or terrorist organizations seeking to acquire nuclear weapons.[53]

In an attempt to make their nuclear industries self-sufficient, several nations have developed reactors that run on recycled plutonium. Some governments have plans to develop a new generation of nuclear plants called "breeder reactors," which produce more plutonium than they consume, thereby creating a continuous supply of nuclear reactor fuel. Such projects would involve shipping large quantities of this deadly material around the globe.[54]

A new agreement between the United States and Japan will allow the unrestricted commercial transfer of large amounts of plutonium for the first time. In the past, Japan needed special permission to recycle plutonium from fuel that originated in the United States. Under the new agreement, Japan could acquire as much as 400 metric tons of recycled plutonium in the next 30 years. Much of this will be processed in France and shipped to Japan by air and by sea, making it more vulnerable to diversion. A U.S. Nuclear Regulatory Commission report has estimated that a project of this size could leave as much as 200 pounds of plutonium unaccounted for each year. A weapon nearly equal in power to the bomb dropped on Hiroshima could be built with just 15 pounds of plutonium.[55]

Vulnerability of Energy Facilities

In 1988, the U.S. Department of Energy released a report advising U.S. companies of a growing threat of terrorist attacks on electric power facilities and other domestic energy installations. The report stressed the vulnerability of the more than 225,000 miles of power transmission lines linking the electric grid that supplies power to other energy facilities such as fuel pipelines, as well as to other electric power users. The report noted that serious power disruptions have already been caused by terrorist action in Western Europe and parts of Africa and South America.[56]

In early 1989, a U.S. Senate Committee held hearings on the vulnerability of telecommunications and energy resources to terrorism and natural disaster. The hearings examined how well U.S. government and industry are organized to repel attacks and to sustain and repair vital services provided by telecommunications, electrical generation and transmission systems, and other energy resources such as oil and natural gas pipelines. The hearings concluded that U.S. energy facilities are quite vulnerable to terrorist action, and that such action could have potentially serious consequences for the nation's energy system.[57]

The complex and interdependent energy systems in most advanced countries are clearly vulnerable to a range of threats. A strategy to achieve a greater degree of decentralization of these systems, for example, by encouraging expansion of small-scale, renewable energy generation, could reduce the level of vulnerability.[58]

Energy and the Environment

Energy production and consumption may have a greater impact on the environment than any other human activity. Nuclear power plant accidents, for example, have caused widespread radiation poisoning. Large-scale hydroelectric projects flood hundreds of square miles, displacing people and disrupting ecosystems. And in many parts of the developing world, the scramble for fuelwood is destroying forests and turning productive lands into desert.

In addition, the combustion of fossil fuels releases vast quantities of chemicals into the atmosphere, including carbon dioxide, sulfur and nitrogen oxides, carbon monoxide, and hydrocarbons, and now threatens to alter the global climate.[59] (See Chapter 12.) The eventual cost of restoring the world's soils, forests and lakes that have been damaged by air pollution from fossil fuel use will be enormous.

Global Warming

The increasing concentration of carbon dioxide (CO_2) in the earth's atmosphere is causing a "greenhouse effect" that many scientists believe is warming the earth. CO_2 molecules allow the sun's rays to pass through but prevent some of the reflected heat from returning to space. In 1988, fossil fuel burning, the primary cause of global warming, added about 5.5 billion tons of carbon to the atmosphere. Deforestation added another 0.4 billion to 2.5 billion tons by releasing the carbon stored in trees and in the soil that supports them. Carbon dioxide levels are now 25 percent higher than before the industrial revolution.[60]

Scientists predict that, if strong action is not taken to control global warming, the average temperature of the

Earth's surface will increase rapidly, as much as 1.5 to 4.5 degrees Celsius (3 to 8 degrees Fahrenheit) by the middle of the next century. Temperatures at mid and upper latitudes could rise twice as fast. Resulting changes in rainfall patterns will threaten agriculture. The cost of adjusting irrigation systems alone could exceed $200 billion.[61] The polar ice caps will begin to melt and warmer oceans will expand, causing sea levels to rise as much as several feet. Beaches will wash away and, without expensive dikes, many coastal cities will be flooded. Ocean salt will pollute fresh water supplies and flood low-lying coastal farmland, wetland areas, and even some island nations.

Acid Rain

Acid precipitation, a serious problem in Europe and North America, results when sulfur and nitrogen oxides combine with water vapor and become airborne acids. The energy sector, particularly coal-burning electric utilities and motor vehicles, is the leading source of these pollutants in many countries.

Acid rain and acidified snow, fog, dew, and dry airborne particles damage nearly everything they touch. The acids cause the death of fish and other aquatic life in lakes and rivers, make healthy forests prone to disease, leach nutrients from the soil, pollute drinking water, and corrode buildings and cars. Sweden has already spent more than $25 million on liming projects to neutralize the acid temporarily in some of its lakes.[62]

Smog

Photochemical smog is a major health hazard in many cities, and also damages crops and forests in rural areas far from the sources of pollution. The smog contains ozone and other chemicals that irritate the lungs of people and animals and cause plant cells to break down. These chemicals are formed when sunlight acts on mixtures of nitrogen oxides and hydrocarbons. Losses to U.S. agriculture from ozone pollution have been estimated at $5 billion per year.[63] (See Chapter 12.)

Oil Spills

It is not only the burning of oil and coal that threatens the environment. The process of extracting these fuels from the earth and transporting them to where they are used also creates problems. Every year, between 3 million and 6 million metric tons of oil pollute the world's oceans; this is roughly one ton for every 1,000 tons produced. Some of the petroleum is spilled accidentally, as in the recent Exxon Valdez disaster off the Alaskan coast, but most of it is discharged in the course of normal operations. Even small concentrations of oil can kill marine animals and seabirds.[64] (See Chapter 8.)

Coal Mining

Underground coal mining is one of the most hazardous industrial occupations. Miners often suffer from coal-induced black lung disease and risk injury and death in mining accidents. Strip-mining of surface coal without expensive land reclamation efforts can leave massive scars on the earth. Mined land is more vulnerable to erosion, landslides, and floods. Acid drainage from coal mining is a significant source of surface and underground water pollution.[65]

Nuclear Radiation

The 1986 disaster at Chernobyl put to rest any notions that serious nuclear accidents are unlikely. The explosion released a massive plume of radiation that contaminated crops in Europe and eventually circled the globe. Dr. Robert Gale, leader of an effort by western scientists to supply medical aid to the Soviets, has estimated that 30,000 additional cancer deaths will occur worldwide as a result of the Chernobyl accident. Some nuclear regulatory officials now estimate that there is a 50-percent chance of a core meltdown at a U.S. plant before the year 2000.[66]

Even the routine, low-level emissions of radiation by nuclear power plants may be hazardous to human life. A recent British study shows a correlation between small increases in radiation levels and increases in childhood leukemia. The U.S. Department of Health and Human Services is now investigating clusters of leukemia cases around the Pilgrim nuclear power plant in Massachusetts and several plants in the United Kingdom.[67]

Nuclear reactors produce a range of radioactive wastes—from highly toxic liquid wastes that will remain radioactive for thousands of years to less toxic intermediate- and low-level wastes. Over the next 30 years, the 10 leading nuclear energy producers will generate over 25,000 cubic meters of high-level waste; the United States will account for over a third of the total.[68] In addition, nuclear power production is creating millions of cubic meters of low-level radioactive wastes, and uranium mining and processing is producing hundreds of millions of tons of low-level radioactive tailings. Some reactor by-products have half-lives roughly five times as long as all of recorded history. No nation has yet implemented a satisfactory program to dispose of radioactive wastes.

Energy Efficiency: An Overview

There are vast differences between nations in how much total energy they use, how much they use on a per person basis, and how efficiently they use it. Many factors affect a nation's energy use, including population size and distribution, geography, climate, and level of development. Figure 11.5 illustrates some of the wide differences in total and per capita energy consumption among nations.

Figure 11.5
Energy Consumption in Selected Countries, 1984

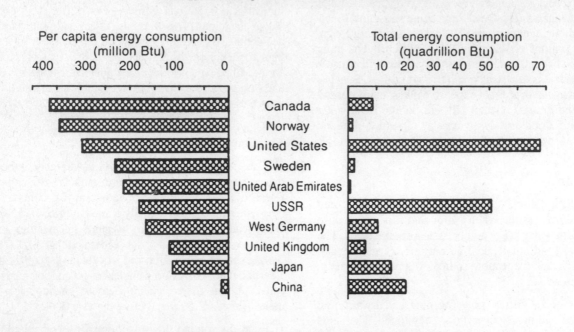

Source: Graphic provided courtesy of The Conservation Foundation. Information provided by The World Bank.

A nation's energy consumption in relation to its economic output, sometimes called energy intensity, reflects the country's economic structure and level of development as well as how efficiently energy is used within that structure. Table 11.2 gives energy consumption per unit of economic output for 12 countries, and also shows how much consumption increased or decreased between 1973 and 1985. All the OECD industrial nations listed in Table 11.2 have lower energy intensities than the United States, and Denmark, Japan, and West Germany use about half as much energy as the United States per unit of economic output.[69]

Other studies show that West Germany's per capita energy consumption is about half that for the United States, and that Japan spends only 5 percent of its gross national product on energy compared with 10 percent for the United States.[70]

In the western industrial countries, since the first OPEC oil price increases in 1973, there have been significant improvements in energy efficiency in all major areas of energy use—including the industrial, transport, service, and residential sectors. Industrial nations used about 1 percent less primary energy in 1985 than in 1980, in contrast to the 26 percent increase in the five years before the 1973 oil price increases.[71] And between 1973 and 1985, per capita energy use in OECD nations dropped 6 percent while per capita gross domestic product rose 21 percent. For the first time,

Table 11.2

Primary Energy Consumption per Unit of Gross Domestic Product, Circa 1985

Country	Energy Consumption*	Average Annual Growth Rate (%)**
Denmark	2.7	-1.7
Japan	2.9	-3.1
West Germany	3.1	-1.7
Italy	3.3	-1.8
United Kingdom	3.5	-2.0
Sweden	4.0	0.6
United States	6.1	-2.2
Brazil	6.8	+2.1
India	7.9	+1.4
Canada	8.0	-0.5
China (PRC)	14.0	-1.3
Venezuela	14.0	+4.6

*Metric tons of oil equivalent per $10,000 U.S. dollars of gross domestic product.
**Average annual growth rate, 1973-85.

Source: Adapted from World Resources Institute and International Institute for Environment and Development, *World Resources 1988-89*, (New York: Basic Books, 1988),p. 114

the link between economic growth and increased energy use was broken.[72]

Unfortunately, in the last few years, momentum toward greater efficiency has been slowed by lower energy prices and the waning commitment of governments to conservation. In the United States, efficiency improvements in industry have leveled off since 1983. The U.S. government has reduced fuel economy standards for new cars and has slashed the budget for research and development of energy efficiency by 50 percent in the last 7 years.[73]

Improving Energy Efficiency: Doing More With Less

Improvements in the efficiency of energy use provide the quickest, least expensive means to alleviate many energy-related problems. An increase in efficiency is usually the cheapest "source" of energy. To recast the old metaphor, a kilowatt saved is a kilowatt supplied. By providing energy for actual or final uses ("end uses") such as lighting and transportation more efficiently, increased demands for such services can be met without more imported oil, coal-fired electric plants, nuclear power, or environmental pollution. In the United States, opportunities have already been identified for boosting energy efficiency that cost one-half to one-seventh the expense of new energy supply.[74]

Regarding environmental benefits, one study suggests that a two-percent annual increase in worldwide energy efficiency could slow the buildup of atmospheric carbon dioxide enough to keep the global average temperature to perhaps within 1 degree Celsius of current levels, thus avoiding the most serious effects of climate change. Another study shows that energy efficiency improvements are a much cheaper way to slow the carbon dioxide buildup than an expansion of nuclear power; improved efficiency displaces nearly seven times as much carbon per dollar invested.[75]

Energy can be conserved in many ways, ranging from turning down thermostats to using more-efficient appliances and vehicles. Through improved efficiency, the amount of energy consumed in buildings, industry, transportation, and other sectors of society can be substantially reduced.

Buildings and Appliances

There is vast potential for improving energy efficiency in the heating, cooling, and lighting of buildings and in the operation of appliances such as stoves, water heaters, and refrigerators. Important savings can be made by reducing heat loss from poorly insulated buildings and replacing inefficient furnaces and air conditioners.

A call to the "house doctor" might be the best way to cut energy consumption in existing homes. In an experiment by a Princeton, New Jersey, research group, house doctors made detailed energy audits of homes. Using infrared scanners and house pressurization techniques, they could quickly identify defects in the shells of houses that were allowing heated or cooled air to escape. After leaks were sealed, fuel use was cut by an average 30 percent, and lower energy bills enabled home owners to gain a 20-percent annual return on their investment in energy conservation.[76] Homeowners can save even more money by finding and fixing energy leaks themselves.

Windows are a major source of energy loss in commercial and residential buildings. Each year, losses through American windows equal the energy that flows through the Alaskan pipeline. Losses can be greatly reduced by installing a "heat mirror" film that lets in light while doubling insulation value, and by removing the air between the glass sheets of double-pane windows.[77]

Some owners of residential buildings and businesses lack the capital to make investments in energy efficiency. The cooperation of public utilities with consumers and with groups called conservation companies can solve this problem. These companies test buildings and make energy-saving improvements free of charge. To recover their investments, such firms charge a fixed monthly fee or a percentage of the money saved in energy costs. The conservation business can become more profitable and much more widely applicable with the cooperation of power companies. Some farsighted public utilities, recognizing that saved energy costs less than building new power plants, are willing to pay conservation companies for kilowatts saved. Such arrangements promote energy efficiency and provide economic benefits for all parties involved.[78]

Improved energy efficiency can easily be incorporated into the design of new buildings. Super-insulated homes can be heated for one-tenth the average cost for conventional homes. Such houses are heated primarily by lights, appliances, and the body heat of the residents, requiring only occasional use of small space heaters. The need for costly central heating systems is often eliminated. Ventilating heat exchangers can simultaneously remove any indoor air pollution and preheat water for washing and cooking.[79]

Compared with the average efficiency of equipment in use in 1985, the most efficient residential appliances and equipment available in the United States use much less energy. The best commercial models of refrigerators, central air conditioners, and electric water heaters use 50 to 60 percent less energy, and the best gas furnaces, water heaters, and ranges use between 26 and 43 percent less energy than 1985 averages. Improved designs now under study or development could increase these energy savings to between 59 and 87 percent compared with the 1985 average.[80] (See Table 11.3 on the following page.)

In the United States, about 20 percent of electricity is used for lighting. In commercial facilities, lighting accounts for between 25 and 40 percent of energy use; part of this is used to remove heat generated by the lighting. Energy used

Table 11.3
Energy Efficiency Improvements and Potential for Residential Appliances and Equipment, United States, 1985

Product	Average of Those in Use	New Model Average	Best Commercial Model	Estimated Cost-Effective Potential[1]	Potential Savings[2]
	(kilowatt-hours/year)				(percent)
Refrigerator	1,500	1,100	750	200–400	87
Central Air Conditioner	3,600	2,900	1,800	900–1,200	75
Electric Water Heater	4,000	3,500	1,600	1,000–1,500	75
Electric Range	800	750	700	400–500	50
	(therms/year)				(percent)
Gas Furnace	730	620	480	300–480	59
Gas Water Heater	270	250	200	100–150	63
Gas Range	70	50	40	25–30	64

[1]Potential efficiency by mid-nineties if further cost-effective improvements already under study are made.
[2]Percent reduction in energy consumption from average of those in use to best cost-effective potential.

Source: Howard S. Geller, "Energy-Efficient Appliances: Performance Issues and Policy Options," *IEEE Technology and Society Magazine*, March 1986, as published in Christopher Flavin and Alan Durning, *Building on Success: The Age of Energy Efficiency*, Worldwatch Paper 82 (Washington, DC: Worldwatch Institute, March 1988), p. 20.

for lighting can be decreased by up to 70 percent while maintaining or improving lighting quality. Newer lighting sources also last longer that conventional ones. Standard incandescent lamps operate for about 750 hours, while newer alternatives can last for 10,000 hours while using up to 75 percent less energy.[81]

A World Resources Institute study, The End Use Global Energy Project, estimates that a family living in a super-insulated home with the most efficient appliances currently available would consume only one-quarter the energy used by a typical household.[82]

In the early 1970s, the town of Osage, Iowa, began an energy conservation program that has cut its natural gas consumption by 45 percent since 1974. Osage residents save energy by plugging leaky windows, insulating walls, ceilings, and hot-water heaters, and replacing inefficient furnaces. In 1988, the program saved an estimated $1.2 million in energy costs. Much of the saving has resulted from initiatives by the local utility company, which offered free building thermograms to locate heat losses, and now gives customers fluorescent light bulbs that are much more energy efficient than incandescent bulbs.[83]

In 1986, the U.S. Congress passed the National Appliance Energy Conservation Act that established energy conservation standards for most major home appliances and central heating and cooling systems. The standards are being phased in over a five-year period and are expected to lower residential energy use 6 percent by the year 2000.[84]

Industry

Industry has made significant gains in energy efficiency. In OECD countries, energy intensity—the amount of energy required per unit of industrial production—has fallen 30 percent since 1973. Japan has led the way, making energy-efficient industry a national priority. Full-time energy managers are required by law in all Japanese companies that use substantial amounts of energy. In Sweden, a new technology called "Plasmasmelt" is being developed that will reduce the energy intensity of their already efficient steel industry by 47 percent. Recycling scrap metals takes much less energy than producing new metals from ore. Fabricating a die-cast machine part takes 95 percent less energy when recycled aluminum is used in place of primary metals. In the United States, the proportion of aluminum produced from recycled metals grew from 25 to 50 percent between 1970 and 1983.[85]

Cogeneration

One of the most promising developments to improve energy efficiency in industry and in cities and towns is cogeneration, a system that simultaneously produces heat and electricity or other forms of energy such as mechanical power.

For the greatest efficiency and lowest cost, the cogeneration facility should be near the site where the heat can be used.[86]

In a typical operation, fuel (usually natural gas, but also wood, plant wastes, coal, or oil) is burned in a boiler to produce steam. The steam turns an electric generator and is then recaptured for heating, refrigerating, or manufacturing processes rather than discarded to the air. The system can more than double the usable energy obtained from each dollar invested. A utility plant producing only electricity is about 32 percent efficient; a cogenerator using the same amount of fuel can approach 80 percent efficiency. The cost of electrical capacity from cogeneration systems is usually less than half the cost of new coal or nuclear power plants.[87]

Cogeneration can be utilized in hospitals, schools, office buildings, and apartment complexes. The amount of electricity provided by cogeneration is increasing rapidly. U.S. production will rise from 13,000 megawatts in 1985 to over 47,000 megawatts (the equivalent of 47 large nuclear plants) when projects already under way are completed.[88]

So far, there is little accurate information about cogeneration activity in the Third World, but the International Cogeneration Society reports that some developing nations have shown substantial interested in small-scale cogeneration applications.[89]

Transportation

Most means of transportation, including cars, trucks, and airplanes, run on petroleum products. In the United States, 63 percent of all oil is used in the transport sector.[90] To reduce environmental pollution and dependence on unstable oil supplies, it is crucial that the transportation sector be as energy efficient as possible.

Considerable progress has been made in automobile fuel efficiency. By 1988, all U.S. cars averaged 19 miles to the gallon (MPG), compared with 13 MPG in 1973. New cars now average over 25 MPG in the United States and more than 30 MPG in Europe and Japan. However, compared with prototype, super-efficient models, most new cars are still "gas guzzlers." Prototypes now exist that are more spacious, more responsive, and safer than many current models and can get 60 to 100 MPG. They use lightweight materials and design improvements such as continuously variable transmissions. Experts have estimated that the cost of purchasing and operating such cars would be roughly the same as for the inefficient cars of today. The higher initial cost would be offset by fuel savings over the life of the car.[91]

In 1975, the United States enacted its most effective energy legislation, the Energy Policy and Conservation Act, which required new car fleets to improve their fuel efficiency to 27.5 miles per gallon (MPG) by 1985. This standard helped sustain efforts to improve fuel efficiency after oil prices collapsed in the early 1980s, and the average efficiency of U.S. passenger cars rose from 13 MPG in 1973 to 25 MPG by 1985. Yet in spite of this improvement, U.S. fuel economy is well below the 30-33 MPG level achieved in other industrial nations.

In 1986, following appeals by luxury-car manufacturers, the U.S. fuel efficiency standard was lowered from 27.5 MPG to 26 MPG, and by late 1987, the major U.S. car manufacturers had stopped much of their research on more fuel-efficient cars. In 1989, however, members of the Bush administration have called for a federal fuel-economy standard of 27.5 MPG for 1990, 40 to 50 MPG for 2000, and as high as 75 MPG by 2025.[92]

A recent study by the American Council for an Energy-Efficient Economy showed that major improvements in the efficiency of energy use in the transportation, industrial, commercial, and residential sectors would yield important benefits for the United States in terms of economic well-being, competitiveness, national security, and environmental quality.[93]

Superconductivity

Looking to the future, superconductive materials are being developed that conduct electricity with no energy loss. Although not ready for commercial applications, superconductors could enable development of more-efficient electric motors and storage systems, and allow dramatic improvements in the overall efficiency and flexibility of energy use.[94]

The Developing World

In most developing countries, rapidly expanding demands for energy are outpacing economic growth.[95] Adequate energy supplies are essential to reducing poverty, but additional oil imports and power stations are costly. Such expenditures can add to a country's debt burden, and they make scarce capital unavailable for other projects that could provide more jobs.

The Third World desperately needs improved energy efficiency as an inexpensive source of power that could also lessen its growing contribution to global environmental problems.[96] Brazil is currently building large, expensive, and environmentally-damaging hydroelectric dams in the Amazon River basin to power its growing economy. The two largest dams have each flooded tropical forest areas the size of Long Island. A quarter of Brazil's $100-million foreign debt stems from such electric power projects, and 80 percent of the money for new projects must be borrowed.[97] In a recent $300 million World Bank energy loan to Brazil, only 0.02 percent was earmarked for improving energy efficiency and conservation. A study for the World Resources Institute concluded that investing $10 billion to make Brazil's electrical sector more efficient would eliminate the need for over $44 billion in new electrical capacity.[98]

The Stirling engine, developed over 170 years ago and recently improved, is a potentially useful source of electrical and mechanical power for Third World applications. The engine can operate on external heat from a variety of sources such as solar radiation and any solid, liquid, or gaseous fuel, including wood and biogas. The engine is especially appropriate for small communities and remote villages, where it can be used to generate electricity and to pump or purify water.[99]

Half the world's people cook with wood and, in developing countries, many use open "three stone" fires that waste most of the heat. Inexpensive, durable stoves already available could cut wood use by half or more and ease deforestation pressures in some areas. But building enough stoves at low cost and getting people with deep-rooted cooking traditions to use them is difficult. The challenge is to improve stove efficiency without straining the household budget or compromising the social and lighting functions of the fire. Projects in several African countries are providing experience that will help improve stove design and production. In one such project, a metal stove design uses up to 35 percent less fuel than traditional cooking methods.[100]

A U.S.-based group, ESTA Inc., has developed a low-cost, high-efficiency metal stove that can easily be mass-produced at the village level. The stove requires less than a quarter of the fuelwood needed for cooking over a three-stone fire, and has been used successfully in Mexico, Egypt, and India.[101] Another U.S. group has developed solar box cookers that can be built and maintained by local people, and used to cook food and pasteurize water without consuming scarce fuelwood or animal waste. The cookers have been tested successfully in Mexico, Guatemala, Bolivia, Sierra Leone, and Kenya.[102]

Between 1980 and 1985, Third World consumption of commercial fuels increased by 22 percent, while commercial energy use by OECD-member industrial nations actually declined by one percent. These statistics underscore the importance for developing countries of adopting energy strategies that are much more efficient than those used in industrial nations, if the world is to avoid the negative consequences of rapid growth in global energy use.

In the End Use Global Energy Project, an international team of researchers showed that developing nations have the potential to improve living standards—to a level comparable to Western Europe—with only a 30-percent increase in per capita energy consumption. This would require a shift from inefficient use of traditional fuels to efficient use of existing modern energy technologies.[103]

Improved energy efficiency in Third World countries could reduce oil imports, save foreign exchange, and ease debt burdens. In particular, efficient use of biomass energy sources would help reduce deforestation, desertification, and soil loss, as well as create jobs, promote technical development, and increase self-reliance.[104]

Incentives for Improving Efficiency

As noted, the collapse of oil prices in 1986 has begun to slow the progress of energy efficiency. Consumers are finding that investments in conservation have longer, less attractive payback periods. To avoid the cycle of "cheap energy" leading to higher consumption and increased pollution followed by future oil price shocks, incentives are needed for continued investment in energy efficiency.

Governments can play a role in providing the necessary incentives. Estimates show that government incentive programs could triple the current $20 billion to 30 billion annual world investment in energy efficiency.[105] Major increases in efficiency programs also could help avoid the enormous potential costs associated with global warming.

The gasoline tax in the United States is far lower than in most industrialized countries, averaging about 29 cents per gallon. In Japan, the tax is about $1.60 per gallon and in Italy, about $3.30. As a result, gasoline in most industrial countries costs 2 to 4 times as much as in the United States.[106] (See Figure 11.6 on the following page.) Not coincidentally, drivers in American cities buy four times as much gasoline as residents of European cities. Many energy experts have recommended that U.S. gasoline taxes be increased as a means of reducing pollution and oil dependency.[107]

A higher gasoline tax would have several related benefits. Most importantly, it would allow cleaner, more efficient means of transportation to compete fairly with motor vehicles. The current U.S. gasoline tax does not begin to cover all the costs associated with cars and trucks. Expenses such as road building and maintenance, lost tax revenues from paved-over land, traffic regulation, accidents and health care, and driver education may require up to $300 billion a year in subsidies by local, state, and federal governments.[108]

U.S. citizens currently support motor vehicle transportation through income, property, and other taxes regardless of how much they drive. A substantial increase in the gasoline tax could distribute these costs more fairly, especially if there were compensation for low-income groups. Public demand would increase for alternative transportation, fuel-efficient cars, and community planning to reduce driving

Figure 11.6
Gasoline Price Levels with Taxes, Selected Countries, 1989

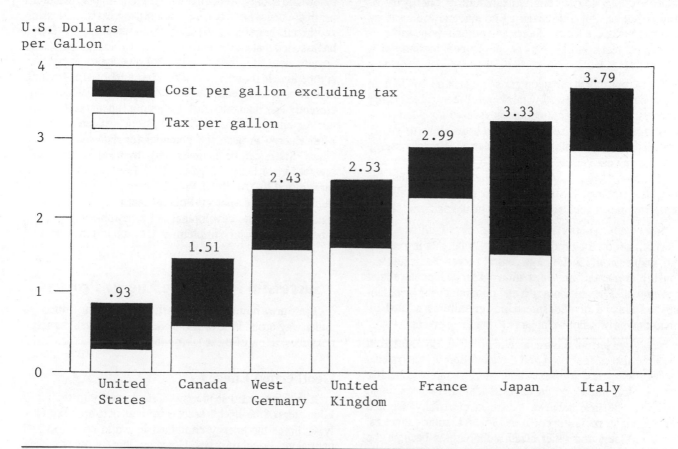

U.S. Dollars
per Gallon

Cost per gallon excluding tax

Tax per gallon

Source: International Energy Agency, *Energy Prices and Taxes, First Quarter 1989* (Paris: Organization for Economic Cooperation and Development, 1989)

distances. Fuel-efficient cars could also be promoted through a "gas guzzler-gas sipper" tax plan in which a tax on relatively inefficient new cars would be used to provide rebates for high-efficiency models.[109]

To combat global warming, a "carbon tax" could be levied on fuels in proportion to the amount of carbon dioxide they create per unit of energy. For example, coal would be taxed almost twice as heavily as natural gas. Clean energy sources would be made more competitive while those that threaten our climate would be discouraged.[110] (See Chapter 12.)

At the state level, incentives for energy conservation and improved energy efficiency can be highly effective in slowing or even reversing the growth in energy use, while the absence of conservation incentives can result in rapid growth. Between 1977 and 1984, California and Texas followed different energy strategies. California set strict conservation standards for all economic sectors, but Texas let energy demand grow without conservation incentives. As a result, during that 7-year period, energy use in Texas grew by 1.7 percent, compared with growth of 1.2 percent

for the United States as a whole, while California actually reduced energy consumption by 0.5 percent. California encouraged conservation and cogeneration in industry, set mandatory building and appliance efficiency standards, and implemented major utility conservation programs, while Texas left energy use entirely to the marketplace. California's conservation initiatives clearly resulted in substantial economic and social savings.[111]

Developing Renewable Energy

By developing renewable energy technologies along with improved energy efficiency, nations may eventually be able to replace fossil fuels and nuclear power. Renewables hold the promise of a long-term solution to many energy problems.

Advantages of Renewable Energy

Renewable energy sources have a number of advantages over coal, oil, and natural gas. In general, they are abundant and more evenly distributed. An estimated 80,000 quads (quadrillion BTUs) of solar and other forms of renewable energy are available annually in the United States alone, roughly a thousand times more than the country has ever used.[112] Collecting all this vast amount of energy is, of course, not feasible or necessary. Enough renewable energy exists to make the United States and other nations self-sufficient, if technologies can be developed to harness it economically.

Numerous developing countries that lack oil and natural gas reserves are endowed with plentiful renewable energy potential. Many tropical countries receive abundant sunshine for generating solar power and are close to warm, tropical oceans suitable for OTEC energy. Biomass energy systems can use various agricultural byproducts that would otherwise be wasted. A recent study calculated that Indonesian rice mills could save $30 million annually by replacing diesel generators with rice husk gasifiers.[113]

Renewable energy sources could also facilitate rural electrification. By producing energy where it is needed, renewables could make large central power plants and extensive transmission lines unnecessary. Decentralized power would benefit both rich and poor countries by reducing the chance of widespread power failures caused by breakdowns or sabotage at large generating plants.

Although serious environmental problems do result from the over-use of fuelwood and the alteration of ecosystems by hydroelectric dams, renewable energy sources generally have far less impact on the environment than conventional sources. The link between increased electricity use and worsening air pollution could be broken if solar collectors, wind turbines, and other clean technologies became the prime sources of power.

Achieving a world powered largely by renewable energy is possible within the next few decades. The U.S. Department of Energy has projected that energy production from a variety of renewable sources could be greatly expanded by the year 2000 (see Figure 11.7 on the following page), and that renewable energy could be competitive in meeting nearly 80 percent of projected U.S. energy demand in the year 2010. The degree to which this potential is realized will depend on how competitive renewables become in the energy market.[114] Changes in public policy could be critical in removing many of the obstacles to this development.

The Role of Governments

Most governments subsidize their conventional and nuclear energy industries, which discourages the development of renewable energy. China, for example, provides its coal industry with $10 billion in subsidies annually. The U.S. nuclear industry has benefited from many types of federal assistance, including billions of dollars spent on nuclear reactor fuel. Nuclear power may cost U.S. taxpayers as much as $11 billion per year. Governments could either eliminate these subsidies or provide comparable support to renewable energy.[115]

Support of basic research and development is perhaps the most important role governments can play in developing renewable energy. Recently, government support in a number of countries has led to substantial progress in harnessing geothermal energy. The U.S. Department of Energy (DOE) has asserted that a sustained government role in research and development (R&D) is essential for renewables to become competitive in the energy market. Regrettably, U.S. funding in this area has been cut 80 percent since 1980 and is currently less than one-sixth the amount spent on R&D for nuclear energy.[116] According to Public Citizen's Critical Mass Energy Project, the potential for renewables in the United States can be achieved only by restoring the DOE budget for R&D to at least the 1985 level of $263 million (in adjusted 1990 dollars). Public Citizen also states that the Public Utility Regulatory Policies Act (PURPA) must be used to encourage development of renewable energy, and other policies must be implemented to support the effort.[117]

Successful Renewable Energy Projects

There are a number of successful or promising initiatives under way around the world to develop renewable energy resources; a few of these are reviewed below.

Geothermal Energy

It is estimated that the geothermal energy in the top 5 kilometers of the earth's crust is equivalent to approximately 40 times the energy contained in world crude oil and natural gas liquid reserves. However, this energy is widely dispersed, and only a tiny fraction can be economically exploited. Because it is difficult to extract, estimates vary on the total contribution geothermal energy can make to the world's energy needs. A recent report by the Electric Power Research Institute estimated that North American capacity alone could exceed 18,000 megawatts by the year 2000. The most promising region of North America is along its western coast. Other areas believed to have important geothermal power potential are parts of the Caribbean, the Himalayan Mountains, East Africa, western Arabia, Central Asia, and along the eastern Pacific Belt.[118]

One of the most promising prospects for tapping geothermal power's potential is in the Hawaiian Islands. Since 1982, a small State-owned geothermal plant on the island of Hawaii has been generating approximately 2 megawatts, enough to supply 2,000 homes with electricity. The State of Hawaii and the Hawaiian Electric Power Company are planning to develop 500 megawatts of electricity from the island of Hawaii's volcanic geothermal resources and transmit it by underwater cable 270 miles to the island of Oahu. Geothermal energy could satisfy a major portion of Oahu's energy demands. Currently, the State of Hawaii is 90 percent dependent on oil imports for its energy, and the island of Oahu accounts for 80 percent of Hawaii's electrical

Figure 11.7
Production of Selected Renewable Energy Sources in the United States, 1984 and 2000

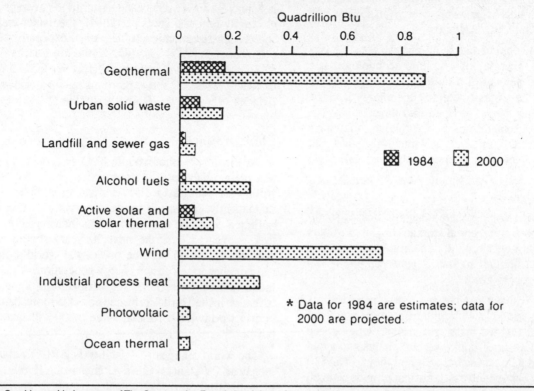

Quadrillion Btu

* Data for 1984 are estimates; data for 2000 are projected.

Source: Graphic provided courtesy of The Conservation Foundation. Information provided by the U.S. Department of Energy.

consumption. Geothermal energy could be available for transmission to Oahu by 1995. When it is installed, the 270-mile deep water cable connecting Hawaii and Oahu will be the longest and deepest electrical power cable in the world.[119]

Wind Power

During the summer of 1987, wind power accounted for more than 5 percent of the power sold by Pacific Gas and Electric, which supplies northern California. During the 1980s, wind power in California has grown rapidly. From 144 generators in 1981, it reached a high of 4,687 machines in 1984, and supplied 398 megawatts in 1985. New designs with new materials, knowledge, and manufacturing techniques have created more reliable products, allowing more electricity to be produced with fewer machines. With the termination at the end of 1985 of the U.S. tax credit for wind energy, however, wind power capacity dropped sharply, as did the sales of turbine suppliers. Sales of wind generators in California are now lower than in 1985, and they are unlikely to return to that level until the nineties, although sales are increasing in many other countries.[120]

By the middle of 1987, Denmark had installed 100 megawatts of wind energy capacity, including a sea-based

plant. The country has plans to increase that capacity greatly over the next few years. In China, the government plans to build 100 megawatts of wind power capacity between 1990 and 1996.

The Netherlands plans to have 150 megawatts of capacity by 1992, and hopes to reach 1,000 megawatts by the year 2000. Smaller programs are planned or under way in Australia, Belgium, Greece, Israel, Italy, Spain, the Soviet Union, the United Kingdom, and West Germany.

The Indian government has an ambitious wind power program. India's Ministry of Energy plans to have 5,000 megawatts of wind power installed by both public and private developers by the year 2000. If India achieves its goal, wind power will supply more electricity there than the country's large nuclear program.[121]

Solar Building Technologies[122]

Residential and commercial buildings can be designed to utilize the sun's heat and light to provide part of their energy needs, including space heating, water heating, and lighting. Building designs can also minimize the absorption of solar heat to keep buildings cool. These designs can displace the need for conventional energy supplies such as oil, gas, and electricity.

Solar building technologies combine the utilization of solar energy with improvements in energy efficiency such as more efficient building insulation.

Residential and commercial buildings use about 37 percent of total U.S. energy. Solar building technologies can provide up to 80 percent of a building's energy requirements, and thus have great potential for energy savings. These technologies can be grouped into three levels. The most basic passive solar design simply involves orienting a building to capture the sun's heat in winter, minimize solar absorption in summer, and facilitate natural ventilation.

Basic passive design can supply 20 to 25 percent of a building's energy needs.

An intermediate stage passive design increases south-facing window area and thermal storage capacity in floors and walls, and improves insulation, natural ventilation and lighting, and heat circulation. Such designs can provide 30 to 40 percent of energy needs.

Finally, an advanced passive solar design maximizes insulation, using windows with high insulating values and large amounts of thermal mass spread over wide surface areas. Advanced designs require careful analysis to avoid unnecessary heat gain during hot weather; they can add 10 to 12 percent to construction costs, but can provide 50 to 60 percent of a building's energy needs.

Passive solar designs have been shown to be effective in producing both comfortable indoor conditions and cost savings.

A study of 48 passive homes by the U.S. Department of Energy (DOE) found that the passive features supplied nearly 40 percent of heating energy needs, and an analysis of small nonresidential passive buildings showed energy savings of 46 percent compared with conventional buildings.

Other DOE studies estimate that buildings using advanced technologies and designs could obtain up to 80 percent of their heating requirements and up to 60 percent of their cooling needs from solar energy.[123]

Solar Thermal Energy

Of the various technologies being developed to produce electricity from the sun's rays, trough collector systems may have the greatest short-term potential to provide significant amounts of renewable energy. A solar thermal plant can be built in less than two years, compared with 5 to 10 years for many fossil-fuel and nuclear power plants.

At its Mojave Desert site in California, Luz Engineering is building five solar collector complexes capable of tripling its current generating capacity to nearly 600 megawatts, enough to meet the residential needs of a city slightly larger than Washington, DC. The company expected that units coming on line in 1989 would produce solar power at about 8 cents per kilowatt-hour, less than the cost of power from new nuclear power plants and closer to the average cost of electricity (about 6 cents per KWH) than solar electricity sources have ever achieved. The cost is expected to decline to 5 cents per KWH for units coming on line in 1994. In addition to the California plant, Luz is producing solar electricity in Israel, and exploring sites for installations in Arizona, New Mexico, Brazil, and India.[124]

Photovoltaics

As the cost of photovoltaic (PV) systems has dropped and their reliability has improved over the last decade, worldwide markets for PV electrical systems have grown dramatically. Industry revenues have risen to almost $400 million a year. The communications industry is the largest user. Advances over the next 20 years should make it possible for PV to become one of the world's preferred technologies for electrical energy generation. Already, PV modules supply electricity for refrigeration, lighting, and irrigation in the Third World, and power to more than 15,000 homes worldwide. In India, some 6,000 villages use PV power.[125]

The world's largest PV facility is ARCO Solar's 6.5-megawatt PV plant on Carissa Plains in California. As part of a joint industry-government program to compare and evaluate current and emerging technologies, the project will determine the cost and technological efficiencies of PV systems and demonstrate their ability to compete successfully with future electric generation options. In September 1988, Chronar Corporation and SeaWest Power Systems of San Diego, California agreed to develop a 50-megawatt, amorphous-silicon PV power station, to be completed in 1992. These efforts could generate increased interest in photovoltaics and initiate a new era in the development of PV technology that will make possible competitive, central-utility power generation.[126]

It is possible that government concerns about global warming and other energy-related environmental issues could override free market ideals and lead to policies that displace fossil fuel use, regardless of cost. Should the United States fully support such a program, photovoltaic energy systems could meet most U.S. electrical power requirements within four to five decades.[127]

Alcohol Fuels

In western industrial nations, transportation accounts for roughly a third of all energy consumption and carbon emissions, and is responsible for a substantial part of acid deposition. Among OECD countries, oil supplies nearly 99 percent of energy used for transportation, and road transport consumes about 80 percent of it.[128]

Because of concerns about cost, supply, security, and environmental damage associated with dependence on petroleum fuels (discussed earlier in the chapter), a number of countries are developing or considering alternative fuels for motor vehicles. Among the various options, alcohol fuels derived from food surpluses and plants grown to produce energy offer much promise. More than a dozen countries are now using alcohol fuels; Brazil and the United States are the largest producers. In 1986, Brazil produced 10.5 billion liters of ethanol (ethyl alcohol) from sugarcane, supplying about 50 percent of the country's automotive fuel. Most Brazilian cars run on a gasoline-alcohol fuel containing 20 percent ethanol, but nearly 30 percent of the cars use pure ethanol.[129]

In 1987, the United States produced 3 billion liters of ethanol, mostly from surplus corn and other grains. The ethanol is added to gasoline and sold as gasohol, usually containing 10 percent ethanol.[130]

Another alcohol, methanol, appears to be a potential automotive fuel and can be produced from wood and other organic substances. Although methanol corrodes some materials and has only about half the energy value of gasoline, its high oxygen content allows efficient combustion. When vehicle engines are properly modified, methanol can serve as an efficient, clean-burning fuel. The State of California has about 500 experimental methanol-powered cars in operation, and the California Energy Commission has an agreement with Atlantic Richfield to market methanol at 25 service stations in the Los Angeles area.[131]

Both ethanol and methanol can be produced from a variety of feedstocks, including sugar, starch, wood, and even organic materials in municipal solid waste. When produced from tree plantations or farms, the use of alcohol fuels can result in no net increase in atmospheric carbon dioxide, since biomass absorbs CO_2. A shift from petroleum to alcohol fuels would thus slow the buildup of greenhouse gases.[132]

Recent advances in the technology of fuel cells (which change the chemical energy of a fuel and an oxidant to electrical energy) offer promise that methanol, compressed natural gas, and probably other liquid fuels could be used to power motor vehicles with twice the energy efficiency of the best conventional cars. Vehicles powered by fuel cells should emit virtually no nitrogen oxides or hydrocarbons, emissions responsible for much of the damage caused by acids, ozone, and other vehicle-related pollutants. With accelerated research and development on fuel cell technology and alternative fuels, a price-competitive fuel cell vehicle could be mass-produced within a decade.[133]

Hydrogen

One of the world's most plentiful chemical elements, hydrogen can serve as a non-polluting, indirect source of energy with promising potential for meeting future energy needs. Since it contains no carbon, when used as a fuel, hydrogen combines with oxygen and produces only water vapor.

Although it is a promising option for the future, hydrogen is not a direct or primary source of energy, but an "energy carrier," or secondary source, like electricity. Much of the world's hydrogen supply is combined with oxygen in the form of water. To obtain hydrogen, these molecules must be split, a process which requires as much energy as it releases as a fuel. There are a number of ways in which water can be split, but one of the most promising is the use of solar energy to generate electricity that can then extract this pollution-free, transportable fuel from water. If research can develop methods to produce and collect it economically, hydrogen could become a practical energy carrier and a valuable method for storing solar energy.[134]

There are many projects currently under way to demonstrate hydrogen's potential. The Canadian province of Quebec and the European Community have together allotted $3.4 million for a study to explore the feasibility of shipping electricity across the Atlantic in the form of hydrogen. Germany's Frauenhofer Institute for Solar Energy Systems and the Chronar Corporation in the United States are developing self-contained systems to produce and store hydrogen. Japanese researchers are working on a wide range of products, including development of liquid hydrogen-powered diesel-type engines for cars and trucks at Tokyo's Musashi Institute of Technology. Researchers at Kogakuin University are experimenting with hydride-driven air conditioning and heating systems that would make conventional compressors obsolete. The Soviet Union has converted a commercial jet to fly partially on liquid hydrogen; it has completed a successful 21-minute flight. In West Germany, industry and government spend an estimated $200 million a year on research, and spending could exceed $500 million by 1992. In the United States, although several government agencies are developing a combination aircraft and space vehicle fueled by hydrogen, the U.S. Energy Department spent only $3 million on hydrogen research in 1988, down from $3.8 million in 1984.[135]

Tree Planting

Planting trees can help in solving several of our most important energy-related problems. Trees can reduce electricity use for air conditioning by blocking the summer sun and lowering air temperature. Tree plantations can provide fuelwood and serve as feedstocks for alcohol fuels. By absorbing carbon dioxide, plantations and new forests can slow the pace of global warming.

Even if wood-burning efficiency can be increased substantially, tree-planting programs will remain essential to solving the fuelwood crisis. The World Bank has estimated that, even with efficiency and other initiatives to reduce fuelwood needs by a fourth by the year 2000, at least 55 million hectares of high-yielding tree plantations would be needed between 1980 and the end of the century to meet growing fuelwood demands—five times the current rate of planting. Unfortunately, many plantations funded by the World Bank and other development agencies are providing pulp for commercial enterprises rather than fuelwood for household use.[136]

Various programs to increase fuelwood supplies are under way around the world. Development agencies have tried, with limited success, to establish community fuelwood plantations in rural areas. However, Third World villagers have generally been reluctant to plant trees for fuelwood alone. Trees are more valuable at the village level for fruit, timber, animal fodder, shade, and other uses. Firewood from trimmings and dead branches is often seen as a secondary benefit.

In Haiti, where over 70 percent of energy demand is met with fuelwood and charcoal, the U.S. Agency for International Development in 1981 initiated an $8 million agroforestry outreach project to promote trees as a cash crop. With the assistance of several development organizations, tree-planting projects were implemented by local residents who received training through agroforestry extension services.

Farmers received trees free of charge and decided what species to plant and where to plant them. Extension agents followed up initial contacts with additional information about care and maintenance of trees and seedlings. By 1986, 39 nurseries were in operation, producing about 5 million seedlings a year, and over 100,000 farmers had planted more than 25 million seedlings with a survival rate of 50 percent. The program has helped stabilize eroding soils in Haiti and meet energy needs, while allowing farmers to diversify their production and increase their income.[137]

In Southeast Asia, intensive energy farms, where fast-growing trees are planted in rows and harvested every few years, can be found in many areas. Some of these projects are quite successful, producing annually up to 50 cubic meters of fuelwood per hectare. Soil quality, however, is often rapidly depleted. Some good results have been achieved though agroforestry, which combines tree planting with traditional farming. Through agroforestry, crops can be protected from excessive sun, wind, and soil erosion. Certain trees can improve crop productivity by adding nitrogen to the soil. The amount of fuelwood produced per tree can be increased up to tenfold through regular pruning by farmers.[138] Community agroforestry programs can provide multiple benefits and sources of income, including food, timber, and animal fodder as well as fuelwood. In addition, agroforestry can improve cropland by increasing water retention, improving soil quality, and preventing soil erosion by wind and water. The most successful forestry programs are generally those that determine the priorities of local people, involve them in project planning, and gain their active support in implementing projects.[139]

Conclusion

Most nations have only begun to exploit the opportunities for improved energy efficiency and the potential of renewable energy sources. In an interdependent world, there are good economic, political, and environmental reasons for the industrial countries, which use most of the world's energy supplies, to take the lead in improving efficiency and developing renewables, and to help Third World countries acquire the means to develop their energy resources in ways that will ensure efficient energy use on a sustainable basis. National and international development agencies, business and industry, and nongovernmental organizations all have vital roles to play in achieving these goals.

What You Can Do

Inform Yourself

Get more information about energy issues by contacting organizations listed in Further Information. Learn about the major sources and uses of energy in your area and find out about community conservation efforts.

- Visit local gas and electric utility companies and find out how they operate, which areas they serve, how efficiently they create energy for industrial and consumer use, and what programs they have to promote energy conservation and efficiency. If there is a nuclear power plant in your area, find out about its operating and safety records, and compare its electricity costs with the costs of electricity generated by fossil and hydroelectric power plants.

- Learn about your town's energy plan and meet with the Town Energy Coordinator or an official who undertakes local energy projects. Look into the growing trend of community energy planning (See *Town Energy Planning: A Framework for Action* in Further Information) and consider the need for such action in your community.

Join With Others

Contact others who are concerned about energy issues and work together to educate your community about the need for conservation, the environmental damage caused by careless energy consumption, and methods of improving energy efficiency.

- Join national or international organizations that address energy issues at the national and global level, and subscribe to newsletters to keep up-to-date on the latest energy legislation.

Review Your Habits and Lifestyle

There is much you can do in your everyday life to promote energy conservation and efficiency. By using energy efficiently and conserving energy resources whenever you can, you will save money and help protect the environment at the same time. Consider what actions you can take as a student, citizen, voter and consumer in your schools, neighborhood, community, or workplace.

- Conserve energy in your daily activities; drive an energy-efficient car, use a thermostat with an automatic night-time setback, recycle metals, glass, and paper, and turn off unneeded lights.

- Improve the energy efficiency of your home by sealing air leaks, adding attic insulation, and installing insulating window shades, storm windows, and doors. Contact your local utility company and ask about obtaining a home energy audit to evaluate energy losses. These are generally offered free or at minimal cost.

- When replacing heating and cooling systems, refrigerators, and other household appliances, check the energy-efficiency ratings of different models and purchase efficient replacements. Consider the purchase of a heat pump that can provide both cooling in the summer and heating in the winter. For guidance, consult publications of the American Council for an Energy-Efficient Economy listed under Further Information.

- Improve the energy efficiency of your diet by consuming fewer animal products and more grains, vegetables and fruits that require less energy and fewer resources to produce.

- Plant as many trees as you can, especially around your house. Trees can help conserve energy by shading the house in the summer. They also absorb carbon dioxide and help counteract the greenhouse effect.

- Drive less. Use public transportation, ride a bicycle, and walk whenever possible. Organize carpools to work and school, and don't make unnecessary trips.

- If you live in a big house, reduce your living area during the winter to only a few rooms and heat only this space. Use space heaters and thermal clothing and blankets with insulating air spaces instead of trying to keep the entire house warm.

Work With Your Elected Officials

Contact your local, state, and national legislators, and find out about their positions and voting records on legislation designed to establish efficiency standards, such as the National Appliance Energy Conservation Act. In brief, clear letters or visits, urge representatives to support legislation that would:

- Assign the highest priority to improving U.S. energy efficiency, and develop an energy policy that simultaneously addresses air quality, the greenhouse effect, energy security, and U.S. economic competitiveness.[140]

- Provide incentives for renewable resource investment and conservation measures, such as low-interest utility loans or rebates, and the federal Solar and Conservation Bank—all of which encourage energy conservation investments to continue even in a period of high inflation.

- Discontinue subsidies for non-renewable energy sources such as the depletion allowance on investments in oil and gas extraction and on power plants that use these fuels.

- Make accurate, reliable, and understandable energy conservation information accessible to all homeowners, businesses, and consumers. Support efforts such as the Residential Conservation Service home audit program, and the Commercial and Apartment Building audit program, that educate consumers on conservation options.

- Promote community programs to enable low-income families to invest in energy conservation measures.

- Supplement, at the state level, the National Appliance Energy Conservation Act of 1987 that establishes energy conservation standards for home appliances and central heating and cooling systems.

- Fund research and development projects involving renewable energy sources such as solar, wind, biomass, and low-head hydroelectric energy, with special attention to small-scale, decentralized energy projects.

- Create controls on emissions of acid rain-forming sulfur and nitrogen compounds from power plants.

- Increase gasoline and diesel fuel taxes to allow the market price for imported oil to reflect its real cost to the U.S. economy and environment. Support an increase in car and light truck fuel economy standards to 45 and 35 mpg, respectively, by the year 2000.

- Promote U.S. government and private technical assistance programs in renewable energy, conservation, and fuelwood replanting in developing countries. Modify the energy planning and lending policies of the World Bank and other multilateral lending institutions so that they stress energy conservation and energy efficiency programs in the Third World.

Publicize Your Views

Use the print and electronic media to inform community leaders, local officials, and the general public about the costs and consequences of inefficient energy consumption in your town, state, and the United States in general.

- Write to the editor of your local newspapers, and create a television and radio public service announcement for a community organization concerned with energy conservation.

- Offer to write articles for local publications about home energy audits, recycling programs, and other energy issues.

Raise Awareness Through Education

Help educate your community about energy-related issues. Find out whether schools, colleges, and libraries in your area have courses, programs, and adequate resources on energy conservation and efficiency. If not, offer suggestions and help supply these resources.

- Urge social studies and science teachers from local schools to include a unit in all courses on the challenges of energy for the future, and the immediate need for conservation and efficiency. Encourage field trips to utility companies, nuclear power plants and solar energy installations; suggest projects that analyze individual energy consumption habits; and sponsor a showing of an energy-related film (See Further Information).

- Promote information-sharing on energy issues community-wide. Organize activities such as meetings and workshops featuring guest speakers or films, energy exhibits and fairs, recycling drives, and van pool and ride-sharing programs.

Consider the International Connections

The United States accounts for one-quarter of global energy consumption and compares poorly with most other industrialized nations in the areas of energy efficiency and conservation.

- If you travel to Japan, West Germany, or other countries that stress energy efficiency, observe gasoline prices and find out about motor vehicle fuel efficiency, government publicity campaigns for energy conservation and recycling, and the prevalence of community recycling programs.

- Consider the ways in which excessive U.S. energy consumption contributes to a wide range of international problems—for example, marine pollution from oil spills such as the 1978 Amoco Cadiz disaster that polluted French beaches, and fossil fuel emissions from U.S. sources that cause acid damage in Canada and contribute heavily to global warming. Consider how a vigorous U.S. effort to improve energy efficiency and develop renewable energy sources could help alleviate global problems.

Further Information

A number of organizations can provide information about their specific programs and resources related to energy.*

Books

American Council for an Energy-Efficient Economy (ACEEE). *The Most Energy-Efficient Appliances.* 1989-90 Edition. Washington, DC, 1989. 28 pp. $3.00. Lists annual energy costs or efficiency ratings for refrigerators, freezers, dishwashers, clothes washers, water heaters, air conditioners, heat pumps, and gas and oil furnaces.

ACEEE. *Saving Energy and Money with Home Appliances.* Washington, DC, 1987. 34 pp. $3.00. An illustrated guide to the purchase and use of energy-efficient appliances. In addition to the appliances listed in the previous entry, the guide also covers ranges, clothes dryers, portable space heaters, and lighting.

---

* These include: Alliance to Save Energy, American Council for an Energy-Efficient Economy, American Gas Association, American Petroleum Institute, American Solar Energy Society, American Wind Energy Association, Conservation Foundation, Critical Mass Energy Project, Energy Conservation Coalition, Electric Power Research Institute, Environmental Action, Environmental and Energy Study Institute (A Joint Committee of the U.S. Congress), Environmental Defense Fund, Environmental Policy Institute, Friends of the Earth, National Audubon Society, National Wildlife Federation, National Wood Energy Association, Natural Resources Defense Council, Passive Solar Industries Council, Public Citizen's Critical Mass Energy Project, Renew America, Renewable Fuels Association, Rocky Mountain Institute, Safe Energy Communication Council, Sierra Club, Solar Energy Industries Association, Solar Energy Research Institute, Union of Concerned Scientists, U.S. Department of Energy, World Resources Institute, and Worldwatch Institute.

Chandler, William U., Howard S. Geller, and Marc R. Ledbetter. *Energy Efficiency: A New Agenda.* Washington, DC: The American Council for an Energy Efficient Economy (ACEEE), July, 1988. 76 pp. Paperback. $8.00.

Clark, Wilson. *Energy for Security: The Alternative to Extinction.* Garden City, NY: Anchor Press, 1975. 652 pp. Paperback. Although outdated, the book still provides a valuable review of renewable and nonpolluting energy options.

Commoner, Barry. *The Poverty of Power: Energy and the Economic Crisis.* New York: Bantam Books, 1977. 297 pp. Paperback. Discusses links between pollution, economic stagnation, and depletion of energy reserves.

Deudney, Daniel and Christopher Flavin. *Renewable Energy: The Power to Choose.* New York: Norton, 1983. 431 pp. $18.95.

Gever, John et al. *Beyond Oil: The Threat to Food and Fuel in the Coming Decades.* A Project of Carrying Capacity, Inc. Cambridge, MA: Ballinger, 1986. 304 pp. Paperback. The first comprehensive effort through mainframe computer simulations to predict U.S. energy needs beyond the year 2000.

Goldemberg, José et al. *Energy for Development.* Washington, DC: World Resources Institute, 1987. 73 pp. Paperback, $10.00. Presents the major findings relating to developing countries of the End-Use Global Energy Project.

Goldemberg, José et al. *Energy for a Sustainable World.* Washington, DC: World Resources Institute, 1987. 120 pp. Paperback, $10.00. Presents the major global findings of the End-Use Global Energy Project; stresses the importance of improving the efficiency of energy use, and of matching energy supply to final, or end uses of energy.

MacKenzie, James J. *Breathing Easier: Taking Action on Climate Change, Air Pollution, and Energy Insecurity.* Washington, DC: World Resources Institute, 1988. 23 pp. Paperback. An excellent summary of actions needed to slow climate change, curb air pollution, and reduce dependence on imported oil.

Munson, Richard. *The Energy Switch: Alternatives to Nuclear Power.* Cambridge, MA: Union of Concerned Scientists, 1987. 65 pp. Paperback.

Ogden, Joan M. and Robert H. Williams. *Solar Hydrogen: Moving Beyond Fossil Fuels.* Washington, DC: World Resources Institute, in press. About 100 pp. Paperback, $10.00.

Rader, Nancy et al. *Power Surge: The Status and Near-Term Potential of Renewable Energy Technologies.*

Washington, DC: Public Citizen Critical Mass Energy Project, 1989. 96 pp. Paperback. Reviews all major renewable energy sources.

U.S. Department of Energy. *Energy Security: A Report to the President of the United States.* Washington, DC: March, 1987, 240 pp. plus appendices. Paperback, $16.00.

U.S. Energy Information Agency. *Annual Energy Review 1988.* Washington, DC, 1989. Approximately 300 pp. Paperback, $15.00. A statistical survey of U.S. and world energy production and consumption.

VanDolemen, Julie. *Power to Spare: The World Bank and Energy Conservation.* Washington, DC: World Wildlife Fund and The Conservation Foundation, 1988. 67 pp. Paperback.

Articles, Pamphlets, and Brochures

Conservation Foundation, The. "Energy," *State of the Environment: A View Toward the Nineties.* Washington, DC, 1987, pp. 234-57.

El-Hinnawi, Essam and Manzur H. Hashmi. "Energy and Transport," *The State of the Environment.* London: Butterworth Scientific, 1987, pp. 70-92.

Flavin, Christopher. "Creating a Sustainable Energy Future." In Lester R. Brown et al., *State of the World 1988.* New York: W.W. Norton, 1988, pp. 22-40.

Flavin, Christopher and Alan Durning. "Raising Energy Efficiency." In Lester R. Brown et al., *State of the World 1988*, pp. 41-61. For an expanded version, see Worldwatch Paper 82.

Gibbons, John H. et al. "Strategies for Energy Use." *Scientific American*, September 1989, pp. 136-43. Solar-based energy sources and nuclear power will help supply energy for growth and development without aggravating the greenhouse effect, but significant improvement in the efficiency of energy use is the real hope to meet the challenge.

Myers, Norman (ed.). "Elements," *Gaia: An Atlas of Planet Management.* Garden City, NY: Anchor Books, 1984, pp. 102-105, 112-117, 124-130.

Odell, Peter R. "Draining the World of Energy." In R.J. Johnston and P.J. Taylor (eds.), *A World in Crisis?* Cambridge, MA: Basil Blackwell, 1989, pp. 79-100.

Renner, Michael. "Rethinking Transportation." In Lester R. Brown et al, *State of the World 1989*, pp. 97-112. For an expanded version, see Worldwatch Paper 84.

Ridley, Scott. "Renewable Energy and Conservation," *The State of the States 1987*, pp. 26-29; "Energy Pollution and Control," *The State of the States 1988*, pp. 36-41. Washington, DC: Renew America, 1987, 1988.

Shea, Cynthia Pollock. "Shifting to Renewable Energy." In Lester R. Brown et al., *State of the World 1988*, pp. 62-82. For an expanded version, see Worldwatch Paper 81.

Sierra Club. *Wasted Energy: The Reagan Administration's Energy Program*. San Francisco, CA, 1988. $2.00.

World Resources Institute and International Institute for Environment and Development. "Energy," *World Resources 1986*, pp. 103-119; *World Resources 1987*, pp. 93-109; *World Resources 1988-89*, pp. 109-126. New York: Basic Books, 1986, 1987, 1988. Paperback, $16.95 each.

Yergin, Daniel. "Energy Security in the 1990s." *Foreign Affairs*, Fall 1988, pp. 110-132.

Your Home Energy Portfolio. Alliance to Save Energy. Outlines over 50 energy-saving home improvements and ways to finance them.

Periodicals

ARCO Solar News. P.O. Box 6032, Camarillo, CA.

The Natural Resource: A Quarterly Newsletter on Energy and the Environment. American Gas Association, 15 Wilson Boulevard, Arlington, VA 22209.

Photovoltaic News. PV Energy Systems, Inc., P.O. Box 290, Casanova, VA 22017.

Solar Today. American Solar Energy Society, 2400 Central Avenue, Unit B-1, Boulder, CO 80301.

Wind Energy Weekly. American Wind Energy Association, 1730 North Lynn Street, Suite 610, Arlington, VA 22209.

A number of the organizations listed at the beginning of the Further Information section above publish journals or newsletters that frequently cover energy-related issues.

Films and Other Audiovisual Materials

Dawn of the Solar Age. Part I, Solar Energy, 29 min.; Part II, Wind and Water Energy, 25 min. 16mm color film. Michigan Media.

Energy: The Fuels and Man. 1978. Describes what energy is and gives its past and present uses. Distinguishes between renewable and nonrenewable fuels. 23 min. Grades 7-12/adult. 16mm color film, $241.50; video, $69.95. National Geographic.

Energy: The Problems and the Future. 1978. Examines renewable energy resources, including tidal, geothermal, wind and solar power, and hydrogen gas. 23 min. Grades 7-12/adult. 16mm color film, $241.50; video, $69.95. National Geographic.

Energy: What Energy Means. 1982. Examines the various forms of energy and how one form can be converted to another. Explains the need to conserve fuels. 15 min. Grades 2-6. 16mm color film, $150.50; video, $59.95. National Geographic.

If You Can See A Shadow. 1979. Introduces usable information on passive solar power by covering 30 different solar projects. Emphasizes the partnership of solar power and conservation. 28 min. 16mm color film, $395; video, $250. Bullfrog Films.

Nuclear Energy: The Question Before Us. 1981. Explains how a nuclear reactor works and objectively examines the pros and cons of nuclear power, focusing on cost, operation, and waste disposal. 26 min. Grades 7-12/adult. 16mm color film, $293.50; video, $69.95. National Geographic.

Solar Energy: The Great Adventure. 1979. 18 min. 16mm color film. Michigan Media.

Solar Promise. 1979. 28 min. 16mm color film. Michigan Media.

Teaching Aids

Christensen, John W. *Global Science: Energy, Resources, Environment*. 2nd Edition. Dubuque, IA: Kendall-Hunt, 1984. Textbook, 355 pp.; Laboratory Manual, 285 pp.; Teacher's Guide, 356 pp. A new edition is in preparation.

Energy, Food and You. An interdisciplinary curriculum guide for elementary school children. 1985. Washington State, Office of Public Instruction, 17011 Meridian Ave. N., Seattle, WA 98133.

Intercom: Handbook for Educators; includes Intercom booklet #98, #102, and #106. Two audiovisuals are also available: *Energy Education* (1980), and *Teaching Against Hunger* (1979). Intercom, 218 East 18th Street, New York, NY 10003.

Air, Atmosphere, and Climate

Acid rain spares nothing. What has taken humankind decades to build and nature millennia to evolve is being impoverished and destroyed in a matter of a few years.

—*The Earth Report*[1]

Damage already done to the ozone layer will be with us, our children and our grandchildren throughout the 21st century.

— British Prime Minister Margaret Thatcher[2]

The five warmest years in the history of instrumental measurements are in the 1980s— 1980, 1981, 1983, 1987, 1988.

— James Hansen, Director, Goddard Institute for Space Studies, National Aeronautics and Space Administration[3]

The greenhouse effect constitutes the most serious disturbance of the biosphere yet caused by humanity.

—International Congress on Nature Management and Sustainable Development[4]

A serious and lasting government commitment to the development and use of energy-efficient and renewable technologies is a prerequisite to stabilizing world climate.

—*State of the World 1989*[5]

Major Points

- Many toxic gases and fine particles entering the air pose hazards to human health; these air pollutants can cause cancer, genetic defects, and respiratory disease, as well as exacerbate existing heart and lung disease.

- Nitrogen and sulfur oxides, ozone, and other air pollutants from fossil fuels are inflicting damage in more than thirty countries. Ozone and the acids of nitrogen and sulfur are damaging forests, crops, soils, lakes, streams, coastal waters, and buildings. Growing use of fossil fuels is spreading damage to industrial and developing countries throughout the world.

- Chlorofluorocarbons and other pollutants entering the atmosphere are depleting the Earth's protective ozone layer, and in some areas this depletion is beginning to increase the amount of harmful ultraviolet radiation that reaches the Earth's surface. Ozone depletion could cause skin cancer and cataracts, damage the immune system, reduce crop yields, disrupt marine food chains, and cause significant climate change.

- Fossil fuel combustion is increasing the amount of carbon dioxide in the atmosphere. This increase, along with growing atmospheric concentrations of other heat-absorbing gases, is already raising global temperatures and could alter weather patterns, worsen storms, destroy natural systems, and disrupt agriculture. Global warming may eventually melt the polar ice caps and raise sea levels, forcing relocation of low-lying coastal populations.

The Issues: Air Pollution, Acid Deposition, Ozone Layer Depletion, and Climate Change

Vast quantities of pollutants are pouring into the atmosphere, posing health threats to humans, damaging the environment and possibly altering the Earth's climate. Historically, the air has renewed itself through interaction with vegetation and the oceans. Today, however, this process is threatened by increasing use of fossil fuels, expanding industrial production, and growing use of motor vehicles.

The most common and widespread pollutants currently emitted by human activities are sulfur dioxide (SO_2), nitrogen oxides (NO_x), carbon monoxide (CO), carbon dioxide (CO_2), volatile organic compounds (hydrocarbons), particulates (tiny solid particles or liquid droplets), and lead. In addition, dozens of toxic chemicals are commonly found in the air surrounding urban areas.

Damage from air pollution often results not from a single pollutant but from several pollutants acting together.

Primary pollutants such as SO_2, NO_x, hydrocarbons, and CO often react with moisture and with one another to form secondary pollutants such as sulfuric and nitric acid, ozone (at ground level, a major contributor to smog), and photochemical oxidants.[6]

In recent years, many industrial nations have controlled air pollution with some success. Europe and North America, however, are now suffering serious damage from acid deposition. Increasing pollution from the growing use of motor vehicles plagues many nations. Car sales in Western industrialized nations rose 71 percent from 1970 to 1986; the total annual mileage driven in these nations reached 2.5 trillion in 1985.[7]

Developing nations with large urban areas are now also suffering from air pollution. Charcoal smoke blankets cities such as Lagos and Jakarta, while vehicle-caused smog looms over Ankara and Mexico City.[8] An estimated 60 percent of Calcutta's residents suffer from pollution-related respiratory diseases.[9]

Because the human body is so efficient in exchanging gases and fine particles between the air and the bloodstream, harmful air pollutants can easily permeate the body. Short-

Table 12.1

Sources and Impacts of Important Air Pollutants

Pollutants and Sources	Health and Environmental Impacts
Sulfur and Nitrogen Oxides: from fossil fuel combustion	Damage to lungs and respiratory tract. Acidification of streams, lakes, and soils; damage to buildings and materials; together with ozone, implicated in death of trees.
Carbon Monoxide: mostly from motor vehicles	Impairs ability of blood to carry oxygen; affects cardiovascular, nervous, and pulmonary systems. Contributes to ozone formation and indirectly to greenhouse warming.
Volatile Organic Compounds: from vehicles and industry	Harmful to human health; some compounds cause mutations or cancers. Contribute to ground-level ozone formation.
Ozone: from atmospheric reactions between nitrogen oxides and organic compounds	Eye irritation; nasal congestion, asthma, reduced lung function, possible damage to lung tissue; reduced resistance to infection. Principal component of smog, important greenhouse gas, injures trees, crops and other plants.

Sources: U.S. Environmental Protection Agency, *Environmental Progress and Challenges: EPA's Update* (Washington, DC, August, 1988), p. 13; James J. MacKenzie, *Breathing Easier: Taking Action on Climate Change, Air Pollution, and Energy Insecurity* (Washington, DC: World Resources Institute, 1988), p. 11.

term exposure to high levels of air pollution, or long-term exposure to low levels, can cause many adverse health effects, including breathing difficulties, increased susceptibility to respiratory infections, development of chronic lung disease, exacerbation of existing heart and lung disease, fetal defects, and cancer. (See Table 12.1 on the preceding page.) The people at greatest risk from air pollution are infants and children, the elderly, and those with respiratory problems.

Major Air Pollutants

Countless pollutants are present in our atmosphere, some of which have been produced naturally throughout time. But today, in addition to natural pollution, anthropogenic (human-caused) pollutants are pouring into the atmosphere in greater quantities than ever before. Globally, nature and people both produce large quantities of air pollutants, but in many urban and industrial areas anthropogenic pollutants greatly exceed those from natural sources. In 1980, human activities released about 110 million tons of sulfur oxides, 69 million tons of nitrogen oxides, 193 million tons of carbon monoxide, 57 million tons of hydrocarbons and 59 million tons of particulates into the atmosphere. The 24 industrial nations belonging to the Organization for Economic Cooperation and Development (OECD) account for about half of these pollutants.[10] (See Figure 12.1.)

Figure 12.1

Global Emissions of Common Air Pollutants from Human Activities, 1980

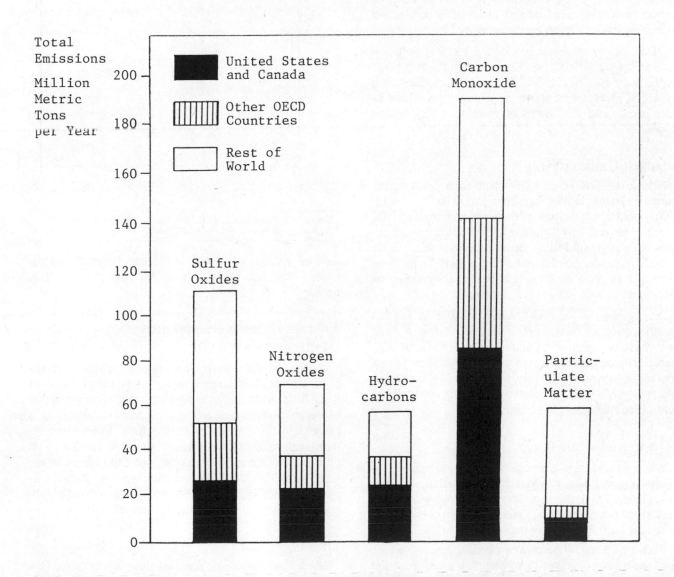

Source: Organization for Economic Cooperation and Development, *OECD Environmental Data Compendium 1989* (Paris, 1989) Table 2.1A, p. 17.

Sulfur Dioxide (SO₂)

Worldwide, nature and human activities produce this corrosive gas in roughly equivalent amounts. Its natural sources include volcanoes, decaying organic matter, and sea spray, while anthropogenic sources include combustion of sulfur-containing coal and petroleum products and smelting of nonferrous ores. Over land, and especially in industrial regions, much more SO_2 comes from human activities than from natural sources. The burning of fossil fuels to generate electricity is the largest single source of SO_2.[11]

Global output of SO_2 has increased sixfold since 1900.[12] However, most industrial nations lowered SO_2 levels by 20 to 60 percent between 1975 and 1984, and several countries have reduced SO_2 pollution in urban areas during the last decade, especially by shifting away from heavy industry and imposing stricter emission standards. Major SO_2 reductions have come from burning coal with lower sulfur content and from using less coal to generate electricity.[13]

Of 54 cities monitored worldwide for SO_2 pollution levels in the early 1980s, most were within the safe range specified by the World Health Organization. However, average SO_2 levels were above the safe range in at least 14 cities and around the homes of more than 625 million people.[14]

Nitrogen Oxides (NOₓ)

Nitric oxide (NO) comes from both natural and human sources and is quickly converted to nitrogen dioxide (NO_2). (NO_x is used to refer to both of these nitrogen oxides.) NO_x is formed naturally by lightning and by decomposing organic matter. Roughly half of anthropomorphic NO_x is emitted by motor vehicles and about a third comes from power plants; most of the rest is produced by industrial operations.

During the 1970s, NO_x emissions rose in several countries and then leveled off or declined. Nitrogen oxide levels have not dropped as dramatically as those of SO_2, primarily because a large part of total NO_x emissions comes from millions of motor vehicles, while most SO_2 is released by a relatively small number of coal-burning power plants from which emissions can be controlled.[15]

Carbon Monoxide (CO)

When inhaled, this gas restricts the blood's ability to absorb oxygen, causing angina, impaired vision, and poor coordination. CO has little direct effect on ecosystems, but it contributes indirectly to the greenhouse effect and depletion of the Earth's protective ozone layer.

Between 60 and 90 percent of global CO emissions are from natural sources, but in some urban areas most CO emissions come from incomplete burning of motor vehicle fuels. Some countries have been successful in reducing CO emissions. The United States, Japan, and West Germany

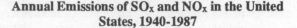

Figure 12.2

Annual Emissions of SOₓ and NOₓ in the United States, 1940-1987

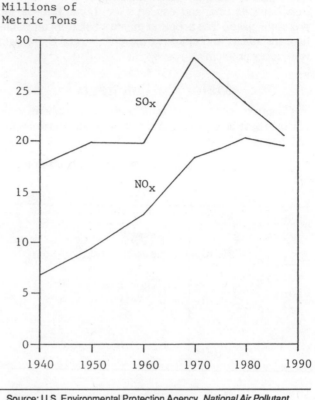

Source: U.S. Environmental Protection Agency, *National Air Pollutant Emissions Estimates* (Washington, DC, 1989)

have all reduced emissions in the last 15 years, but the United Kingdom has shown an increase over the past decade.[16]

Volatile Organic Compounds (Hydrocarbons)

Emissions of volatile hydrocarbons from human resources are primarily the result of incomplete combustion of fossil fuels. Natural sources are fires and the decomposition of matter. Some hydrocarbons are known or suspected to cause cancer, mutations, or birth defects. In the lower atmosphere, sunlight causes hydrocarbons to combine with other gases, such as NO_2, oxygen, and CO to form ozone, peroxyacetyl nitrate (PAN) and other types of photochemical oxidants. These active chemicals react with other substances and can damage human health and vegetation.[17]

Ozone

Each ozone molecule contains three atoms of oxygen. Ozone is not emitted directly from human activities, but is formed when volatile organic hydrocarbons and nitrogen oxides react with oxygen in the presence of sunlight.[18] At ground level, ozone is a serious air pollution problem throughout the industrialized world, posing threats to human health and damaging trees, crops and building material. While average ozone concentrations in the United States dropped from 0.16 parts per million (ppm) in 1977 to 0.12 ppm in 1986, as of 1988, 150 million Americans lived in areas that exceeded the EPA's maximum safe level for ozone at least once during the year. In 1988, 96 cities, counties, and other areas failed to meet EPA standards for ozone.[19] Especially vulnerable to ozone are more than 16 million Americans who already suffer from emphysema, asthma and other chronic respiratory problems.[20]

During the latter half of the 1970s and the early 1980s, the ozone concentration in Japan and Europe was 75 percent higher than levels considered safe by WHO. Australia and the United States exceeded this limit by 400 percent.[21]

Between 1984 and 1986, 62 U.S. metropolitan areas exceeded the federal ozone standard of 0.12 ppm. Los Angeles had the highest average ozone level with 0.35 ppm, and exceeded the federal standard, on average, 154 days each year—more than five times as often as the next highest city. The Los Angeles basin has all the conditions that promote severe pollution, including 12.5 million people with 8 million cars, a dozen oil refineries, a sunny climate, and surrounding mountains. In November 1988, the EPA issued a report outlining options for controlling the region's serious air pollution, including the possibility of a ban on gasoline-fueled vehicles.[22]

The average ozone concentrations in industrialized cities of Europe and North America are up to 3 times higher than the level at which damage to crops and vegetation begins. Ozone harms vegetation by damaging plant tissues, inhibiting photosynthesis, and increasing susceptibility to disease, drought, and other air pollutants. The U.S. National Crop Loss Assessment Program estimates that damage to corn, wheat, soybeans, and peanuts due to ozone results in annual losses of $1.9 to $4.5 billion. Other studies estimate that ozone causes U.S. crop losses of $5.4 billion each year.[23]

A 50 percent reduction in rural ozone levels, from 0.050 to 0.025 ppm, would increase yields of corn, wheat, peanuts, and soybeans by an estimated 6 to 30 percent, and a 40 percent drop in surface ozone levels could increase the value of eight U.S. food crops by $3.5 billion annually.[24]

Smog

A common air pollutant in urban areas, smog causes extensive environmental damage and is a serious health hazard. Smog contains a number of chemicals, notably ozone and peroxyacetyl nitrate (PAN), and is formed when strong sunlight acts on a mixture of nitrogen oxides and volatile organic compounds. Because the sun plays an important role in the formation of smog, cities are more affected by this form of air pollution during summer months.

The U.S. Environmental Protection Agency reported that during some summer days in 1988, about 110 million Americans breathed air with smog levels considered to be unhealthful. The city of Los Angeles has been called the smog capital of the nation. Smog has also become common in many large cities around the world such as Ankara, New Delhi, Melbourne, Mexico City, and Sao Paulo.[25]

In Mexico City, where 18 million people now live, school officials extended the 1988-89 Christmas holiday by a month in an effort to reduce children's exposure to the winter smog. Some environmentalists scoff at this action, saying more dramatic and effective steps must be taken such as eliminating some of the three million cars that clog the streets. Cars are responsible for 80 percent of the 5.5 million tons of contaminants entering the air every year in Mexico City. Because of its geographic layout, smog may be trapped over the city for weeks. It is feared that each year thousands may die as a result of air pollution.[26]

Motor vehicle exhaust is a major source of the nitrogen oxides and volatile organic compounds that create smog. A recent report indicated that despite vehicle emission controls and other technological improvements in the United States during the past two decades, U.S. efforts to control smog are losing ground because of the 25 percent increase in motor vehicles between 1977 and 1988 (from 147 million to 183 million), and the 40 percent increase in trucks over the same period.[27]

Carbon Dioxide and Chlorofluorocarbons

Produced mainly by the burning of fossil fuels, atmospheric carbon dioxide (CO_2) absorbs radiant heat from the Earth and is a major contributor to the greenhouse effect and global warming. Chlorofluorocarbons (CFCs) are synthetic chemicals produced for use as aerosol propellants, refrigerants, solvents, and as foam-blowing agents. CFCs are a major cause of ozone depletion in the stratosphere and also contribute to global warming.[28] Ozone depletion and global warming are discussed later in this chapter.

Particulates

Solid and liquid material suspended in the air may vary in size from fine aerosols to larger grit. The health effects of particulates depends on the size: larger particulates reduce visibility but have minor health effects, while smaller ones can cause eye and lung damage.[29]

Dust, spray, forest fires, and the burning of certain types of fuels are among the sources of particulates in the atmosphere. Emission controls have reduced the amount of particulates being released by several industrial nations. Nevertheless, the U.S. Office of Technology Assessment estimates that current levels of particulates and sulfates in ambient air may cause the premature death of 50,000 Americans every year, accounting for 2 percent of annual mortality.[30]

Lead

In high concentration, lead can damage human health and the environment. In humans and animals, it can affect the neurological system and cause kidney disease. In plants, lead can inhibit respiration and photosynthesis as well as block the decomposition of microorganisms. Once lead enters an ecosystem, it remains there permanently.

Most lead enters the environment from human sources; it has been used for thousands of years in piping, roofing, and coins. Most lead emissions come from burning leaded gasoline. Since the 1970s, however, stricter emission standards have caused a dramatic reduction in lead output. In 1978, the European Community allowed 0.4 grams of lead per liter; in 1989, this level will drop to 0.013 grams per liter. In Japan, 99 percent of all gasoline sold is lead-free. And in the United States, a shift to unleaded gasoline produced a 90-percent drop in lead emissions between 1970 and 1985.[31]

Acid Deposition

Acidic fallout has become one of the most damaging and controversial forms of air pollution in the industrialized world. Acid deposition can occur in various forms including rain, sleet, snow, fog, and dry particles. The primary sources of acid deposition are sulfur and nitrogen oxides released from electrical power plants, industrial boilers, mineral smelting plants, and motor vehicles that burn fossil fuels.[32] Sulfur and nitrogen oxides combine with moisture in the atmosphere and return to Earth as sulfuric and nitric acids.

Levels of Acidity

The acidity of water and other substances can be measured on the pH scale, which ranges from 0, the most acidic, to 14, the most alkaline. Since the pH scale is logarithmic, a change of one unit represents a tenfold change in acidity. Thus a solution of pH 4 is 10 times as acidic as one of pH 5, and 100 times as acidic as one of pH 6. While distilled water has a neutral pH level of 7 (neither acidic nor alkaline), "natural" rainfall is slightly acidic, with a pH between 5 and 6. In some heavily populated industrial areas of the United States and Europe, the pH of precipitation averages around 4. In 1987, rainfall in Norway was found to be as acidic as lemon juice with a pH level of about 2. In the United States, Kane, Pennsylvania experienced precipitation as acidic as vinegar, while Wheeling, West Virginia once reported rain nearly as acidic as battery acid, which has a pH of 1.[33]

Acidic Tolerance

Acid precipitation does not affect all ecosystems in the same way. Soil and water types vary a great deal; some can tolerate acids better than others. The American Midwest, for example, has alkaline soils that can buffer acid fallout.

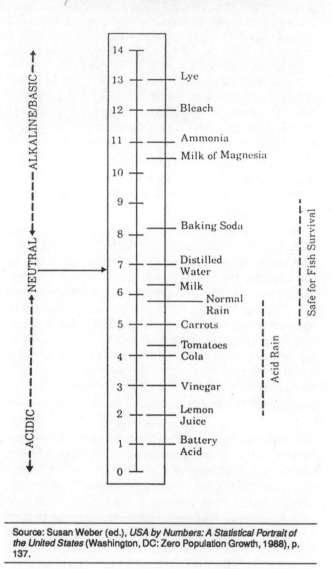

Figure 12.3
The pH Scale of Acidity and Alkalinity

Source: Susan Weber (ed.), *USA by Numbers: A Statistical Portrait of the United States* (Washington, DC: Zero Population Growth, 1988), p. 137.

Likewise, some lakes lie on limestone, sandstone, or other alkaline formations that help neutralize the acid. On the other hand, regions where lakes and soils lie on granite or thin glacial tills have lower pH values and are thus more susceptible to acid damage. Not surprisingly, regions suffering most from acid rain have soils and water with relatively low pH levels.[34]

Affected Regions

Damage from acid rain is worst in Scandinavia, Central Europe, and eastern North America. Yet signs of its destruction are now apparent in less industrialized parts of the world, and have been detected in more than thirty countries. The rain in Brazil's Sao Paulo state has an average pH level of 4.5, while precipitation in Guiyang, China averages 4.02.

The Taj Mahal in India is now being damaged by airborne acids from local oil refineries. In industrial countries, air pollution and acid deposition have damaged an estimated 31 million hectares of forests, an area the size of New Mexico.[35]

European nations have been especially hard hit by acid rain. Fish have completely disappeared from 4,000 lakes in Sweden, while 20,000 other lakes have been damaged by acid precipitation. The situation is even more dramatic in Norway, where 80 percent of the lakes have been declared technically "dead" or placed on the critical list. In the central alpine region of Switzerland, 43 percent of the conifers are already dead or seriously damaged, and 52 percent of all trees in West Germany are suffering from acidic deposition. The United Kingdom suffers from the highest percentage of forest damage, with 67 percent of its trees suffering from acid rain. In the heavily industrialized Upper Silesian region of Poland, acid deposition is eroding iron railroad tracks,

forcing trains to travel at only 40 kilometer per hour. Acid rain and other pollutants are dissolving ancient monuments such as the Parthenon in Athens and the Cathedral in Cologne.

In North America, the damage from acid precipitation is equally serious. In Canada's Ontario province, sulfur and nitrogen oxides from International Nickel's Sudbury plant have killed all vegetation and resulted in extensive soil erosion in a 20-mile area east of Sudbury. Over 300 Ontario lakes have a pH level below 5.0, while in Nova Scotia, nine rivers have pH levels below 4.7, making them incapable of supporting salmon and trout reproduction. In the United States, thousands of lakes on the eastern seaboard have been labeled "fish graveyards" due to their acidic content. At least 10 percent of the lakes in the Adirondack mountains have a pH level below 5.0. Acid rain is implicated in tree damage reaching all the way down the Appalachian mountains to Georgia.[36]

Figure 12.4
The Global Scope of Acid Damage

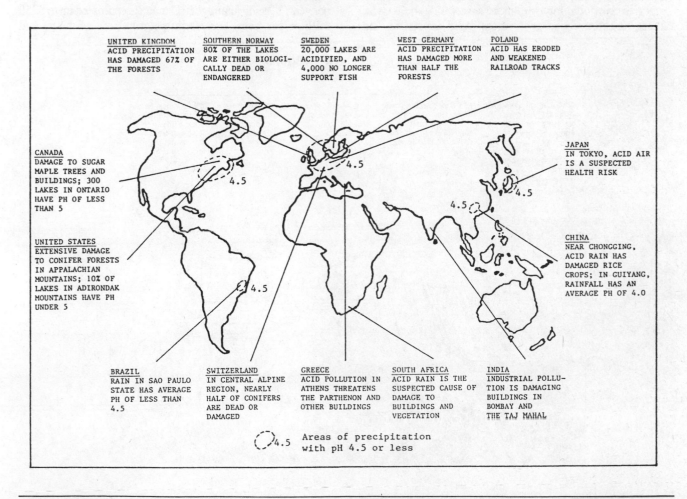

Source: Don Hinrichsen, "Acid Rain and Forest Decline," in Edward Goldsmith and Nicholas Hildyard (eds.), *The Earth Report* (Los Angeles: Price Stern Sloan, 1988), p. 67; D.M. Whelpdale, "Acid Deposition: Distribution and Impact," *Water Quality Bulletin*, Vol 8, 1983, p. 72.

Ironically, one reason acid rain has become such a widespread problem is because of earlier pollution control efforts. In recent decades, smokestacks from power plants and industries have been made much taller to reduce local effects of pollution. As a result, winds now carry pollutants over vast areas; SO₂ emissions have traveled up to 2,000 kilometers in a few days. Acid emissions have become a source of political controversy, since one country's pollution is often carried into another country, as is the case with U.S. pollution in Eastern Canada. Sweden claims that 80 percent of the SO₂ it receives comes from other countries; Norway puts its claim at 90 percent.[37]

Acidic Damage to Ecosystems

The effects of acid fallout can be seen throughout ecosystems. Acid deposition damages leaf surfaces, preventing some tree species from retaining water. Acidic water can leach minerals such as calcium, magnesium, and potassium from leaves and from the soil, depriving plants of vital nutrients. Acids also release aluminum from the soil, which can damage tree roots, block nutrient absorption, and impair water transport, making trees more susceptible to drought, insects, and other sources of stress. Acid deposition and acid-catalyzed releases of toxic metals such as aluminum can filter into streams and lakes, contaminating fish and threatening public water supplies.[38]

Acidified water itself can kill many species of fresh water fingerlings and larvae. That disrupts the food chain. In saltwater, nitrates from acid deposition can boost the nitrogen content of coastal estuaries, creating algae blooms that cause oxygen depletion and the suffocation of fish and other aquatic plants.[39]

Economic Costs

The economic costs of acid deposition are increasingly recognized. Damage to metals, building facades, and paint alone costs the 24 OECD nations an estimated $20 billion a year.[40] The total environmental cost of acid fallout is difficult to estimate, since it would include damage to agriculture, lakes, fisheries, vegetation, human and animal health, tourism, and numerous other factors. The European Community estimates that acidic deposition harms at least $1 billion worth of crops annually.[41] In West Germany alone, damage to timber and agriculture is estimated at $1.4 billion per year.[42] Acidic damage in Canada is estimated up to $250 million per year, threatening agriculture, fishing, and

Figure 12.5
Understanding Acid Deposition

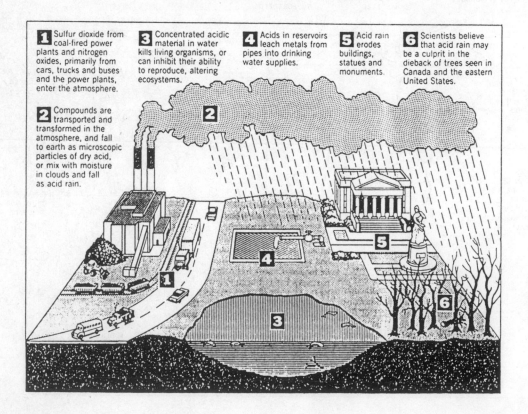

1. Sulfur dioxide from coal-fired power plants and nitrogen oxides, primarily from cars, trucks and buses and the power plants, enter the atmosphere.

2. Compounds are transported and transformed in the atmosphere, and fall to earth as microscopic particles of dry acid, or mix with moisture in clouds and fall as acid rain.

3. Concentrated acidic material in water kills living organisms, or can inhibit their ability to reproduce, altering ecosystems.

4. Acids in reservoirs leach metals from pipes into drinking water supplies.

5. Acid rain erodes buildings, statues and monuments.

6. Scientists believe that acid rain may be a culprit in the dieback of trees seen in Canada and the eastern United States.

Source: *The Washington Post*, March 19, 1987, p. A25.

tourism.[43] In the United States, total harm to forests, agriculture, and aquatic ecosystems may exceed $5 billion annually.[44]

Efforts to Limit Acid Damage

Some European nations are spreading limestone, caustic soda, sodium carbonate, and other alkaline compounds over soils and waters in an effort to neutralize acidic fallout. By 1986, Sweden had limed some 3,000 lakes at a cost of about $25 million. West German scientists have fertilized damaged trees to replace minerals leached from the soil by acidic water. While treated trees have recovered in a few weeks, the treatment only works for trees with severe nutrient deficiencies. Such programs are merely quick-fix remedies; the only long-term solution to acid damage is to reduce emissions of sulfur and nitrogen oxides.[45]

In 1984, Norway and Sweden helped form the "30 Percent Club," initially consisting of 10 members that received more airborne pollution than they emitted. Using 1980 emissions as the baseline, the group called for a 30-percent reduction in sulfur dioxide levels by 1993. However, the United States and Great Britain, major producers of both SO_2 and NO_x, refused to join the group, claiming there is insufficient evidence of benefits from reducing emissions.

By 1995, the European Community plans to reduce emissions of SO_2 by 60 percent and NO_x by 40 percent. In addition, the Community proposes that all members have unleaded gas available by 1989 and that all new cars must have catalytic converters by 1995.[46]

Air Pollution and Damage to Vegetation

Acid deposition and ground-level ozone can each impair plant growth and damage forests. In combination, they can be deadly. Separately and together, acid deposition and ozone are the main suspected causes of forest decline in the United States, Canada, and Europe. High levels of these

Figure 12.6
States Where Air Pollution is Known or Suspected to Cause Major Damage to Trees or Crops

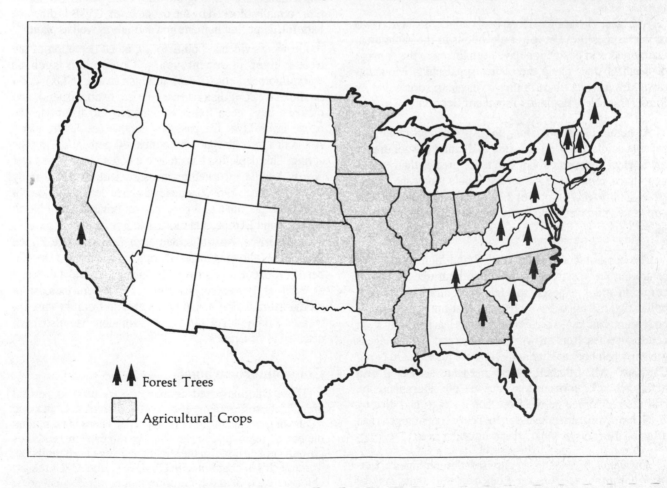

Source: James J. MacKenzie and Mohamed T. El-Ashry, *Ill Winds: Airborne Pollution's Toll on Trees and Crops* (Washington, DC: World Resources Institute, 1988), p. 11.

pollutants are found where damage to U.S. forests and crops occurs.[47] (See Figure 12.6 above.)

Forest damage from air pollutants appeared in the northeastern United States as early as 1962. Damage in this region is greater at higher elevations, where pollutants are more concentrated. At elevations greater than 900 meters, 20 percent of all spruces alive in 1982 had died by 1987. More than 70 percent of standing trees covering 14,500 acres of spruce-fir forests in Virginia, Tennessee, and North Carolina were dead in 1987.[48]

For some time, high concentrations of air pollutants have been blamed for killing trees and crops surrounding industrial areas. Sulfur dioxide and nitrogen oxide emissions were thought to be the primary cause. But scientists observed similar tree damage in the San Bernardino Forest near Los Angeles, where experiments detected high ozone concentrations from the city's automobile emissions. Surface ozone now has been linked to the damage or death of 87 percent of ponderosa and Jeffrey pines in southern California.[49]

At higher elevations, ozone concentrations may be two or three times those at lower elevations in the same area. Ozone and acid deposition have been linked to the damage or death of trees along the entire Appalachian Mountain range, from white pines in the Northeast to spruce and fir forests from New England to southern Georgia.

At higher elevations along the Appalachian range, the average acidity of cloud moisture is 10 times greater than at lower elevations, and about 100 times greater than that of unpolluted rain water. At several points along the range, cloud acidity reaches a pH of 2.3 — 2,000 times higher than unpolluted rain water and equivalent to the acidity of lemon juice.[50]

The combination of high acidity and high ozone concentration has a synergistic effect on both trees and food crops. In areas exposed to high concentrations of both pollutants, scientists have observed that nutrients such as potassium, calcium, nitrates and sulfates leach rapidly from conifer needles. Root growth diminishes and leaf damage is greater when both acidity and ozone are present. On North Carolina's Mt. Mitchell, where trees are subjected to a highly acidic, high-ozone fog for up to 3,000 hours annually, half the red spruce on the west-facing slopes had died by 1987. No significant infestations of insects or pathogens had affected the roots, trunks, or crowns of the trees.[51]

An overall assessment suggests that the combined effects of airborne acids, ozone, and other air pollutants may be reducing agricultural and forest productivity over large areas in industrial countries by 5 percent to 15 percent below historic levels.[52]

Depletion of the Ozone Layer

Ozone, while a hazard at ground level, acts as a filter in the upper atmosphere, screening out harmful ultraviolet (UV) radiation from the sun. Small quantities of ozone—only a few molecules to each million molecules of air—exist in the stratosphere, a layer of the atmosphere between 15 and 50 kilometers (9 and 30 miles) above the Earth's surface. Ozone concentrations at different altitudes also affect air temperature, air movements, and other factors that influence climate.

Ozone forms in the upper atmosphere when oxygen molecules (O_2) are split by ultraviolet radiation. The two free oxygen atoms resulting from this reaction quickly bind to other oxygen molecules to form O_3. This process is reversible; UV also breaks down ozone, forming O_2 and O and creating a balance between O, O_2, and O_3. When other reactive substances such as chlorine and nitrogen oxides are present in the upper atmosphere, they can upset this balance and reduce the amount of O_3. Single reactive chlorine, bromine, or nitrogen molecules can eliminate thousands of ozone molecules. Among the different types of UV radiation normally blocked by the ozone layer, UV-B is the most harmful, damaging humans and animals as well as plants.[53]

Growing evidence points to a gradual depletion of the ozone layer in recent years. Compounds such as chlorofluorocarbons (CFCs), nitrogen oxides (NOx), carbon dioxide (CO_2), halons (used in fire extinguishers), and methane have been found responsible for disrupting the ozone layer. Over the past 20 years, ozone levels above Antarctica have dropped by almost 40 percent during the spring. This depletion has created a hole in the ozone layer clearly visible by satellite measurement. A 1988 study shows that since 1969, ozone levels have declined 2 percent worldwide, as much as 3 percent over urban areas of North America and Europe, and more than 3 percent over parts of South America, Australia, and New Zealand. (See Figure 12.7 on the following page.) During several days in December 1987, ozone levels over Melbourne, Australia dropped as much as 10 percent, causing a 20 percent increase in ultraviolet radiation reaching the ground from the sun. Increased ground-level ultraviolet radiation has also been detected in Antarctica.[54]

Chlorofluorocarbons

These chlorine-based compounds are used as aerosol propellants, refrigerants, coolants, sterilizers, solvents, and as blowing agents in foam production. There is no natural method of removing CFCs from the atmosphere, and they are only broken down in the stratosphere by UV-B radiation. Through this interaction with UV-B radiation, CFCs release chlorine; each of these chlorine atoms can destroy some 100,000 ozone molecules. Because CFCs are stable, they could remain intact for over 100 years, continuing to destroy the ozone layer, even if their production is ceased immedi-

Figure 12.7
Atmospheric Ozone Levels: A Global Decline

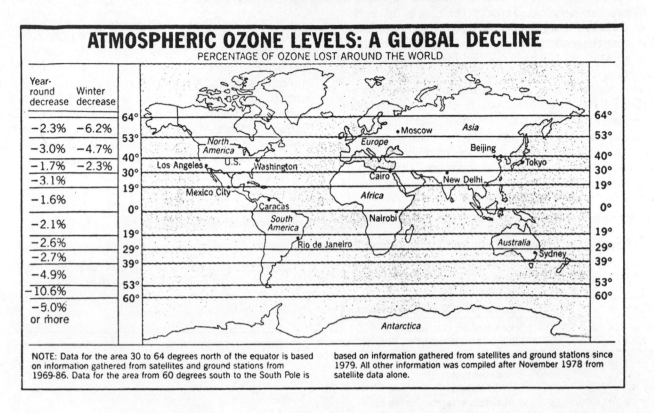

Source: Ozone Trends Panel, National Aeronautics and Space Administration, as published in *The Washington Post,* March 6, 1988, p. A8.

ately. The United States has already banned the use of CFCs as aerosol propellants, but they continue to be used in other products.[55] As described later in the chapter, the 1987 Montreal Protocol specifies reductions in CFC production over the next decade.

Nitrogen Oxides, Carbon Dioxide, and Methane

These gases also affect the ozone layer, but in opposite ways. Methane and carbon dioxide build up ozone in the stratosphere; the former directly affects atmospheric chemistry while the latter indirectly affects atmospheric temperature and reaction rates. Nitrogen oxides, on the other hand, work to break down atmospheric ozone. All three of these compounds are important in industrial and agricultural production, so that a significant reduction in their emissions is quite difficult to achieve.[56] In addition to these gases and CFCs, several other compounds are known to affect the ozone layer. (See Table 12.2.)

Table 12.2
Major Ozone-Modifying Substances Released by Human Activities

Chemical	Source
CFC-11, CFC-12	Aerosol propellants, refrigeration, foam blowing, solvents
CFC-22	Refrigeration
CFC-113	Solvents
Methyl Chloroform	Solvent
Carbon Tetrachloride	CFC production and grain fumigation
Halon 1301, 1211	Fire extinguishers
Nitrous Oxides	Industrial activity
Carbon Dioxide	Fossil fuel combustion
Methane	Agriculture, industry, and mining

Source: Alan S. Miller and Irving M. Mintzer, *The Sky Is the Limit: Strategies for Protecting the Ozone Layer* (Washington, DC: World Resources Institute, 1986), p. 5.

229

Damage from Ultraviolet Radiation

A continued thinning of the stratospheric ozone layer will have grave ecological effects. The EPA estimates that over the next century, 800,000 additional cancer deaths and an increased number of cataracts may be attributed to depletion of the ozone layer and the resulting increase of UV-B radiation. This radiation is thought to be linked to more deadly forms of skin cancer, which annually kill an estimated 100,000 people worldwide. By increasing UV-B radiation, each 1 percent drop in stratospheric ozone may increase the rate of skin cancer by 4-6 percent. Between 1969 and 1986, the average ozone concentration worldwide dropped by 2 percent. Ozone depletion could cause 3 million to 15 million new cases of skin cancer in the United States by the year 2075.[57]

Table 12.3
Effects of Ultraviolet Radiation on Human Health

Acute	Sunburn Thickening of the skin
Chronic	Aging of skin, thinning of epidermis
Carcinogenic	Nonmelanoma skin cancer Malignant melanoma
Eye disorders	Cataracts (probable) Retinal damage Corneal tumors Acute photokeratitus ("snow blindness")
Immunosuppression (possible)[58]	Infectious skin disease
Conditions Aggravated by UV Exposure	Genetic sensitivity to sun-induced cancers Nutritional deficiencies Infectious diseases Autoimmune disorders

Source: Miller and Mintzer, *The Sky Is the Limit*, p. 12

UV-B radiation can destroy individual cells in plants and animals and kill microorganisms. In plants, UV exposure can decrease photosynthesis, stunt leaf growth, damage seed quality, and reduce crop yields. Even small changes in stratospheric ozone or UV-B exposure can kill large numbers of aquatic organisms, which form the base of the ocean's food web. Many species of phytoplankton (small plants) and fish larvae cannot tolerate increases in UV-B. A 15-day exposure to UV-B levels 20 percent higher than normal can kill anchovy larvae to a water depth of 10 feet.[59]

UV-B exposure greatly accelerates the breakdown of some paints and plastics. Without additional chemical stabilizers in plastics to mitigate the effects of UV radiation, damage to polyvinyl chloride alone could approach $5 billion by the year 2075.[60]

Global Warming and Climate Change: The Greenhouse Effect

Evidence is growing that emissions of carbon dioxide and other gases are accumulating in the atmosphere. These gases allow sunlight to penetrate to the Earth's surface, but trap radiant heat and prevent its escape into space. This "greenhouse effect" is beginning to raise the average global temperature. Although a recent study indicates that in the United States (which only covers two percent of the Earth's surface), temperatures have not increased during the last 100 years, many researchers agree that the global average air temperature at the Earth's surface has risen about 0.5 degree Celsius (nearly 1 degree Fahrenheit) over the last century. There is evidence that the increase in air temperature may be accelerating; the five warmest years in recorded history occurred in the 1980s.[61]

Figure 12.8
Observed Global Average Temperatures, 1880-1988, With Projections to 2040

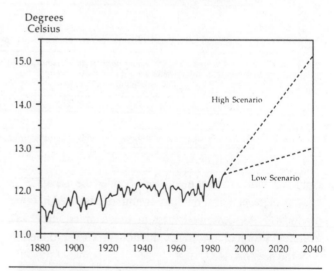

Source: Data based on James E. Hansen, "The Greenhouse Effect: Impacts on Current Global Temperature and Regional Heat Waves," in Dean Edwin Abrahamson (ed.), *The Challenge of Global Warming* (Washington, DC: Island Press, 1989); and Richard A. Warrick and Philip D. Jones, "The Greenhouse Effect, Impacts and Policies," *Forum for Applied Research and Public Policy*, Fall 1988; as published in Christopher Flavin, *Slowing Global Warming: A Worldwide Strategy*, Worldwatch Paper 91 (Washington, DC: Worldwatch Institute, October 1989), p.18.

In addition to the data indicating that surface air temperatures are rising, a recent analysis of sea surface temperatures monitored by satellite concluded that between 1982 and 1988, ocean temperatures increased about 0.1 degree Celsius per year, providing additional evidence that global warming is under way.[62]

Coastal communities, food production, water supplies, forest products industries, fisheries, and entire ecosystems will all be at risk if global warming accelerates during the next several decades. The huge quantities of oil, coal, and natural gas that fueled the modern age could lead to its decline.[63]

Carbon Dioxide Buildup

Carbon dioxide is the single most important "greenhouse gas," accounting for about half of all current additions to the greenhouse effect. Small amounts of CO_2 exist naturally in the atmosphere, holding life-supporting warmth at the Earth's surface.

Prior to the Industrial Revolution, the concentration of CO_2 in the earth's atmosphere was 275 to 285 ppm. Since then, humans have burned vast quantities of fossil fuels — coal, oil and natural gas — releasing carbon gases and other emissions. As a result, atmospheric CO_2 concentration is now 25 percent higher than before the Industrial Revolution, and increasing rapidly. In the last 30 years, the CO_2 level has risen from 316 ppm to about 350 ppm, the highest concentration in the past 160,000 years. Data from ice cores and ocean sediments show a clear link between atmospheric carbon dioxide and global temperatures over this period; higher CO_2 concentrations have coincided with higher temperatures.[64]

In 1988, at least 6 billion tons of carbon were added to the atmosphere—about 5.5 billion tons from fossil fuel combustion and between 0.4 and 2.5 billion tons from burning or clearing forests, primarily in tropical areas. Of the portion attributed to fossil fuel use, the United States contributes about 24 percent of total emissions, the Soviet Union 20 percent, and Western Europe 15 percent. The United States, the Soviet Union and Europe together produce two-thirds of all carbon emissions from fossil fuels. Of the carbon emissions from deforestation, in 1980 Brazil accounted for roughly 20 percent.[65]

In 1986, U.S. consumption of fossil fuels released nearly 1.3 billion tons of carbon. Texas led the country in carbon emissions, accounting for about 12 percent of the total, far ahead of second-place California, which emitted about 7 percent of the total. Ohio, Pennsylvania, Illinois, and Louisiana together released another 19 percent of U.S. carbon emissions.[66]

If no action is taken to slow carbon emissions, continued population, economic, and energy growth could result in global warming of 1.6 to 4.7 degrees Celsius above pre-industrial temperatures by the year 2030. Over the same period, if strong policies were implemented to discourage fossil fuel use, improve energy efficiency, encourage solar energy use, and stop deforestation, global temperatures might increase by only 1.1 to 3.2 degrees C.[67]

Other Greenhouse Gases

Gases such as methane (CH_4), chlorofluorocarbons (CFCs), ozone, and nitrous oxide also contribute to global warming. Although trace gases make up only a small percentage of the total atmosphere, some trap heat a thousand times more effectively than CO_2. Their effect is compounded because they absorb different wavelengths of infrared radiation.[68]

The concentration of trace gases in the atmosphere is increasing even faster than that of CO_2, and will cause about half of the additional greenhouse effect in the 1990s. There is considerable uncertainty about the relative contributions of the various greenhouse gases to global warming, but estimates are available. During the past decade, methane has probably accounted for about 16 percent of global warming, with much of the methane coming from rice and cattle production, decomposition of organic wastes, and the burning of fossil fuels and forests. The release of CFCs and other halocarbons, the same gases that are destroying the ozone layer, may account for about 20 percent of greenhouse warming.[69] (See Table 12.4.)

Table 12.4
Greenhouse Gases: Sources and Estimated Relative Contributions to Greenhouse Warming

Gas	Major Sources	Percent Contribution
Carbon dioxide	Fossil fuels, deforestation	50
Chlorofluorocarbons, other halocarbons	Refrigeration, solvents, insulation, foams, aerosol propellants, other industrial and commercial uses	20
Methane	Rice paddies, swamps, bogs, cattle and other livestock, termites, fossil fuels, wood burning, landfills	16
Tropospheric ozone	Fossil fuels	8
Nitrous oxide	Fossil fuels, fertilizers, soils, burning of wood and crop residues	6

Source: Based on 1989 estimates by World Resources Institute and information from other sources. The estimates of relative greenhouse gas contributions involve significant scientific and data uncertainties.

Rising Temperatures and Changing Climate

A 1988 study by the American Association for the Advancement of Science predicted that an "equivalent doubling" of CO_2 will raise the average global temperature 2 to 5 degrees C (4 to 9 degrees F). (A temperature increase of 1 degree C is equivalent to a rise of 1.8 degrees F.)[70] As long as greenhouse gases are added to the atmosphere, global warming will increase. The earth's middle and upper latitudes, which include most of North America, Europe, and Asia, are expected to warm at twice the average global rate. The average temperature in these areas could increase 1 degree C or more per decade.[71]

Global temperature increases may appear small until they are compared with previous climate change. In the thousands of years since the last Ice Age, global temperatures gradually increased by 3 to 5 degrees C. The predicted warming in this century would be 10 to 40 times as rapid. Temperatures could soon be higher than at any time in human history. A 5 degree C (9 degree F) average increase would make the earth hotter than it has been in the last 2 million years.[72]

While many climate scientists agree that the Earth will warm as carbon dioxide and other greenhouse gases accumulate in the atmosphere, there is little agreement on how much warming will occur. Much of the uncertainty stems from the difficulty of predicting the temperature effects of the increased cloud cover that is expected to result from higher temperatures and greater evaporation of water from the Earth's surface. When the effects of clouds were removed from computer simulations by 14 different climate models, all the models produced similar predictions of global warming. But when clouds were included, the predictions of warming differed greatly. In some models, cloud cover caused a net warming (for example, by trapping more outgoing radiation than the incoming solar radiation it blocks); in other models, clouds caused overall cooling. In a recent analysis using one of the climate models with a more sophisticated and perhaps more realistic cloud modeling scheme, the predicted warming from an equivalent doubling of carbon dioxide was only half of that predicted by the same model using a simpler cloud scheme—less than 5 degrees Fahrenheit instead of 9 degrees.[73]

While the Earth's average temperature is rising, regional and seasonal variations exist. One recent study (mentioned earlier) indicates that the United States has not become warmer during the last 100 years, and there is evidence of a trend toward cooler winters in some parts of the Northern Hemisphere.[74] Other studies indicate that a buildup of greenhouse gases in the atmosphere may increase cloud cover in some regions, which could block more incoming solar radiation than the outgoing radiation it traps, causing some regional cooling.[75] An overall assessment suggests that as greenhouse gases accumulate and the Earth's average temperature rises, local climates will become more variable—warmer and drier in some areas, cooler and wetter in others. Some regions could even experience both hotter summers and colder winters.[76]

The greatest warming could take place at middle and upper latitudes during winter, but the number of uncomfortable summer days will also increase. An equivalent doubling of CO_2 could increase the average number of days over 90 degrees F from 15 to 56 in Chicago and from 36 to 87 in Washington, D.C. By the middle of the next century, summer temperatures in many U.S. cities will exceed 100 degrees F far more often than they do now.[77] (See Table 12.5.)

Table 12.5
Predicted Number of Days Temperature Will Exceed 100°F for Selected U.S. Cities by Middle of Next Century

City	Days Above 100° in 2050	Days Above 100° at Present
Washington, DC	12	1
Omaha	21	3
New York	4	0
Chicago	6	0
Denver	16	0
Los Angeles	4	1
Memphis	42	4
Dallas	78	19

Source: James J. MacKenzie, *Breathing Easier: Taking Action on Climate Change, Air Pollution, and Energy Insecurity* (Washington, DC: World Resources Institute, 1988), p. 7.

Global warming may increase precipitation in some areas. Regions in higher latitudes, including parts of Canada and the Soviet Union, could receive more rain and snow during the winter. Some wet tropical areas may receive additional rainfall. Rising ocean surface temperatures may increase the frequency and severity of hurricanes and other tropical storms.[78]

The interior of large land masses could receive less rain. Even without drought conditions, soils would be drier during the summer. Less snow would accumulate during mild winters and snow would melt earlier. Lack of snow cover in spring would allow the sun to begin drying the soil earlier, increasing the chance that soils will be too dry to support crops through the summer. The combination of higher temperatures and lower soil moisture in the U.S. grain belt would be likely to lower agricultural production.[79]

Rising Seas

Because rising temperatures melt polar ice and continental glaciers, and warm and expand ocean waters, global warming is projected to raise sea levels between one and five feet by the year 2050, which could cause extensive flooding of coastal regions. Ocean levels along the northeastern U.S. coast have already risen about a foot during the past century. The U.S. Environmental Protection Agency estimates that sea levels could rise 2.2 meters (7 feet) by the year 2100.[80]

The world's coastal lands are densely populated and extremely valuable. A relatively small rise in sea level would contaminate groundwater supplies with salt, flood coastal roads, buildings, and wetlands, and drive millions of people from their homes.

A half meter (1.5 foot) rise would displace 16 percent of Egypt's population. Rice, the staple food of Asia, is grown on low-lying river deltas and flood plains. A 1 meter (3 ft.) rise would inundate many rice-growing areas and substantially reduce harvests.[81]

Most of the Maldives, a nation of 1,196 islands in the Indian Ocean, would disappear if sea levels rose 2 meters (6.5 feet). Such a sea level rise would flood 28 percent of Bangladesh, along with large portions of Louisiana and Florida.[82]

To avoid losing major cities, nations could be forced to hold back the sea at tremendous cost. Charleston, South Carolina would have to invest an estimated $1.5 billion to protect itself from a 1 meter rise in sea level.[83]

Threats to Plants and Animals

In the past, organisms have adapted to climate change by migrating and evolving, but the speed of the expected warming will make adaptation difficult.

Forests Tree species are sensitive to climate variation; many species can survive only within a narrow band of temperature and moisture. A warming of only 1 degree C (1.8 degrees F) could shift some forest zones by more than 200 kilometers (125 miles). In North America, an equivalent doubling of carbon dioxide could shift the ranges of birch, sugar maple, hemlock, and beech trees by 500 to 1,000 kilometers (300 to 600 miles).

If warming proceeds as rapidly as some scientists expect, entire forests across the southern United States could be threatened. Trees at the warmer, drier limits of a species range would become more susceptible to disease, insects, and other stresses. Tropical species, more suited to the new climate, could eventually move in to replace them. Yet no type of tree has ever migrated faster than 200 kilometers per century.[84]

Agriculture Crops are also climate-sensitive. The hotter, drier conditions predicted for the interior of North America would lower yields of Midwestern corn and wheat. A temperature increase of 2 degrees C combined with a 10 percent decrease in precipitation could reduce U.S. corn yields by one-fifth. Climate change will require costly alterations in agricultural practices. Adjusting the world's irrigation systems alone could cost $200 billion.[85]

Higher levels of carbon dioxide will benefit plants by accelerating the process of photosynthesis, although some plant species will adapt better than others. Weedy plants would thrive in a carbon-rich atmosphere, crowding out food crops and other plants and depriving them of nutrients.[86]

Global warming will favor warm-weather insect species with short life spans that can adapt and evolve quickly. Insect pests, parasites, and pathogens are expected to flourish under rapid warming conditions. Crop losses due to insects are expected to increase, and mid-latitude countries could experience an influx of tropical diseases.[87]

Wildlife Animal species that can migrate quickly will face obstacles that did not exist during previous climate changes. As global warming progresses, wildlife may become "civilization locked" in refuges and wilderness areas that are no longer suitable habitats. Cities, roads, farmland, and other human barriers will make migration difficult. To preserve animal life, it may become necessary to create networks of "migration corridors" or to undertake vast relocation efforts.[88]

Summary of Environmental Effects

For major air pollutants, Table 12.6 (on the following page) lists the most significant human-caused sources and summarizes the more important harmful environmental effects, including greenhouse warming, stratospheric ozone depletion, acid deposition, smog, and damage to trees and crops. In Table 12.6, a plus sign indicates a contribution to the effect; a minus sign signifies a lessening of the effect. In the case of ozone-layer depletion, the impact of pollutants may vary at different altitudes; this is indicated by a dual sign (+/-).

As the table suggests, most environmental damage caused by air pollution results from more than a single pollutant. A comprehensive strategy to limit damage by airborne pollutants will require reductions in harmful emissions from a variety of sources, especially transportation, energy production, industry, agriculture, and deforestation.

Protecting the Atmosphere

Air pollution, damage to living organisms, and global climate change are complex and diverse problems. Yet they all share a common root—energy consumption. To slow damage to plants and animals and to avoid destructive climate change will require fundamental changes in energy policies within the next decade.

Carbon dioxide now accounts for about half of global warming. To curb CO_2 emissions will require integrated energy policies that are less dependent on fossil fuels, especially coal and oil, and that stress energy efficiency and renewable energy sources.[89]

To achieve widespread use of the least polluting energy sources on a global scale is a massive but necessary task. Initiatives to improve energy efficiency, tax fossil fuels, expand renewable energy sources, reverse deforestation, limit emissions of sulfur dioxide and nitrogen oxides, eliminate chlorofluorocarbons, encourage sustainable

Table 12.6
Major Air Pollutants, Human-Caused Sources, and Environmental Effects

Air Pollutant	Major Human-Caused Sources	Greenhouse Warming	Stratospheric Ozone Depletion	Acid Deposition	Smog	Damage to Vegetation
Carbon Dioxide (CO_2)	Fossil fuels, deforestation	+	+/-			
Methane (CH_4)	Rice Fields, cattle, landfills, fossil fuels	+	+/-			
Nitric Oxide (NO), Nitrogen Oxide (NO_2)	Fossil fuels, biomass burning		+/-	+	+	+
Nitrous Oxide (N_2O)	Nitrogenous fertilizers, deforestation, biomass burning	+	+/-			
Sulfur Dioxide (SO_2)	Fossil fuels, ore smelting	-		+		+
Chlorofluoro-carbons (CFCs)	Aerosol sprays, refrigerants, solvents, foams	+	+			
Ozone (O_3)	Fossil fuels	+			+	+

Source: Thomas E. Graedel and Paul J. Crutzen, "The Changing Atmosphere," *Scientific American*, September 1989, p. 62; and James J. MacKenzie and Mohamed T. El-Ashry, *Ill Winds: Airborne Pollution's Toll on Trees and Crops* (Washington, DC: World Resources Institute, 1988).

agriculture, and curb population growth can help us curb harmful pollution and reduce overall energy use.

Improving Energy Efficiency

Improving energy efficiency in transportation, industry, offices, and homes is the quickest, most cost-effective way of reducing carbon emissions and other gases that are altering the atmosphere and harming the environment.[90]

The transportation sector absorbs two-thirds of all oil consumed in the United States, and contributes a third of all carbon dioxide emissions. In OECD member countries, cars, trucks and buses generate 75 percent of carbon monoxide emissions, 48 percent of nitrogen oxides, 40 percent of hydrocarbons, 13 percent of particulates, and 3 percent of sulfur oxides. The average American car emits its own weight in carbon every year.[91]

Greater use of public and mass transit would reduce the number of vehicles on the road, improving urban air quality and relieving traffic congestion. Raising fuel efficiency standards for new automobile and truck fleets from the current 26.5 miles per gallon to 45 mpg—feasible with commercially available technology—also would reduce carbon emissions and oil imports.[92]

Fossil fuels used to generate electricity, run industrial processes, and heat buildings contribute two-thirds of U.S. carbon dioxide emissions.[93] (See Table 12.7.) Sixty-four percent of the world's electricity is produced by burning fossil fuels (chiefly coal), accounting for 27 percent of global carbon emissions from fossil fuels (1.5 billion tons annually). Energy consumption in buildings, industries, homes, and workplaces could all be reduced by improving the efficiency of lighting, space heating, motors, and appliances.[94]

In 1987, after years of delay by the Reagan administration, the U.S. Congress passed laws establishing national efficiency standards for 13 types of appliances. In 1988, it also enacted a second tier of efficiency standards for lighting ballasts. These measures are expected to save more than $100 billion and 22,000 megawatts of electricity by the year 2000.[95] Worldwide, lighting absorbs 17 percent of electricity use, accounting for annual carbon emissions of 250 million tons.

Table 12.7
U.S. Sources of Carbon Dioxide Emissions

Electric Utilities	33%
Transportation	31%
Industry	24%
Buildings	12%

Source: MacKenzie, *Breathing Easier*, p. 10.

Cogeneration, the simultaneous production of steam and electricity, is common in many countries. Waste heat can be used directly to run industrial processes or heat buildings, or it can be recycled for more electricity production, boosting the output of the original fossil fuel up to 30 percent and cutting carbon emissions.[96]

Non-fossil fuel energy sources such as nuclear power have been proposed to lower greenhouse gas emissions from electrical generation. But fossil fuel consumption to produce uranium fuel for nuclear power is substantial, and nuclear power itself is both environmentally dangerous and more expensive than other energy options. (See Chapter 11.) In the United States, improvements in end-use efficiency— techniques to reduce the energy required at the point of use—cost between 0.5 cents and 2 cents per kilowatt hour, compared with more than 13 cents per kwh for nuclear energy. Efficiency improvements can displace 2.5 to 10 times more carbon dioxide per dollar invested than a comparable investment in nuclear power, and improved efficiency can reduce carbon output over the entire range of fossil-fuel uses in addition to electricity generation. Full use of available efficiency improvements in the United States could cut energy consumption in half and save consumers an additional $220 billion.[97]

Taxing Fossil Fuels

Fossil fuels vary greatly in their carbon content and therefore in their contribution to the greenhouse effect. To produce the same amount of useful energy, oil releases about 40 percent more carbon dioxide than natural gas, and coal emits up to 95 percent more. Synthetic fuels produced from coal would give off two to three times as much CO_2 as natural gas.[98]

Levying a "carbon tax" on various fuels according to their carbon content would increase the price of energy, and encourage investments in energy efficiency by government, industry, and individual consumers. As the price of fossil fuels increased, renewable energy sources would become more competitive.[99]

Imposing a gasoline tax at the pump would encourage the purchase of energy-efficient vehicles and reduce overall vehicle use. Revenues from both taxes could be used to develop renewable energy sources or to reduce air pollution. Substituting less carbon-intensive fuels such as compressed natural gas would also cut carbon emissions from vehicles.[100]

Expanding Renewable Energy Sources

Hydro, solar, wind, and geothermal power are all under development around the world. Hydropower currently supplies about 21 percent of global energy supply. Although renewable energy sources have much less short-term potential to replace fossil fuels than advances in energy efficiency, renewable sources are rapidly improving in performance and dropping in cost.[101]

Some renewable energy sources produce greenhouse gases. For example, large hydropower projects flood vegetation and create methane. Wood and other biomass fuels release carbon dioxide, but this carbon dioxide can be offset by replanting the biomass crops. Renewable energy sources generally release much less CO_2, SO_2, and NO_x than other fuels, and their use reduces the need for coal, oil, and natural gas in power production and industry.[102]

Reversing Deforestation

Because trees absorb CO_2 from the atmosphere while growing and release it when burned, curbing deforestation and encouraging replanting, especially in the tropics, would slow the buildup of atmospheric CO_2 as well as provide many other environmental benefits. (See Chapter 7.) An additional 150 million hectares of tree cover could be needed by the year 2000 to meet developing countries' growing demands for fuel and wood products and to rehabilitate deteriorating ecosystems. While scientists estimate that additional tree cover with an area about the size of Australia would be needed to absorb the CO_2 released from fossil fuels, large-scale tree planting may be the lowest cost initial measure to slow the greenhouse effect.[103]

Limiting Sulfur Dioxide and Nitrogen Oxides

Emissions of sulfur dioxide and nitrogen oxides, the leading sources of acid deposition, can be reduced by implementing four broad strategies: increasing energy efficiency and conservation, using less-polluting fuels, preventing the formation of pollutants during combustion, and capturing pollutants before they are released.[104] Burning coal and oil with lower sulfur content is a simple approach, but supplies of such fuels are fairly limited and their use can create economic conflicts between producing regions. Sulfur can be removed from coal by washing and from oil by a chemical process, but these procedures are expensive. Coal can be burned in fluidized beds with lime to prevent harmful sulfur emissions. Scrubbers and filters can trap SO_2 and NO_x emissions before they are released. However, all emission control technologies increase the final cost of electricity. In addition, the disposal of trapped pollutants can pose environmental hazards.

Since motor vehicle exhausts are primary sources of nitrogen oxide emissions, improved vehicle efficiency, stricter controls on vehicle emissions, the use of cleaner-burning automotive fuels such as methanol and compressed natural gas in urban areas, and expanded mass transit systems are among the priorities for reducing acid deposition and other damage caused by NO_x.[105]

Phasing Out Chlorofluorocarbons

In the 1970s, when the connection between CFCs and ozone depletion was first realized, several nations, including the United States, took action to ban CFCs from being used in spray cans.[106]

In 1987, delegates from 24 countries signed the Montreal Protocol, a historic agreement to cut CFC production in half by the end of the century. The agreement, now signed by more than 40 nations, will come into force in 1989. Under its terms, CFC production gradually will be reduced to 20 percent below 1986 levels by 1994 and to 50 percent below 1986 levels by 1999. The Protocol leaves it up to each nation to determine the way it will cut CFC production.[107]

In March 1988, Du Pont Chemical Company, the world's leading producer of CFCs, announced plans to phase out production of these chemicals. In early 1989, Du Pont reported development of a gas to replace CFC-12, the most common of the CFCs. Du Pont said the new gas, which it hopes to introduce commercially by 1993, is 97 percent less destructive to the ozone layer than CFC-12.[108] Unfortunately, many potential CFC substitutes that are less destructive to the ozone layer have other disadvantages: some are toxic, less energy efficient to produce, contribute to smog formation, or add to the greenhouse effect.[109]

More immediate steps must also be taken to reduce CFC emissions. For example, all nations and industries should act to ban aerosol propellants containing CFCs, eliminate the rapid evaporation of CFC solvents, and capture the blowing agents used to inflate plastic foams. Sealing refrigerators that are leaking CFC coolants and recovering coolants from home and automobile air conditioners that are being repaired or discarded would also reduce ozone-depleting emissions.[110]

Encouraging Sustainable Agriculture

World fertilizer use per unit area of cropland has risen nearly 40 percent in the last decade, and petrochemical fertilizers are a major source of nitrous oxide emissions that contribute to greenhouse warming. In addition to other environmental benefits from adopting less energy- and chemical-dependent agricultural practices, the use of lower-input, more sustainable methods in agriculture would reduce greenhouse gas emissions from oil consumption and fertilizers. Multiple cropping and intercropping techniques can improve yields and decrease the need for chemical fertilizers and pesticides, thus reducing air pollution. Expansion of agroforestry practices that combine trees and food crops would help absorb carbon dioxide.[111]

Because livestock production is highly resource-intensive, shifting from a human diet based on meat to one based on grains can conserve energy, water, and soil while improving air quality. To produce one pound of steak with 500 calories of food energy requires 20,000 calories of fossil fuel, expended mostly in producing the crops fed to livestock. Compared with beef, grains and beans require only 2 to 5 percent as much fossil fuel to produce, per calorie of protein. Livestock production also consumes prodigious quantities of water as well. In the U.S. Pacific Northwest, meat production accounts for more than half the water consumed. The energy losses to the region, in pumping this water and diverting it from potential hydropower production, amount to 17 billion kilowatt hours of electricity each year.[112] (See Chapter 5.)

Curbing Population Growth

In 1987, world population grew by 83 million people, passing the 5 billion mark and adding to the burden on existing resources. Programs to control population growth can benefit environmental quality by stabilizing demands for energy, food, and durable goods. Yet curbs on population alone cannot guarantee a lessening of atmospheric pollution and other environmental pressures. Even with a population growth rate approaching zero, U.S. demand for energy grows nearly 2 percent a year. Without improved energy efficiency, expanded use of renewable energy, increased reforestation, and sustainable agriculture, existing populations in industrial and developing countries will continue to pollute the atmosphere and accelerate damage from acid deposition, ozone layer depletion, and climate change.[113]

Industrial and developing countries both must adopt energy and pollution control strategies to limit atmospheric damage, but because industrial and Third World countries have differing development and energy needs, those strategies will vary from country to country. Debt-burdened developing countries may be unable to afford low-pollution energy technology without substantial international assistance.

Recent Responses by Governments and Organizations

Governments and organizations are devoting more attention to air pollution, acidic damage, and the threats of global warming and ozone depletion, and are beginning to respond to these threats with a range of initiatives to protect the atmosphere.

Urban Air Pollution

Of the world's heavily-polluted urban areas, the Los Angeles region may have developed the most comprehensive program to control air quality. The South Coast Air Quality Management District, covering a four-county area including Los Angeles, has prepared a 20-year plan for meeting air standards in the region. Provisions of the plan include:

- Employer incentives to persuade employees to join car pools, take buses, or ride bicycles.

- Restrictions on how and where future businesses and housing developments may locate, with industrial growth limited to areas close to affordable housing.

- Construction of more car-pool lanes on freeways, and a ban on heavy-duty diesel trucks from freeways during rush hours.

- Enforced use of vapor-recovery nozzles on gasoline pumps, and curbs on excessively-smoking vehicles.

- A "clean fuels" program to require all buses, rental cars, and fleet vehicles purchased after 1993 to be powered by electricity or low-polluting methanol. Methanol-fueled cars and buses are already on the region's roads, and an electric van with a top speed of 60 mph and a range of 120 miles soon will be in use. [114]

At the national level, the U.S. Congress recently passed legislation with incentives for motor vehicle manufacturers to produce vehicles that can operate on non-polluting natural gas, or on methanol, ethanol, or alcohol-gasoline mixtures. The legislation also authorizes the Department of Energy to purchase and test cars, trucks, and buses that use alternative fuels.[115]

However, there is concern that this bill will encourage the production of methanol from coal, which would increase carbon dioxide emissions. Methanol can also be produced from trees, which can be replanted to avoid a net CO_2 increase.

Acid Deposition

In 1979, the International Convention on Long-Range Trans-Boundary Air Pollution began work on measures to reduce sulfur and nitrogen oxide emissions that cause acid deposition. By June 1988, 21 European nations and Canada had agreed to reduce sulfur emissions by at least 30 percent of 1980 levels within 5 years. Unfortunately, some of the largest sulfur emitters have not signed the agreement, including the United States, the United Kingdom, and Poland.

In October 1988, 12 Western European nations agreed to cut nitrogen oxide emissions by 30 percent over the next decade. Thirty other European nations have agreed to freeze NO_x emissions at current levels.[116]

Global Warming

In 1987, the World Commission on Environment and Development called for international agreement on policies to reduce greenhouse gases.[117] In November 1988, representatives of 30 nations and 16 international organizations met to form the Intergovernmental Panel on Climate Change

in Geneva, organized by the World Meterological Organization and the United Nations Environment Program. Three working groups were established to develop specific proposals by mid-1990 for an international convention to regulate carbon emissions. There appears to be some support for having international lending agencies make development assistance conditional on actions to protect the atmosphere.[118]

The Global Greenhouse Network, which includes public interest organizations and legislators from 35 nations on 6 continents, has been organized by environmentalist Jeremy Rifkin to mobilize public opinion on the global warming issue. At its international conference in October 1988, more than 100 activists agreed on several recommendations for action to slow greenhouse warming. These include cuts in carbon dioxide emissions of 20 percent by the year 2000 and 50 percent by 2030, incentives to encourage energy efficiency and conservation, a rapid transition to solar and renewable energy technologies, and Third World debt reduction to encourage tropical forest conservation. At a recent meeting with Los Angeles Mayor Tom Bradley and other community leaders, Rifkin noted that California ranks among the top 10 countries in the world in greenhouse emissions. Rifkin and Bradley announced the start of a sister city campaign to persuade city mayors and councils around the world to plant trees, which absorb CO_2, and to raise awareness of the importance of energy conservation.[119]

During 1988, several legislative initiatives were introduced in the U.S. Congress to slow the buildup of greenhouse gases. These include the National Energy Policy Act in the Senate, and the Global Warming Prevention Act in the House of Representatives. In January 1989, the World Environment Policy Act, with measures to curb greenhouse gas emissions, was also introduced in the Senate.[120]

In late 1988, the American Agenda, chaired by former Presidents Ford and Carter, recommended that the next U.S. president set global warming as a major environmental priority and work for international agreements to confront the issue. At about the same time, 18 U.S. environmental organizations issued *Blueprint for the Environment*, which urged the incoming president to make global warming a top priority, to work with other nations toward a global treaty for reducing CO_2 emissions through increased energy efficiency and reliance on renewable energy, and to propose an international program to halt deforestation and promote reforestation.[121]

In its January 1989 global environmental agenda for the Bush Administration, Worldwatch Institute called for aggressive action on an international agreement to cut global carbon emissions by 20 percent, a U.S. gasoline tax of $1.00 a gallon, and U.S. aid to reforestation efforts in the Third World. Finally, in its "Planet of the Year" issue highlighting the Earth's endangered environment, *Time* magazine called on nations to tax carbon dioxide emissions to encourage energy conservation, increase funding for alternative energy sources, help build high-efficiency power plants in developing nations, and launch major tree planting programs. *Time* urged the United States to raise the gasoline tax by 50 cents per gallon, tighten fuel efficiency requirements for cars, and promote natural gas usage.[122]

After years of official U.S. inaction on the global warming issue, in January 1989 the new Secretary of State James Baker called for action to counter the warming trend, including cuts in greenhouse gas emissions, increased energy efficiency, and reforestation.[123]

Finally, in March 1989, the EPA presented to the U.S. Congress the results of a two-year study of ways to counter global warming. The report recommended radical steps to be taken by the year 2000, including greatly improved motor vehicle efficiency and home insulation, a carbon tax on fossil fuels, major reforestation programs, and a phase-out of CFCs.[124]

Ozone Depletion

In 1988, new evidence of rapid ozone depletion triggered a spate of international conferences that seek greater and faster reductions of CFCs. In March 1989, 123 nations gathered in London to discuss accelerating the phaseout of CFCs. A week prior to the London conference, the 12-nation European Community agreed to end the use of all CFCs by the year 2000. A day after the agreement was announced, the United States joined in the turn-of-the-century CFC ban. Industrialized nations have pressured other countries to speed cuts in CFCs, but large developing nations, including China and India, have balked at CFC reductions. They claim to need CFCs to promote economic development, particularly for the manufacture and sale of cheap consumer goods and refrigerators. Nations participating in the London conference reaffirmed their willingness to allow CFC manufacturers to continue exporting the chemicals, while halving domestic production in order to accommodate Third World growth. Other financial incentives, such as debt remission favoring environmental protection, and technical and financial assistance from industrial nations and the World Bank could help developing nations switch to CFC substitutes while maintaining economic growth.[125]

In May 1989, representatives of the European Economic Community and 81 countries that adhere to the Montreal Protocol met in Helsinki and agreed to a total phaseout of ozone-depleting CFCs, by the year 2000 if possible. They also declared that as soon as it was feasible, they would phase out other ozone-depleting chemicals such as methyl chloroform, carbon tetrachloride, and halons.[126]

Many nations are responding to the threats of global warming, ozone depletion, and other consequences of atmospheric pollution. Following the London conference on CFCs, leaders of 15 nations and ranking envoys from nine others held an environmental summit conference at The Hague and called for increased United Nations authority to protect the Earth's atmosphere, either by strengthening existing institutions or by creating a new institution.[127] An effective response to pollution of the atmosphere will require concerted global action at all levels by a wide range of nongovernmental and intergovernmental organizations. Such action should be supported by legal instruments declaring the atmosphere to be the common heritage of humankind.[128]

What You Can Do

Inform Yourself

Get more information about issues related to air pollution by contacting some of the organizations listed under Further Information. Among the services these groups can provide are issue updates, suggestions for citizen action, newsletters, speakers, seminars, and membership programs. The National Clean Air Coalition includes nearly 30 national organizations and serves as a network of individuals, state, and local organizations concerned with the environment, health, labor, parks, and other resources threatened by air pollution.

- Find out about major sources of air pollution in your area. They might include motor vehicles, power plants, industrial operations, and municipal waste incinerators. Where are the largest sources located? Are streams and lakes in your area affected by acid precipitation? Are trees or crops affected by surface ozone? What are the ozone levels in your community or in nearby cities? Find out about your state's air quality standards and determine whether or not your region is in compliance with federal regulations.

- If you live in a rural area that is relatively unaffected by smog, visit a city like Los Angeles or New York and observe firsthand how serious urban air pollution can be. Learn what cities in your state are doing (or not doing) to combat air pollution.

- Consult the report by Renew America on the states' role in reducing global warming (listed under Further Information) and find out how much your state contributes to the greenhouse effect. See how much of your state's total carbon emissions come from oil, coal, and natural gas.

Join With Others

Contact others in your area who are concerned with the hazards of air pollution, and work together to spread your message to the community. Join existing groups or start your own action network.

- Become active in your local civic association and find out what agencies in your local government have responsibility for monitoring air quality and enforcing air pollution standards. Are these agencies effective in their methods? Are they adequately staffed and funded?

- Join or become a volunteer for organizations that focus on air pollution. Subscribe to their newsletters and action bulletins to keep up-to-date on the latest developments in legislation and clean air enforcement.

- Share your knowledge and concern about air pollution issues through local action campaigns. Arrange talks and workshops with community schools, churches, synagogues, unions, businesses, and other organizations.

Review Your Habits and Lifestyles

Consider what actions you can take to affect air quality in your everyday life—in your family, home, neighborhood, community, and school or workplace; as a student, citizen, voter, and consumer.

- Eliminate or cut down your use of products that release harmful chlorofluorocarbons (CFCs) into the atmosphere and encourage recycling of CFCs wherever possible. CFCs contribute to the greenhouse effect as well as to ozone layer depletion, and though their use was banned from spray cans in 1978, they are still found in several consumer products.

  » Use cellulose fiber instead of foam to insulate your house. Rigid foam insulation contains large amounts of CFC-11 that eventually will be released into the atmosphere. Fiberglass, fiberboard, gypsum, and foil-laminated board provide effective alternatives.

  » Do not buy household items that contain CFCs, including fire extinguishers filled with Halon-1211, liquids and aerosols used to clean electronic equipment, and polystyrene foam products. Buy plates and cups made of paper instead of styrofoam, and petition local fast-food chains to reduce their use of styrofoam in packaging.

  » When your home or car air conditioner is serviced, insist that the ozone-depleting liquid refrigerant be drained into a closed container to be cleaned and recycled—not evaporated into the atmosphere.

  » Have your car serviced regularly at licensed service stations, and replace air conditioner hoses to prevent harmful leaks of CFC-laden coolant.

- Conserve energy. Acid rain and the greenhouse effect are directly linked to the burning of fossil fuels, airborne industrial pollutants, and automobile exhaust. Since many power plants create energy by burning coal, gas, or oil, any measure you take to reduce energy consumption or use energy more efficiently will help lessen environmental damage.

  » Save electricity by turning off unnecessary lights; turn down the air conditioner in the summer and the heat in the winter; use ceiling fans; install storm windows and insulation; and insulate your water heater. Saving energy can also save you money on fuel bills—contact your local utility company for more suggestions.

  » Buy appliances that are energy-efficient. Manufacturers of refrigerators, air conditioners, hot water heaters, furnaces, and stoves all sell "low energy use" models. Drive a fuel-efficient car with a high mileage-per-gallon rating.

  » Drive less. Use public transportation, ride a bicycle, and walk whenever possible. Organize carpools to work and school and avoid unnecessary trips.

- Plant as many trees as you can, especially around your house. Trees can help conserve energy by shading the house in the summer. They also absorb carbon dioxide and help counteract the greenhouse effect.

- Eat fewer animal products, and consume more grains, vegetables, and fruits that require less energy to produce and therefore create less air pollution.

Work With Your Elected Officials

Contact your local, state, and national representatives, and find out where they stand on the issues of global warming, acid rain, and ozone depletion. In brief, clear letters or visits, urge them to support or introduce legislation to:

- Develop an energy policy that simultaneously addresses air quality, the greenhouse effect, energy security, and U.S. economic competitiveness; halt the release of all chlorofluorocarbons to the atmosphere by 1999; and initiate international and national forestry programs that absorb carbon dioxide.[129]

- Reduce growth in carbon dioxide emissions through energy conservation, improved energy efficiency, and the development of energy technologies such as solar and wind power that lessen dependence on fossil fuels.

- Create a commission in your state to consider actions to improve energy efficiency and limit greenhouse gas emissions; the State of Missouri has recently considered such legislation.[130]

- Strengthen the Clean Air Act to improve air quality and limit damage from acid deposition, surface ozone, and other air pollutants. Needed measures include reduction of sulfur and nitrogen oxide emissions from industrial operations and power plants, strengthened auto emission standards, and expansion of mass transit systems.

- Reduce the emission of CFCs by establishing mandatory CFC recycling projects at service stations and at refrigerator and automobile disposal sites, and by requiring industry standards to prevent ventilation of CFCs during repair of air conditioners and retail food refrigerators.

- Ban new vehicles in your state with air conditioners containing CFCs; the State of Vermont has enacted such a ban to take effect in 1992.

- Limit use of CFCs in hospital sterilization procedures, home insulation, household and automobile coolants, and fire extinguishers. Persuade foreign nations to adopt additional controls on CFC use.

- Encourage international cooperation on measures to ban all harmful CFCs, reduce global use of fossil fuels, halt the destruction of tropical forests, and launch a major international reforestation effort.

Publicize Your Views

Use local print and electronic media to express your views on air pollution, and the views of the organizations you represent, to community leaders and the general public.

- Create public service announcements for local radio and television stations, and write letters to the editor of town and city papers. Suggest article topics and offer to cover action campaigns yourself. Write about the causes and consequences of air pollution, and relate both to situations in your own community. Urge others to write to their elected representatives and to get involved in efforts to improve air quality.

Raise Awareness Through Education

Find out whether schools, colleges, and libraries in your area have courses, programs, and adequate resources related to acid rain, ozone depletion, and global warming; if not, offer to help develop or supply them.

- Meet with teachers of biology, ecology, and social studies in your local schools and urge them to include a unit on environmental, social, and political aspects of air pollution. Suggest field trips to lakes and forests to look for acid rain damage, and outline projects that link individual habits to acid rain, ozone depletion, and global climate changes. Offer to speak to classes and lead projects yourself.

Consider the International Connections

If you travel to other countries, observe the global effects of air pollution first-hand. Heavy smog in Mexico City, forest death in Central Europe, acidified lakes in Norway and Sweden: each case demonstrates air pollution's international implications. Learn about U.S. groups that have links with organizations in other countries that are actively combatting air pollution, and work to strengthen those ties.

- If you visit Canada, find out about the damage done to lakes and forests in southeastern Canada by sulfur and nitrogen oxides that originate in the midwestern United States.

- Consider locating a foreign "sister city" for your city or town; assist with efforts to reduce air pollution in both cities by organizing exchanges of visitors and technical experts.

- Because tropical deforestation contributes to the greenhouse effect, consider how you can help slow deforestation and encourage tree planting in tropical countries. (See Chapter 7.)

Further Information

A number of organizations can provide information about their specific programs and resources related to air pollution and climate change.*

Books

Abrahamson, Dean Edwin (ed.). *The Challenge of Global Warming.* Washington, DC: Island Press, 1989. 358 pp. Paperback, $19.95. Contains 21 reports by scientists and environmental policy experts explaining global warming

---

* These include: Climate Institute, Conservation Foundation, Global Greenhouse Network, Environmental Defense Fund, Environmental Policy Institute, Friends of the Earth, National Audubon Society, National Clean Air Coalition, National Wildlife Federation, Natural Resources Defense Council, Renew America, Sierra Club, Union of Concerned Scientists, World Resources Institute, and Worldwatch Institute. Addresses for these organizations are listed in the appendix.

and presenting recommendations for action to counter the threat.

Environmental Defense Fund. *Protecting the Ozone Layer: What You Can Do*. A Citizen's Guide to Reducing the Use of Ozone Depleting Chemicals. New York: 1988. 32 pp. Paperback, $12.00.

MacKenzie, James J. *Breathing Easier: Taking Action on Climate Change, Air Pollution, and Energy Insecurity*. Washington, DC: World Resources Institute, undated. 23 pp. Paperback, $5.00.

MacKenzie, James J. and Mohamed El-Ashri. *Ill Winds: Airborne Pollution's Toll on Trees and Crops*. Washington, DC: World Resources Institute, 1988. 74 pp. Paperback, $10.00.

McCormick, John. *Acid Earth: The Global Threat of Acid Pollution*. Washington, DC: Earthscan and International Institute for Environment and Development, 1985. 190 pp. Paperback, $7.75.

Miller, Alan S. and Irving M. Mintzer. *The Sky Is the Limit: Strategies for Protecting the Ozone Layer*. Research Report #3. Washington, DC: World Resources Institute, November, 1986. 38 pp. Paperback, $7.50.

Mintzer, Irving M. *A Matter of Degrees: The Potential for Controlling the Greenhouse Effect*. Research Report #5. Washington, DC: World Resources Institute, April, 1987. 60 pp. Paperback, $10.00.

National Research Council (NRC). *Ozone Depletion, Greenhouse Gases, and Climate Change*. Washington, DC: National Academy Press, 1989. 122 pp. Paperback, $20.00. Proceedings of a 1988 joint symposium sponsored by NRC's Board on Atmospheric Sciences and Climate, and the Committee on Global Change.

Regens, James L. and Robert W. Rycroft. *The Acid Rain Controversy*. Pittsburgh, PA: University of Pittsburgh Press, 1988. 228 pp. Paperback, $12.95. Covers the origins of acid rain, its ecological and human health effects, and measures to prevent these effects.

Renew America. *Reducing the Rate of Global Warming: The States' Role*. Washington, DC: November, 1988. 33 pp. Paperback, $7.00.

Roan, Sharon L. *Ozone Crisis: The 15 Year Evolution of a Sudden Global Emergency*. New York: Wiley, 1989. 270 pp. Hardback, $18.95.

Rosenberg, Norman J. et al. (eds.). *Greenhouse Warming: Abatement and Adaptation*. Washington, DC: Resources for the Future, 1989. 196 pp. Paperback, $18.95.

Schneider, Stephen H. *Global Warming: Are We Entering the Greenhouse Century?* San Francisco: Sierra Club Books, 1989. 317 pp. Hardback, $18.95. Examines the

causes and likely consequences of climate change, and considers what can and should be done about it.

Topping, John C., Jr. (ed.). *Coping With Climate Change*. Proceedings of the Second North American Conference on Preparing for Climate Change: A Cooperative Approach. Washington, DC: The Climate Institute, 1989. 696 pp. Paperback, $35.00 plus $2.50 postage.

United Nations Environment Program (UNEP). *The Greenhouse Gases*. UNEP/GEMS Environment Library No. 1. Nairobi: UNEP, 1987. 40 pp. Paperback. Available from UNEP, 1889 F Street, N.W., Washington, DC 20036.

UNEP. *The Ozone Layer*. UNEP/GEMS Environment Library No. 2. Nairobi: UNEP, 1987. 36 pp. Paperback. Available from UNEP, 1889 F Street, N.W., Washington, DC 20036.

Articles

"Acidification." A Special Issue on Acid Deposition and Its Effects. *Ambio*, Vol. 18, No. 3, 1989.

El-Hinnawi, Essam and Manzur H. Hashmi. "Air Quality and Atmospheric Issues." In El-Hinnawi and Hashmi, *The State of the Environment*. London: Butterworth Scientific, 1987, pp. 5-34.

Flavin, Christopher. *Slowing Global Warming: A Worldwide Strategy*. Worldwatch Paper 91. Washington, DC: Worldwatch Institute, October 1989. 94 pp. $4.00.

Graedel, Thomas E. and Paul J. Crutzen. "The Changing Atmosphere." *Scientific American*, September 1989, pp. 58-68. Emissions from fossil fuels, manufacturing, and farming are causing deleterious effects that may become worse.

Hinrichsen, Don. "Acid Rain and Forest Decline." In Goldsmith and Hildyard (eds.), *The Earth Report*. Los Angeles: Price Stern Sloan, 1988, pp. 65-78.

Houghton, Richard A. and George M. Woodwell. "Global Climatic Change." *Scientific American*, Vol. 260, No. 4, April 1989, pp. 36-44.

Myers, Norman (ed.). "Elements." In Myers, *Gaia: An Atlas of Planet Management*. Garden City, NY: Anchor Books, 1984, pp. 116-119, 126-132.

Organization for Economic Cooperation and Development (OECD). "Air." In OECD, *The State of the Environment 1985*. Paris: OECD, 1985, pp. 17-46.

Postel, Sandra. "Stabilizing Chemical Cycles." In Lester R. Brown, et al., *State of the World 1987*. New York: W.W. Norton, 1987, pp. 157-176. For an expanded version, see Worldwatch Paper 71.

Ridley, Scott. "Air Pollution Reduction." In El-Hinnawi and Hashmi, *The State of the States 1987*. Washington, DC:

Fund for Renewable Energy and the Environment, 1987, pp. 6-9.

Schneider, Stephen H. "The Changing Climate." *Scientific American*, September 1989, pp. 70-79. Rising atmospheric concentrations of carbon dioxide and other gases threaten to intensify the greenhouse effect. The danger of global warming is serious enough to warrant prompt action.

Shea, Cynthia Pollock. *Protecting Life on Earth: Steps to Save the Ozone Layer*. Worldwatch Paper 87. Washington DC: Worldwatch Institute, December 1988. 46 pp. $4.00.

Smith, Kirk R. "Air Pollution: Assessing Total Exposure in the United States." *Environment*, Vol. 30, No. 8, October 1988.

"Special Report on the Greenhouse Effect." *Newsweek*, July 11, 1988, pp. 16-23.

Udall, James R. "Turning Down the Heat." *Sierra*, July-August 1989. Includes a "climate checklist" of what you can do to help avert climate change.

U.S. Environmental Protection Agency. "The Greenhouse Effect: How It Can Change Our Lives." *EPA Journal*, January-February 1989. Contains 19 articles on greenhouse warming and its projected impacts.

World Resources Institute and International Institute for Environment and Development. "Atmosphere and Climate." *World Resources 1986*, pp. 161-181; *World Resources 1987*, pp. 143-161; *World Resources 1988*, pp. 138-169. New York: Basic Books, 1986, 1987, 1988 respectively.

Zero Population Growth. "Global Warming: A Primer." *ZPG Activist*, Summer 1989.

Periodicals

Climate Alert. Published quarterly by the Climate Institute, 316 Pennsylvania Avenue, S.E., Suite 403, Washington, DC 20003.

A number of organizations listed at the beginning of this section publish journals or newsletters that regularly cover air pollution and related issues.

Films and Other Audiovisual Materials

Charlie Brown, Clear the Air. This "Peanuts" cartoon, produced by the American Lung Association, examines pollution caused by autos, factories, home heating, and trash burning, and suggests ways to clear up the air we breathe. 7 min. 16mm film. Film Distribution Center.

Cooperation Across Boundaries: The Acid Rain Dilemma. 1987. Looking at the United States and Canada, the film explores acid precipitation as a problem calling for diplomatic and political solutions. 30 min. 16mm film ($495), Video ($395). Umbrella Films.

On the Road to Clean Air. Produced by the American Lung Association, the film shows how a one-minute tailpipe test for carbon monoxide and hydrocarbon emissions can clean up the air and save motorists money by increasing the efficiency of cars. 15 min. 16mm film. Film Distribution Center.

Pollution of the Upper and Lower Atmosphere. Looking at fixtures in our world such as the internal combustion engine and jet aircraft, this film analyzes the effects of these technologies on the Earth's atmosphere. 17 min. 16mm film ($300), video ($220); rental, $75. Coronet/MTI Film and Video.

Hazardous Substances

The most alarming of all man's assaults upon the environment is the contamination of air, earth, rivers, and sea with dangerous and even lethal materials. This pollution is for the most part irrecoverable; the chain of evil it initiates not only in the world that must support life but in living tissues is for the most part irreversible. In this now universal contamination of the environment, chemicals are the sinister and little-recognized partners of radiation in changing the very nature of the world—the very nature of life.

—Rachel Carson, *Silent Spring*[1]

Disposing of hazardous waste is expensive and risky. It is better to stop producing waste in the first place.

—Joel Hirschhorn, "Cutting Production of Hazardous Waste"[2]

How to handle society's toxic chemical waste now ranks among the top environmental issues in most industrial countries. Without concerted efforts to reduce, recycle, and reuse more industrial waste, the quantities produced will overwhelm even the best treatment and disposal systems.

—Sandra Postel, "Defusing the Toxics Threat"[3]

Major Points

- Hazardous substances—including toxic waste, industrial chemicals, pesticides, and nuclear waste—are entering the workplace, the marketplace, and the environment in unprecedented amounts.

- The United States produces over 260 million metric tons of hazardous waste each year—more than one ton for every person in the country. Some states generate as much as three tons per person each year.

- Through pollution of the air, the soil, and water supplies, hazardous substances pose both short- and long-term threats to human health and environmental quality.

- For most hazardous substances, there are already cost-effective methods of preventing human and environmental exposure by reducing the amount of toxic substances used and hazardous waste produced at the source.

The Issue: Proliferation of Hazardous Substances

In the 1950s, a major disaster struck the city of Minamata, Japan. The first sign of impending doom was the death of the city's cat population, as the animals underwent convulsions and often leaped into the sea. Over time, some 700 people were killed directly, and 9,000 others were crippled. The city's economic base was ruined as fish died and the fishing industry shut down. Minamata's population gradually declined, and now only 35,000 remain of the 50,000 who once lived there. The disaster was traced to a single cause—the dumping of a mercury-based compound into the bay by a chemical company.[4]

The Minamata incident has become a landmark as the first widely publicized confrontation between civilization and manufactured hazardous substances poured into the environment. Many other incidents have followed it. Frequent reports of abnormally high numbers of deformities, "cancer clusters," prolonged illnesses, miscarriages, and the disappearance of local animal and insect life have come from many countries. Often, such health problems can be traced to manufactured hazardous chemicals released into the air, ground, or water and inhaled, eaten, or drunk by people.

Simply put, "hazardous substances" include any substance that poses a threat to human health or to the environment. The term usually encompasses a broader range of materials than does the term "toxic," which generally denotes substances directly poisonous to humans. Substances might be classified as hazardous because they are corrosive, flammable, or explosive, in addition to toxic.

Hazardous substances in the environment are one of the most insidious and personal of all environmental threats because they often strike families or neighborhoods in the places where people feel safe: at work, in the home, or at the dinner table. Hazardous substances have disrupted and separated communities, as neighbors have been forced to scatter after learning that their houses had been built over a toxic waste dump. Hazardous substances also have generated feelings of betrayal, as workers have discovered that the company that provided them jobs had poisoned their families with toxic fumes.

The problems caused by hazardous substances are twofold. First, they cause a wide range of harmful effects on human health, some of which are summarized in Table 13.1 on the following page.

Second, these substances can cause long-term or permanent damage to ecosystems, such as Minamata Bay. Since manufactured toxic chemicals can accumulate in the environment over time, the long-term danger is especially great. The use of some chemicals, such as toxic synthetic organic compounds, has increased about fifteenfold since World War II. (See Figure 13.1.)

Figure 13.1
Synthetic Organic Chemicals Production, United States, 1945-85

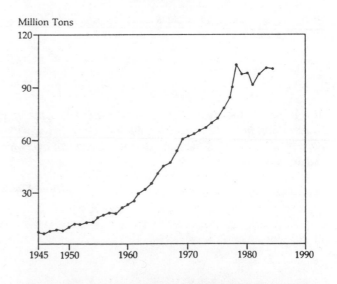

Source: U.S. International Trade Commission, as published in Sandra Postel, *Defusing the Toxics Threat: Controlling Pesticides and Industrial Waste*, Worldwatch Paper 79 (Washington, DC: Worldwatch Institute, September 1987), p. 8.

The number of chemicals in existence is also increasing, with 70,000 chemicals now used regularly, and 500 to 1,000 new ones added each year. Much is understood about the damage caused by those chemicals, but the greatest harm of all may come from what we do not understand about them. No data are available on the toxic effects of 79 percent of all chemicals, and complete data exist for only 2 percent.[5]

Table 13.1
Health Effects of Selected Hazardous Substances

Chemical	Source	Health Effects
Pesticides DDT	Insecticides	Cancer, damages liver, embryo, bird eggs
BHC	Insecticides	Cancer, embryo damage
Petrochemicals Benzene	Solvents, pharmaceuticals and detergent production	Headaches, nausea, loss of muscle coordination, leukemia, linked to damage of bone marrow
Vinyl chloride	Plastics production	Lung and liver cancer, depresses central nervous system, suspected embryotoxin
Other Organic Chemicals Dioxin	Herbicides, waste incineration	Cancer, birth defects, skin disease
PCBs	Electronics, hydraulic fluid, fluorescent lights	Skin damage, possible gastro-intestinal damage, possibly cancer-causing
Heavy Metals Lead	Paint, gasoline	Neurotoxic, causes headaches, irritability, mental impairment in children brain, liver, and kidney damage
Cadmium	Zinc processing, batteries, fertilizer processing	Cancer in animals, damage to liver and kidneys

Sources: World Resources Institute and International Institute for Environment and Development, *World Resources 1987*, (New York, Basic Books, 1987), pp. 205-06; OECD, *The State of the Environment 1985*, p. 39; and other sources.

Dioxin is an example of a highly toxic chemical. A common contaminant of herbicides and pesticides, it is one of the most toxic substances known. A tiny amount of dioxin can cause cancer, immune system depression, and birth defects in laboratory animals.[6]

Chemicals can be carried hundreds or thousands of miles in air and water currents, and their effects may not become apparent until years or decades after exposure. Pinpointing and eliminating the sources of toxic contamination is often a monumental task.

Hazardous Waste

When hazardous substances are released into the environment as waste, they may contaminate the air, soil, surface water, or groundwater, and they may harm people and ecosystems.

There are no reliable estimates of how much hazardous waste the world produces. One estimate is 375 million metric tons each year; another estimate is nearly 500 million tons for just 19 countries.[7]

The United States alone produces over 260 million metric tons each year—more than one ton for every man, woman, and child in the country.[8] Louisiana generates over three tons per person each year, and West Virginia, Tennessee, and Texas each produce more than two tons per person. (See Figure 13.2 on the following page.)

The chief producers of hazardous wastes are the chemical and petrochemical industries, which contribute nearly 70 percent of these wastes in industrialized countries.[9] In the United States, about two-thirds of hazardous wastes is disposed of on land, into injection wells, surface impoundments such as pits and ponds, or landfills. These disposal sites are all subject to leaks that can contaminate groundwater. Another 22 percent of U.S. hazardous wastes is discharged into sewers or directly into streams and rivers, and only about 11 percent is recycled or processed to eliminate or reduce its toxicity before discharge.[10] (See Table 13.2 on the follwoing page.)

Some industrial countries process a larger portion of their hazardous waste to reduce its toxicity before discharging it. In West Germany, for example, about 35 percent of the waste receives detoxification treatment, 15 percent is incinerated, and 50 percent is discharged into landfills.[11]

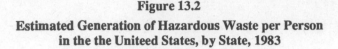

Figure 13.2

Estimated Generation of Hazardous Waste per Person in the the Uniteed States, by State, 1983

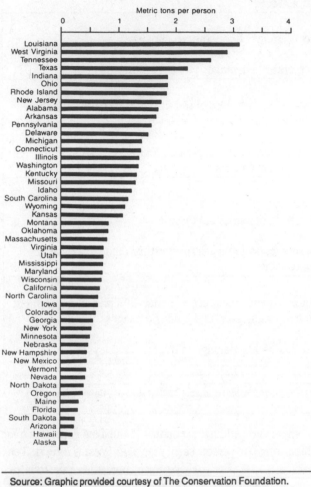

Metric tons per person

Source: Graphic provided courtesy of The Conservation Foundation. Information provided by the U.S. Department of Commerce and Congressional Budget Office.

Table 13.2

Hazardous Waste Management Methods, United States, 1983

Management Method	Share of Total Waste Managed, percent
Land disposal	67
Discharge to sewers, rivers, streams	22
Distillation for recovery of solvents	4
Burning in industrial boilers	4
Chemical treatment by oxidation	1
Land treatment of biodegradable waste	1
Incineration	1
Total	100

Source: Sandra Postel, *Defining the Toxics Threat*, Worldwatch Paper 79 (Washington, DC: Worldwatch Institute, September 1987), p. 13.

Disposal in the United States

The inadequacy of toxic waste disposal in the United States was revealed with brutal clarity in 1978, when the Love Canal neighborhood of Niagara Falls, New York, was declared a federal disaster area. Thirty years before, a chemical company had dumped almost 40,000 metric tons of waste (much of it carcinogenic) into the canal, filled in the open site, and then sold the land to the local Board of Education. By 1977, the dump was leaking and endangering the lives of the people who had built houses on the land. More than 200 families were evacuated from the area. In 1988, the chemical company was ruled liable for the cleanup of Love Canal, estimated to cost $250 million.[12]

Estimates of the number of U.S hazardous waste disposal sites vary, but at least 15,000 uncontrolled hazardous waste

landfills have been identified in the United States, along with 80,000 contaminated surface lagoons.[13]

The problem is so widespread that the Council on Economic Priorities estimates that 8 out of 10 Americans live near a hazardous waste site. The Centers for Disease Control report that in 1980, nearly half of all U.S. residents lived in counties containing a site classified among the most dangerous in the United States. In 1986, EPA reported contaminated surface waters at 44 percent, and airborne toxic pollutants at 15 percent of these sites.[14]

Except for Nevada, every U.S. state has identified at least one hazardous waste site for inclusion on EPA's National Priority List, making them eligible for cleanup using federal funds allocated under the Superfund legislation (discussed later in this chapter). By 1988, 1,177 sites had been identified. New Jersey listed 110 sites, Pennsylvania had 97, and California, Michigan, and New York each had more than 75 sites.[15] (See Figure 13.3 on the following page.)

In 1986, EPA reported that hazardous materials had leaked into groundwater at about three-fourths of all sites on the National Priority List.[16] Groundwater is the principal water supply for half of U.S. residents, and many are accordingly threatened by the contamination.[17] Sadly, every one of the commonly used disposal systems for hazardous wastes risks polluting groundwater, and some pose an extremely high risk. Seventy percent of the pits, ponds, and lagoons used to store hazardous wastes have no liners, and as many as 90 percent may threaten groundwater.[18] The dangers of such poor disposal methods have become evident in Woburn, Massachusetts, where carcinogenic industrial solvents seeped into the town's water supply. Residents believe

Figure 13.3
U.S. Hazardous Waste Sites, 1988

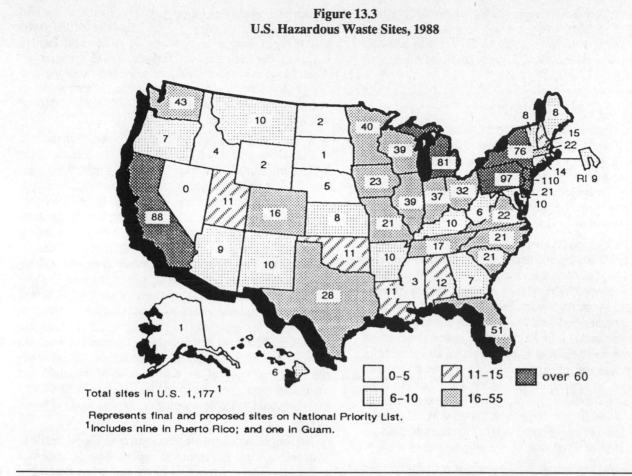

Total sites in U.S. 1,177 [1]

☐ 0-5	▨ 11-15
▦ 6-10	▨ 16-55
	■ over 60

Represents final and proposed sites on National Priority List.
[1] Includes nine in Puerto Rico; and one in Guam.

Sources: U.S. Environmental Protection Agency, as published in U.S. Bureau of the Census, *Statistical Abstract of the United States, 1989* (Washington, DC, 1989), p.202.

that the leakage explains why child leukemia rates in Woburn are eight times that of the national average.[19]

A recent EPA report released results of the first U.S. inventory of toxic substances discharged into the environment. The report states that in 1987, U.S. industry released some 10.5 billion pounds of toxic substances into the air, land, and water, including 550 million pounds dumped into the nation's waters. (See Table 13.3.)

Table 13.3

Toxic Substances Discharged by U.S. Industry, 1987

Destination	Millions of Pounds
Air	2,700
Lakes, Rivers, and Streams	550
Landfills and Earthen Pits	3,900
Treatment and Disposal Facilities	3,300
Total	10,450

Source: Environmental Protection Agency, reported in *The Washington Post*, April 13, 1989, p. A33.

The EPA report also revealed that over half of the compounds discharged into streams, lakes and rivers are not covered by EPA regulations. As a result, 10.5 million pounds of carcinogenic arsenic compounds were released into U.S. waters in 1987.[20] These figures became available to EPA as a result of the Community Right to Know Act on toxic substances, described later in this chapter.[21]

The federal government's chief tool for identifying and cleaning up the most dangerous toxic dumps in the United States (see Figure 13.3) is the "Superfund" legislation. Superfund was revised and reauthorized in 1986 after five years of activity. The revised law strengthened the public's right and ability to participate in cleanup programs. People living near one of the priority sites may now be eligible for technical assistance grants from EPA to help them pay for expert assistance.[22]

Many observers were disappointed with the initial results of the Superfund program. Despite years of effort and the expenditure of $1.6 billion, EPA cleaned up only 13 sites from the National Priority List, and the quality of the cleanups was questionable.[23] Some feel that the new $9 billion reauthorization is too far below the $100 billion dollars that the Office of Technology Assessment estimated would be needed over as many as 50 years to clean up 10,000 priority waste sites.[24] The National Priority List contained only 1,177 sites as of 1988.

Disposal in Other Countries

In other industrial countries, the situation is also alarming. West Germany faces an estimated $10 billion bill for detoxifying its hazardous waste sites, while The Netherlands expects to pay $1.5 billion. Some time ago, Denmark initiated a program to detoxify hazardous waste before disposing of it, and because of this foresight has estimated cleanup costs of only $158 million.[25]

Most developing countries lack the regulations and technology, and often the political will, to minimize output or dispose safely of hazardous waste.[26] In parts of the Third World, the absence of regulations on the dumping of waste has created a toxic nightmare. Mexico City discharges wastewater contaminated with heavy metals and toxic organic compounds directly into the city sewer system, from which the wastes pass virtually untreated into agricultural areas for irrigation. The same heavy metals and organic compounds have begun to appear in vegetables and other crops. In China, the 400 million tons of industrial waste and tailings produced annually are dumped in the outskirts of cities or released into surface waters; 60,000 hectares of land are believed to be covered with potentially hazardous waste.

The 400-kilometer stretch between Rio de Janeiro and Sao Paulo, Brazil, is one of the most industrialized areas in the world, but the dumping of waste by several hundred metallurgical and other factories is unregulated. Sepetiba Bay, Rio's primary source of shellfish, is contaminated with chromium, zinc, and cadmium.[27] These easily observed cases are only the tip of the iceberg. The absence of worldwide data on the subject underscores how little is known about the production and disposal of hazardous substances across the globe.[28]

International Shipments

Because of the increased expense and regulation associated with hazardous wastes in Western nations, some industries are finding it increasingly appealing to export these toxic materials. Unfortunately, much of the exported waste goes to areas less capable of handling it properly than in the industry's home country. The Center for Investigative Reporting has found that notifications by U.S. companies for overseas shipments of hazardous waste to all countries jumped from 30 in 1980 to more than 400 in 1987, while notifications for shipments to Third World nations increased from 4 in 1984 to 19 in 1986. These figures probably represent only a small portion of the wastes actually shipped, since they exclude illegal shipments or shipments of potentially harmful waste not classified as hazardous.[29]

In the past two years, approximately 3 million tons of hazardous waste have been shipped from the United States and Western Europe to countries in Africa and Eastern Europe.[30] The environmental organization Greenpeace estimates total international shipments of hazardous waste at three million tons a year, and the United Nations Environment Program says that about 800,000 tons are shipped within Europe each year.[31]

Sometimes the waste is willingly received, as in the case of East Germany, which accepts waste for disposal from West European nations such as Denmark and West Germany, who pay for this service. Often, however, the dumping is conducted secretly and illegally, and evokes indignation from the unwilling host countries. In the summer of 1988, Nigeria discovered that Italy was dumping hazardous materials in one of Nigeria's ports, and the enraged government threatened the perpetrators with a firing squad. Similarly, an arm of the U.N. Environment Program uncovered a plan to ship millions of tons of U.S. and European hazardous wastes to Africa and Latin America. Greenpeace has reported two illegal shipments of nuclear waste from France to Benin, and one shipment of toxic ash from Philadelphia to Guinea.[32] As industrialized countries run out of economical ways to dispose of industrial and municipal waste, both toxic and non-toxic, efforts to send the unwanted trash to developing nations are likely to intensify.[33]

In recognition of growing concern over illegal traffic in toxic waste, legal and technical experts from 50 countries worked for more than a year under the auspices of the United Nations Environment Program to draft a treaty on international trade in hazardous wastes, which was adopted in March of 1989. The treaty defines illegal traffic as any transboundary movement of waste without notification by the exporter, without the consent of the importer, or with documents that do not conform to the shipment.

The treaty also requires that illegally shipped wastes be disposed of in an environmentally sound manner within 30 days by the exporter, the importer, or both cooperatively.[34]

In another effort to limit international traffic in hazardous waste, The World Bank recently announced that it will not finance any projects that involve the disposal of hazardous wastes from another country.[35]

Airborne Toxic Pollutants

The 85-mile stretch of the Mississippi River between Baton Rouge and New Orleans, Louisiana, is known as the "Petrochemical Corridor," because of the large number of petrochemical manufacturers on either side of the river. Part of the corridor town of Chalmette has been dubbed "Cancer Alley." Though only two blocks long, this area has claimed 15 cancer victims, a record approached only by the one block with seven cancer victims a half-mile away. In another town, problems with miscarriages, nausea, headaches, and vomiting are frequent, and six of the river parishes rank in the top 10 percent nationwide for lung cancer deaths. In the corridor town of St. Gabriel, 26 petrochemical companies pour an annual total of 400 million pounds of toxic pollutants into the atmosphere, including carcinogens like benzene and carbon tetrachloride, and fetal toxins such as toluene, ethylene oxide, and chloroform.[36]

The Cancer Alley of Chalmette is one example among many that reflect the critical health effects of exposure to airborne toxic chemicals, which are released even more freely into the air than solid and liquid wastes are dumped into the ground. A national inventory of toxic air pollution by a House of Representatives subcommittee has found that 2.4 billion pounds of hazardous pollutants, including tons of toxic chemicals that cause cancer and damage the nervous system, were emitted in 1987.[37] EPA has estimated that 2,000 people develop cancer each year as result of hazardous airborne chemicals, and more pessimistic figures indicate that as many as 7 million American workers face an increased risk of illness from three common air pollutants encountered on the job: benzene, asbestos, and arsenic.[38]

The EPA recently released results of the first national survey of cancer risks from toxic air pollutants released by U.S. industry. The survey found that more than 200 plants located in 37 states pose threats at least 1,000 times higher than levels considered acceptable by federal standards. Although the U.S. Congress has directed the EPA to limit emissions of such toxic air pollutants, the agency has regulated only seven of 200 air pollutants deemed hazardous.[39]

Airborne toxic pollutants can come from a number of sources. Most easily recognized as poisonous are mercury, lead, and arsenic—a few of the toxic metals. Produced as fine particles or vapors by high-temperature processes such as copper smelting or fossil fuel combustion, these metals can cause cancer, neurological problems, and damage to kidneys and the central nervous system. Currently, worldwide atmospheric emissions of lead from industrial sources total at least 2 million tons per year, while annual emissions of mercury and arsenic are about 11,000 tons and 78,000 tons, respectively—20 to 300 times higher than natural emission rates.[40]

Toxic fibers are another common airborne pollutant. Naturally occurring asbestos is widely recognized for both its usefulness and toxicity. The high tensile strength and resistance to heat of asbestos fiber led to wide use in over 3,000 industrial and commercial applications before discoveries that inhaled asbestos fibers could cause lung cancer and asbestosis. Asbestos is still used, though with much greater caution. Even so, fibers often escape into the atmosphere from vehicle brakes, old insulation, and mining operations. Like many of the atmospheric pollutants, asbestos can travel hundreds or thousands of miles before settling. Traces of asbestos fibers have been found in regions as remote as the Greenland ice cap.[41]

A large number of airborne compounds, including benzene, formaldehyde, some nitrates, and radionuclides, are emitted by sources ranging from automobiles to chemical factories; many of these are harmful to human health.[42] In addition, a number of airborne compounds containing fluorine and chlorine are toxic. One of these is vinyl chloride, a common ingredient of plastics that is a known carcinogen. In 1987, more than 500,000 pounds of this toxic substance were released from the "Petrochemical Corridor" town of St. Gabriel.

Everyday Hazardous Substances

Seemingly more benign but actually as dangerous as many toxic industrial chemicals are the hazardous materials that consumers encounter every day. These can be found in items such as mothballs and roach sprays, batteries, drain cleaners, and bleaches.[43] They are also present in office furnishings, carpets, and dry-cleaned clothes. Hazardous substances are often poured into septic tanks, flushed down sinks, or thrown out with the trash. People fail to realize that the toxic substances these materials contain can easily enter the air or water and threaten people or wildlife.

The alarming results of this usage and disposal include contaminated surface water and groundwater, poisoned wildlife, damaged municipal sewer systems, blinded refuse collectors, burned landfill tractor operators, and sick office workers. As evidence of how severe the problem is, half of all designated Superfund sites are municipal landfills.[44]

Rodent bait is lethal to humans and pets; as little as one taste can cause death. Drain cleaners are poisonous, can cause serious burns, and may contain carcinogens. Some window cleaners contain harmful chemical compounds and sometimes carcinogens, and may cause birth defects.[45] Paint, certain building materials, and even some ordinary household and office goods contain neurotoxins that can cause headaches, irritability, impairment of mental functioning, brain damage, and liver and kidney damage. Canadian government studies have shown that highly toxic dioxin can leach into milk from bleached paper milk cartons.[46]

EPA recently took regulatory action on one common source of toxic leakage—underground storage tanks such as those used by gas stations, taxi companies, fire departments, and marinas. There are about 1.4 million of these tanks in the United States, and at least 15 percent leak petroleum products and cancer-causing substances into the ground.[47] A quart of gasoline or oil can contaminate several million gallons of drinking water; thus any leak from a gasoline tank can be a serious threat.

Pesticides

The San Joaquin valley in California is one of the most productive agricultural regions in the world. It is also one of the farming areas most heavily treated with pesticides. Though constituting only 1 percent of U.S. farmland, the San Joaquin valley accounts for 7 percent of all U.S. pesticide use. In 1986, 60 million pounds of pesticides were used on the fields of the valley, much of which either drifted and evaporated into the air, or seeped into groundwater. Pesticides, including one probable carcinogen, have turned up in over 2,000 wells, including 125 public water systems. Although definite links are difficult to establish, many observers suspect that the pesticides are responsible for the excessive cancer rates in the valley. McFarland, a small town in the area, has a childhood cancer rate eight times the national average, and birth defect and stillbirth rates which are also substantially higher than normal.[48] A recent study by the Natural Resources Defense Council states that pesticides pose a greater cancer risk to children than to adults because of the youngsters' higher vulnerability to toxic chemicals and their relatively high consumption of fruits and vegetables.[49]

Between 400,000 and 2 million pesticide poisonings occur worldwide each year, and 10,000 to 40,000 of these result in death.[50] Most are among farmers in developing countries who are exposed to the pesticides without taking proper safety precautions, but many also occur among people living in towns like McFarland, near places where pesticides are heavily applied.

Often less than 0.1 percent of pesticides applied ever reach the target pests; the rest enter the air, soil, and water.[51] In the United States, groundwater supplies drinking water for 95 percent of rural residents, yet routine agricultural practices have contaminated this groundwater with more than 50 different pesticides in at least 30 states.[52]

Many pesticides also contaminate the food they are intended to protect, remaining on the surface or being absorbed into the plant. In the summer of 1985, a thousand people in the western United States and Canada were poisoned by residues of the pesticide Temik in watermelons. They experienced symptoms ranging from nausea, vomiting, and blurred vision to grand mal seizures and cardiac irregularities. Data from a recent National Academy of Sciences report suggests that pesticides in food may cause as many as 1 million additional cases of cancer in the United States through the lifetime of the present generation.[53]

Despite mounting evidence of the damage from these chemicals, pesticide use in the United States has climbed from a yearly application of 300 million pounds in 1966 to over a billion pounds in 1987. Usage worldwide is currently increasing at more than 12 percent annually. But as the amount of pesticides applied has risen, their effectiveness has declined. In fact, since the introduction of modern pesticides in the 1940s, crop losses attributed to pests have increased from 32 percent to 37 percent.[54]

The reason for the change is a mixture of complex factors, including the increasing size of farms, cultivation of monoculture crops that are more vulnerable to pests, and a reduced labor force. But pesticides themselves are also to blame. In 1938, scientists knew of only seven species of insects and mites that had become resistant to pesticides. But by 1984, at least 447 insect and mite species, encompassing most of the world's major pests, had acquired resistance to pesticides. Ironically, many of their natural predators had been destroyed by the very chemicals aimed at eliminating these pests.[55] As time goes by, a "pesticide treadmill" develops, where more and more pesticides are required to control infestations effectively. (See Chapter 5.) Many of those additional pesticides will either reach humans directly, through the produce they consume or the water they drink, or will enter the food chain through plants, animals, and water.

Pesticides Overseas

Africans fishing on Ghana's Lake Volta discovered an ingenious way to increase their daily catch. When they dumped the insecticide Gammalin 20 (imported for use by cocoa farmers) into the lake, many fish died and floated to the surface for retrieval. Many of these fish were eaten by the villagers, and others were sold. People in the community began suffering from dizziness, headaches, vomiting, and diarrhea—the first symptoms of poisoning by lindane, the active ingredient of Gammalin 20, and a prelude to more serious problems such as convulsions, brain disturbances, and liver damage. The fish population of the lake declined by up to 20 percent. However, the fishers did not link the pesticide to the damage done to their health and livelihood until a private aid agency noted the connection.[56]

Those living beside Lake Volta are not the only people to fish with insecticides, nor is fishing the only abuse of these highly toxic substances. While Third World nations use only 10 to 25 percent of the world's pesticide production, they suffer up to 50 percent of the acute poisonings and between 73 and 99 percent of the fatalities among pesticide applicators.[57] One 1985 survey found that 6 out of every 10 Third World farmers using pesticides had suffered acute poisonings, mostly because of inadequate training or inability to read label instructions.[58] The ignorance of farmers concerning the hazards of these chemicals is sometimes appalling. There are reports of empty pesticide containers being used to carry drinking water, and one instance of children using a large container as a bathtub.

Aggravating the problem are the pesticide export policies of industrial nations. Because of legal loopholes, manufacturers in industrialized nations can often sell pesticides overseas that are banned or severely restricted at home. As an example, both DDT and benzene hexachloride (BHC) are banned from use in the United States and much of Europe, but they account for three-fourths of all pesticide use in India. Both of these chemicals accumulate in the body and can cause cancer and other life-threatening illnesses.

Some 30 percent of India's cereals, eggs, and vegetables have residues of these chemicals exceeding the World Health Organization's tolerance limits, and residues of DDT and BHC were found in all 75 samples of breast milk taken from women in India's Punjab.[59]

U.S. government reports have stated that about 25 percent of all pesticides sold overseas by U.S. companies are banned, restricted, or unregistered for use in the United States. Government studies have also concluded that the EPA does not properly inform importing countries about U.S. restrictions affecting exported pesticides, as required by law.[60]

The export of banned or severely restricted chemicals, mostly pesticides, will be controlled under a new United Nations system of Prior Informed Consent (PIC). Instead of being linked to individual shipments of restricted chemicals, the PIC system will be based on a notification plan that will encourage developing nations to make decisions about whether to accept shipments after they receive information on a product's potential risks and benefits.[61]

The Circle of Poison

Merely shipping hazardous pesticides overseas does not remove the threat to people living in Western nations, because much of the produce grown in the Third World is ultimately sold to the industrial countries. In the United States, less than 1 percent of imported fruits and vegetables are inspected for pesticides that are banned domestically, despite evidence that pesticide-contaminated food is entering the country. In 1983, the Natural Resources Defense Council independently tested Latin American coffee beans sold in New York City markets, and discovered that each of four samples contained residues of DDT, BHC, and other pesticides. The Brazilian beans that are most popular in the United States contained residues of five suspected carcinogens.[62] In a similar case, the U.S. Department of Agriculture blocked the marketing of 30 million pounds of Australian beef because it contained illegally high pesticide residues of DDT and dieldrin.[63]

Accidents Involving Hazardous Substances

Since World War II, a number of serious accidents have occurred involving both toxic substances from chemical manufacturing and storage facilities, and radioactive materials from nuclear power plants and weapons facilities.

Chemical Accidents

In November 1986, a fire at a chemical warehouse in Basel, Switzerland, released 30 tons of pesticides into the Rhine River. The accident created a 60-kilometer toxic slick that moved downstream. A 120-mile stretch of the river was left biologically dead for several years, and the fishing industries and drinking water of communities in West Germany and The Netherlands were threatened. An estimated 500,000 fish and 150,000 eels were killed, posing a threat to the livelihoods of commercial fishers. The long-standing efforts of the International Commission for the Protection of the Rhine to revitalize the river from previous pollution were destroyed in a single day.[64]

With so many manufactured chemicals produced and stored, it is inevitable that some will be released accidentally, whether on the minor scale of a leaking storage tank or on a large scale, like the Rhine River spill. On just one day in May 1988, three industrial accidents in three different U.S. states killed nine people, injured more than 250, and exposed thousands to toxic chemicals that may eventually cause chronic illness. These incidents are only a small sample from the incomplete EPA data base for acute hazardous events in the United States, which lists nearly 7,000 such events between 1980 and mid-1985.[65] If production of hazardous substances continues to grow, the number of significant accidents is also likely to increase.

Hazardous substances can cause serious problems for many reasons, ranging from technical failure—such as the Rhine River spill—to ignorance and carelessness. One of the most notorious careless mishaps in the United States took place in Times Beach, Missouri, when a waste oil dealer combined oil with waste sludge, which was given to him for disposal by a chemical manufacturer, and then sprayed the resulting mixture on the streets of the town to help settle the summer's dust. After a flood several years later, it was discovered that the sludge had been contaminated with dioxin, one of the most toxic synthetic substances. In 1983, with dioxin levels in the town 300 times greater than considered safe for human habitation, the federal government bought out all 2,400 residents and began the $64-million process of removing the toxin.[66] As costly as such cleanups are, their expense may be only a fraction of the costs incurred in lives lost and long-term health care for those exposed.

The potential for accidental releases of hazardous substances is even greater in developing countries. Some international manufacturers have realized that regulations are even less restrictive and less well enforced in the Third World than they are in industrial nations, and have chosen to move some of their most dangerous facilities abroad where they can operate with less expense. This strategy roughly parallels the overseas shipment of pesticides banned in the Western nations, and the results are similar.

The rapidly growing squatters' settlements of the Third World are often located near industrial facilities, where illegal tenants have less chance of eviction from the less desirable land. As a result, large populations are near the most dangerous areas. Exemplifying the hazards was the release in 1984 of toxic methyl isocyanate gas from a Union

Carbide plant in Bhopal, India, which has killed more than 3,300 people and left at least 20,000 others suffering from ailments including lung damage, internal bleeding, abdominal pains, reproductive disorders, and damaged eyesight.[67]

Nuclear Accidents

The toxic dangers of the nuclear age were starkly revealed in April 1986, when one of the four reactors at Chernobyl in the Ukraine republic of the Soviet Union exploded, releasing between 50 million and 100 million curies of radioactive material into the environment. There was no doubt about the immediate health consequences of Chernobyl, in contrast to the Three Mile Island accident in Pennsylvania almost a decade earlier. The amount of radiation released at Chernobyl was a thousand times greater than in the Pennsylvania accident.[68]

Twenty-nine people died of radiation poisoning within the first few months, and 200 others were believed to have poor long-term prospects with the probability of developing cancer later in life. One hundred million people in Europe fell under mandatory or voluntary food restrictions, as the radioactive cloud spread over the continent and contaminated fruits, vegetables, and grass for grazing livestock. The eventual toll of the nuclear accident has been estimated as high as 135,000 additional cancer cases and 35,000 deaths.[69]

Recent reports indicate that more than half the children near Chernobyl have developed thyroid illness that may be linked to the disaster, and that among middle-aged people exposed to radiation from the accident, some cancer rates have doubled. At a collective farm near the Chernobyl site, the number of deformed pigs and cows born since the accident has soared.[70]

Chernobyl demonstrated that radiation endangers human health and can threaten entire ecosystems. The effects are often subtle and indirect, as in the near meltdown at Three Mile Island in 1979, which may have contributed to the premature deaths of elderly people between 1979 and 1982 by damaging their immune systems.[71] Though the U.S. Department of Energy (DOE) has claimed that an accident like the one at Chernobyl could not happen in the United States, contradictory reports by independent organizations have demonstrated that U.S. plants are vulnerable—particularly federal reactors not governed by the strict requirements that apply to commercial facilities.[72]

In addition to the accidents at Chernobyl and Three Mile Island, there have been at least two other serious nuclear accidents since World War II, notably those in 1957 at Kyshtym in the Soviet Union and at Windscale in the United Kingdom.[73]

Toxic Nuclear Substances

The 14-state complex of weapons-related nuclear reactors operated by the U.S. DOE has proven that a nuclear explosion or a meltdown is not necessary for hazardous substances from reactors to contaminate the environment. Called a "creeping Chernobyl" by the Physicians for Social Responsibility, the storage of radioactive materials and wastes has already caused a number of problems at these facilities.

An investigation by the U.S. General Accounting Office discovered radioactive materials in the groundwater at DOE nuclear weapons facilities at Hanford, Washington, and at the Savannah River plant in Aiken, South Carolina. The radiation level of the water was over 400 times greater than the proposed drinking water standard at both sites. One off-site drinking well near a nuclear facility in Cincinnati was found to be contaminated with uranium, one of a number of problems that prompted Ohio's governor to call for closure of the plant in 1988.[74]

Much of the radiation may have seeped gradually into groundwater during the 1950s and 1960s, when DOE dumped 200 billion gallons of low-level waste into shallow ponds and burial pits. But accidental discharges have also occurred. At the Hanford site, 500,000 gallons of highly radioactive materials has escaped from cracked storage tanks. There are 100 million gallons of waste still stored in tanks at federal facilities, many of which are cracked and leaking. In 1984, the Oak Ridge Associated Universities Center for Epidemiological Research found that 9 out of 12 health studies of areas near weapons plants reported excessive cancer deaths.[75]

One great hazard facing the entire nuclear industry, both commercial and federal, is that no long-term solution for waste disposal has been developed. The world has more than 400 nuclear reactors, but not one long-term disposal program is in place.[76] Nearest to implementation are plans for the disposal of low-level radioactive wastes—including medical fluids, power reactor fluids, contaminated clothes, and plastics—which remain hazardous for 60 to 300 years. In the past, commercial sites have been used to bury these wastes. However, the last three such sites—at Barnwell, South Carolina; Beatty, Nevada; and Hanford, Washington—are closing either for safety reasons or because they are nearly full. To deal with the imminent crisis, Congress passed the Low-Level Radioactive Policy Act (LLRPA) in 1980, which encourages states to form compacts to dispose of radioactive waste. LLRPA mandated that all states be members of a compact or build their own disposal facilities by 1990.[77]

Mill tailings, produced in uranium extraction, and transuranic waste, generated by reprocessing plutonium-bearing fuel and manufacturing nuclear weapons, are far more dangerous than low-level waste. Nuclear industries simply abandoned the tailings in uranium mills until 1983, when

EPA ordered a cleanup to be completed by 1990. Transuranic waste has been in temporary storage, primarily at the Hanford Reservation and at Idaho National Engineering Laboratory in Idaho Falls, until a permanent depository opens at Carlsbad, New Mexico.[78]

But the most serious disposal problem concerns the high-level waste and spent fuel rods from nuclear reactors. The rods account for 90 percent of the radioactivity emanated by nuclear waste, and 15,000 tons of this fuel are stored in cooling ponds at 106 reactor sites in the United States.[79] Because the half-life of radioactive elements in high-level waste is so long, this material must be isolated from the environment for thousands of years. The Nuclear Waste Policy Act specifies that these substances must be stored in a "geological repository"—a geologically stable location 1,000 to 4,000 feet underground.[80] Current plans call for the use of a site at Yucca Mountain, Nevada, which will hold 70,000 tons of waste for 10,000 years.[81]

Disposing of waste from the nuclear industry is becoming increasingly expensive. The Department of Energy estimates costs of $100 billion just to clean up its nuclear weapons facilities, which is more than was spent to create the fissionable materials in the first place. The Yucca Mountain site is also expected to cost $100 billion, and neither of these figures includes the costs for non-military, low-level waste, transuranic waste, mill tailings, or temporary storage for high level waste while the Nevada facility is constructed. For the DOE, 45 cents of every dollar invested in production of nuclear weapons goes to waste management, an illustration of the economic problems plaguing the nuclear industry.[82]

Solving the Toxics Dilemma

Several different strategies are available to curb the spread of hazardous substances. One approach is treatment of hazardous wastes to neutralize them or make them less toxic. However, a better strategy is to reduce or eliminate the use of toxic substances and the generation of hazardous waste. This section reviews waste treatment and waste reduction programs, regulatory legislation designed to control and limit hazardous wastes, programs that reduce the use of toxic substances, the role of government incentives, the use of integrated pest management in agriculture to lessen dependence on hazardous pesticides, and methods to improve the storage of nuclear wastes and limit their future production.

Hazardous Waste Treatment

Industries are pursuing several strategies to render toxic substances less hazardous. Technologies currently in use include incineration, settlement, evaporation, air flotation, and filtration.[83] Chemical treatment, in which hazardous substances are combined with other chemicals to produce relatively harmless waste, is also widely used. Among the

most innovative technological solutions to the hazardous waste problem are genetically engineered bacteria that can digest toxic substances. Already, special strains of bacteria can break down PCBs, creosote, pentachlorophenol, and 4-ethylbenzoate.[84]

Legislation

Regulatory legislation is an important part of the effort to combat the hazards of toxics waste. In 1976, Congress enacted both the Resource Conservation and Recovery Act (RCRA) and the Toxic Substances Control Act to control the threat of hazardous waste. RCRA is the primary legislation governing the management and disposal of hazardous waste. More recent legislation includes the 1977 amendments to the Clean Air Act, and the Clean Water and the Safe Drinking Water Acts, all designed to help prevent pollution. In 1980, to deal with the massive project of cleaning up the hundreds of already-contaminated sites, Congress passed the Comprehensive Environmental Response, Compensation, and Liability Act known as "Superfund."

Superfund is the federal government's best-known mechanism for combatting hazardous waste, and was created as a five-year, $1.6 billion project to identify and design cleanup programs for the nation's worst waste sites. EPA worked with state governments to devise a work plan and budget, and Superfund provided the money to implement actual cleanup. Superfund was reauthorized in 1986 and now incorporates the Emergency Response and Community Right-To-Know Act. Under this section of the law, some industries must provide lists of certain hazardous chemicals produced, used, or stored at their facilities to EPA, the state, and to local emergency planning committees. Local committees, composed of emergency services personnel, industry representatives, and citizens use the information to design an emergency response plan in the event of an accident involving toxic materials.[85]

While many environmentalists call for the strengthening of controls on pollution and the allocation of more funds for cleaning up existing hazardous waste sites, there is general recognition that costs of environmental cleanup must be included in actual costs of production. Legislation embodying this idea might include laws that obligate industry to prove that a waste or product is safe, rather than making the government prove it is harmful. Laws should also reflect the "polluter pays" principle by requiring the industry that created a toxic hazard to clean it up.

Hazardous Waste Reduction and Toxics Use Reduction[86]

The best strategy for dealing with toxic substances is to reduce or eliminate their production. According to the National Toxics Campaign, waste reduction alone, though a positive step toward eliminating the hazardous waste problem, is not a comprehensive strategy to reduce all of the hazards associated with the use of toxic substances. The Campaign maintains that the use of toxic substances in industry can be reduced or eliminated in the following ways:

- Substituting non-toxic or less-toxic raw materials for highly toxic ones. For example, Cleo Wrap, a producer of gift wrapping paper, switched from organic solvent-based inks to water-based inks to prevent the release of hazardous waste.

- Changing the design of a product so that its production involves fewer toxic substances or less-toxic materials. For instance, the Minnesota Mining and Manufacturing (3M) Company replaced a metal alloy in a product and eliminated a waste stream containing toxic cadmium.

- Modifying the production process to avoid the use of toxic substances. For example, a U.S. Air Force base in Utah switched from chemical solvents to sandblasting for removing paint from aircraft.

- Improving the monitoring of the production process to better identify the origin of toxic materials and eliminate any release of toxic substances. For instance, Exxon Chemical installed floating roofs on tanks of volatile organic chemicals to reduce their release to the atmosphere.

- Recycling materials within a production process to reduce waste streams and recover valuable materials by creating closed loops within the process. This kind of internal recycling is preferable to out-of-process recycling which is more likely to result in worker exposure, transportation hazards and release of toxic substances.[87]

Initiating and maintaining the momentum of a hazardous waste reduction program in industry is difficult unless sufficient motivation, technical assistance, and other resources are part of the program. According to Joel Hirshhorn of the U.S. Office of Technology Assessment, several measures are necessary for a successful program:

- » Transfer the economic motivation for waste reduction to those engaged in the manufacturing process.

- » Conduct and maintain a waste-reduction audit.

- » Make waste reduction a lasting part of the corporate culture: rewrite job descriptions to reflect a waste-reduction ethic, and transfer waste-reduction knowledge throughout the company.

- » Motivate employees by setting specific waste-reduction timetables.

- » Utilize technical assistance from outside sources to obtain a fresh view of an operation and external support for reduction goals.[88]

Benefits of Toxics Use Reduction

The National Toxics Campaign has identified a number of benefits that could be achieved by reducing all exposures to toxic substances:

- Fewer accidents and spills resulting from transporting and storing toxic chemicals and waste.

- Fewer routine and accidental occupational exposures to toxic materials.

- Less hazardous waste to contaminate air, water, and land; and therefore less damage to food chains and natural ecosystems.

- Less reliance on costly, nonproductive pollution control systems and on hazardous waste treatment, storage, and disposal facilities.

- Fewer new Superfund sites that require lengthy and costly cleanups.

- Fewer consumer exposures to products containing toxic substances.

- Less household hazardous waste from discarded consumer products entering municipal waste streams.

The reduction or elimination of toxic substances does not necessarily cost industry any additional money. In fact, as the 3M Company has shown, reducing hazardous waste at the source can save millions of dollars. Through a series of measures introduced between 1975 and 1985, including recycling, substitutions in the chemical process, and equipment changes, 3M cut its waste generation in half and eliminated 10,000 tons of water pollutants, 140,000 tons of sludge, 90,000 tons of air pollutants, and 3.7 billion liters of wastewater. These measures also conserved the equivalent of 250,000 barrels of oil a year and saved the company $300 million over the 10 year period.[89]

The DuPont Corporation has used innovative techniques that both reduce waste and save money. One example is a Freon plant that converts anhydrous hydrogen chloride, a toxic by-product, into the chlorine and hydrogen used to produce the Freon.[90]

Riker Laboratories in California achieved a one-time saving of $180,000 in pollution-control equipment, and now saves $15,000 a year by using a water-based solvent instead of an organic solvent to coat medicine tablets. Chevron USA saves $50,000 a year in waste management costs by using a high-pressure hot water cleaning system to replace a system that used toxic chemicals. And Borden Chemical saves $48,000 annually in waste disposal costs by using filter rinsing and tank cleaning procedures to lessen the release of organic solvent wastes.[91] A number of other companies have had success in reducing waste and saving money, both in the United States and elsewhere.[92] Table 13.4 on the following page contains several examples of successful industrial waste reduction that have saved money.

In an EPA study of 28 firms engaged in waste reduction, 54 percent of the companies reported full investment return in a year or less, 21 percent were repaid in one to two years, and only 7 percent of the cases took more than four years to regain investment expenditure.[93]

The EPA has estimated that the total U.S. industrial waste stream could be reduced by 15 to 30 percent if industry would merely expand its use of existing techniques.[94] And the 3M Company has proved that waste reduction can be profitable if reduction programs are given a high priority.

Waste reduction programs can yield multiple benefits. For example, an OECD survey of 200 French companies with reduction programs found than more than half had conserved energy, nearly half had saved raw materials, and 40 percent reported improved working conditions.[95]

Government Incentives

Governments at both the national and state level can encourage source reduction in various ways. France and Denmark both supply grants to promote the development of clean technology. The results not only help reduce hazardous waste, they also provide new technology for export. In the United States, North Carolina's "Pollution Prevention Pays" program is similar, but in addition to grants, the state offers on-site technical assistance. Though North Carolina funds the program with only $650,000 a year, the annual savings documented in 60 case studies were $16 million in 1987.[97]

Other types of government-promoted source reduction include a waste-end tax, already enacted in 30 states, under which companies pay a tax based on the volume and toxicity of the waste produced. Environmental groups have suggested that EPA incorporate waste reduction into its other programs by allowing states to substitute verifiable waste reduction for the construction of new storage and treatment facilities, thereby lessening the need for expensive new waste facilities.[97]

A Model State Program

In the absence of a strong U.S. national program to reduce the use of toxic substances, the National Toxics Campaign has proposed a model state program with the following elements:

- Reporting and planning regulations requiring that large users of toxic substances report the type, amount, and ultimate destination of those substances.

- Education and technical and financial assistance, including a clearinghouse of toxic use-reduction methods, research and technical assistance programs, training and certification of use-reduction planners, provision of loans, and availability of insurance rates that reward toxic use reduction.

- Coordination of law enforcement, including a "multiple environment" approach (air, water, land, workplace) to regulation, and regulation procedures that monitor toxic releases into all environments and encourage toxic use reduction.

- A program to phase out the use of particularly dangerous toxic chemicals, in order to eliminate their use while giving industry time to develop substitute chemicals, products, and processes.

- Citizen involvement, including access by citizens, residents, and workers to chemical use data and toxics use reduction plans; and citizens' rights to request public hearings, inspect toxics users' facilities, petition to have substances considered for phase-outs, and sue to enforce environmental laws.

- A toxics use tax, to be levied on the first sale or use of toxic substances in the state, with the tax rate for a substance dependent on its toxicity, and with proceeds from the tax used to support the state's toxics use reduction program.[98]

Integrated Pest Management

Since pesticide use generally does not reduce the threat of pests over the long term, and may even increase it, alternative pest control methods are clearly needed. Recent agricultural research indicates that integrated pest management (IPM) may be the key. Instead of relying solely on chemicals, IPM utilizes a broad spectrum of biological controls, genetic manipulations, and planting practices to limit pest damage and improve growth conditions.

For example, cotton growers in Jiangsu Province, China, plant sorghum between cotton plants to attract the natural enemies of cotton pests, while changing tillage practices and applying chemicals at much lower doses. Their pesticide use is down 90 percent while the cotton yield has increased. In the United States, farmers in the corn belt have begun using crop rotation to replace soil insecticides that can leach into groundwater. Crop rotation discourages large numbers of pests, and soil insecticide use has dropped 45 percent.[99]

Table 13.4
Selected Successful Industrial Waste Reduction Efforts

Company/ Location	Products	Strategy and Effect
Astra Södertälje, Sweden	Pharmaceuticals	Improved in-plant recycling and substitution of water for solvents cut toxic wastes by half.
Borden Chem. California, United States	Resins; adhesives	Altered rinsing and other operating procedures cut organic chemicals in wastewater by 93 percent; sludge disposal costs reduced by $49,000/year.
Cleo Wrap Tennessee, United States	Gift wrapping paper	Substitution of water-based for solvent-based ink virtually eliminated hazardous waste, saving $35,000/year.
Duphar Amsterdam, The Netherlands	Pesticides	New manufacturing process cut toxic waste per unit of one chemical produced from 20 kilograms to 1.
Du Pont Barranquilla, Colombia	Pesticides	New equipment to recover chemical used in making a fungicide reclaims materials valued at $50,000 annually; waste discharges were cut 95 percent.
Du Pont Valencia, Venezuela	Paints; finishes	New solvent recovery unit eliminated disposal of solvent wastes, saving $200,000/year.
3M Minnesota, United States	Varied	Companywide, 12-year pollution prevention effort has halved waste generation, yielding total savings of $300 million.
Pioneer Metal Finishing New Jersey, United States	Electroplated metal	New treatment system design cut water use by 96 percent and sludge production by 20 percent; annual net savings of $52,500; investment paid back in three years.

Source: Sandra Postel, *Defusing the Toxics Threat: Controlling Pesticides and Industrial Waste*, Worldwatch Paper 79 (Washington, DC: Worldwatch Institute, September 1987), pp. 42-43.

Other IPM techniques include intercropping, where a nitrogen-fixing legume is grown between rows of wheat in order to crowd out the weeds and add nitrogen to the soil for use in the next growing season. Fungi that destroy weeds, one of a number of bioherbicides, are often used. And one of the most popular methods is seeding an area with natural predators. The *Trichogramma*, a wasp that eats the eggs of some pests, has been used on 17 million acres worldwide. Recently introduced in El Salvador, the wasp eliminated the need for 10 pesticide applications every season.[100]

Integrated pest management techniques have been used successfully in North and South America, Africa, and Asia. (See Chapter 5.) If every nation adopted IPM strategies, the need for pesticides would decline considerably around the world, along with the dangers to people in industrial and developing countries alike. Once again, reducing and ultimately ceasing production of hazardous substances would be the ideal solution.

Nuclear Waste

Curtailing the production of hazardous radioactive substances may be the most critical of all hazardous waste challenges, as many of these substances have no known technological detoxifiers and will remain dangerous for thousands of years. Some progress has been made in curbing radioactive waste production. Federal nuclear facilities have been shut down under protest, no new orders for commercial reactors have been placed in the United States since 1974, and countries like Sweden—after a public referendum—have begun to phase out nuclear power use entirely.[101] Yet many nations are still expanding their nuclear power capability, and in the United States the danger of new sources of nuclear contamination remains, as U.S. defense agencies appear unlikely to cease their demand for weapons. Political pressures for a complete halt in production of nuclear waste, both at home and abroad, may be the only path toward a solution until proper disposal methods have been developed.

What You Can Do

Inform Yourself

Get more information about toxic chemical use and hazardous wastes so you can speak and write with authority on the issue. Consult sources and contact organizations listed under Further Information.

- Find out about hazardous waste production and disposal in your community. Pinpoint local industries with the potential to create hazardous waste and learn about their disposal procedures. How safe are your community's waste treatment facilities? If your community incinerates its waste, how much toxic material is released in the air, and how much remains in the ash? Is there a hazardous waste collection site in your area, and if there is, do people use it?

- Find out whether your community has complied with the federal Community Right-To-Know Law, requiring that communities prepare local emergency response plans and that citizens be allowed to participate in local emergency planning efforts. It also requires that companies respond to any official requests for information about chemical storage or use relevant to emergency planning.

- Find out about the planning committee and emergency response plan for hazardous substances in your area. As an individual citizen, get involved in the local committee, either by becoming a member or monitoring the process.

- Contact national organizations active on the issue of hazardous waste listed under Further Information and subscribe to newsletters to keep up-to-date on recent developments and legislation.

Join With Others

Combine efforts with others by joining an existing organization or starting a grass roots education effort in your own community. Remember, there is power in numbers.

- Meet with other members of your community concerned with toxic chemical use and waste, and organize an action network. A number of organizations such as Environmental Action and the National Toxics Campaign provide technical, organizational, and legal support for grassroots campaigns against hazardous substances.

- Organize a town meeting in your community to discuss hazardous waste and other environmental issues. Organize meetings with local industries; ask them to sign a "good neighbor" agreement promising to obey all environmental and safety laws, committing to perform an environmental audit of their facility, and agreeing to reduce toxic chemical use.

- Reach out to employees and unions in your area to support each other in efforts to improve safety and working conditions in local industries.

- If your community does not have a household hazardous waste collection program, work with others to organize one.

- Organize a group to walk along local waterways to identify unlawful dumping by industries. State chapters of Public Interest Research Groups (PIRGs) can provide instruction booklets.

- Join a national or international organization working to protect the environment and control hazardous waste. Volunteer your services and tell others about the group.

- Form a consumer group with your neighbors and buy pesticide-free organic foods in bulk from your local natural foods store. Ask local grocery stores and supermarket chains to carry produce free of toxic chemicals.

Review Your Habits and Lifestyle

There is much you can do within your own household to reduce toxic damage to the environment.

Obtain a copy of the *Household Hazardous Waste Wheel* listed under Further Information.

In general:

- Handle household chemicals with extreme care. Make sure you read the label to know exactly what you are buying and what the potential hazards are.

- Use alternative, less harmful products wherever you can. Ask retailers about alternatives. You may be able, for example, to control a pest problem without a pesticide. Contact your county Cooperative Extension Service for effective alternatives. If you must use a hazardous substance, select the least toxic product you can find (those labeled CAUTION are less toxic than those marked WARNING).

- Dispose of your unwanted household chemicals safely.

  » Pour liquids such as cleaning fluids into a plastic container that is filled with kitty litter or stuffed with newspaper, and close the top securely.

  » Take used motor oil and antifreeze to a gas station with an oil recycling program.

- Reduce or eliminate the use of pesticides on your garden or lawn.

  » Introduce natural predators like ladybugs, lacewings, mantises, toads, and garter snakes to eat insect pests.

  » Buy plants that are relatively pest-and disease- resistant.

  » Regulate planting and harvesting to avoid those times when insects are most abundant and damaging.

- Grow your own fruits and vegetables without toxic chemicals, and patronize natural food stores that sell organic produce.

- If you do use pesticides, minimize potential hazards.

  » Make sure people and pets are out of the way during application.

  » Never apply near wells, streams, ponds, or marshes unless instructions allow.

  » Don't apply if rain is forecast, unless otherwise specified on the label.

Work With Your Elected Officials

Contact your local, state, and national legislators, find out where they stand on hazardous waste issues, and examine their voting records on legislation such as the Resource Conservation and Recovery Act (RCRA) and Superfund. In brief letters or personal visits, express your concerns and urge them to:

- Request a community inspection tour of local industrial facilities, to include citizens and local regulators. At each facility, ask for a list of chemicals stored on site and released from the facility, including a description of their health effects, and request a copy of their "worst case" accident scenario and emergency response plan.

- Broaden the definition of "hazardous waste" to include all wastes that are damaging to human health or the environment, without exceptions based on economics or politics.

- Broaden public participation in the establishment, management, and cleanup of hazardous waste facilities.

- Provide stronger incentives to industry for the source reduction of hazardous waste.

- Require industry to clean up sites that they have polluted in order to avoid future publicly-financed cleanups under Superfund.

- Provide greater support to Superfund so the government can do a better job of identifying and cleaning up hazardous waste sites.

- Press the EPA to complete re-evaluation and testing of chemical pesticide agents under 1972 safety legislation.

- Support research to develop alternatives for pesticides and common household toxic substances.

- Halt U.S. exports of banned or restricted hazardous products to Third World countries.

- Help developing countries establish their own technological capabilities and regulations for the use and disposal of hazardous substances.

- Cooperate with recent international efforts to develop uniform regulations to control the export of hazardous substances.

Publicize Your Views

Disseminate your views on the issue of hazardous waste to the print and electronic media. Organize community activities stressing the dangers of toxic waste.

- Write letters to the editor of local newspapers to alert the community about hazardous waste sites or industrial polluters in the area. Endorse and publicly congratulate legislators who are active on toxic waste issues. Create public service announcements for local radio and television stations. Call in and discuss environmental issues on radio talk shows.

- Sponsor a showing and discussion of a film on toxic waste at a library, community center, or school. (See Further Information for suggestions.) Invite a guest speaker from a local environmental organization.

Raise Awareness Through Education

Find out whether schools, colleges, and libraries in your area have courses, programs, and adequate resources about hazardous waste. If not, offer to help develop or supply them.

- Meet with biology, chemistry, ecology, and social studies teachers from local schools and discuss the importance of addressing toxic waste issues. Encourage field trips to industrial plants and hazardous waste disposal sites, and suggest projects to evaluate individual household habits affecting waste disposal.

Consider the International Connections

The problem of hazardous waste is not confined to the United States alone. If you travel to another country, find out about the toxic waste issues there firsthand. Investigate the operations and waste disposal practices of American industries abroad. Bring your observations home and inform local interest groups. Learn about U.S. organizations that are actively combatting toxic waste on a global level and work to strengthen ties between these groups and those in developing countries.

Further Information

A number of organizations can provide information about their specific programs and resources related to hazardous substances.[*]

Books

Baybook: A Guide to Reducing Water Pollution at Home. 1986. 32 pp. Paperback. Available from Alliance for Chesapeake Bay, 6600 York Road, Baltimore, MD 21212. Single copies free, multiple copies $1 each. Explains how to prevent damage to inland and coastal water resources from soil erosion, sewage, pesticides, household chemicals.

Concern, Inc. *Hazardous Waste.* Washington, DC: Concern, Inc., 1981. 22 pp. Paperback.

Dover, Michael and Brian Croft. *Getting Tough: Public Policy and the Management of Pesticide Resistance.* Washington, DC: World Resources Institute, November, 1984. 80 pp. $3.50.

Elkington, John and Jonathan Shopley. *Cleaning Up: U.S. Waste Management Technology and Third World Development.* Washington, DC: World Resources Institute, 1989. 80 pp. Paperback, $10.00.

Environmental Action. *Making Polluters Pay.* Washington, DC: Environmental Action Foundation, 1987. 135 pp. Paperback, $15.00.

Gold, Steve (ed.). *Eating Clean: Overcoming Food Hazards.* Washington, DC: Center for Responsible Law. 155 pp. Paperback, $8.00.

Institute for Local Self-Reliance. *Proven Profits From Pollution Prevention.* Washington, DC, 1986.

Lave, Lester B. and Arthur C. Upton (eds.). *Toxic Chemicals, Health, and the Environment.* Baltimore, MD: Johns Hopkins University Press, 1987. 304 pp. Paperback, $16.50.

League of Women Voters. *Nuclear Waste Primer.* Publication No. 448. Washington, DC: 1985. 90 pp. $5.95.

Mott, Lawrie and Karen Snyder. *Pesticide Alert: A Guide to Pesticides in Fruits and Vegetables.* A publication of the Natural Resources Defense Council. San Francisco: Sierra Club Books, 1987. 179 pp. Paperback, $6.95.

National Research Council. *Regulating Pesticides in Food.* Washington, DC: National Academy Press, 1987. 272 pp. Paperback, $19.95.

The National Toxics Campaign. *Citizen's Toxic Protection Manual.* Boston, MA: 1988. 560 pp. plus appendices. Loose-leaf binder, $30.00 plus $5.00 postage. Includes chapters on the toxics crisis, its solution, and health effects of wastes in the environment.

National Wildlife Federation. *Reducing the Risk of Chemical Disaster.* Washington, DC: 1989. 75 pp. $7.00. Examines the recently enacted Right-to-Know Act and its importance as a tool for awareness. Available from NWF, Dept. TD, 1412 16th St., NW, Washington, DC, 20036.

---

[*] These include: Citizen's Clearinghouse for Hazardous Waste, Clean Sites, Inc., Coalition for the Superfund, Conservation Foundation, Environmental Action Foundation, Environmental Defense Fund, Environmental Hazards Management Institute, Environmental Policy Institute, Friends of the Earth, National Center for Policy Alternatives, National Wildlife Federation, Natural Resource Defense Council, Renew America, Sierra Club, The National Toxics Campaign, Southwest Research and Information Center, U.S. Public Interest Research Group, Working Group on Community Right-To-Know, World Resources Institute, and Worldwatch Institute.

Regenstein, Lewis. *How to Survive in America the Poisoned*. Revised Edition. Washington, DC: Acropolis Books, 1986. 432 pp. Paperback, $18.95.

Sarokin, David et al. *Cutting Chemical Wastes*. New York: INFORM, 1988. $47.50. A comprehensive report of hazardous waste reduction in the U.S. organic chemical industry.

U.S. Congress, Office of Technology Assessment. *Serious Reduction of Hazardous Waste: For Pollution Prevention and Industrial Efficiency*. Washington, DC: U.S. Government Printing Office, 1986.

Weir, David and Mack Schapiro. *Circle of Poison: Pesticides and People in a Hungry World*. San Francisco: Institute for Food and Development Policy, 1981. 101 pp. Paperback, $3.95. Documents the global scandal of corporate and government exportation of pesticides and reveals the threat this poses to consumers and workers throughout the world.

Articles, Pamphlets, and Brochures

El-Hinnawi, Essam and Manzur H. Hashmi. "Chemicals and Hazardous Waste." In El-Hinnawi and Hashmi, *State of the Environment*. London: Butterworth Scientific, 1987, pp. 93-121.

Environmental Action. "The Toxic Waste Crisis—A Solution Is In Sight." Washington, DC: 1984. 8 pp. Single copy free.

Environmental Hazards Management Institute. *The EMHI Household Hazardous Waste Wheel*: A practical guide to safe management of hazardous household products.

The EHMI Water Sense Wheel: Explains how to investigate and improve your own water quality. Single copies of each wheel, $2.75; order from EHMI, 10 Newmarket Road, P.O. Box 932, Durham, NH 03824.

Environmental Protection Agency (EPA). *Citizen's Guide to Pesticides*. Washington, DC: EPA, 1987. 16 pp. Contains tips for safe pesticide use. Explains how to determine correct dosage, how to store and dispose of pesticides, how to choose a pest control company, and how to reduce your exposure to pesticides.

EPA. "Hazardous Waste Management." In EPA, *The Waste System*. Washington, DC: November 1988.

Frosch, Robert A. and Nicholas E. Gallopoulos. "Strategies for Manufacturing." *Scientific American*, September 1989, pp. 144-52. Creative engineering, characterized by dematerialization and closed-system manufacturing, can help maintain the industrial way of life without exhausting resources, generating excessive waste, and poisoning the environment.

Hirschhorn, Joel. "Cutting Production of Hazardous Waste." *Technology Review*. April 1988, pp. 53-61.

Myers, Norman (ed.). "Hazardous Chemicals—A Growing Problem," In *Gaia: An Atlas of Planet Management*. Garden City, NY: Anchor Books, 1984, pp. 122-124.

The National Toxics Campaign. *Toxics Use Reduction: From Pollution Control to Pollution Prevention*. A Policy Paper. Boston, MA, February 1989. 9 pp.

Organization for Economic Cooperation and Development (OECD). "Atmospheric Trace Pollutants," "Hazardous Waste Disposal" and "Industry." In OECD, *The State of the Environment 1985*. Paris, 1985, pp. 37-43, 166-170, 217. Available from OECD, 2001 L St., N.W., Washington DC 20036.

Postel, Sandra. "Controlling Toxic Chemicals." In Lester R. Brown, et al., *State of the World 1988*. New York: Norton, 1988, pp. 118-136. For an expanded version, see Worldwatch Paper 79.

Ridley, Scott. "Hazardous Waste." In Ridley, *The State of the States 1987*. Fund for Renewable Energy and the Environment. Washington, DC: Renew America, 1987, pp. 22-25. Paperback.

Robbins, John. "America the Poisoned." In Robbins, *Diet for a New America*. Walpole, NH: Stillpoint Publishing, 1987, pp. 308-349.

World Resources Institute (WRI) and International Institute for Environment and Development (IIED). "Managing Hazardous Wastes: The Unmet Challenge." In WRI and IIED, *World Resources 1987*. New York: Basic Books, 1987, pp. 201-209.

WRI and IIED. "Pesticide Use and Health" and "Waste Generation in Selected Countries." In WRI and IIED, *World Resources 1988-89*. New York: Basic Books, 1988, pp. 28-31, 314-316.

Periodicals

Community Plume. A Right-To-Know newsletter published by Environmental Policy Institute.

Hazardous Materials Intelligence Report. Published by World Information Systems, P.O. Box 535, Cambridge, MA 02238.

Hazardous Waste News and *Right-to-Know News*. Both published by Thompson Publishing Group, 1725 K St. N.W., Suite 200, Washington, DC 20006.

Rachel's Hazardous Waste News. Published by Environmental Research Foundation, P.O. Box 3541, Princeton, NJ 08543.

Toxic Times. A quarterly newsletter published by The National Toxics Campaign, 37 Temple Place, Boston, MA 02111.

Films and Other Audiovisual Materials

Hazardous Waste: Who Bears the Cost? 1987. Explores the problem of hazardous waste—what it is, where it comes from, and how it affects us—using Woburn, Massachusetts (America's oldest toxic waste dumpsite) as a case study. 28 min. 16 mm film, $475; video, $375. Umbrella Films.

Toxic Chemicals: Information is the Best Defense. 1985. Part I: Who Needs to Know? Part II: Developing a Community Right to Know Law. This two-part documentary focuses on the importance of toxic waste awareness within the community. Describes how business leaders, local officials, and concerned citizens in one community worked together to create a model hazardous materials disclosure ordinance. Each part 26 min. Video purchase, $95 each part; rental, $30 each part. Bullfrog Films.

The Toxic Trials. 1988. NOVA investigates the controversial story of one community affected by hazardous waste and its lawsuit against a local chemical company. 60 min. Video purchase, $350; rental, $125. Coronet/MTI Film & Video.

Solid Waste Management

The United States generates more and more municipal solid waste every year...but there are fewer and fewer landfills to absorb it.... There is no single, simple solution. Recycling, incineration, landfills, manufacturers and municipalities, private carters, and consumers—all are part of the solution.

—James Cook, "Not in Anybody's Backyard," *Forbes* Magazine[1]

Waste stream reduction is the answer. We must start with the manufacturer, to curb production of what will become unnecessary waste. We must educate the purchaser not to buy what may become unnecessary waste. We must educate the user not to be wasteful and not to create unnecessary waste.

—Cedric Maddox, Director of Sanitary Services, Atlanta, Georgia[2]

Major Points

- Each year, the United States generates about 10 billion metric tons of non-agricultural solid waste. Municipal solid waste alone accounts for at least 140 million metric tons each year. An average U.S. citizen discards about 3.5 pounds of waste each day.

- In 1978, the United States had about 20,000 landfills; by 1988, the number had dropped to about 6,000. Eighty percent of U.S. solid waste is now dumped in landfills. In the past 5 years, 3,000 dumps have been closed; by 1993, some 2,000 more will be filled and closed.

- Solid waste disposal is causing serious pollution of ground-water and surface water in some areas.

- To manage our growing volume of solid waste, we will need to rely on a combination of waste reduction, recycling, composting, use of landfills, and incineration. A serious effort to reduce the volume of waste will require major changes by consumers and manufacturers.

- Waste recycling saves energy and materials and reduces air and water pollution. The United States recycles only about 11 percent of its waste, while Western Europe recovers 30 percent and Japan reuses more than 50 percent of its waste.

The Issue: Municipal Solid Waste Volume Is Exceeding Disposal Capacity

Each year, the United States generates more than 150 million tons of municipal solid waste, and the amount continues to grow. Landfills, which now receive about 80 percent of this waste, are rapidly reaching capacity. Many landfills have already closed, and few new sites are available.[3] (See Figure 14.1.) The situation already is critical in many urban areas. Additional ways to handle the growing waste stream must be adopted.

Since both population size and the volume of waste materials are growing every year, the urgency to act is great. The Environmental Protection Agency (EPA) estimates that U.S. municipal solid waste will increase 20 percent by the year 2000.[4]

Figure 14.1

Growth of U.S. Municipal Solid Waste and Decline in the Number of Landfills

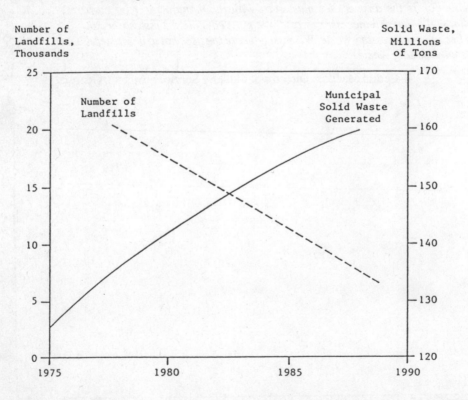

Source: U.S. Environmental Protection Agency

A recent study of recycling potential for Montgomery County, Maryland indicates that within 20 years, waste generation will grow so much that even if 30 percent of that waste were recycled, there would still be more waste to be disposed of than exists at present.

The *Wall Street Journal*, the *Washington Post*, and the *Christian Science Monitor* have all called the garbage disposal situation in America a crisis.[5] *Scientific American* wrote that "America is a throwaway society." Total U.S. per capita waste generation is twice that of any other country.[6] Americans perceive waste as a matter of convenience, and treat even some expensive products as disposable. Convenience has a high value in consumer goods, and the waste stream associated with it has therefore been generally accepted. The packaging on food and merchandise constitutes about 30 percent of the weight and 50 percent of the volume of household waste.[7] At the same time, Americans already pay between $4 billion and $5 billion a year to dispose of waste in landfills and by incineration and recycling.[8]

Each year, the United States generates about 10 billion metric tons of solid waste. About two-thirds of this total is categorized as non-hazardous industrial waste. Almost all is disposed of in private facilities separate from municipal landfills.[9] Another 20 percent of U.S. waste is associated with the production and use of oil and gas, and 12 percent comes from mining. Municipal solid waste (MSW) makes up about 1 percent of the total, yet it amounts to about 140 million metric tons (154 million U.S. "short" tons) each year, or about 3.5 pounds per person each day.[10] Less than 11 percent of this waste is recovered, but recovery is up from less than 7 percent during the 1960s. (See Table 14.1.)

Table 14.1
Municipal Solid Waste Generation and Recovery, United States, 1960-86

	1960	1970	1980	1986
Gross waste generated, millions of tons	88	121	143	158
Gross waste per person per day, pounds	2.7	3.2	3.4	3.6
Resources recovered, percent of total	6.6	6.6	9.3	10.7

Source: U.S. Bureau of the Census, *Statistical Abstract of the United States 1989* (Washington, DC, 1989), Table 349, p. 202.

Worldwide, municipal solid waste generation tends to increase with the level of economic development. Large cities in low-income countries, such as Calcutta and Jakarta, produce 1.1 to 1.3 pounds of waste per person each day. Cities in middle-income countries, such as Manila, Tunis,

and Singapore, generate 1.1 to 1.9 pounds. Large cities in industrial countries produce more waste; New York City tops the list of cities included in a World Bank study with 4 pounds of waste per person a day. (See Table 14.2.)

Table 14.2
Waste Generation Rates in Selected Cities, circa 1980

	Pounds per Person per Day
Industrial Cities	
New York, United States	4.0
Tokyo, Japan	3.0
Paris, France	2.2
Singapore	1.9
Hong Kong	1.9
Hamburg, West Germany	1.9
Rome, Italy	1.5
Lower-Income Cities	
Jakarta, Indonesia	1.3
Lahore, Pakistan	1.3
Tunis, Tunisia	1.2
Medellin, Colombia	1.2
Calcutta, India	1.1
Manila, Philippines	1.1

Sources: Cynthia Pollock, *Mining Urban Wastes: The Potential for Recycling*, Worldwatch Paper 76 (Washington, DC: Worldwatch Institute, 1987), p. 9; World Resources Institute (WRI) and International Institute for Environment and Development (IIED), *World Resources 1988-89* (New York: Basic Books, 1988), p. 46.

What Is Waste?

Wastes are generated at every stage in our use of materials, from their extraction and processing to their abandonment as used items. These leftovers are variously termed garbage, junk, waste, trash, scrap, and sewage — depending on what and where they are and on the viewpoint of the speaker. Some of this waste can be recovered and reused as secondary resources — materials that can replace primary (virgin) resources, which would otherwise have to be extracted from the earth or obtained from other sources, such as forests, at a greater cost.

Solid Waste: A Broad Definition

The Resource Conservation and Recovery Act (RCRA), passed by the U.S. Congress in 1976, defines solid waste as including garbage, refuse, and sludge; and solids and liquids from industrial, commercial, mining, agricultural, and community activities; but excluding solid or dissolved material in domestic sewage.[11]

Municipal Solid Waste

As used here, municipal solid waste refers to those wastes that:

1) originate in households, commercial establishments, small industries, institutions, and governmental entities that regulate or organize waste collection systems, and

2) are either collected by public or private haulers and delivered to landfill, incineration, or processing sites, or disposed of at other specified sites.

Municipal solid waste is creating serious problems for local governments, and has been in headlines nationwide for several years. The magnitude and the cost of the waste problem were illustrated for the American public by the Mobro, the garbage barge from Islip, New York that went on a 6,000-mile voyage in search of a dump-site, only to return after five futile months to its point of origin.

Other Consumer Waste

Many discarded materials bypass the municipal waste collection system and are handled by specialized collectors or dealers. These include scrap dealers, auto dismantlers, and scrap-metal processors, often represented and assisted by large trade organizations. Their work has become more sophisticated as the items they collect have become more complex.

Items in this type of waste tend to be large and have many parts, such as automobiles and "white goods" (large household appliances such as refrigerators and washing machines). When compared to newspapers and cans, the number of items is small, but the weight, volume, and value of each is substantial. This type of waste also includes electrical and plumbing materials, aluminum siding, lawn furniture, and some small tools, appliances, and machinery.

Saleable parts are removed and sorted, materials are sorted, and the parts and materials are sold to appropriate markets for reuse or recycling. Dealers must pay for disposal of remaining unacceptable materials for which no market exists. Their destination depends on their composition, and on whether they contain hazardous material. The increased use of plastics in automobiles and appliances is raising disposal costs and forcing some handlers out of business.

The U.S. Department of Energy estimates that 240 million auto and truck tires are disposed of each year, and there may be as many as 2 billion discarded tires in U.S. dumps and along roadsides.[12] Except for retreads, markets for discarded tires are limited. Waste and scrap dealers do not want them, municipal landfills avoid them, and huge mounds of used tires therefore appear off back roads in the countryside. There are numerous potential uses for used tires including asphalt additives, fuel, rubber mats, bumpers, roofing material, and playground equipment that could be developed if it were economically beneficial to do so.

Non-Municipal Wastes: Industrial and Agricultural

The amount of waste generated by industry and agriculture is many times that discarded by consumers; however, consumers cannot disclaim all responsibility for industrial waste. It is generated in producing the goods demanded by citizens of a modern industrial society.

Industrial processes and waste generation begin at the raw material stage, whether the materials are minerals to be mined and processed, trees for wood and paper products, petroleum products and coal for plastics, or latex for rubber. There are leftovers at every stage, whether the final product is to be a highway, a house, a pin, a can of soup, or heavy machinery for a manufacturing plant.

Increasingly, wastes generated by basic industries in production processes are considered "secondary resources" and are immediately reused — fed back into the furnaces of a steel plant, collected from a lumber mill and used for fuel or for particle board. It is advantageous for industries to recycle as much as possible. It saves raw materials and processing costs, reduces energy and water requirements, lowers pollution control costs, and lessens waste disposal problems.

Despite waste reduction efforts, the quantities of unused industrial waste are enormous. Copper, for example, typically comprises only about 0.5 percent of the ore in which it is found. U.S. copper production exceeded one million tons in 1988, and was accompanied by large amounts of waste.[13] Huge quantities of water are used in industrial processes (60 to 150 tons per ton of steel production, for example). The waste water from all industrial plants (textile, food processing, paper, and others) is classified as "solid waste" by the Resource Conservation and Recovery Act (RCRA).

The volume of agricultural waste is also enormous, mostly due to runoff from croplands. It is classified as "solid waste" by RCRA because the runoff contains dissolved and suspended substances. Agricultural runoff has changed greatly with the introduction of mechanized techniques and the expanded use of chemical fertilizers, insecticides, and herbicides. The soil carried away can hardly be classified as waste since it represents a serious loss to the farmer; it also causes harmful siltation of streams and rivers.

Our habits as citizens affect the amount of industrial and agricultural wastes generated, but methods for handling the wastes vary for different industries. Control and regulation of wastes rest more with Federal and state governments than with local jurisdictions.

Environmental Hazards of Waste Disposal

methods for managing waste have some environmental impact. Waste disposal in landfills can cause pollution groundwater and surface waters when rainfall leaches hazardous substances from materials in the waste. And as organic waste decays, methane accumulates, creating the potential for explosions.

When waste in incinerated, effluent gases may contain dioxins and other hazardous air pollutants. The ash from incineration usually is disposed of in landfills, where heavy metals and other toxic substances may leach into groundwater.

Recycling of waste can also cause air or water pollution if chemicals used in reprocessing materials are not properly handled. The environmental impacts of waste management are discussed further under the various management strategies reviewed below.

With proper use of new and existing technologies, however, pollution from waste can be reduced to tolerable levels. Eliminating all hazard from waste disposal is impossible. But old disposal facilities can be improved or replaced with new facilities. This is expensive, but it is one of the costs we must pay to maintain a high material standard of living.

Municipal Waste Management Strategies

The three primary methods available for managing municipal waste are recycling, incineration, and landfills

In the United States, our overwhelming reliance on landfills has placed us in a precarious situation. Development of integrated programs that make appropriate use of all three is essential. As they develop waste management strategies, states, counties, and communities must choose a combination of management methods that is appropriate to local conditions.

The major components of U.S. municipal solid waste are paper and paperboard, yard wastes, metals, glass, food wastes, plastics, rubber, leather, textiles, and wood.[14] (See

Table 14.3
Composition by Weight of U.S. Municipal Solid Waste, 1986*

	Percent
Paper and paperboard	35.6
Metals	8.9
Glass	8.4
Plastics	7.3
Wood	4.1
Rubber and leather	2.8
Textiles	2.0
Yard wastes	20.1
Food wastes	8.9
Other wastes	1.8

*Net discards after materials recovery and before energy recovery

Source: Franklin Associates, Characterization of Municipal Solid Waste in the United States, 1960 to 2000 ([Prairie Village, KS, March 1988).

Table 14.3.) All of these materials are at least partially recoverable by recycling, but in 1986 only about 11 percent were recycled; 9 percent were incinerated (6 percent with energy recovery), and the remaining 80 percent were disposed of in landfills or by other means.[15]

To reduce the effort and expense required to manage waste through recycling, incineration, and disposal in landfills, it is essential to limit waste generation at the source as much as possible—using a strategy known as *source reduction*, discussed later in this chapter.

Recycling

The term recycling is commonly applied to the processing of materials into new products that may or may not resemble the original material. Recycling not only reduces waste; it also lowers energy, water, and primary raw material requirements and reduces both air and water pollution. (See Table 14.4.) Recycling can also create jobs and opportunities for small business, and reduce dependence on foreign mineral imports.[16]

Table 14.4
Environmental Benefits Derived from Substituting Secondary Materials for Virgin Resources

(percent)

Environmental Benefit	Aluminum	Steel	Paper	Glass
Reduction of:				
Energy Use	90-97	47-74	23-74	4-32
Air Pollution	95	85	74	20
Water Pollution	97	76	35	—
Mining Wastes	—	97	—	80
Water Use	—	40	58	50

Source: Robert Cowles Letcher and Mary T. Scheil, "Source Separation and Citizen Recycling," in William D. Robinson (ed.), The Solid Waste Handbook (New York: Wiley, 1986), as published in Cynthia Pollock, *Mining Urban Wastes: The Potential for Recycling*, Worldwatch Paper 76 (Washington, DC: Worldwatch Institute, April 1987), p. 22.

In January, 1988, J. Winston Porter, Assistant Administrator for Solid Waste and Emergency Response of the U.S. Environmental Protection Agency, said "It is in recycling that I see the most early promise for improved solid waste management." He recommended a national goal of 25 percent recycling of solid waste within four years, saying some localities would do better, others worse.

Porter said that we must be realistic, and pointed out that the Japanese experience is a useful benchmark. Japan, even with a highly sophisticated source reduction and recycling program developed over many years, recycles only about 50 percent of its waste and does not expect to be able to recycle much more.[17]

Porter's statements are significant because he emphasized the importance of recycling and at the same time, with his example of Japan, cautioned against being overly optimistic that recycling could eliminate reliance on other methods to handle municipal solid waste.

Nevertheless, Japan's example demonstrates the potential for recycling. As noted above, the country recycles or reuses about 50 percent of its solid waste, compared with only 11 percent for the United States. And, after burning 23 percent of its trash in waste-to-energy facilities, only 27 percent of Japan's waste remains to be disposed of in

Table 14.5

Solid Waste Management in the United States, Japan, and West Germany (percentages)

	United States	Japan	West Germany
Recycled or Reused	11	50	15
Waste-to-Energy	6	23	30
Landfilled or Other	83	27	55
Total	100	100	100

Sources: U.S. Environmental Protection Agency, *The Waste System* (Washington, DC, 1988), p. 2-3; Allen Hershkowitz and Eugene Salerni, *Garbage Management in Japan* (New York: INFORM, 1987); Federal Environment Agency, Berlin, Federal Republic of Germany, 1987.

landfills and by other means, compared with 83 percent for the United States.[18] (See Table 14.5.)

In 1987, Japan was reported to have some 360 resource recovery plants serving its population of 120 million people, while the United States had only about 70 plants for 240 million people.[19]

In 1988, Japan recycled 50 percent of its waste paper, 55 percent of its glass bottles, and 66 percent of its beverage and food cans, and converted much of the remaining trash into fertilizers, fuel gases, and recycled metals. With few exceptions, residents in Japan's communities separate their trash into six classifications to simplify recycling. In 1986, the United States recovered only 23 percent of its paper products, 9 percent of its glass, and 25 percent of its aluminum.[20]

Municipal solid waste is being recycled in many European cities. Recovery plants in Rome, Vienna, and Madrid produce metals, glass, paper, plastics, and fibers, and other products. Glass recycling in Europe is growing rapidly; the amount of glass recovered more than doubled between 1979 and 1984.[21]

In developing countries, composted household waste is being used to produce methane gas, fuel pellets, fertilizers, and animal feed. In many cities in India, more than a third of urban wastes is being composted and sold.

In China, the city of Shanghai, with a population of about 12 million, produces about 13 million metric tons of solid waste annually. The city processes and sells more than 10 percent of the waste for biogas production, fertilizers, brick and cement manufacture, and other uses. Shanghai also reprocesses more than a thousand materials, including metals, rubber, plastics, paper, rags, glass, and waste oil.[22]

Recyclable Materials

In the United States, many components of the solid waste stream are separated and processed for reuse, sometimes in products similar to the original product, sometimes into other goods.[23]

Paper About 40 regular grades and 30 special grades of paper, each with carefully defined specifications, are listed by the Paper Stock Institute of America. Paper comprises between 35 percent and 41 percent of the municipal waste stream. Which papers can be recycled and how they can be used are limited by carefully defined material requirements for each grade. Even minor contaminants can make paper unusable.[24]

Newspapers Nationally, newspapers comprise about 7 percent of municipal solid waste by weight. They are easily separated and are the most frequently recycled materials. In 1987, about 36 percent of recycled newspapers were made into newsprint; another 31 percent was converted to paperboard, mainly for packaging, while much of the remainder was used for construction purposes such as roofing and insulation.[25]

Newspapers to be recycled for newsprint must be carefully sorted to remove all contaminants (including clay-filled glossy sections), and baled before shipping to a paper de-inking mill. Steps to create recycled newsprint involve mixing the old paper with water and beating it to separate fibers, addition of chemicals to remove ink, washing, cleaning to remove remaining contaminants, removing water, pressing, drying, and rolling. This process can only be carried out in specially-designed facilities; most paper mills are equipped to handle wood pulp fibers and do not have the equipment required to process secondary fibers.

Each time paper is recycled, the fibers become shorter and weaker; therefore, there is a limit to the number of times it can be reused for paper. In most processes, some virgin fiber must be added. In Japan's newsprint industry, the fiber is about 50 percent new and 50 percent recycled. However, a new process being used by the Southeast Paper Manufacturing Company in Georgia reportedly can produce newsprint using 95 percent recycled newspapers and only 5 percent wood pulp.[26]

Some publishers hesitate to use recycled newsprint because of quality problems. Gannett News, which publishes *USA Today*, uses 100 percent virgin fiber because of the high quality required for adequate color reproduction, and because recycled paper can cause problems with high-speed presses. Gannett's papers are printed at several different locations that obtain paper from different mills; the company says it cannot obtain recycled newsprint of a consistently adequate quality. In general, the quality of paper made of 50 percent new and 50 percent recycled fiber can equal that of paper made from all new fiber, but the demand for recycled newsprint does not yet justify the expense of setting up the mills.[27]

The export market for old newspapers has grown rapidly in the past few years, both in European and East Asian countries, where wood for fiber is scarce. Prices reached as much as $40 a ton in 1983. By late February 1989, however, prices in New York and Philadelphia had dropped to under $5 a ton, partly because of the glut of papers collected by new recycling programs, and partly because overseas markets had accumulated large inventories.[28]

The other major use for old newspapers is for making paperboard or cardboard, used for products such as cereal boxes, shoe boxes, backing for tablets, and tubes for rolled paper products. Since ink is not removed from the fibers, the products tend to be gray.

While use of old newspapers in newsprint and paperboard has been increasing, their use in construction materials such as insulation and fiberboard has changed little in recent years. Cushioning material, mulch, and cat litter are among other uses.

Corrugated Cartons   This material makes up almost as large a part of the U.S. municipal waste stream as newspapers. The American Paper Institute estimates that about 42 percent of corrugated cartons are recycled and made into new corrugated boxes or cardboard products.

Enough cartons are disposed of by households, small businesses, and offices to make their separation from the municipal waste stream worthwhile. Most, however, are discarded by large businesses, supermarkets, retail stores, and factories. These usually go directly to processing mills, bypassing municipal collection systems.

There is a large export market for corrugated containers; shipments from Baltimore, a major port of embarkation, exceed export shipments of old newspapers.[29]

High Grade Office Papers   These paper products have good recycling potential. There are many types of office papers and specifications for their reuse. Used office paper often bypasses municipal waste collection because large businesses and government offices tend to sort their own papers and sell them to recyclers.

Benefits of Paper Recycling   Benefits go beyond reduction of waste. Recycling can dramatically reduce the number of trees that have to be cut to make paper. According to the Southern Forest Institute, a 20-year-old pine tree contains 500 pounds of wood that can yield 120 pounds of paper, enough to produce 3,600 12-pound grocery bags. Paper manufacturing requires large amounts of energy, water, and chemicals. Recycled paper requires up to 74 percent less energy and more than 50 percent less water than paper made from virgin timber. Recycling reduces air and water pollution and pollution abatement costs.

Aluminum Cans   Aluminum is the second most used metal after iron. In 1988, about 972,000 metric tons of aluminum (equivalent to 18 percent of U.S. consumption) were recovered from scrap, mostly from used cans. Packaging accounted for about 30 percent of U.S. aluminum consumption. In 1988, the average U.S. market price for aluminum ingot was $1.10 a pound, up from $0.71 in 1987, but by March 1989 the price had fallen back to $0.70 a pound. Aluminum scrap recyclers paid $0.70 a pound for carload lots in 1988.[30]

As with paper, there are detailed specifications for different uses of aluminum. Beverage cans are easily recycled; they are melted and reprocessed to make new cans. In cases of contamination, the metal can be used for lower grade products. About 50 percent of U.S. beverage cans were recycled in 1987. There is relatively little recycling of other aluminum products from municipal waste.

Energy requirements for producing aluminum from bauxite ore are the highest of any major metal. Recycling, on the other hand, requires between 90 percent and 97 percent less energy than that needed to produce the metal from ore. Recycling of aluminum also reduces air and water pollution by comparable amounts and decreases water requirements as well.

Steel Cans: Tin and Bi-metal   Tin cans are made of steel with a thin coating of tin to prevent corrosion of the steel; bi-metal cans are tin-coated steel with one or both ends of aluminum. Because of their composition, recycling of these cans is much more complicated and therefore less common than recycling of aluminum cans. Both aluminum and tin are contaminants in iron- and steel-making processes, as is lead, which often is used to solder cans. Only very small amounts of those metals, if any, can be allowed in a steel furnace charge.

Clean tin cans can be detinned, which leaves high quality steel for recycling. About one-fourth of the U.S. tin supply comes from used material. Most of the rest is imported, because the United States has almost no tin resources.

Steel from cans supplied only a tiny part of old steel scrap, which comes mostly from used machinery, transportation equipment (including ships and boats), and construction materials. The U.S. exported almost 11 million tons of iron and steel scrap in 1988.[31]

Glass About 2 billion pounds of glass containers are collected annually in the United States. When properly designed and handled, many of these need only to be washed and sterilized before reuse; most handled this way are recyclable glass bottles collected in states that have bottle bills.

Most of the glass containers collected are crushed to form cullet (pieces less that a quarter-inch across), which are then fed into furnaces with new raw materials and melted to produce new containers. The cullet not only saves raw materials; it also allows the use of lower furnace temperatures, thereby reducing fuel consumption and air pollution.

Most recyclers require that glass for containers be sorted by color (clear, amber, or green) and be free of contaminants such as metals and ceramics. In addition to its use in containers, small amounts of recycled glass are used to produce fiberglass, construction materials, road paving, and abrasives.

Plastics Measuring the contribution of plastics to the municipal waste stream by weight is deceptive because plastics are very light. By weight, they account only for about 7 percent; but by volume, they makeup 25 to 32 percent of the total.[32]

Recycling of plastics is limited because there are many different types of plastics and because it is difficult and sometimes impossible to identify and separate them according to type. The plastics industry has devised a voluntary container coding system to enable recyclers to identify and separate the different types of plastics for recycling. So far, 12 states have adopted this coding system, and others are expected to follow.[33]

Two kinds of plastic containers that can be easily identified and recycled to a limited extent are PET (polyethylene terephthalate) and HDPE (high-density polyethylene). When recycled, these plastics are not remade into beverage containers, because they melt at relatively low temperatures, and complete neutralization of all contaminants cannot be assured.[34]

PET soft drink bottles comprise about 20 percent of all molded plastic containers; HDPE containers make up about 70 percent. PET scrap is successfully used to make fiber fill for jackets, pillows, and sleeping bags; it is also used as a liner in upholstery and as a fiber in carpet construction. PET scrap is also recycled to produce industrial strapping, wall tile, flooring, and tail-light lenses. HDPE plastic can be used to make boards suitable for boat piers and garden furniture. Other products include flower pots, toys, trash cans, and various kinds of plastic containers.[35]

State and local laws regulating disposable containers have stimulated research and development on ways to recycle plastics. Biodegradable plastics are high on the list of research priorities, but there is some concern about the impact of such plastics on the stability of landfills, about potential toxicity of the plastics as they degrade, and about whether they can be successfully recycled.[36]

The Mobil Corporation has just announced the first commercial plastic-foam recycling plant in the United States, to be built in Leominster, Massachusetts. Initially it will recycle polystyrene-foam products and plastic cutlery used in school lunchrooms and industrial cafeterias. Later it is planned to include plastics from fast-food outlets.[37]

The Amoco Corporation and McDonald's Restaurants in New York City are initiating a joint effort to recycle foam plastics and plastic cutlery. Browning-Ferris Industries (BFI), one of the biggest U.S. waste haulers, and Wellman, Inc., the world's largest recycler of rigid plastics, are combining in a nation-wide program to remove rigid plastic containers from the municipal solid waste stream.[38]

Yard Wastes Leaves, branches, brush, and shrub and grass clippings comprise most of yard waste. In the summer and fall, grass clippings, leaves, and other yard wastes can constitute a major part of municipal solid waste, especially as leaf-burning often is prohibited or restricted in many urban and suburban areas.

Composting is the most satisfactory way to handle yard wastes. Grass is more difficult to compost than leaves, but leaves and grass can be readily composted together, especially if they are well mixed to prevent clumping. Montgomery County, Maryland operates a composting facility that has become an important source of soil conditioner for landscaping companies, and provides income to help support the county's program.

There are a number of ways that households can set up home composting systems to provide their own mulch. In addition to yard and garden wastes, home composting can include household vegetable wastes and other plant material. Information on composting is available in gardening books.

Recycling: Setting Up a System

It takes time, public education, and political will to implement a large-scale recycling program. Fortunately, limited programs of different types have been under way for some time in various parts of the United States. We can learn from the recycling experiences of others, both successful and unsuccessful, and both here and abroad.

No two locations have identical conditions. Communities vary in size from a few thousand to several million people. Regions may be largely industrial or mainly residential. They may be dominated by single-family homes or may have a mixture of housing types. Climate may be warm or cold, wet or dry. The community may have many families with small children, or consist mostly of senior citizens.

Assessing the Situation

To develop an effective recycling strategy, a jurisdiction must have a clear picture of its own waste situation. In addition to those mentioned above, factors to be considered include:

- waste materials — types, amounts, and sources;

- waste producers — kinds, numbers, and distribution;

- markets for recycled materials — kinds, capacities, locations, and prices.

Many states can provide help and guidance to local communities. They can be especially helpful in providing information on markets.

With the above information in hand, the process of making decisions can begin. Should all recyclables be removed from the waste stream? Should a program be started with just newspapers? Or just aluminum cans?

How will recyclables be separated from the rest of the waste? Where? By whom? How and where should they be collected? In what kind of containers? How much processing will be necessary before recyclables can be sold? What will be the costs? What should be the responsibilities of government, corporations, schools, and individual citizens?

Before undertaking a large-scale program, pilot projects are advisable to test the feasibility of the proposed system and to work out difficulties. The two major issues that must be resolved before recycling can be feasible are: identification of markets and choice of collection systems.

Markets

Without assurance of markets for recycled materials, collecting recyclables is futile. They may well have no destination but a landfill or incinerator. Viable markets exist when recycled materials are competitive in price and quality with primary (new) materials, and when the system for converting the waste to usable materials is in place.

The market cycle begins with the producer of goods (who is also the user of raw materials). With some variations, the cycle proceeds on to wholesalers, to retailers, and to consumers. Recycling involves a series of material handlers, each requiring equipment, and each contributing to the conversion of waste to usable material. These waste handlers include collectors, waste processors, refiners, and material processors. Expensive equipment and skilled personnel are essential all along the way to prepare the recycled raw material for the manufacturer of goods (the user of raw materials). Final prices for recycled materials are set by the producer or manufacturer, who decides how much he can pay for his raw materials, whatever the source. Consumers of goods pay a price that includes all costs of production and delivery. The amount paid for waste by collectors depends on the progressive increase in costs on the way back to the manufacturer.

Prices and markets fluctuate widely. Competition for markets will increase as more and more jurisdictions undertake recycling. The present glut of newspapers, mentioned above, is an example. Jurisdictions in New Jersey, trying to comply with new state requirements to recycle, are finding that a shortage of markets is a major problem.

Materials Collection and Sorting

This is the other major problem. The program devised must consider the pros and cons of requiring citizens to separate all or part of the recyclables included in the program into separate containers, or to collect all mixed waste to be sorted at a separation center. The goal should be to recycle as much as possible at the lowest cost. Two long-time voluntary means of collecting recyclables have been drop-off and buy-back centers, sponsored either by community organizations or, especially in rural areas, by local jurisdictions. Success with these centers, which has depended to a large extent on convenience, has been limited.

The third option, curbside collection, has been much more effective in gaining citizen participation, especially when it is mandatory. If this method is chosen, factors that must be considered include collection methods, costs, time and equipment involved, and contamination of recyclables.

Important questions include: To what extent will sorting be done by waste producers? How much separating are they willing or able to do? How much will waste separation increase hauling costs? Each additional container dumped adds time. Trucks must be designed to keep different materials separated to avoid contamination. Will this be done by separate bins, which will add time and labor and may result in inefficient use of space? Or will separate trucks and separate pick-ups be preferable? Can separate pick-ups be on the same day?

Will it be more efficient and less costly to collect newspapers separately and have one container for other mixed recyclables, with final sorting to be done at a sorting center? What are the requirements for a sorting center? What methods will be used?

In a single jurisdiction, collection plans will have to vary for single-family residences, multi-unit dwellings (and within that category, condominiums, small garden apartments, high-rise structures, and special housing such as nursing homes) and for small businesses and commercial establishments.

Each jurisdiction must assess its own waste situation; incentives or disincentives may be necessary. Ongoing education to encourage citizen cooperation and support is an essential part of any successful program.

Incineration

The use of incineration to dispose of municipal waste has been erratic since the first systematic tests in England in 1874. Incineration's major purpose has been to reduce the volume of municipal refuse. Another benefit, only recently achieved, is the use of the energy released by burning to produce either electricity or steam for heating buildings. The problems with these waste-to-energy processes have been their costs, concern about air pollutants, and hazardous components in the ash. The stimulus for waste incineration has been the increasing difficulty of finding land-disposal sites for urban waste.

Growing populations and changing life-styles, especially since World War II, have produced growing amounts of waste. This has led to increasing pressure on local dumps, periodic burning at many dump sites, and construction of incinerators to reduce the volume of waste to be dumped. The growing shortage of landfill sites has focused attention on incineration, which can reduce the volume of discards by as much as 90 percent.[39] Reduction in weight is less, usually about 70 percent.[40]

Opponents of incineration cite environmental effects such as gaseous and particulate emissions, including highly toxic substances such as dioxins, toxic components in both bottom and fly ash, possible water pollution, and siting problems. They also emphasize the high costs of construction and maintenance of facilities, and the potential for incineration to obstruct the establishment of recycling programs or to weaken their effectiveness. How valid are those objections?

The choice is not between incineration and recycling but between incineration and landfills. Even if a recycling rate of 50 percent, equal to that of Japan, can be achieved in the United States, the remaining municipal waste cannot be accommodated in landfills. Some incineration is essential. The challenge is to use state-of-the-art technologies to expand both recycling and incineration—and to continue to refine those technologies. The understanding and cooperation of citizens is essential.

Air Pollution from Emissions

Concerns about pollution and health effects, although certainly valid for early incinerators that lacked emission controls and exhausted combustion gases directly to the atmosphere, are much less valid today. Pollution will continue to decrease as new state-of-the-art facilities are built, old ones are upgraded, and incinerator emissions are monitored.

Since enactment of the Clean Air Act in 1962 (amended in 1970 and 1977), a new generation of incinerators has been developed that includes scrubbers and filters to remove pollutants before they are released. EPA requires continuous monitoring of gases for several components, including carbon monoxide and carbon dioxide; acid gases such as hydrogen chloride, sulfur dioxide, and nitrogen oxides; particulates; and hydrocarbons. Other emissions, including dioxins and furans, lead, mercury, cadmium, chromium, beryllium, nickel, arsenic, and zinc are monitored periodically.

If not removed from incinerator effluents, acid gases can cause respiratory disease, harm plants, corrode metals, and contribute to acid rain. Heavy metals carried by particulate emissions can harm health even in small amounts. Lead and mercury can cause neurological disorders, and cadmium and arsenic may cause cancer.[41] In general, the toxicity of air pollutants is measured in terms of the estimated number of cancer deaths they would produce over time.

Of particular concern are highly toxic compounds known as dioxins, which can be formed by low-temperature combustion of plastics, as well as by the burning of pesticides, bleached paper, and wood preservatives. Dioxins are extremely poisonous substances strongly suspected of causing cancer and birth defects.[42] A 1987 research project at the Vicon Incinerator Facility in Pittsfield, Massachusetts tested different types of waste under duplicate sets of varying operating conditions. Wood and cardboard wastes alone formed dioxins under certain operating conditions, and the addition of plastics did not increase the amounts. Dioxin emissions were greater at lower temperatures and when carbon monoxide levels were high. However, extensive operating experience with incinerators clearly indicates that if operating temperatures are kept high enough, and carbon monoxide levels low enough, there is virtually no risk of dioxin emissions occurring.[43]

The EPA has set standards for all significant gases emitted by waste incinerators, and the 1989 reauthorization of the Clean Air Act will strengthen these standards. Many states and communities already have standards more strict than the EPA regulations, and incinerator technology now available can perform better than these strict standards.

Ash Problems

Bottom ash, which remains in incinerators after burning, and fly ash, the lightweight particles in gaseous effluents that are trapped by emission control devices, both contain hazardous materials. The amounts of ash and the kinds and amounts of hazardous components depend largely on the composition of the trash. EPA requires that ash be tested and disposed of as hazardous material if it contains hazardous components. However, such testing is rare and ash is now usually disposed of in landfills, where it increases leachate pollution and the possibility of groundwater pollution.

Since most of the hazardous components of incinerator ash are metallic elements, removal of metals from the processes can largely eliminate their presence in ash. Alternatively, incinerator ash can be mixed with lime and other additives to produce a solid aggregate that immobilizes

Figure 14.2
Diagram of a Mass-Burn Incinerator

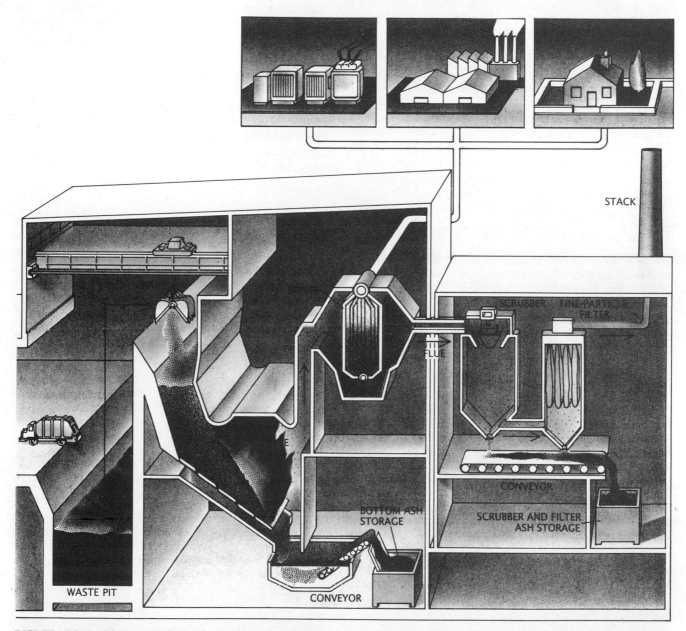

INCINERATORS with systems for pollution control and energy recovery are a viable option for waste management. The mass-burn incinerator diagrammed here reduces 1,000 tons of solid waste into 250 tons of incinerator ash in a single day. Collection trucks deposit solid waste in a pit. A crane transfers the waste to the furnace, where it is burned at a high temperature. The furnace heats a boiler that produces steam for generating electricity, heating buildings and other industrial processes. Ash collects at the bottom of the furnace and is removed by conveyors. Smoke flows through a flue at the top of the furnace and into two pollution-control devices. The scrubber sprays a wet or dry calcium compound into the smoke. The calcium compound reacts with heavy metals and toxic organic compounds and regulates them. The fine-particle filter removes tiny ash particles by passing them through either a porous bag or an electric field that attracts charged particles.

From Philip R. O'Leary, Patrick W. Walsh and Robert K. Ham, "Managing Solid Waste," *Scientific American*, Vol. 259, No. 6, December 1988, p. 40.
Copyright © 1988 by Scientific American, Inc. All rights reserved.

hazardous heavy metals. The aggregate can then be used for road construction and other purposes. The process is now being tested in the United States.[44]

High Costs

The expense of building an incinerator can be as high as $150 million or more, depending on size and type. Operating costs also are high, perhaps $15 million or more annually for a 2,000 ton-per-day incinerator. Income from sales of energy can offset operating costs, and must be deducted from those costs to compare the net cost of incineration to the rapidly rising expense of waste disposal in landfills. In some cases, waste heat from incinerators is being used to heat nearby buildings; the value of such heat can also be used in calculating the net cost of incineration.

Resource Recovery (Mass Burn) Facilities

The mass burn furnaces, which accept almost all kinds of mixed municipal waste, burn at temperatures from 1300 to 2400 degrees Fahrenheit, depending on the technology used. The larger plants can process 500 to 3000 tons per day; small plants handle from 40 to 100 tons a day.

In a typical mass-burn facility, trucks dump unsorted waste into a large pit, from which it is moved by crane to a hopper that feeds the furnace. Furnace temperature and airflow are monitored and adjusted depending on composition of the waste. Bottom ash is collected. The hot gases heat water in a boiler to produce steam, which is used for heat or to produce electricity. Then the gases pass through scrubbers and filters that remove most of the pollutants before they escape through the stack.[45] (See Figure 14.2 on the preceding page.)

More than half of these incinerators have energy-recovery systems that generate electricity and/or produce steam for heating, thereby reducing requirements for traditional fossil fuels (coal, oil, gas). The ash residue occupies only 15 to 20 percent of the original volume of the furnace charge. If recyclable metals and glass are removed before burning, the unit will produce less residual ash with fewer hazardous components.

Refuse-Derived Fuel Facilities

Several types of refuse-derived fuel facilities also receive mixed municipal waste, but they sort it and remove nonburnable portions, chiefly metals and glass. The remaining burnable waste is processed into a form suitable for fuel that can reduce demand for coal, oil, or gas. Some facilities have furnaces to use their own burnable waste as fuel to produce steam or electric power. Others sell the fuel for use elsewhere. In either case, when the waste is burned, ash must be disposed of. But with metals and glass removed, there is less ash and it is less toxic.

An effective means of sorting waste relies on magnetic separation to remove ferrous metals, then shredding and screening to remove non-ferrous metals and glass. These materials are sold for recycling. The process does not sort the non-ferrous material, much of which is exported to countries where the availability of low-cost labor makes hand sorting possible.

Waste with high energy content can be an efficient source of power, and can be burned in facilities designed primarily to produce electricity. In California's San Joaquin Valley, for example, a 15-megawatt power plant produces electricity by burning about 7 million tons of discarded vehicle tires a year, drawing on a stockpile of over 40 million tires.[46]

Landfills

The destination of most waste is the land. Four-fifths of U.S. municipal solid waste—including the waste we produce in our homes—ends up in landfills, and landfill sites in our country and many other countries are filling rapidly.

In 1978, the United States had about 20,000 landfills; by

Figure 14.3
Remaining Life of U.S. Municipal Solid Waste Landfills

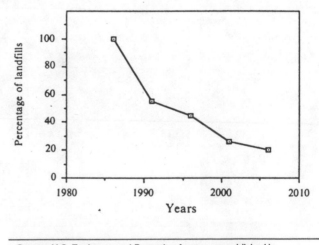

Source: U.S. Environmental Protection Agency, as published in Concern, Inc., *Waste: Choices for Communities* (Washington, DC, 1988), p.2

1988 the number had dropped to about 6,000. In the past 5 years, 3,000 landfills have been closed; by 1993, some 2,000 more will be filled and closed. And by the year 2005, about 75 percent of those remaining are expected to close.[47] (See Figure 14.3.)

Creating new landfills is difficult. In the United States, only half as many landfills have been created in the last five years as in the preceding five-year period. Landfill disposal

Figure 14.4

Tipping Fees at Selected U.S. Landfills, 1984-87

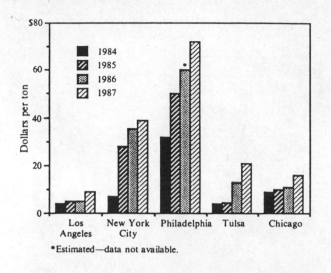

*Estimated—data not available.

Source: *Beyond the Crises: Integrated Waste Management,* Community Environmental Council, 1987, as published in Concern, Inc., *Waste: Choices for Communities* (Washington, DC, 1988), p. 10.

costs, called tipping, or dumping fees, are climbing rapidly (see Figure 14.4), reflecting the growing expense of insurance, audits, permits, fees, improvements, and the future cost of closure and construction of a new landfill. For example, in 1987, tipping fees paid by Boston, Massachusetts jumped from $23 to $70 per ton.[48]

Only within the past two decades has the problem of land disposal of waste become widely recognized. In addition to the long-lamented problems of litter, odors, and rodents are the recently-discovered serious problems of leachate and methane produced in landfills.

Leachate is the solution formed as rainwater percolates down through the landfill. The water is slightly acidified by biochemical processes so that it more readily dissolves or leaches various elements and compounds present in the waste. Some of these, particularly some of the metallic elements, are toxic and pollute groundwater when the leachate flows out of the landfill.

Methane (CH_4), a colorless, combustible, and potentially explosive gas, is produced by bacterial decomposition of the waste. It permeates the landfill and frequently seeps into adjacent land and buildings where fires and explosions have occurred. Methane also can kill vegetation by reducing oxygen in the soil.

The state-of-the-art landfill today is a completely new structure—the product of interdisciplinary scientific research and technological development. It is far more sophisticated that those of the early 1980s and, except for size and shape, has little resemblance to those built before 1970. (See Figure 14.5.)

This new generation of landfills has been carefully designed to eliminate the environmental problems that plague older ones. Siting preferably is restricted to geologically suitable areas. To prevent leachate and methane from leaving the landfill, the bottom is covered with an impermeable liner usually made of layers of clay and synthetic material.

A leachate collection system is installed above the liner. It consists of a permeable layer of sand to trap the leachate and a series of pipes through which it is pumped to a storage tank. Samples from the tank are tested regularly and, depending on the results, the leachate is either sent to a regular sewage treatment plant or diverted to a special plant, usually on-site, for special treatment. To be sure that no leachate is escaping, monitoring wells are located around the site.

Methane is monitored with a series of probes inside and outside the landfill. Wells are drilled to collect the methane, which is either flared or burned to generate steam or electricity. Methane can also be recovered successfully from ordinary landfills not originally designed for its recovery.[49]

Unfortunately, only a small percentage of the 6,000 municipal landfills in the United States have state-of-the-art capabilities. Only about 15 percent are lined, 5 percent collect leachate, and 25 percent monitor groundwater. There are only about 100 sites where methane is recovered.[50]

Technologically, modern landfills can handle municipal solid waste safely. Available space for landfills, however, is almost exhausted, and four-fifths of our waste is still headed for landfills. Time is short to develop alternatives. Fortunately, rapidly rising costs of landfill disposal, due largely to increasingly strict federal and state landfill requirements, will encourage the shift to other strategies for waste management.

Waste Management Strategies: Summary

Great progress has been made in recent years in the quality and effectiveness of technologies for managing solid waste by recycling, incineration, and landfills. Improvements will continue. New technologies have only begun to replace old facilities with their many defects and weaknesses, but the next few years will see many changes as facilities comply with new regulations issued at local, state, and federal levels.

Source Reduction

The foregoing sections review ways to manage the waste we have. They do not discuss the increasingly vital question of how to reduce the amount we generate.

Source reduction involves strategies to decrease the amount of materials to be disposed of, both by eliminating unessential components and by producing more durable,

Figure 14.5
Diagram of a Modern Landfill

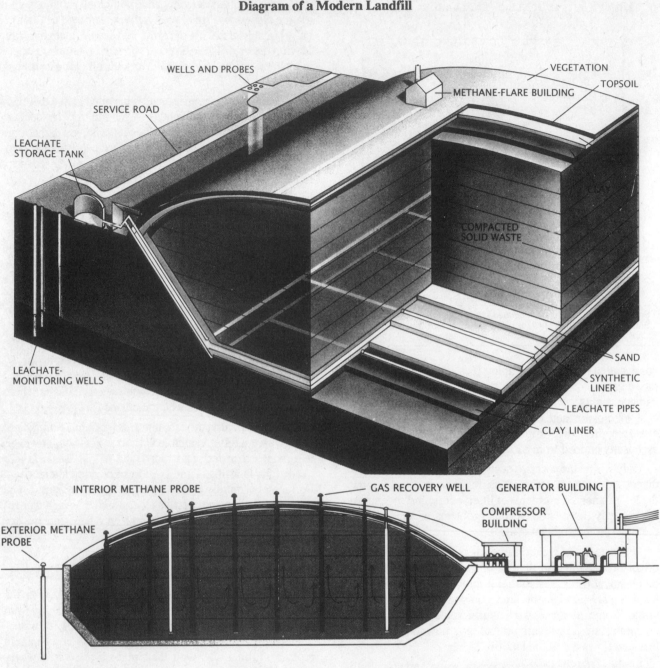

NEW LANDFILL TECHNOLOGIES (*top*) can protect valuable resources from pollutants. Water from weather and other sources leaches pollutants out of the garbage, forming a solution known as leachate. To protect groundwater from leachate, layers of low-permeability clay and synthetic material line the landfill. The liners contain the leachate, which is pumped out through pipes. Every day layers of waste are dumped, compacted and covered with soil. When the landfill is full, it is covered with clay, sand, gravel, topsoil and vegetation to reduce leachate production and minimize erosion. Another landfill pollutant produced by bacteria decomposing waste is the volatile gas methane. Wells and probes are placed at the perimeter of the landfill to detect leakage of leachate or methane. Methane can be burned off by a flare. An alternative to burning methane is to install a recovery system (*bottom*). The system collects the methane and fuels a turbine to generate electricity.

From Philip R. O'Leary, Patrick W. Walsh and Robert K. Ham, "Managing Solid Waste," *Scientific American*, Vol. 259, No. 6, December 1988, p. 41.
Copyright © 1988 by Scientific American, Inc. All rights reserved.

longer-lasting goods. Durable goods designed to allow easy repair, at prices competitive with their replacement, can contribute significantly to waste reduction. Packaging makes up about 50 percent of the volume and 30 percent of the weight of municipal waste. The bag inside the box inside the wrapper is unnecessary. More economical use of materials can reduce both municipal waste, and waste resulting from the production of raw materials.

Cedric Maddox, Director of Sanitary Services for Atlanta, says "...waste stream reduction is the answer. We must start with the manufacturer, to curb production of what will become unnecessary waste. We must educate the purchaser not to buy what may become unnecessary waste. We must educate the user not to be wasteful and not to create unnecessary waste."[51]

A condition that has been developing for many years and has been encouraged by society cannot be changed dramatically in a short time — just as a large ship sailing at 30 knots cannot stop or change course immediately.

To some extent, we can change our own habits—use fewer disposable materials, wear our clothes longer, drive our cars longer, or repair instead of replacing appliances. Yet how many of us are willing to take these steps, especially when we are constantly bombarded with advertisements for appealing goods?

Will we form community organizations to insist on changes in the products offered for sale—and in their packaging? Will we participate in citizen-education programs? Will we write letters to our elected representatives at all levels to support legislation that would discourage waste? All of the above and more will be necessary if consumers are to make a difference.

On the production side, we must understand that industry similarly cannot change overnight. It will take time and money to make the necessary changes in plant and equipment. Changes will affect employees; some jobs may be eliminated, new ones will be created. Production will have to decrease in some areas while it increases in others. Economic, political, social, and technical problems will appear and will have to be solved.

Integrated Waste Management: Nothing is Simple

This chapter has described various methods being used to manage municipal solid waste. Source reduction is absolutely indispensable for the long term. A final solution to waste disposal problems will require the use of all four possibilities —source reduction, recycling, incineration, and landfills.

Federal Legislation

In 1965, the U.S. Congress passed the Solid Waste Disposal Act to provide support for alternatives to incineration and unregulated dumping. In 1970, this Act was amended by the Resource Recovery Act, which directed the EPA to investigate the potential for waste reduction.

The Resource Conservation and Recovery Act (RCRA), passed in 1976, was designed to address the problems of hazardous wastes (see Hazardous Substances chapter), but it also empowered the EPA to help states develop solid waste management programs that would give priority to conservation of material and energy resources.

RCRA sets standards for solid waste disposal facilities and provides financial and technical assistance. In the late 1970s, support for waste reduction and recycling was undercut by oil price increases and interest in incinerating waste for energy. By 1981, budget cuts had eliminated EPA's solid waste grant program for states, and now only 18 states have completed solid waste management plans or regulations.[52]

There is as yet no other Federal legislation that deals specifically with solid waste management. During recent sessions of Congress, bills that address various municipal waste problems have been introduced but have failed to pass.

In the Senate, legislation has been proposed to control municipal waste combustion by amending the Clean Air Act, and to prohibit interstate disposal of solid waste. In the House, bills have been introduced to establish a National Packaging Institute to promote reduced packaging and to ensure that waste disposal will not endanger health and the environment; to promote recycling by improving collection, analysis, and dissemination of information; and to encourage development of technology to aid recycling and the use of degradable materials.

The increasing concern in Congress about the waste issue is encouraging, even if new legislation fails to pass again in this session. Informed citizen pressure can be very helpful in generating needed support.

Current EPA Proposals

In 1988, EPA's Municipal Solid Waste Task Force released *Solid Waste Dilemma: An Agenda for Action*, a report on MSW problems facing the nation, containing goals and recommendations. The report's stated goal is to "find a safe and permanent way to eliminate the gap between waste generation and available capacity in landfills, incinerators, and in secondary material markets."[53]

The Task Force recommends that a waste management hierarchy be instituted, including, in descending order of preference, waste reduction, recycling, incineration, and landfilling. Thus, the report states that source reduction and recycling are the preferred options for reducing the amount and toxicity of waste that must be placed in landfills or incinerated.[54]

The EPA Task Force proposed four objectives to meet its goal of closing the gap between waste volume and disposal capacity: increase available information, encourage greater planning, promote source reduction and recycling, and reduce the risks associated with incineration and landfills.[55]

The State Role in Waste Management Planning

Each state is responsible for developing a solid waste management plan and implementing a management program.[56] The plan must contain the state's overall strategy for protecting human health and the environment, specify efforts to encourage resource conservation and recovery, and provide for adequate disposal capacity. Oregon, Washington, Illinois, Michigan, and Vermont are among the states that have used the waste management hierarchy as a guideline in developing their plans.

Washington, among other states, requires communities that request disposal facility permits first to analyze reduction and recycling options before allowing a permit to be issued. Similarly, Minnesota requires solid waste plans that consider reduction, recycling, and waste-to-energy options. Minnesota's overall goal is to reduce the waste stream by 3 percent, recycle 25 percent of the waste, and compost 12 percent for a total diversion from disposal of 40 percent.[57]

Some states have chosen to implement recycling through mandatory recycling and/or bottle bills. Oregon, Wisconsin, and Washington require disposal facilities to provide recycling opportunities for the public. In 1986, Rhode Island introduced a comprehensive recycling law requiring communities to divert 15 percent of their discards from the waste stream by 1989, with recycling containers to be provided by the state for every household. New Jersey's mandatory recycling law requires each county to recycle 25 percent of its waste by 1989. Residents are required to sort at least three of four usable materials—glass, paper, plastic, or aluminum. The law also requires counties to prepare a marketing strategy for recycling waste.[58] There is no assurance, however, that markets for recycled materials can be found.

Legislation requiring a deposit for beverage containers, popularly known as a "bottle bill," has been adopted in 10 states. The laws are designed to provide economic incentives for consumers to recycle beverage containers, and for retailers, distributors, and manufacturers to reuse or recycle them. Oregon and Michigan charge a lower deposit fee for refillable bottles that can be used interchangeably by different bottling companies. Bottlers and distributors generally do not favor this strategy, citing increased handling and higher costs. Nevertheless, states with bottle bills have realized an average waste diversion from disposal of 5 percent.[59]

Many states make funds available for local waste management planning and project start-up; local programs have grown rapidly. Michigan established a $10 million fund for comprehensive waste management assistance to local governments. Pennsylvania included provisions in its SWM plan for grants to begin recycling programs. California has provided grants to private corporations to encourage diversion of waste from landfills. A privately owned and operated, $8 million material sorting and recovery facility in Marin County, California runs a highly successful private pick-up program. Curbside collection of recyclables is combined with regular weekly trash pickups. Sixty percent of the county's 255,000 people participate. The program diverts about 22 percent of the waste stream, and it was initiated with a $575,000 state grant.[60]

Funds to pay for SWM programs can be raised in ways that complement program goals. Products, packaging, and disposal methods that contribute heavily to the waste stream can be taxed or made subject to surcharges, making these methods less attractive and internalizing social and environmental costs. Illinois, New Jersey, Vermont, Maine, Iowa, and the District of Columbia tax waste that is dumped at disposal facilities. The Environmental Defense Fund estimates that New Jersey's tax raises $15 million annually for the state. In the District of Columbia, a $2 per ton surcharge on waste handled by private waste haulers is expected to raise over $1 million to support that city's recycling program.[61] Iowa gives local governments the option of assessing their own fee to support waste management options. Washington, New Jersey, Nebraska, and Ohio all levy taxes to raise money for their waste management and litter control programs.

Local Waste Management Programs

As mentioned earlier, community governments are responsible for making most municipal waste management decisions. Although each government will have to choose the balance between recycling, incineration, and disposal in landfills that best satisfies local needs, the problems faced by small rural or suburban communities, where daily waste generation may total perhaps forty tons, have little in common with large urban centers where several thousand tons of waste may be generated each day.

Communities of all sizes are beginning to develop waste-reduction ordinances, but major reductions will depend on changes in the goods that are produced for consumers. Governments at all levels will have to be involved. Local citizen pressure can help spur them to action. As an example of what local initiatives can accomplish, Minneapolis and several other cities have effectively eliminated plastic food packaging by banning its use unless it can be recycled.[62]

More and more frequently, local jurisdictions are mandating recycling programs. Except for very small, closely-knit communities, legislation is much more effective than depending on voluntary efforts.

Cities are still experimenting to develop optimal waste management programs. In Seattle, for example, two different recycling programs are being tested. In the northern part of the city, residents are given three-compartment containers for newspaper, mixed paper, and mixed bottles and cans. The materials are collected weekly. Residents in the southern part of the city receive one large bin for all recyclable materials, which is collected monthly and then processed at a materials recovery facility.[63]

In 1982, when Berkeley, California, adopted a recycling ordinance with the goal of recycling 50 percent of its waste stream by 1991, it already had a city-sponsored voluntary recycling program that included curbside newspaper, glass, and can pickup. A voluntary curbside pickup program is still operated by the non-profit Ecology Center, and subsidized by the city. Private sector recycling enterprises also flourish. Backyard composting is popular. Berkeley encouraged recycling instead of incineration by passing a 5-year moratorium on garbage burning in 1982.[64]

The Role of Citizens

Public understanding of the waste management problem and the various components of its solution is essential. To gain the public's understanding and cooperation will require extensive, balanced, and realistic education programs. One approach must not be promoted to the exclusion of others, since a combination of management methods will be required for successful programs.

Government and industry will eventually respond to citizen pressure for changes in the production of goods and packaging, for markets to process and reuse recyclable materials, for support of up-to-date technologies for incineration and landfills, and for continued research and development.

What You Can Do

Inform Yourself

Get more information about solid waste management and recycling. Learn the facts about the latest incinerator and landfill technologies. Remember that source reduction and recycling alone cannot handle all municipal solid waste. Consult the sources and organizations listed in the Further Information section.

- Contact your local authorities dealing with sanitation, public works, or environmental resources and learn about solid waste management in your community. Here are some questions to ask:

» What is the quantity and composition of waste in your community?

» How is your present solid waste management system organized?

» Who is responsible for waste management?

» Is more than one government agency involved?

» What does your community pay per ton to dispose of its waste?

» What is the local budget for solid waste management?

» How much has it increased recently?

» How much of the waste is placed in landfills?

» Where is the landfill that serves your community?

» What technology is used: liners to prevent groundwater contamination? Leachate monitoring and control? Methane monitoring and control? Is the landfill covered?

» What is the remaining life-expectancy of your landfill?

» Is waste hauling done by government or private haulers?

» How much is the tipping fee?

» Has the tipping fee increased recently or is an increase anticipated?

» Does your community have an incinerator?

» What is its capacity? How old is it?

» What technology does it use? Resource recovery, mass burn, or resource-derived fuel?

» Is the energy used to generate electricity and/or industrial steam? What is the income from such sales?

» What are the operating costs? What landfill costs are avoided?

» Does your community have a recycling program?

» Is it voluntary or mandatory?

» Does it have drop-off centers? Buy-back centers? Curbside collection?

» If curbside, what separation is required? Are containers provided?

» How frequent is collection?

» What collection equipment is used?

» Is there a separation center for recyclable materials?

» How does it operate?

» What are collection costs? Separation costs?

» How much is saved by avoiding landfilling?

» Are markets adequate? Where are they?

- If your community has no recycling program, contact your state government and ask for the location of the nearest recycling centers for paper, metals, glass, plastic, and other materials.

Join With Others

Contact and join a local group or national organization with a regional chapter that works on solid waste management issues. Find out if the group is supporting a viable waste management program that integrates recycling, incineration, and landfilling as well as source reduction into its goals. Volunteer your time. Develop a network of contacts. Attend meetings of your city council or county board. Arrange to visit waste disposal facilities in your jurisdiction and elsewhere. Arrange visits for citizen and student groups.

- If no local recycling program exists, try to help create one. Contact nearby recycling groups, private recyclers, and national recycling organizations. Find out whether there are nearby markets and collection points for paper, glass, metals, and plastics. For information about recycling plastics, call 1-800-542-7780.

- Organize roadside walks to collect cans, bottles, and other recyclable trash. Organize community pickups of recyclable materials. Examine potential markets for recycled materials. Establish a public education program. Invite experts to speak at meetings.

- Try to involve various sectors of your community in recycling programs. As an example of how this can be done, a recycling center in New York City initiated a successful collection program using homeless people as paid gatherers.

Review Your Habits and Lifestyle

Individuals can make a real difference in their community's production and management of solid waste. Home composting is practical for suburban and rural dwellers, but less so for urban residents. If recycling is convenient, use it; if it is an hour drive to recycle, however, your time and energy might be better spent working on the development of a local recycling program.

- If you have a yard or garden, consider composting your yard and vegetable wastes.

- Remember the importance of reducing your contribution to the waste stream. Try to reduce the amount of materials you use and buy. Purchase items with the least packaging and with packaging that is recyclable and non-toxic. Reuse containers and any other items that you can.

- If you can locate an organization that will accept used paper, organize a collection system in your school or workplace, and arrange for periodic pickups.

- Try to obtain recycled printing and writing paper for your office or home through your local printer or paper distributor. Buying recycled paper locally through ordinary sources tends to support local recycling efforts and helps create markets for products. Don't be surprised if you can get recycled paper only from the Midwest or the West Coast; that is where most of it is made. For information about obtaining recycled paper, contact the following companies: Conservatree Paper Company (California, wholesale); Earth Care Paper Company (Michigan, retail); Prairie Paper (Nebraska, retail); or Recycled Paper Outlet (Oregon, retail); see the Appendix for addresses.

- When buying recycled paper, find out how much secondary (waste) fiber it contains; how much of it was derived from post-consumer waste paper; and whether it was de-inked, bleached, or dyed.

- Buy in bulk — this saves packaging and allows you to reuse package containers you already have. Minimize the use of paper whenever cloth can substitute. Take a reusable bag when shopping, or ask for paper bags instead of plastic.

Work With Your Elected Officials

Find out what legislation, if any, has been passed or proposed to improve the solid waste management situation. Urge legislators to take constructive action to reduce waste generation and encourage recycling. Work for legislation to reduce the amount of packaging material on consumer products and make it readily recyclable.

- Support federal initiatives for an explicit waste reduction policy and creation of a national task force to study and recommend effective ways to change wasteful and extravagant production and consumption habits, and to develop markets for recyclables.

- Encourage your state government to provide technical and economic assistance to communities for source reduction and recycling programs. Support state tax incentives for businesses engaged in recycling, and for businesses that use recycled materials or purchase production equipment that uses recycled material. Support taxes, fines, and enforcement procedures to curb industrial waste.[65]

- Work for the appointment of a full-time recycling coordinator in your state and in your community.

- Discuss with community officials and leaders the possibility of charging households a trash-collection fee based on the amount of refuse produced.[66]

- If your community has plans to incinerate solid waste, examine the economics of the proposal. Be sure that plans for recycling and incineration are compatible. The amount of recycling anticipated should be considered in determining incinerator capacity.

Publicize Your Views

Write letters to local newspapers and public officials stressing the need for waste reduction and recycling. Let them know that incinerators and landfills will be necessary parts of the solution, and that these must meet the highest technical standards that can be achieved. Volunteer to write articles for local publications. Help local organizations create public service messages for local television and radio stations.

Raise Awareness Through Education

Find out whether schools and libraries in your area have courses, programs, books, films, and other resources related to solid waste management. If they do not, volunteer to help develop some. Offer to speak to classes or organize projects on the issue. Help teachers arrange field trips to local waste management facilities such as landfills, incinerators, and facilities that reprocess used paper, glass, and metal.

- Encourage local schools and colleges to organize their own recycling programs.

Consider the International Connections

Compare the amount of recycling in your community to that in Japan and Western Europe (see Recycling section), and consider how your community could recycle more of its waste. Consider how the export of garbage from one country to another can create international tensions.

Further Information

A number of organizations can provide information about their specific programs and resources related to solid waste management and recycling.[*]

Books

The Biocycle Guide to Composting Municipal Wastes. Emmaus, PA: Biocycle, 1989. Paperback, $49.95. Edited by the staff of *Biocycle*, the Journal of Waste Recycling (see listing under Periodicals).

Blumberg, Louis and Robert Gottlieb. *War on Waste: Can America Win Its Battle With Garbage?* Washington, DC: Island Press, 1989. 325 pp. Paperback, $19.95. Analyzes the historical, political, and sociological roots of the waste crisis, and reviews a range of possible solutions.

Cohen, Levin et al. *Coming Full Circle: Successful Recycling Today*. New York: Environmental Defense Fund, 1988. 162 pp. Paperback, $20.00. The finest one-stop guide to successful state level recycling programs; discusses technical and financial feasibility.

Concern, Inc. *Waste: Choices for Communities*. Washington, DC: 1988. 30 pp. Paperback, $3.00.

Eldred, Kate. *Waste Choices for Communities*. Washington, DC: Concern, Inc., 1988. A general treatment of the waste issue.

Hershkowitz, Allen et al. *Garbage Burning: Lessons from Europe: Consensus and Controversy in Four European States*. New York: INFORM. Documents Europe's more sophisticated practices and policies in the areas of in-plant monitoring, ash handling, regulatory standards, penalties, and mandatory formal worker training.

Hershkowitz, Allen and Eugene Salerni. *Garbage Management in Japan: Leading the Way*. New York: INFORM,

---

[*] These include: Association of State and Territorial Waste Management Officials, Center for the Biology of Natural Systems, Citizens Clearinghouse for Hazardous Wastes, Coalition for Recyclable Waste, Concern Inc., Conservation Foundation, Council for Solid Waste Solutions, Environmental Defense Fund, Environmental Task Force, Friends of the Earth, INFORM, Institute of Scrap Recycling Industries, Institute for Local Self-Reliance, Keep America Beautiful, National Association for Plastic Container Recovery (NAPCOR), National Audubon Society, National Center for Policy Alternatives, National Recycling Coalition, National Recycling Congress, National Solid Waste Management Association, National Wildlife Federation, Natural Resources Defense Council, Pollution Probe Foundation, Renew America, Resource Policy Institute, Sierra Club, Wellman, Inc. (plastic recycling), World Resources Institute, Work on Waste, Worldwatch Institute. Addresses for these organizations are listed in the appendix.

1987. 131 pp. Paperback, $15.00. Reviews aspects of Japan's advanced waste management system, including state-of-the art plant technology, worker training, siting, and recycling.

Keep America Beautiful, Inc. *Overview: Solid Waste Disposal Alternatives.* An Integrated Approach for American Communities. Stamford, CT, 1989. 25 pp. Paperback. An excellent review of source reduction, recycling, composting, landfills, and waste-to energy systems.

Kirshner, Dan et al. *To Burn or Not to Burn.* New York: Environmental Defense Fund, 1988. An economic comparison of incineration and recycling in New York City.

Morris, David et al. *Garbage Disposal Economics: A Statistical Snapshot.* Washington, DC: Institute for Local Self-Reliance, 1987. A comparison of U.S. case studies of recycling, processing and incineration programs.

Newsday. *Rush to Burn: Solving America's Garbage Crisis?* Washington, DC: Island Press, 1989. 269 pp. Paperback, $14.95. Written for citizens and community leaders who need to understand the extent of the problem and what they can do to help solve the garbage crisis.

Platt, Brenda et al. *Garbage In Europe: Technologies, Economics, and Trends.* Washington, DC: Institute for Local Self-Reliance, 1988. European case studies in waste management technologies with country overviews.

U.S. Environmental Protection Agency, Office of Solid Waste. *State and Local Solutions to Solid Waste Management Problems.* Washington, DC, January 1989. 52 pp. Paperback.

U.S. Environmental Protection Agency, Office of Solid Waste. *The Waste System.* Washington, DC, November 1988. 72 pp. plus appendices. A review of hazardous and solid waste management issues.

Wirka, Jeannie. *Wrapped in Plastics.* Washington, DC: Environmental Action, 1988. 159 pp. The environmental case for reducing plastics packaging.

Articles

Chandler, William. "Recycling Materials." In Lester R. Brown et al., *State of the World 1984* (New York: Norton, 1984), pp. 95-114. For an expanded version, see Worldwatch Paper 56.

Conservation Foundation. "Solid Waste," and "America's Waste: Managing for Risk Reduction, " *State of the Environment: A View Toward the Nineties* (Washington, DC, 1987), pp. 106-15, 407-77.

Cook, James. "Not in Anybody's Backyard." *Forbes*, November 28, 1988, pp. 172-82.

Elkington, John and Jonathan Shopley. *Cleaning Up: U.S. Waste Management Technology and Third World Development.* Washington, DC: World Resources Institute, 1989. 80 pp. Paperback.

Environmental Action. *Fact Packet on Bottle Bills*, *Fact Packet on Recycling, Legislative Summary of Statewide Recycling Laws.* Washington, DC.

Organization for Economic Cooperation and Development. "Solid Waste," *The State of the Environment 1985* (Paris: 1985), pp. 159-172.

O'Leary, Philip R. et al. "Managing Solid Waste." *Scientific American*, December, 1988.

Pollock, Cynthia. *Mining Urban Wastes: The Potential for Recycling*, Worldwatch Paper 76. Washington, DC: Worldwatch Institute, 1987. $4.00. An expanded version of the next listing.

Pollock, Cynthia. "Realizing Recycling's Potential." In Lester R. Brown et al., *State of the World 1987* (New York: Norton, 1987), pp. 101-121.

"Recycling in Third World Cities," *The Futurist*, March-April 1989, pp. 50-51.

Ridley, Scott. "Solid Waste and Recycling. In Ridley, *The State of the States 1987* (Washington, DC: Fund for Renewable Energy and the Environment, 1987). An overview and rating of state solid waste programs.

Schwartz, Anne. "Drowning in Trash, We Begin to Discard Our Wasteful Ways." *Audubon Activist*, May-June 1989.

World Resources Institute and International Institute for Environment and Development. "Recovery of Selected Materials," *World Resources 1987*, pp. 299, 308-309, 311-312; "Urban Solid Waste Disposal" and "Waste Generation in Selected Countries," *World Resources 1988-89*, pp. 45-47, 314-316. New York: Basic Books, 1987, 1988.

Periodicals

Biocycle. Journal of Waste Recycling. Monthly, $43 a year. JG Press, Inc., P.O. Box 351, Emmaus, PA 18049.

Recycling Times. National Solid Waste Management Association. Biweekly, $90 a year.

Resource Recycling. P.O. Box 10540, Portland, OR 97210.

Hazardous Waste News. (Also covers solid waste management.) Weekly, $18 a year. Environmental Research Foundation, Box 3541, Princeton, NY 08543. Free access to computer database is available.

Waste Not. Weekly, $25 a year. Work on Waste, 82 Judson St., Canton, NY 13617.

Waste Age. (An industry journal.) National Solid Waste Management Association, 1730 Rhode Island Ave., Washington, DC 20036.

Films and Other Audiovisual Materials

Building a Homegrown Economy. Describes resource flows in the city of St. Paul, Minnesota. Looks at local enterprises and reasons for the pursuit of local self-reliance. 12 min. Video tape, VHS or Beta, $30. Institute for Local Self-Reliance.

Citizens Training Conference on Solid Waste Management. More than 100 scientists, citizen organizers, and local officials participated in this conference held in Washington, D.C. to review waste management strategies. 52 min. VHS video tape, $27.50. Institute for Local Self-Reliance.

Recycling. An EPA film that samples some of the methods for treating municipal solid waste. 21 min. 16 mm color. Film Distribution Center.

The Trash Monster and *The Wizard of Waste.* 12 minute film strips. California Solid Waste Management Board, 1020 9th St. Suite 300, Sacramento, CA 95814.

Uncle Smiley Goes Recycling. A trip to a recycling station shows children how cans, bottles, and paper can be reused. After the visit, the kids start their own neighborhood collection project. 13 min. 16 mm film, $240; video, $120; rental, $50. Coronet/MTI Film & Video.

The Village Green. 1975. An EPA training film describing many of the issues and considerations involved in organizing and running a successful, self-sustaining recycling center in Greenwich Village, New York City. 16 min. 16 mm film. Film Distribution Center.

What About Waste? (Also available as a school curriculum unit.) Michigan Department of Natural Resources, P.O.Box 30028, Lansing, MI 48909.

Work on Waste offers numerous VCR tapes on the hazards of incineration. Address: 82 Judson Street, Canton, NY 13617.

Teaching Aids

Contact your state department of education or the agency that regulates solid waste management for educational materials designed for use in your state.

A-Way With Waste: A Waste Management Curriculum for Schools. 1984. Kindergarten through Grade 12. Washington State Dept. of Ecology, Litter Control and Recycling Program, 4350 150th Ave. NE, Redmond, WA 98052.

Recycling Study Guide. 1988. Designed to help teachers and students understand what solid waste is, where it comes from, why it's a problem, and what can be done about it. For grades 4-12. 31 pp. Bureau of Information and Education, Wisconsin Department of Natural Resources, P.O.Box 7921, Madison, WI 53707.

Recycling Activities for the Classroom 1978. ERIC/SMEAC Center, 1200 Chambers Rd., 3rd floor, Ohio State University, Columbus, OH 43212.

Global Security

Great Powers in relative decline instinctively respond by spending more on "security," and thereby divert potential resources from "investment" and compound their long-term dilemma.

—Paul Kennedy, *The Rise and Fall of the Great Powers*[1]

If we examine defense expenditures around the world...and measure them realistically against the full spectrum of components that tend to promote order and stability within and among nations—it is clear that there is a mounting misallocation of resources.

—Former U.S. Defense Secretary and Former World Bank President Robert S. McNamara[2]

It is difficult to find among today's pressing issues any that lend themselves to military solutions. In this sense military power seems to be irrelevant to national and global security. Behind a facade of "defense" the arms race, in its extravagant technological triumphs, has become a serious threat to civilization's survival.

—Ruth Leger Sivard, *World Military and Social Expenditures*[3]

International economic security is inconceivable unless related not only to disarmament but also to the elimination of the threat to the world's environment.

—Soviet President Mikhail Gorbachev[4]

It will be a great day when our schools get all the money they need and the air force has to hold a bake sale to buy a bomber.

—Women's International League for Peace and Freedom, U.S. Sector[5]

Major Points

- At $940 billion a year, global military spending absorbs a disproportionate share of critical resources and technical skills in most nations. Low-income countries spend $200 billion a year on their military establishments, more than on health and education combined.

- While an arms race often reflects an unresolved underlying conflict, it may well gain a momentum and life of its own.

- Conflict often is related to population pressures and to environmental and economic decline. Military solutions to these non-military problems are inappropriate and ineffective.

- In addition to the military and economic aspects of national security, recent global trends point to the need to broaden the definition of security to include natural resource, environmental, and demographic issues.[6]

- A reallocation of resources from the military to programs such as reducing poverty and Third World debt, slowing population growth, protecting soils and forests, improving energy efficiency, and developing renewable energy may be the highest priority for reversing global ecological decline, providing true security, and moving toward a sustainable future.

The Issue: Excessive Military Spending Diverts Resources from Other Security Priorities

The world now spends more than $900 billion each year to support military activities. In 1985, military spending was greater than the total income of the poorest half of humanity, and equivalent to almost $1,000 for each of the world's 1 billion poorest people.[7]

National security is an emotional priority for the vast majority of the world's people. Personal security is perceived to be linked closely to national security; threats to a nation are seen as threats to one's own well-being. Since World War II, military establishments everywhere increasingly have claimed prerogative over defining national security interests, and national priorities have been ordered accordingly. In this approach to security, the resources allocated to military institutions have expanded dramatically, diverting money, materials, and personnel away from other priorities. Indeed, military spending has grown so much that it has become a detriment to achieving true national security. In many cases, increasing military spending in the pursuit of unilateral military advantage has resulted in greater overall insecurity.

The preoccupation with military concerns has led to neglect of pressing social and economic problems. Government spending on health and on education, two critical aspects of personal security, falls short of military spending in most nations, especially in the Third World, and expenditures for international peacekeeping are tiny compared with military spending. Third world countries spend nearly 50 percent more on the military than on education, and nearly as much on arms imports as on health.[8] (See Table 15.1.) And according to World Bank President Barber Conable, low-income countries spend $200 billion a year on their military establishments, or more than on health and education combined.[9]

Table 15.1
Public Expenditures for Selected Priorities, 1984
(Million U.S. Dollars)

	Military	Arms Imports	Int'l Peace Keeping	Education	Health
World	768,833	41,425	265	659,352	565,503
Developed Countries	618,849	9,360	258	558,718	527,096
Developing Countries	149,984	32,065	7	100,634	38,407

Source: Ruth Leger Sivard, *World Military and Social Expenditures 1987-88* (Washington, DC: World Priorities, 1987).

On a per person basis, military expenditures range from more than $2,000 for Saudi Arabia to less than $10 for some developing countries such as India and Zaire. Public spending for education and for health both vary from around $1000 to less than $10. Table 15.2 (on the following page) illustrates the wide differences in spending between nations in these sectors, and includes data for the world (142 countries), developed countries (29 in number), developing countries (113 in number), and selected individual nations.[10]

Large military expenditures divert money from programs that would strengthen national economies or protect and conserve natural resources such as air, water, soil, and energy supplies upon which all nations ultimately depend. A former U.S. State Department official once lamented the irony of the situation in which military strength is viewed as the only form of security, noting: "If a foreign power seized several hundred square kilometers of Sudanese territory, the government would not hesitate to call out the army. Yet every year hundreds of square kilometers of valuable land are irrevocably lost to that nation through bad land management, and the government accepts it meekly."[11]

Table 15.2
Public Military and Social Expenditures per Person
for Selected Countries, 1984

(in U.S. Dollars)

	Military Expenditures Per Person	Education Expenditures Per Person	Health Expenditures Per Person
World	$163	$140	$120
Developed Nations	550	497	469
Developing Nations	42	28	11
United States	1002	771	674
West Germany	359	506	891
Soviet Union	816	332	227
Hungary	130	293	164
Saudi Arabia	2091	753	486
Lebanon	133	105	22
Angola	133	49	11
Ethiopia	10	3	2
El Salvador	43	25	13
Costa Rica	0	82	20
Peru	73	31	12
Brazil	11	57	23
South Korea	114	101	5
Japan	103	529	474
China	21	8	2
India	8	8	2

Source: Sivard, *World Military and Social Expenditures 1987-88.*

By stressing the existence of an "external enemy" and building up military forces to counter an alleged external threat, a nation's leaders can use such an "enemy" as a scapegoat to blame for or divert attention from their country's own domestic problems, or to help justify internal repression of dissident groups in the name of "national security."

Intense concentration on the military sector can mislead people into believing that their security and well-being are adequately insured by their nation's military forces, when in fact the root causes of poverty, insecurity, and conflict are often found in environmental, economic, and social problems. By providing nations with military means to augment economic and political efforts to achieve their objectives, high levels of military spending can increase the chance of conflict. The nations of the world drastically need a new definition of national security—one that incorporates the concept of sustainable development.

The scale on which the military sector now consumes resources is unprecedented. In 1985, global military spending exceeded $940 billion, more than the combined gross national product (GNP) of China, India, and African countries south of the Sahara.[12]

Over 70 million people worldwide are believed to be engaged in military activities, including about 29 million in regular armed forces (not including paramilitary organizations and reserves, which would add at least 10 million more), and at least 11 million workers directly engaged in the production of weapons and other specialized military equipment.[13]

Technological research and development (R&D) has increasingly become dominated by military industry. The military sector currently accounts for a third of global R&D expenditures and involves 3 million scientists and engineers. The world's military also accounts for 5 percent of the total world consumption of petroleum (up to 50 percent for developing nations) and between 3 and 12 percent of major minerals such as aluminum, copper, iron ore, lead, silver, and tin.[14]

Seventy percent of the world's military expenditures can be traced to just six nations: China, France, the United Kingdom, the United States, the Soviet Union, and West Germany.[15] Prodigious defense spending by these countries has assured the military a prominent place in their economies, accounting for 7 percent of U.S. GNP in 1985, and 14 percent of the Soviet Union's (much smaller) GNP.[16] For the United States, spending for military purposes grew tremendously during the eight years of the Reagan administration, amounting to over $2 trillion for the period, or more than $21,000 for every U.S. household.[17] More than a quarter of all federal taxes in the United States, and roughly an equivalent amount in the Soviet Union, are levied in the name of military security. In 1987, 29 percent of the U.S. federal budget was spent for defense. During the rapid U.S. defense buildup during the 1980s, the growth of military funding has outpaced all other budget increases, and has been a major factor in the doubling of the U.S. national debt during this period.[18]

Military security dominates the thinking of other countries as well, and a significant fraction of military output is exported, in a multibillion-dollar arms trade with developing nations lacking an indigenous arms-manufacturing capacity.[19] The United States and Soviet Union together account for 53 percent of the world's arms exports, but are joined by France, the United Kingdom, West Germany, and Italy, and more recently by Brazil, China, South Africa, India, and Israel. For the United States, arms exports totalled $7.7 billion in 1984, just under 4 percent of total exports, while the Soviet Union's $9.4 billion came to almost 12 percent of total exports.[20] For exporters, the arms trade has become a significant source of foreign trade earnings. However, this income will be short-lived if the economies of developing countries cannot support this investment in militarization, as appears to be the case: since the mid-1980s, arms exports have declined substantially.[21]

Indeed, for many weapons importers the arms trade has already become an economic drain. Third World military expenditures grew an average of 7 percent per year between 1960 and 1981, compared with an average growth of 3.7 percent for industrial nations. The Third World's share of global military spending rose from one-tenth to one-fifth in the same time period.[22] The repercussions of this dramatic increase include stalled socioeconomic development. South Korea stands almost alone as a nation that has managed to achieve a higher level of development despite high military spending.[23] Meanwhile, many developing countries remain trapped in regional arms races, trying to match the advanced missiles and aircraft purchased by would-be adversaries instead of devoting those funds to development.

Unfortunately, high levels of military spending have not bought security for nations, nor for the people to whom it matters the most. The average civilian is less secure now than at any time in human history. Since World War II, some 150 armed conflicts, primarily in developing countries, have killed about 20 million people and turned millions more into refugees.[24] In the 1950s, there were an average of 9 outbreaks of conflict each year; by the 1980s, the figure was 14 outbreaks each year.[25] In the 1950s, 52 percent of wartime casualties were civilian, while by the 1980s civilians accounted for 85 percent of casualties.[26] Finally, although they have remained unused thus far, the 50,000 weapons in the nuclear arsenals of industrial nations represent the greatest potential threat to the life and livelihood of every person on the planet.

Security, Resources, and the Environment

In its report, *Our Common Future*, the World Commission on Environment and Development concluded that "the whole notion of security as traditionally understood—in terms of political and military threats to national sovereignty—must be expanded to include the growing impacts of environmental stress—locally, nationally, regionally, and globally. There are no military solutions to 'environmental insecurity'."[27]

A preoccupation with military factors distracts attention from the underlying causes of insecurity and conflict. Tracing the sources of human conflict is difficult, and it is rare that a single reason can be isolated as the cause of any violent confrontation. Ideology, religion, politics, social injustice, and individual personalities can all contribute to wars and civil disturbances. However, today's decision-makers rarely even consider one of the most fundamental reasons for conflict— depleted or degraded resources.

Scarce and Degraded Natural Resources

Threats to natural resources can pose as great a danger to national security as military threats. Scarcity, depletion, and degradation of critical resources such as fresh water, ocean fisheries, energy supplies, and agricultural land are often a source of intense competition and conflict.

Fresh Water   All countries depend on fresh water supplies for drinking and agricultural production. Yet as world populations grow, fresh water supplies are dwindling. Between 1940 and 1980, global water use doubled; it is expected to double again by the year 2000, with two-thirds of projected water use going to agriculture. Already, 80 countries, with 40 percent of the world's population, suffer from serious water shortages.[28] (See Chapter 9.) All nations try to ensure the flow of fresh water to people and crops. Unfortunately, many underground aquifers and more than 200 river systems are shared by at least two countries, and increasing demands for those resources provide the seeds of conflict.[29] Both the scarcity of water supply and the pollution of available water supplies can lead to conflict over water resources.

Arab-Israeli competition in the water-poor Middle East has often been intensified by conflict over water supplies. In 1965, Israel diverted 100 million cubic meters of water from watersheds and rivers originating in neighboring Arab states after negotiations on water rights failed. In retaliation, the Arab nations attempted to divert the headwaters of the affected rivers to rivers that did not flow through Israel. The Israelis launched preemptive strikes on Syrian construction equipment to halt the Arab diversion; the action involved military aircraft on both sides.

While this dispute was only one aggravating factor among many in the 1960s, water supplies could soon become a more significant element in Middle-East foreign relations. Forecasts indicate that Israel is likely to face a water deficit of between 300 million and 400 million gallons yearly in the 1990s.[30] Egypt's foreign minister warned in 1985 that "The next war in our region will be over the waters of the Nile, not politics."[31] Similar disputes have occurred over the waters of the Mekong and Ganges in Asia, the Rio de la Plata and Parana in South America, and the Rio Grande in North America. The Middle East probably has the greatest potential for water-related conflicts, due to the scarcity of water, the many additional bases for conflict, and the region's high degree of militarization.[32]

Ocean Fisheries   Marine fisheries are nearly as important as fresh water to seafaring nations that depend on fish for food and as export commodities. Overfishing has depleted this resource in recent years. The world fish catch in 1982 reached 75 million tons, with 70 percent of the total destined for human consumption.[33] By the year 2000, the world demand for fish is expected to reach 110 million tons, which would exceed the maximum sustainable yield estimated by the United Nations Food and Agriculture Organization to be 100 million tons.[34]

The United Nations Convention on the Law of the Sea (UNCLOS), completed in 1982, tried to define fishing rights and regulations for coastal nations. But overlapping national fishing zones, the inability of some nations to enforce their rights, and the migration of fish species across national zone boundaries have undermined the success of the Convention. The conflicts over fishing rights that preceded UNCLOS are likely to continue and may grow worse, as human populations grow and mismanagement of fisheries continues. In this century, marine fishing rights have been a source of confrontations between Japan and South Korea, the United Kingdom and Argentina, and China and Japan.[35]

The "cod wars" between Britain and Iceland illustrate the divisiveness and potential for confrontation over scarce marine resources. Between the 1950s and 1970s, Iceland's efforts to extend its exclusive fishing zone interfered with Britain's long-distance fishing fleet, which depended primarily on Iceland's waters. When Britain sent ships from the Royal Navy to protect British trawlers in Icelandic waters, Iceland—despite its shared membership with Britain in the NATO alliance—excluded British military planes from its territory. A few years later, Iceland also initiated steps to remove U.S. military forces, in hopes of pressuring the United States and other NATO countries to take its side in the cod war. In 1975, the situation boiled over with Icelandic gunboats firing blanks and live rounds near British trawlers, the severing of diplomatic relations, and the collision of Icelandic gunboats with British frigates deployed to uphold British interests.[36] The volatility of resource issues, and the national security priority assigned by Iceland to resource protection, were both made clear. Iceland perceived its fisheries as more essential to national security than even its military alliances.

In Southeast Asian waters, overfishing led to violent conflict between trawlers and peasant fishers in the late 1960s, after modern trawling was first introduced there. The catches of the large trawlers quickly surpassed the sustainable yield of the fishery, and both the quantity and quality of the fish catch declined. The peasant fishers of Malaysia and Thailand, with living standards already below the poverty line, were hurt badly by the decline. Many gave up their way of life and moved to the cities, but others began attacking the fishing trawlers. Forty vessels were rammed or set ablaze in the late 1960s and early 1970s, 100 violent clashes took place, and 28 people were killed. In addition, some of the peasant fishers turned to piracy and pillaged boats filled with Vietnamese refugees on the Gulf of Thailand.[37]

Energy Supplies   Demands for energy resources have already increased the danger of United States involvement in regional conflicts, as exemplified by the U.S. Navy's presence in the Persian Gulf during the Iran-Iraq war to ensure the flow of oil to Western nations. The United States spent nearly $50 billion in 1987 to "protect" the Gulf area.[38] Energy-related conflicts will probably multiply in the future as the supply of nonrenewable energy resources continues to decline. (See Chapter 11.) As long as the scarcity of energy and other critical resources is an underlying cause of tension, military confrontations are likely.

Agricultural Land   When inequitable land distribution and government neglect of rural areas is combined with the loss of natural resources such as topsoil, guerrilla uprisings and other forms of organized conflict can develop. In Peru's Andean region, deforestation and poor agricultural policies fill the rivers with some 600 million cubic meters of sediment each year, which could otherwise provide productive topsoil for at least 300,000 hectares of land. The "Shining Path" guerrilla movement in that area has forced the Peruvian government to use growing numbers of troops to maintain order. The conflict resulted in at least 5,000 deaths between 1980 and mid-1984.[39]

In some cases, agricultural decline leads to tensions severe enough to precipitate the overthrow of governments. In Ethiopia, traditional farming areas in the highland regions lost a billion tons of topsoil each year in the early 1970s. Combined with an outdated economic structure in the countryside, soil loss led to a drop in agricultural production and then to food shortages. The ensuing disorders culminated with the overthrow of Emperor Haile Selassie in 1974. Civil war and the degradation of soils both continued in the 1980s, along with some of the worst famines in recent history. Ethiopia spent $225 million on the military between 1976 to 1980, and $500 million on relief efforts during the 1985 famine. In contrast, the United Nations Anti-Desertification Plan estimates that timely expenditures of only $50 million a year could have effectively addressed the underlying causes of soil erosion in Ethiopia.[40]

The degradation of natural resources that are already scarce has the potential to inflame existing tensions. The pollution of a river by one nation can anger other countries that border the river; when a Swiss chemical manufacturer accidentally released pesticides into the Rhine River—a source of water and fish for West Germany and the Netherlands—Swiss relations with these countries were strained. Likewise, relations between the United States and Canada have been hurt by Canadian anger over acid rain generated by U.S. industrial production and energy use. There are also tensions over acid rain between the United Kingdom and Scandinavia. The links between environmental degradation and political conflict are readily apparent in some developing nations where natural resources have been severely degraded or depleted.

Population Pressures and Other Complicating Factors

Finding connections between conflict and the environment is easy, but to propose a direct causal link between the two would obscure many other dimensions of the problem and oversimplify the nature of the conflict. Environmental degradation is never likely to be the sole cause of strife, and even if environmental factors can be pinpointed as a principal source of conflict, governmental priorities and factors such as population growth, migration, and land distribution will always play a role.

El Salvador provides a vivid illustration of the destruction that can result when natural resources are depleted. Overpopulated, suffering from the worst soil erosion in Central America, and maintaining a large base of poor, mostly landless people, El Salvador went to war with neighboring Honduras in 1969 primarily to obtain more land.[41] A decade later, a protracted civil war began in El Salvador which even today shows no signs of resolution. By the mid-1980s, 60,000 citizens were dead, 200,000 displaced people roamed the country, and thousands had fled overseas. One of every two workers had no satisfactory means of sustenance.[42] According to a report by the U.S. Agency for International Development, "the fundamental causes of the present conflict are as much environmental as political, stemming from problems of resource distribution in an overcrowded land."[43]

Although conflict can be a result of misuse and depletion of natural resources (and poor resource management could therefore be considered a primary security threat), environmental concerns rarely play a role in policy decisions about conflict. In 1984, a commission on Central America led by Henry Kissinger recommended a number of strategies for bringing peace to the region, but failed to include a single proposal for addressing the root problems of overpopulation and environmental degradation. Reflecting this attitude, the United States provided $605 million in aid to El Salvador in 1987—over half the country's national budget—but $3 went to the military effort for every $1 directed at the causes of the conflict.[44]

Environmental Refugees

An estimated ten million people have been forced to leave their homes since World War II because they could not obtain enough food, water, or other basic necessities to survive. Such environmental refugees both suffer from insecurity and cause it. They are the intermediate step between poor resource management and conflict—an active, moving force of discontent that can threaten the security of governments and nations.[45]

Bangladesh has long been a source of environmental refugees. More than 104 million people there are crowded into a country the size of Wisconsin, trying to eke out a living on land that is rapidly deteriorating. Bangladesh is almost completely deforested, leaving its inhabitants with vulnerable topsoil and little firewood. As a result, many desperate Bengalis seek livelihoods in the neighboring Indian state of Assam. The inhabitants of Assam resent the new competitors for land and other resources, and their anger often flares into violence. In 1983, 3,000 people were killed in armed conflicts frequently characterized as a religious struggle between Moslem Bengalis and Hindu Assamese. But the fighting would never have taken place if the Bengalis had not been forced by environmental deterioration to flee their own homes.[46]

Environmental refugees often move to cities, where they become a disaffected underclass that can destabilize governments. In Iran, a combination of overcultivation and inequitable land distribution in the 1970s made it impossible for 78 percent of the rural peasants to acquire sufficient land for subsistence living. Accelerated erosion led to declining yields, and by 1977 the cost of producing a ton of wheat or rice exceeded its value on the urban retail market. Migration to the cities increased tremendously, and, once there, many landless laborers joined revolutionary movements and participated in the overthrow of the Shah in 1979.[47] Massive rural-to-urban migration has become a standard feature of most Third World nations, making many of them vulnerable to the same kind of instability experienced by Iran. Egypt, Morocco, Tunisia, Zambia, Pakistan, Brazil, Bolivia, and Haiti have all experienced major urban riots.[48]

Environmental refugees also can be a problem in industrial nations not suffering debilitating environmental deterioration. Many people fleeing their impoverished homelands head to wealthier nations where economic prospects look brighter. More than 500,000 El Salvadorans currently reside in the United States, either legally or illegally, amounting to 8 percent of the total Salvadoran population.[49] The exodus is reminiscent of more than a million Haitian boat people who fled their island where soils in some places had eroded down to bedrock, to seek refuge and opportunity in the United States.

Such mass migrations may be seen as a security threat by the United States, which has increased spending to protect its borders. Of course, it is often difficult to distinguish between political refugees fleeing wars and environmental refugees trying make a living, especially since military conflict is frequently linked to environmental deterioration. Those fleeing from environmental destruction, however, may never find their lands habitable, while those fleeing from war may be able to return once peace is reestablished. This problem of migration away from untenable, impoverished regions of the world underlines the stake of wealthier nations in preserving the environmental well-being and resource productivity of these regions.

Environmental Threats to Industrial Nations

Both developing countries and industrial nations face environmental threats to their security. Many industrial countries are threatened by the prospect of environmental damage. For example, global warming caused by the build-up of industrial gases in the atmosphere could cause billions of dollars of damage from rising sea levels alone. (See Chapter 12.) The Netherlands currently spends a larger percentage of its GNP to maintain dikes and sea walls than the United States spends on military defense; just one four-kilometer barrier completed there in 1986 cost $3.2 billion.[50] If the threat of global warming materializes, the United States would be severely pressed to afford protection for coastal cities such as Boston, New York, Miami, Los Angeles, and San Francisco. Furthermore, if global warming reduced rainfall in the U.S. farm belt, the United States might be confronted with urban food shortages and its own class of environmental refugees. And because the United States is the world's largest food exporter, a climate-related disruption of U.S. agriculture could have vast security implications worldwide.

Security and the Economy

Current emphasis on military security not only ignores the underlying causes of insecurity and conflict, it diverts resources from important components of long-term security, especially economic productivity. Economic well-being is considered the best measure of national well-being, and an unstable or faltering economy can weaken a country politically and lead to civil disturbances—even to the overthrow of governments. Heavy investment in a nation's military machine drains resources away from other segments of the economy, leaving it vulnerable to the type of internal violence and rebellion that military forces often have to combat.

Third World Debt

Developing nations now owe more than $1.3 trillion to developed nations, and that debt is growing by about $60 billion a year. With annual interest payments approximating $80 billion, the net capital flow from rich nations to poor has reversed, as developing countries now send $30 billion more to their creditors each year than they receive. (See Chapter 4.) Economic policies are geared to paying these debts instead of to achieving socioeconomic development. World Bank President Barber Conable recently pointed to one sign of the deterioration: middle-income developing nations reduced their imports between 1980 and 1985 from $165 billion to $110 billion; normally, imports would have been expected to rise over that period to $220 billion.[51]

Because of the large military expenditures by many Third World nations, and because military spending contributes relatively little to economic growth and, except for arms exports generates little foreign exchange, a developing country's outlays for its own military establishment can be an important source of its foreign debt. And as Michael Renner of Worldwatch Institute has noted, to the extent that military expenditures contribute to indebtedness, they actually undermine national security.[52]

For many Third World nations, debt is becoming harder to pay back as each year passes. Sudan is one of 14 nations in which farmland productivity is lower today than it was a generation ago, forcing the nation to import food and decreasing its ability meet its $9 billion debt. The collapse of Peru's anchovy fishery, the source of its most important export commodity, combined with heavy topsoil loss has made it far more difficult for the nation to contend with its debt burden of $14 billion. According to Lester Brown of Worldwatch Institute, nations such as Sudan and Peru, and larger debtors such as Brazil and Mexico, have crossed a threshold beyond which debt payments can only be met by consuming basic stocks—natural resources and other economic inputs. That is equivalent to spending capital and sacrificing both its current income and its potential for future use.[53]

Along with the deterioration of natural resources, debt has contributed to political instability in many developing nations. Mexico and Brazil—under pressure from the International Monetary Fund—have tried to impose economic austerity measures to help cope with their debt burdens, only to find the associated economic decline worsened economic and social conditions. A dissatisfied populace reduces national stability. Observing the unrest that plagues Peru, President Alan Garcia Perez declared, "We are faced with a dramatic choice: it is either debt or democracy."[54]

Even though many Third World economies are declining under the pressure of external debts, most find the funds to supply their armed forces. Ethiopia, for example, has assembled one of the largest armies in sub-Saharan Africa, spending 42 percent of its budget for military purposes. Only the oil-rich nations of the Middle East have been able to build vast military establishments without incurring crippling debts. However, they too have made sacrifices by trading the opportunity to build a sustainable economic infrastructure for the prestige of being a regional military power. One *New York Times* reporter noted the "technological incongruity" in many Third World nations where soldiers receive guns and bullets while farmers lack hoes and seeds, and jet fighters fly over fields plowed by oxen.[55]

Second World Economic Stagnation

While the Soviet Union and Eastern bloc countries have generally lagged behind Western Europe in technological advancement, their economies have not always appeared as sluggish as they do today. From the early 1950s until the late 1970s, the Soviet economy grew at a healthy average rate of 5 percent a year. But in the 1980s, industrial growth slowed dramatically, with production of principal commodities such as wheat and oil falling off or stagnating. Part of the decline can be traced to environmental problems such as land degradation and growing water scarcity. Another portion can be attributed to failures of the Soviet centrally planned, state-controlled economy, which does not work as well in a modern, diversified, increasingly consumer-oriented society as it did earlier for the development of heavy industry.[56] However, much of the decline can be linked directly to massive military spending.

Soviet military budgets rival those of the United States in absolute size, but constitute an even larger portion of the Soviets' smaller economy. This heavy emphasis on the military has diverted leadership, innovation, and resources from other sectors of the economy. Even though the Soviets produce some of the more advanced military technology available, they depend on western imports for auto manufacturing, computers, and oil extraction, and many of their consumer goods are so shoddy that they cannot compete on the world market.[57] This state of affairs contrasts sharply with the promise of the 1950s when the Soviet Union was a potential world leader in the production of machine tools—a promise that never materialized. The Soviet Union's continuous diversion of resources to the military has condemned it to fall behind newly industrialized countries like Brazil and South Korea in technological advancement.[58]

First World Economic Decline

Beginning at the end of the Carter Administration, and especially during the Reagan Administration, the United States experienced an unprecedented surge in military spending. With more than $2 trillion invested in the course of eight years, the Reagan military buildup was the largest the nation had ever seen during peacetime. While the program was criticized for lack of direction and mismanaged procurement, the most significant complaints centered on its effects on the national economy.

The major increases in military spending, combined with tax reforms dramatically increased the U.S. national debt. The debt doubled in just the first five years of the Reagan Administration, rising from $914 billion in 1980 to $1,841 billion in 1985. Over the same five years, annual military spending increased by $110 billion, while other major federal programs such as health and agriculture increased by only $11 billion and $15 billion, respectively. The debt has been called a mortgage on the future of the U.S. economy, and it required interest payments of $143 billion in 1986. Interest payments continue to rise as federal budget deficits remain high. The debt has also led to an overvaluation of the dollar, hurting U.S. exports on the global market and contributing to record trade deficits.[59]

Figure 15.1 on the following page shows the increase in U.S. military spending beginning in the late 1970s and the growth in the U.S. federal and trade deficits during the 1980s.

Military spending contributes to the trade deficit by siphoning capital away from investment in new industrial facilities. U.S. equipment and manufacturing techniques have become outdated and caused a decline in areas of traditional economic strength. The United States is hardly able to produce such basic goods as cars, steel, and machine tools efficiently enough to compete abroad, or even at home. Two million workers lost jobs in these areas between 1981 and 1984—a period of moderate economic expansion.[60]

The U.S. emphasis on military spending has also drawn brainpower and financial resources away from new high-technology industries. Between 25 and 30 percent of U.S. scientists and engineers are involved in military research and development, creating occupational shortages in the civilian sector. In 1985, the National Science Foundation found that 28 percent of the firms they surveyed reported shortages of scientists and engineers, despite that year's excess economic capacity and high unemployment.[61] By 1986, the United States showed a trade deficit in high-technology goods of $2.6 billion, in contrast to a surplus in 1980 of $27 billion.

A good example of the impact of military spending on high-technology investment is its effect on technology to improve energy efficiency. The U.S. Department of Energy (DOE) devotes 80 percent of its budget to the military, and only $89 million—less than 1 percent—to energy conservation. But DOE research and development (R&D) for energy conservation has repeatedly provided a return to taxpayers of as much as $4,400 for each $1 invested. Further, because military interests receive priority, the United States is losing its preeminence in R&D on energy efficiency, and countries such as Japan and West Germany have become world leaders in areas of commercial technology initially developed in America.[62] West Germany and Japan now use energy about twice as efficiently as does the United States. (See Chapter 11.)

Defense contractors and military officials have often claimed that military R&D creates technological spin-offs that can be marketed commercially. The claim is true to an extent, but much military technology remains classified, and spin-offs become less common as military technology grows more and more exotic. Fewer than 1 percent of the 8,000 patents produced from navy-sponsored R&D are licensed commercially, while 13 percent of the U.S. Department of Agriculture's patents are licensed for commercial use. Such figures suggest that investing in military R&D is not an effective way for the federal government to strengthen high-technology industry.[63] Even Simon Ramo, a co-founder of TRW, Inc., a major U.S. military contractor, maintains that if the United States had spent its last 30 years of military R&D funds in areas of science and technology that promise the greatest economic progress, the United States already would have reached an economic level that it will not now attain until the year 2000.[64]

Military expenditures are inherently inflationary, since military industry employs people and pays them, but

Figure 15.1
U.S. Military Expenditures and Economic Deficits

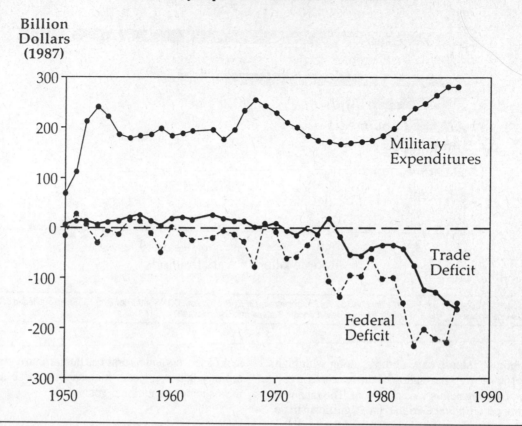

Source: *Economic Report of the President* (Washington, DC: U.S. Government Printing Office, 1988), as published in Michael Renner, *National Security: The Economic and Environmental Dimensions,* Worldwatch Paper 89 (Washington, DC: Worldwatch Institute, May 1989), p. 23.

provides no new goods for them to buy. Instead, they buy existing consumer goods, and as the demand for those goods rises, prices also rise. This is true especially when increased military spending is financed by borrowing, since tax hikes will not absorb any of the new money generated in salaries. Even if the government seeks to provide jobs through military programs, military spending is not an efficient way to do so. Many studies have shown that military spending creates significantly less employment than alternative forms of federal spending.[65]

Military spending diverts resources from investments in the nation's economic future. After spending $2 trillion on defense in the 1980s, the United States has little new infrastructure, no renovated education system, and no radical innovations in R&D designed to meet pressing world needs. Instead, the country has financed submarines, aircraft carriers, missiles, personnel carriers, and the first phases of the Strategic Defense Initiative ("Star Wars"), which threatens to consume more national resources than any previous military programs.

The ultimate trade-off for the United States is the gradual abdication of world economic leadership to Japan, which allocates only about 1 percent of its GNP to military programs, compared with about 7 percent for the United States. In 1986, only 0.7 percent of Japanese public and private expenditures for research and development went to military projects, while 32 percent of U.S. R&D spending was in the military sector. (See Table 15.3.)

Table 15.3

Military and Civilian R&D Expenditures in Selected Countries, 1986

Country	Military billion $	Civilian billion $	Military Share %
United States	37.3	79.4	32.0
United Kingdom	3.5	8.5	29.3
France	3.5	14.6	19.5
West Germany	1.4	26.1	5.1
Japan	0.3	44.0	0.7

Source: Michael Renner, "Enhancing Global Security," in Lester R. Brown et al., *State of the World 1989* (New York: Norton, 1989), p. 140.

Figure 15.2
Number of People Killed in Wars, 1945-85

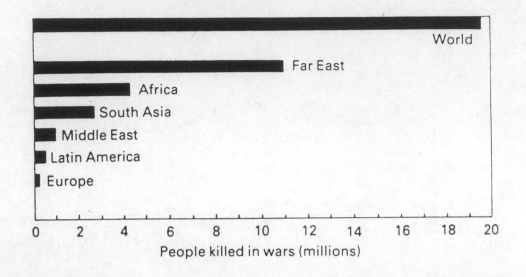

Source: Data from Ruth Leger Sivard, *World Military and Social Expenditures, 1985* (Washington, DC: World Priorities, 1985), as published in Essam El-Hinnawi and Manzur Hashmi, *The State of the Environment* (London: Butterworth Scientific, 1987), p. 143.

Japan's minimal defense expenditures, along with high rates of domestic savings, have allowed the nation to invest heavily in modernizing plants and equipment. The Japanese thus gain a competitive advantage and run a significant trade surplus despite being forced to import all of their oil and most of their raw materials. The United States, with an economy twice as large as Japan's and an immensely superior resource base of land, fuels, minerals, and forest products, may be unable to withstand Japan's challenge for leadership of the global economy.[66]

Military Threats to Security

The redefinition of national security to include environmental and economic threats is not meant to deny the existence of military threats. Such threats do exist and justify a prudent response. Since the end of World War II, in which at least 35 million people were killed, there have been some 150 armed conflicts, mostly in the Third World. This fighting has taken nearly 20 million lives (see Figure 15.2), and at least 12 of these conflicts have caused substantial environmental damage.[67]

But a new trend is developing. Aggression and military action are becoming less successful as means of achieving national goals. In the twentieth century, aggressors have won only about 4 out of every 10 wars, and in the 1980s that ratio has dropped to 1 in 10.[68] The U.S. experience in Vietnam, the Soviet presence in Afghanistan, and the grueling war between Iran and Iraq have accomplished little more than extreme devastation and loss of human life. Given its lack of success and its drain on human and natural resources, military action appears increasingly to be an ineffective means to pursue national goals.

Conventional Warfare

Military action often destroys economic and natural resources that nations need for their security. Economic production has been limited by conflict at least since the ancient Romans damaged Carthaginian agriculture by salting their fields. In recent times, the destructive potential of military action has grown enormously and spread over vast areas. World War II resulted in an average 38 percent reduction in agricultural productivity in 10 nations, though the loss was later recovered. The remnants of war, such as unexploded land mines, barbed wire, residues of chemical defoliants, and wrecked vehicles and aircraft, can affect ecological balances by disturbing soil, harming flora and fauna, and contaminating water supplies.[69]

Modern chemical and biological weapons can cause death indirectly, by infecting air or water, or they can kill or immobilize people directly. Iraq has been accused of using nerve gas and mustard gas repeatedly against the Iranians and against their own separatist Kurdish population. Chemical weapons, called the "poor man's atomic bomb," require relatively little technical know-how and financial investment. They are appealing to nations such as Libya, which has allegedly been constructing a large chemical weapons facility. In spite of an international ban on the production of chemical and biological weapons, the United States resumed production of these weapons in 1986.

Economic deterioration, military destruction, and environmental degradation are mutually reinforcing. Wars destroy fields and forests that then—as in Ethiopia and Sudan—can no longer provide food or fuelwood. Wars destroy cities and factories, which can no longer produce industrial goods. Iran and Iraq have frequently attacked each other's oil terminals and tankers to cripple production capacity. Wars occupy the productive energy of a nation's people and kill many of them, including younger and more skilled individuals and workers. An estimated 1 million people died in the Iran-Iraq War, and another 1.7 million were injured.

Since 1945, almost all military conflict has taken place in the Third World. In a frantic effort to ensure security against regional rivals, countries are acquiring ever more sophisticated and lethal weaponry. Laser-guided "smart" weapons including missiles, bombs, and torpedoes are now entering arsenals in the Middle East, Asia, and Africa, in addition to the chemical weapons mentioned earlier. State-of-the-art missile guidance systems and multiple-stage rockets may not be far behind.[70] If these trends in weaponry continue, Third World nations will face the threat of even greater devastation.

Environmental Warfare

Potentially more dangerous than conventional warfare, environmental warfare's goal is the systematic destruction of natural resources on which the enemy depends. This type of warfare could include contamination of drinking water or destruction of dams and dikes to flood arable land and deplete water supplies.

The use of environmentally dangerous chemicals in warfare is increasing. Chemical defoliants used by the United States in South Vietnam devastated crops, destroyed 1,500 square kilometers of forest, and damaged another 15,000 square kilometers. People in exposed areas ingested toxic herbicides, which have since led to cancer, spontaneous abortions, and birth defects.[71]

Advanced technology makes other kinds of environmental damage possible, such as cloud seeding to increase rainfall and cause flooding of enemy territory. In the future, it might even be possible to alter the directions of hurricanes, or reduce the stratospheric ozone over specific land areas to expose designated populations to harmful ultraviolet radiation.[72]

Nuclear War

A major nuclear exchange would be the ultimate example of environmental warfare. There is no doubt that a full-scale nuclear war would cause environmental devastation on an unprecedented global scale, kill hundreds of millions or even billions of people, and create an unlivable world for months or years.[73] Currently, the superpowers and their allies have assembled nuclear arsenals of 50,000

nuclear warheads, with a combined destructive potential of about 20,000 megatons, nearly 70 times the firepower needed to destroy all the world's large and medium-sized cities. For comparison, it would take 80 bombs like the one dropped on Hiroshima in 1945 to create 1 megaton of explosive power, and all the firepower expended in World War II amounted to only 3 megatons.[74]

The immense power involved in a full-scale nuclear exchange between the superpowers combined with the side-effects of nuclear explosions would produce a wide spectrum of environmental damage. Studies of different scenarios for a large-scale nuclear war, involving the use of weapons with a yield between 5,000 and 10,000 megatons, estimate that some 750 million people could be killed by the effects of blast alone. The combined effects of blast, fire, and radiation could kill about 1.1 billion, and another 1.1 billion could suffer injuries requiring medical attention.[75]

Also deeply disturbing are the probable climatic effects of nuclear war. The ash and soot rising from burning cities following a nuclear holocaust would be carried into the atmosphere, blocking 80 percent or more of the sunlight reaching the mid-latitudes of the Northern Hemisphere. This would result in an average temperature decrease of 5 to 20 degrees Celsius within two weeks. The loss of solar energy would reduce rainfall over the temperate and tropical latitudes by up to 80 percent. The combination of cold temperature, dryness, and lack of sunlight in this "nuclear winter" would cripple agricultural production across the northern half of the globe and destroy its ecosystems. Resultant food shortages could put a majority of the world's population at risk of starvation.[76] (See Figure 15.3 on the following page.)

It is also likely that the nitrogen oxides produced by a large-scale nuclear war would deplete the Earth's protective ozone layer by up to 50 percent, exposing the survivors of the war to damaging amounts of ultraviolet radiation.[77]

Many recent studies confirm that the indirect effects of a nuclear war could be greater than the direct effects, and that noncombatant and combatant countries alike would suffer unprecedented damage.[78]

Most of the scenarios studied assume that the nuclear exchange would take place between the Soviet Union, the United States, and their allies. But a nuclear winter with dire consequences could also be triggered by the explosion of a much smaller number of warheads.[79] The greatest danger of nuclear war may lie in the Third World, where several nations are seeking to obtain their own nuclear weapons despite the existence of the Non-Proliferation Treaty banning the transfer of nuclear-arms technology. Only five nations—the United States, the Soviet Union, Great Britain, France, and China—are declared nuclear powers, but several other countries, including Israel, South Africa, India, and Pakistan, are believed to have built weapons or to have the capability to do so. In 1987, Brazil announced it

Figure 15.3

Vulnerability of Human Population to Loss of Food Production Due to Nuclear War

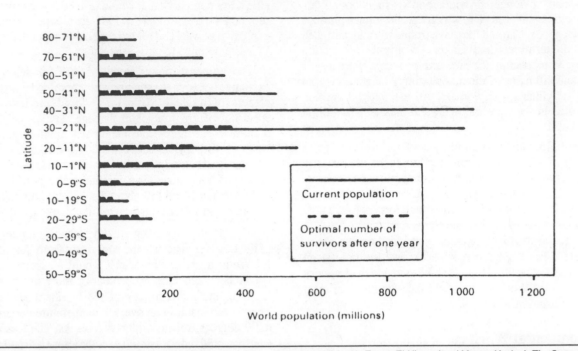

Source: Data from L. Dotto, *Planet Earth in Jeopardy* (Chicester: J. Wiley, 1986), as published in Essam El-Hinnawi and Manzur Hashmi, *The State of the Environment* (London: Butterworth Scientific, 1987), p. 150.

was capable of enriching uranium but would not build nuclear weapons. Other states, including Argentina, Iran, Iraq, and possibly Libya, are on the nuclear threshold.[80] The potential for proliferation of nuclear weapons and other powerful forces of destruction around the world threatens the security of all nations for the foreseeable future.

Arms Control

Since 1946, numerous approaches to limit the arms build-up have been made by East and West, and 16 international agreements relating to nuclear weapons have been signed. These include agreements to prevent the spread of nuclear weapons, to reduce the risk of nuclear war, to limit nuclear testing, and to limit stocks of nuclear weapons themselves.

Though these agreements may have had some moderating effect on the arms race, they have not stopped the production of nuclear weapons, or noticeably curbed it. For example, the "ceilings" on warheads included in the Strategic Arms Limitation Treaty (SALT) have allowed enormous growth in weapon stockpiles, creating at best a regulated arms race.[81] In spite of three decades of negotiations on nuclear weapons, there has been no agreement on a comprehensive test ban or on measures to end the nuclear arms race. The three agreements negotiated by the United States and the Soviet Union between 1974 and 1979 to limit

strategic nuclear weapons have not been ratified, but have essentially been adhered to in practice.

Recently, however, there has been encouraging progress. The U.S.-Soviet summit meeting of December 1987 marked a significant achievement in arms reduction, in contrast to earlier efforts to control or limit arms production. The superpowers not only reached an accord to eliminate an entire class of nuclear weapons, but took encouraging steps towards negotiating a substantial reduction in strategic nuclear weapons.[82] Continued improvements in U.S.-Soviet relations could enhance chances for meaningful arms reduction and diversion of resources to sustainable development priorities.[83]

Security and Sustainable Development

Since sources of conflict are often related to environmental and economic deterioration, and since conflict itself causes environmental and economic deterioration, a cyclical cause-and-effect process can begin when war breaks out or resources are degraded. Such a cycle can feed on itself and move toward a collapse in which people try to wrest from their neighbors the scarce resources they need. The only way out of the cycle is to seek solutions to the underlying problems.

Table 15.4
Rough Estimates of Additional Expenditures to Achieve Sustainable Development, 1990-2000

(billion dollars)

Year	Protecting Topsoil on Cropland	Reforesting the Earth	Slowing Population Growth	Raising Energy Efficieny	Developing Renewable Energy	Retiring Third World Debt	Total
1990	4	2	13	5	2	20	46
1991	9	3	18	10	5	30	75
1992	14	4	22	15	8	40	103
1993	18	5	26	20	10	50	129
1994	24	6	28	25	12	50	145
1995	24	6	30	30	15	40	145
1996	24	6	31	35	18	30	144
1997	24	6	32	40	21	20	143
1998	24	7	32	45	24	10	142
1999	24	7	32	50	27	10	150
2000	24	7	33	55	30	0	149

Source: Lester R. Brown and Edward C. Wolf, "Reclaiming the Future," in Lester R. Brown, et. al., *State of the World 1988* (New York: Norton, 1988), p. 183.

To find such solutions, we must redefine security to encompass the sustainable development of economic and natural resources. People have not always regarded military security as the most critical national priority; there is no inherent reason why other aspects of security should not receive comparable attention. The price to be paid for sustainable management of global resources is not as high as one might expect, at least not when compared with the accelerating costs of the global military budget.

Resource Reallocation

Worldwatch Institute has proposed a budget for the 1990s that devotes funds to sustainable development efforts in six priority areas: protecting topsoil, reforestation, slowing population growth, raising energy efficiency, developing renewable energy, and retiring Third World debt. Surprisingly, the programs needed in these areas could cost less than a sixth of the current global military budget.[84] (See Table 15.4.)

The problem of soil erosion is a serious concern for rich and poor nations alike. In the early 1980s, the U.S. Department of Agriculture (USDA) and American farmers were spending a billion dollars each year to control erosion on U.S. croplands, but were still losing 2 billion tons of soil more than the loss that could be tolerated without a reduction in productivity. In 1985, as part of the Food Security Act, the U.S. Congress established a Conservation Reserve to control farm production and reduce soil loss by converting 16 million hectares (40 million acres) of highly erodible cropland to woodland or grassland. USDA pays farmers, on

average, $48 an acre as compensation for crops not grown. Combined with the conservation measures already in place, the total cost of stabilizing soils in the United States will come to $3 billion a year once the program is fully implemented in 1990. Applying the U.S. program worldwide would cost an estimated $24 billion each year to stabilize topsoil. That is less than the U.S. government paid farmers to support crop prices in 1986, and far less than the $940 billion spent worldwide on military budgets.[85]

A complementary program of global reforestation would also help to stabilize soils in countries such as India and Malaysia, where excessive deforestation is leading to massive erosion and floods. In addition, reforestation would provide fuelwood for the more than 1 billion people currently experiencing shortages of this vital resource, and well-managed forests could be used to satisfy demand for lumber, paper, and other forest products. Worldwatch estimates that to accomplish these objectives would require reforestation of 150 million hectares by the year 2000, at a cost of less than $60 billion over a ten-year period, assuming villagers and farmers would do most of the planting.[86]

Reforestation would slow the buildup of atmospheric carbon dioxide and help stabilize the earth's climate, as would development of renewable energy sources to replace fossil fuels and expansion of programs to improve energy efficiency. The energy improvements could be made for about $500 billion over the ten-year period.[87]

The pressures of population growth, which underlie many environmental and economic crises, could be tackled through education and literacy campaigns, provision of contraceptive services, efforts to reduce infant mortality, and financial incentives to limit family size. These programs to stabilize population would cost less than $300 billion over the period 1990-2000.[88]

Finally, Third World debt could be reduced to manageable levels with an investment of $300 billion in a debt retirement fund over the next decade, coupled with forgiveness of the part of the debt that creditors are unlikely to recover.[89]

In the first year, Worldwatch Institute's global security budget for the 1990s would divert $46 billion from military expenditures to sustainable development priorities. This investment would increase for the next few years and then level off at annual expenditures of between $140 and $150 billion toward the end of the decade, leaving between $755 and $760 billion for military needs. (See Table 15.4.)

Worldwatch suggests that such a shift from military to sustainable development priorities is not without precedent, noting that in less than ten years China cut its military budget by 10 percent and substantially increased investments in food production, reforestation, and family planning. Coupled with economic reforms, these shifts have helped China raise per capita food production by half and dramatically lower its birth rates. Worldwatch suggests that such a reallocation of resources throughout the world could yield the greatest progress toward achieving true global security.[90]

Practical Politics

Any radical reallocation of resources away from the military sector will be opposed by communities, businesses, organizations, and individuals that profit economically and politically from the present system, or by those whose livelihood is at stake. Opponents can be expected to include military and civilian personnel whose careers and prestige are linked to the established military institutions and system, workers and investors in the defense industry whose salaries and profits come from weapons manufacturing, and politicians whose constituencies profit from military spending. Such constituencies should not be ignored, but conversion to non-military economic activity need not be painful. The Defense Economic Adjustment Act introduced in the U.S. Congress by Representative Ted Weiss would mandate the formation of local alternative-use committees in U.S. military plants, bases, and laboratories. The committees would draw up blueprints for civilian product development and marketing, as well as for the retraining of workers.[91] The change from a wartime to a peacetime economy has been accomplished quickly in the past. At the end of World War II, 30 percent of the U.S. gross national product was transferred from military production to civilian industries. Under today's different economic conditions, however, conversion presents more of a challenge, and competent and extensive advance planning is a necessity.[92]

Other recent legislation that would direct human resources away from military defense includes a bill that would provide educational benefits and training for Peace Corps candidates who agree to serve three years in the Peace Corps after graduation. A successful candidate who has completed two years of study at an institute of higher education would be eligible to receive benefits covering the costs of the remaining two years of study as well as appropriate Peace Corps training in return for a three year commitment to service in the Corps. If passed, this legislation would offer students an alternative to the existing Army and Navy Reserve Officers Training Corps (ROTC), and a chance to contribute to sustainable development and global security through Peace Corps service. While the Peace Corps works within developing countries to enhance self-sufficiency and promote social and economic development, the United Nations offers a forum for airing grievances between nations, and can provide the means to help mediate disputes, resolve conflicts, and preserve peace. A strengthened peacekeeping role for the United Nations could lessen national concerns about vulnerability to hostile neighbors and encourage demilitarization. The United Nations has gained credibility as a peacemaker over the past several years, especially through its roles in the Iran-Iraq cease-fire agreement and the Soviet withdrawal from Afghanistan. Its peacekeeping troops have proved to be financially efficient—annual outlays for seven current missions are currently only $380 million. They could rise to $2 billion if forces are sent to Kampuchea, Western Sahara, and southern Africa, but would still amount to less than one day's expenditure on the arms race.[93] A strengthened United Nations could help all nations seek effective political and diplomatic solutions rather than military ones.

Successful arms reduction agreements, for both conventional and nuclear weapons, can make cuts in military spending politically more attractive. A treaty limiting sales of conventional arms to the Third World, for instance, would accomplish such a goal, and could stimulate economic development while keeping destabilizing weapons such as long-range missiles and advanced aircraft out of the Third World.[94]

New satellite technology would make possible the creation of an international satellite monitoring agency to independently verify arms agreements, discourage clandestine missile tests, and deter surprise attacks. Such a monitoring system could greatly increase international stability and make arms races less tempting. Start-up and operating costs for the agency would be less than one percent of the world's current military expenditures. Unilateral initiatives, such as the Soviet Union's recent 18-month moratorium on nuclear testing, can help expedite negotiations and set precedents for future agreements.[95]

Hopeful Horizons

Despite the grim presence of an ongoing arms race and excessive military expenditures around the world, there are reasons to hope that perceptions of the military's position among national priorities is changing. In Argentina, military governments increased defense expenditures from 1.5 percent of GNP to 4 percent in the late 1970s and early 1980s. But a new president elected in 1983 capitalized on the ill-fated Falkland Islands war to cut arms outlays to half their 1980 peak, shifting the resources to social programs. Peruvian President Garcia, taking office in 1985, issued a call to halt the regional arms race and announced his commitment to reduce military spending, demonstrating his resolve by cancelling half of an order for 26 French Mirage fighter planes.[96]

China has been the most vigorous in redirecting resources away from the military sector. Between 1972 and the present, China has lowered the military share of the economy from about 14 percent of GNP to around 4 percent. In 1985, the government cut the 4 million member armed forces by one-fourth, which reduced the number of people in the military worldwide by 4 percent. About 38 percent of China's military industrial capacity has been shifted to the manufacture of civilian goods, and that transfer should rise to 50 percent by the year 2000.[97]

Leadership on significant redistribution of global resources to the cause of sustainable development must come from the superpowers and their allies. Those countries together are responsible for most of the world's military spending. Here, too, there is reason for hope. In December 1988, Soviet President Mikhail Gorbachev announced to the United Nations his decision to withdraw several thousand troops from forward positions in Eastern Europe, declaring that "one-sided reliance on military power ultimately weakens other components of national security."[98] This dramatic announcement followed a 1987 Intermediate-Range Nuclear Forces Treaty with the United States, which removed from Europe both superpowers' stocks of an entire class of missiles. The much-needed redefinition of security seems to have begun.

International Organizations and Security

Many international organizations, both intergovernmental and nongovernmental, play important roles in various aspects of global security, including defense, international law, agriculture, health, population, social and economic development, energy, and the environment. Major intergovernmental organizations include the United Nations and its specialized agencies,[99] and a number of regional organizations. The security-related activities of some of these organizations are summarized below.

United Nations Security Council has primary responsibility under the U.N. Charter to maintain international peace and security. The Council can authorize and deploy regional peacekeeping forces, which have played a significant role in the Congo (now Zaire), Cyprus, and the Middle East. In many situations, Cold War confrontations have prevented the Council from taking effective action.

International Court of Justice, the principal judicial organ of the United Nations, renders judgments on international disputes submitted to the Court in areas such as territorial rights, interpretation of treaties, and law of the sea. The Court has rendered final judgments in over 20 of the more than 50 cases submitted to it by states. In addition, international organizations have requested nearly 20 advisory opinions from the Court. In actual practice, most nations limit their adherence to Court rulings.

United Nations agencies and related bodies are active in several other areas important to global security.

Food and Agriculture Organization (FAO) helps countries improve nutrition and standards of living, increase food production, and improve food distribution.

International Fund for Agricultural Development (IFAD) grants loans to support agricultural development in Third World countries.

World Health Organization (WHO) assists governments in public health programs, sets international drug and vaccine standards, and promotes medical research.

United Nations Children's Fund (UNICEF) works to improve the health and welfare of children in developing countries.

United Nations Population Fund (formerly the U.N. Fund for Population Activities) promotes family planning programs and helps developing countries deal with their population problems.

United Nations Development Program (UNDP) helps developing countries increase the wealth-producing capabilities of their natural and human resources by providing experts and training local people.

United Nations Environment Program (UNEP) monitors changes in global ecological systems and promotes sound environmental practices by governments and in development activities. UNEP programs include "Earthwatch," an international monitoring network; an international register for toxic chemicals, an environmental law unit, action plans to curb desertification and other environmental deterioration; and regional seas programs to combat marine pollution.

United Nations Industrial Development Organization (UNIDO) provides assistance for the development, expansion, and modernization of industry.

The World Bank, an autonomous specialized agency related to the United Nations, works to raise living standards in developing countries by channeling financial resources to those countries from industrialized countries. The bank itself includes three institutions: the *International Bank for Reconstruction and Development* (IBRD), the *International Development Association* (IDA), and the *International Finance Corporation* (IFC). The IBRD promotes economic development through loans and technical advice. The IDA provides funds for development projects in the poorest countries on much easier terms than the IBRD, and the IFC provides risk capital, without government guarantee, for the growth of productive private enterprise. In addition to the World Bank group, there are several regional banks outside the United Nations framework which provide development assistance in Latin America, Africa, and Asia.

International Monetary Fund (IMF), also a specialized agency of the United Nations, seeks to promote international monetary cooperation and facilitate the expansion of trade and employment. Some IMF policies may actually weaken the economic security of Third World nations; the Fund's "structural adjustment" programs often require cuts in social spending without insisting on reductions in military expenditures.[100]

While the United Nations and related organizations work on security issues around the world, several regional alliances have been formed to strengthen the security of member states.

The *North Atlantic Treaty Organization* (NATO) was created in 1949 and now has 16 members that seek to strengthen their individual and collective capacity to resist armed attack.

The *Warsaw Pact*, created in 1955 as a mutual defense alliance, now includes the Soviet Union and six Eastern European nations. The Pact provides for a unified military command; if one member is attacked, the others will provide necessary aid, including military force.

Other international organizations with regional security objectives include the *Organization of American States* with 32 member states, the *League of Arab States* with 21 members, and the *Organization of African Unity* with 51 member countries.[101]

In addition to these intergovernmental organizations, there are many international and national nongovernmental organizations that work to strengthen various aspect of global security, including many working primarily in the United States. Some of these are listed under Further Information.

What You Can Do

Inform Yourself

Consult the sources and organizations listed in the Further Information section for more information on issues concerning all aspects of global security, including environmental, economic, social, and demographic factors as well as political and military factors. Consider the political,

social, and environmental impacts of excessive military spending.

- Learn about the risks and consequences of nuclear war, and what you can do to help prevent it. Obtain a copy of the *Nuclear War Prevention Kit* published by the Center for Defense Information and listed under Further Information.

- Learn about the activities of the United Nations and its related agencies that help maintain and strengthen global security in areas such as international peacekeeping, agriculture, health, family planning, social and economic development, and the environment. Consult *Basic Facts About the United Nations* listed under Further Information. Read about current issues as they are discussed at the U.N. General Assembly. Follow U.S. activity at the United Nations, including U.S. positions on issues before the General Assembly and the Security Council.

- Study the history of U.S. foreign and economic policy and its effect on countries around the world. Follow news developments in countries where the United States has a military presence. Study the relationship between the United States and the Soviet Union. Learn about the Soviet people—their music, traditions, common concerns, and culture in general—and work to increase U.S.-Soviet understanding.

Join With Others

Concentrate your efforts to educate the community, influence policy making, and promote general awareness of security issues by working with existing peace and security organizations or by starting a group in your region.

- Meet with others in your area who are concerned about escalating military spending in the face of social, environmental, and economic degradation. Share ideas and concerns in forums open to the community.

- Coordinate a Cooperative Alliance Forum in your community to build stronger cooperation for sustainable development among local peace, social justice, religious, environment, development, population, and resource groups. Contact the Global Tomorrow Coalition for further information.

- Join local, national, and international organizations that work to promote global security through sustainable development. Volunteer to work with their publicity campaigns and spread the word.

Review Your Habits and Lifestyle

Consider what actions you can take to affect the issue of global security in your everyday life—in your family, home, neighborhood, community, and school or workplace; as a student, citizen, voter, and consumer.

- Find out which businesses in your area have defense contracts. Encourage local studies of economic impacts of military spending such as those assisted by the Jobs With Peace Campaign. Work with local industry to draw up plans for conversion to alternative, civilian production. Publicize your findings.

- Be aware of the amount of money the federal government spends on defense—nearly 30 cents of each tax dollar—in relation to other programs for social and economic development, resource conservation, and environmental protection. Calculate how much you contributed to military expenditures through your federal taxes last year.

- Consider how you can help protect and conserve critical resources such as fresh water, cropland, and energy supplies on which the security of all people depends.

Work With Your Elected Officials

Contact your local, state, and national legislators, find out their stands on issues of global security, and express your opinions in clear, short letters or personal visits.

- Work to decentralize foreign affairs legislation down to the state and local levels. More than 1,000 state and local governments of all political persuasions are participating in foreign affairs with initiatives such as nuclear-free zones, comprehensive test ban treaty support, and sanctuary declarations for refugees. Work with your city council and mayor on similar initiatives in your community.

- Encourage representatives on local, state, and national levels to introduce legislation, or support existing efforts to:

  » decrease military spending and recognize environmental, social, and development programs as effective contributions to international security;

  » increase voluntary contributions to United Nations Environment Program and restore support of United Nations Population Fund;

  » reduce both nuclear and conventional weapons between the superpower blocs through treaties, national initiatives, and international dialogue;

  » encourage work on an international treaty regulating and limiting the arms trade;

  » limit U.S. military aid to developing nations in favor of economic and other developmental aid;

  » put in place a comprehensive framework for planning and implementing the conversion of military production to civilian use, based on the Defense Economic Adjustment Act discussed in the Practical Politics section of this chapter;

  » strengthen the Peace Corps as an alternative to military service in promoting global security; expand the program to provide free or subsidized college educations for those who work for three years after graduation in social, environmental, and economic areas of sustainable development.

Publicize Your Views

Use the local print and electronic media to express your opinions on security legislation, inform the public of the many varying opinions surrounding the military and global security, and generate support for social, environmental, and economic in nature. Write letters to the editor of the local newspaper, and create public service announcements for local radio and television stations.

Raise Awareness Through Education

Find out whether schools, colleges, and libraries in your area have courses, programs, and adequate resources related to the issue of global security; if not, offer to help develop or supply them.

- If your state has a peace institute or peace studies centers affiliated with area universities or colleges, visit one and see what programs it offers. Encourage the discussion of topics such as "redefining national security," which bridge the interests and agendas of a variety of organizations and community groups. If your state does not have such an institution, consider working to establish one.

- Encourage local high schools and universities to host Model United Nations Conferences. Suggest the use of other simulations like the World Game to introduce students to the complexity of national security, resource allocation, and sustainable development on a global scale. (See Further Information—Teaching Aids.)

- Encourage the use and application of dispute resolution techniques in classroom activities and the teaching of non-competitive sports and games at school and youth clubs.

- Visit local schools and work with teachers to expand their collection of peace studies curricula, libraries, and video resources.

Consider the International Connections

Work to establish links with countries around the world at the individual, community, and national level.

- Consider exchanging letters and artwork with a family or student in the Soviet Union to increase mutual cultural awareness and understanding. Several U.S. organizations coordinate exchange efforts, including Children's Art Exchange, International Friendship League/ Russian Pen Pal Project, and US-USSR Youth Communications Initiative.

- Try to get your city involved in international trade, cultural exchange, and global politics by hosting foreign student exchanges, organizing cross-cultural festivals and celebrations, and sponsoring global awareness weeks.

- Involve yourself and your community with citizen diplomacy with nations around the world through a "sister city" program. Over 800 U.S. communities have sister city relationships with cities abroad—15 with Soviet cities and 25 with Chinese cities. If your city has such a program, broaden the structure to include the concept of sustainable development. To start a program in your area, contact the Center for Innovative Diplomacy and subscribe to its publications (see Further Information—Periodicals).

- Join or lead a tour to the Soviet Union. Many exchange programs and sister city initiatives have begun this way.

- Join an organization that supports and seeks to strengthen U.S. support for the United Nations and its many specialized agencies. The United Nations gains credibility with each individual who gives it active support.

Further Information

A number of organizations can provide information about their specific programs and resources related to national and global security.[*]

Books

Barnaby, Frank (ed.). *The Gaia Peace Atlas: Survival into the Third Millennium.* New York: Doubleday, 1988. 271 pp. Paperback, $18.95. Presents options for peace and a sustainable future based on research and contributions from numerous international authorities and organizations. Lavishly illustrated with many effective graphics.

Conetta, Carl (ed.). *Peace Resource Book 1988/89: Comprehensive Guide to Issues, Groups, Literature.* Institute for Defense and Disarmament Studies. Cambridge, MA: Ballinger, 1988. 440 pp. $14.95. A reference book on peace issues, groups, and materials; gives details on 7,000 national and local groups.

Ehrlich, Paul, Carl Sagan, Donald Kennedy, and Walter Orr Roberts. *The Cold and the Dark: The World After Nuclear War.* New York: Norton, 1984. 229 pp. $12.95.

Fahey, Joseph and Richard Armstrong (eds.). *A Peace Reader.* Essential Readings on War, Justice, Non-Violence and World Order. New York: Paulist Press, 1987. 477 pp. Paperback, $4.95.

The Fund for Peace. *Working for Peace: Annotated Resource Guide.* 1989-1990 Edition. New York, 1989. 95 pp. Paperback, $5.95.

Goldring, Natalie. *A Concerned Citizen's Introduction to National Security.* Mothers Embracing Nuclear Disarmament. 1987. 26pp. Free copy available from MEND, P.O. Box 2309, La Jolla, CA 92038. A clear and concise guide to a more balanced understanding of national security and a broader perspective on global security.

Kennedy, Paul. *The Rise and Fall of the Great Powers.* Economic Change and Military Conflict from 1500 to 2000. New York: Vintage Books, 1989. 677 pp. Paperback. Sheds light on the consequences of military spending for economic development.

---

* These include: American Peace Network, Arms Control Association, Center for Defense Information, Center for Economic Conversion, Center for Innovative Diplomacy, Committee for National Security, Council on Economic Priorities, Council for a Livable World, Employment Research Associates, Environmental Policy Institute, Federation of American Scientists, Friends of the Earth, The Fund for Peace, Institute for Defense and Disarmament Studies, Institute for Policy Studies, Institute for Soviet-American Relations, International Friendship League, Jobs With Peace Campaign, National Audubon Society, National Commission for Economic Conversion and Disarmament, Natural Resources Defense Council, Parliamentarians Global Action, PeaceNet, SANE/FREEZE (Committee for a Sane Nuclear Policy and Nuclear Weapons Freeze Campaign), United Nations Association of the USA, United States Institute of Peace, Union of Concerned Scientists, World Federalist Association, World Priorities, and Worldwatch Institute. Addresses for these organizations are listed in the appendix.

Moore, Melinda and Laurie Olsen. *Our Future at Stake: A Teenager's Guide to Stopping the Arms Race*. Citizens Policy Center-Nuclear Action for Youth Project. New Society Publishers. 72 pp. Paperback, $7.95. An informative and illustrated resource for education and action, complete with personal statements by teenagers themselves.

Mueller, John. *Retreat from Doomsday: The Obsolescence of Major War*. New York: Basic Books, 1989. 327 pp. $20.95. Argues that resort to major war as a serious choice for serious political leaders is becoming out-of-date.

Myers, Norman. *Not Far Afield: U.S. Interests and the Global Environment*. Washington, DC: World Resources Institute, 1987. 73pp. Paperback, $10.00.

Sivard, Ruth Leger. *World Military and Social Expenditures, 1987-1988*. Washington, DC: World Priorities, 1987. 56 pp. Paperback, $6.00. A unique book of facts and analysis of the grossly unbalanced allocation of global resources. Contains data on resource allocations for 142 countries, and a historical review of war casualty figures.

Union of Concerned Scientists. *Presidential Priorities: A National Security Agenda for the 1990s*. Cambridge, MA, 1988. 42 pp. Paperback, $4.95.

United Nations. *Basic Facts About the United Nations*. New York: United Nations, 1987.

World Commission on Environment and Development. *Our Common Future*. New York: Oxford University Press, 1987. 383 pp. Paperback, $10.95. Calls for a unity of economics and ecology so that governments and people can take responsibility not just for environmental damage, but for the policies that cause the damage. See especially Chapter 11, "Peace, Security, Development, and the Environment."

Zuckerman, Lord Solly. *Nuclear Illusion and Reality*. New York: Vintage, 1982. 154 pp. Paperback, $2.95. Explains why there can be no such thing as a "limited" nuclear war, why no one side can have a nuclear advantage, and why nuclear weapons have no bearing on national defense; urges that we seek security by other means.

Zuckerman, Peter A. *A New Covenant: Blueprint for Survival*. Washington, DC: American Peace Network (APN), 1988. 131 pp. Paperback, $7.50. APN is an educational organization dedicated to promoting human survival through world peace and shifting savings from military spending to the reduction of human suffering.

Articles

Brown, Lester R. "Redefining National Security." In Brown et al., *State of the World 1986*. New York: Norton, 1986, pp. 195-211.

Brown, Lester R. and Edward C. Wolf. "Reclaiming the Future." In Brown et al., *State of the World 1988*. New York: Norton, 1988, pp. 170-188.

Center for Defense Information. "Nuclear War Prevention Kit." Washington, DC, 1985. 36 pp. Single copy, $5.00; 10 copies, $12.00. Lists many resources and over 200 organizations involved in security issues.

El Hinnawi, Essam and Manzur H. Hashmi. "Military Activity." In El Hinnawi and Hashmi, *The State of the Environment*. Surrey, England: Butterworth Scientific, 1987, pp. 143-154.

Feldman, Jonathan et al. "Criteria for Economic Conversion Legislation." Briefing Paper No. 4. Washington, DC: National Commission for Economic Conversion and Disarmament, December 1988.

Jacobson, Jodi L. *Environmental Refugees: A Yardstick of Habitability*, Worldwatch Paper 86. Washington, DC: Worldwatch Institute, November 1988. 46 pp. $4.00.

Mathews, Jessica Tuchman. "Redefining Security." *Foreign Affairs*, Spring 1989. Maintains that global developments suggest the need to broaden the definition of national security to include resource, environmental, and demographic issues.

Myers, Norman (ed.). "Crisis: The Threat of War." In Norman Myers (ed.), *Gaia: An Atlas of Planet Management* (Garden City, NY: Anchor Books, 1984), pp. 242-51.

Renner, Michael. "Enhancing Global Security." In Lester R. Brown et al., *State of the World 1989*. New York: Norton, 1989, pp. 120-136.

Renner, Michael. *National Security: The Economic and Environmental Dimensions*. Worldwatch Paper 89. Washington, DC: Worldwatch Institute, May 1989. 78 pp. $4.00. An expanded version of the previous entry.

Renner, Michael G. "Swords into Consumer Goods." *World Watch*, Vol. 2, No. 4, July-August 1989, pp. 17-25.

Shapiro, Charles S. "Radiological Effects of Nuclear War." *Environment*, Vol. 30, No. 5, June, 1988, pp. 39-41.

Warner, Frederick. "The Environmental Effects of Nuclear War: Consensus and Uncertainties." *Environment*, Vol. 30, No. 5, June, 1988, pp. 2-7. This issue also contains articles on the biological, climatic, and health effects of nuclear war.

Periodicals

Defense and Disarmament Alternatives. Institute for Defense and Disarmament Studies, 2001 Beacon Street, Brookline, MA 02146.

The Defense Monitor. Center for Defense Information, 1500 Massachusetts Avenue, N.W., Washington, DC 20005. Sent 10 times a year to contributors of $25 or more.

FAS Public Interest Report. Federation of American Scientists, 307 Massachusetts Avenue, N.E., Washington, DC 20002. Sent monthly to contributors of $25 or more.

Journal. United States Institute of Peace, 1550 M Street, N.W., Suite 700, Washington, DC 20005-1708. Published six times a year by the Institute as part of its congressionally-mandated responsibility to expand knowledge and provide public education about peace, war, and international conflict management.

Municipal Foreign Policy. Center for Innovative Diplomacy, 17931 Sky Park Circle, Suite F, Irvine, California 92714, (714-250-1296). Published quarterly. Covers city involvement in international trade, cultural exchange, and global politics.

Nucleus. Union of Concerned Scientists, 26 Church Street, Cambridge, MA 02238. A quarterly newsletter. Covers the hazards of U.S. nuclear policy and energy policy.

Surviving Together: A Journal on Soviet-American Relations. Institute for Soviet-American Relations, 1608 New Hampshire Avenue, NW, Washington, DC, 20009. Reports on current events in the Soviet Union, exchanges, joint projects, trade, legislation, public education programs, and media coverage.

Thinkpeace. San Francisco Study Group for Peace and Disarmament, 2735 Franklin Street, San Francisco, CA 94123. A bimonthly newsletter.

Films and Other Audiovisual Materials

A Call for Survival. Four portraits of average people working to prevent nuclear war: a physicist, an 80 year old surveyor, a mother, and a Trident submarine worker. Examines motivation and explores the psychological, spiritual, ethical, and political issues underlying the current debate on nuclear weapons. 25 min. Video ($49 VHS or Beta). Bullfrog Films.

The Day After. ABC's nationally televised incineration of Lawrence, Kansas. Embassy Home Video.

The Freeze. 1983. An overview of the arms race, combining excerpts from five films on the nuclear war issue. Commentary from Paul Warnke, Robert McNamara, Dr. Helen Caldicott, and others. 25 min. 16mm film: sale, $250; rental, $45. Video: sale, $100. Direct Cinema, Ltd.

Gods of Metal. 1982. Analyzes the arms race from a moral perspective, showing economic and social effects in the United States and the developing world. Includes interviews filmed on the B-1 bomber assembly line. 27 min. 16mm film: rental, $50. Icarus Films.

In Our Defense. Exposes planning for a first-strike capability and shows how the quest for security through nuclear weapons has actually made us less secure, while undermining social and human-oriented programs. Interviews Pentagon officials, business leaders and average citizens. 26 min. 16mm film ($435), video ($265); Rental, $65. Public Media, Inc.

What About the Russians?. 1982. Uses the testimony of military, scientific, and political leaders to explore the relative strength of the United States and the Soviet Union. The film considers how to reconcile the desire for national security with the threat of nuclear war. 26 min. 16mm film: sale, $300; rental, $50. Video: sale, $55: rental, $35. Educational Film & Video Project.

Teaching Aids

Wien, Barbara J. (ed.). *Peace and World Order Studies: A Curriculum Guide.* Fourth Edition. New York: World Policy Institute, 1984. 742 pp. $14.95. Includes over 100 model syllabi in areas such as Global Problems, Hunger, International Organization and Law, Peacemaking, Regional Studies, Militarism, Women, Ecology, and Alternative Futures. Also lists organizations, films, and periodical resources.

The World Game has sessions and workshops available for universities, corporations, and government leaders, as well as simpler programs designed for primary and secondary schools. The object of the game is, according to R. Buckminster Fuller, "to make the world work for 100 percent of humanity in the shortest possible time through spontaneous cooperation without ecological offense or the disadvantage of anyone." For information, contact The World Game, University City Science Center, 3508 Market St., Philadelphia, PA 19104, telephone 215/387-0220.

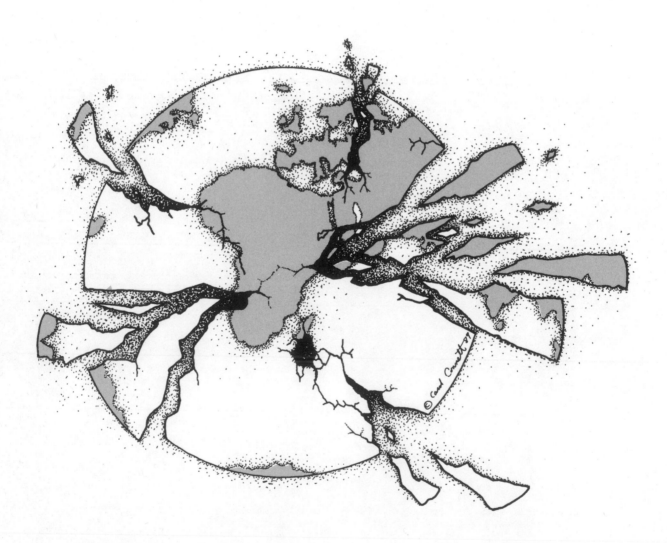

Toward a Sustainable Future:
Priorities and Progress

Successful economic development and environmental protection go hand in hand. You cannot have one without the other.

—U.S. President George Bush[1]

[Development must focus on] human potential and the sustainable management, restoration, and enhancement of ecosystems on which all economic systems rely.

—Hazel Henderson[2]

Achieving sustainable economic growth will require the remodeling of agriculture, energy use, and industrial production after nature's example....

—Jessica Tuchman Mathews, World Resources Institute[3]

Our global future depends upon sustainable development. It depends upon our willingness and ability to dedicate our intelligence, ingenuity, and adaptability—and our energy—to our common future. This is a choice we can make.

—*Our Common Future*, the report of the World Commission on Environment and Development[4]

Sustainable Development: The Overall Objective

A major focus of the recent work of the World Commission on Environment and Development (summarized in Chapter 1) was to develop an understanding of the concept of sustainable development. As defined in the Commission's report, *Our Common Future*, sustainable development is a dynamic process designed to meet today's needs without compromising the ability of future generations to meet their own needs. It requires societies to meet human needs by increasing productive potential and by ensuring equitable economic, social, and political opportunities for all. Sustainable development must not endanger the atmosphere, water, soil, and ecosystems that support life on Earth. It is a process of change in which resource use, economic policies, technological development, population growth, and institutional structures are in harmony and enhance current and future potential for human progress.[5]

To achieve sustainable development, a society must employ a variety of economic and political measures and achieve a careful balance between free market mechanisms and judicious public management to prevent excessive or damaging use of natural resources. Successful sustainable development must also include a thorough understanding of cultural values and natural resource management systems that have proven successful in the past.[6]

There seem to be grounds for cautious optimism about prospects for achieving successful sustainable development worldwide. There is evidence that technological changes can result in better control of environmental pollution and more efficient use of resources in manufacturing; there are also indications that social and economic changes can lead to higher prices for natural resources and greater conservation and recycling of resources.[7]

Progress toward sustainable development will require careful consideration of a range of specific social, economic, and environmental goals, the setting of priorities among the goals based on public discussion and debate, allocation of resources to those priorities, and a high level of political leadership supported by public understanding of the issues and commitment to needed action.

Priority Goals for Sustainable Development

Among a growing number of compilations, one of the best and most concise listings of priorities was the Worldwatch Institute's global security budget in its *State of the World 1988* report.

As noted in the previous chapter, the report called for reallocating some of the world's military spending over the next decade to seven priority areas for achieving sustainable development: stabilizing the Earth's climate, protecting top-soil, reforesting the Earth, slowing population growth, raising energy efficiency, developing renewable energy, and retiring Third World debt.

International Priorities

In addition to *State of the World*, several other recent publications have included international or global agendas specifying priorities for curbing environmental deterioration and making development more sustainable. These include the World Commission on Environment and Development's *Our Common Future*, published in 1987; the World Resources Institute's *The Global Possible*, published in 1985; "Agenda 2000," a special report by *The Christian Science Monitor* published in July 1988;[8] *Time* Magazine's "Planet of the Year: Endangered Earth," published in January 1989; *The Gaia Peace Atlas*, published in 1988; and Ambassador Richard Benedick's article, "Environment: An Agenda for the Next G-7 Summit," published in July 1989. Table 16.1 summarizes the major environment-related priority recommendations of these seven international agendas, organized into 12 categories.[9]

The agendas from *Our Common Future*, *The Global Possible*, "Agenda 2000," and *Time* reflect the views of diverse groups of policymakers and experts from 21, 20, 12, and 11 nations, respectively. *Our Common Future* covered all 12 priorities listed in Table 16.1, *The Global Possible* included 11 priorities, and "Agenda 2000" and *Time* included 10 of the 12. All seven agendas included population, poverty and debt, forests and species, energy efficiency, renewable energy, and greenhouse gases and other air pollutants as areas of priority concern. Six of the seven included sustainable agriculture, and five of the seven included a shift of military spending to sustainable development programs, and protection of the ozone layer. Only four of the international agendas referred to waste reduction or water quality and efficiency as priorities, and only three included ocean and coastal resources.[10]

At its meeting in May 1989, the 58-nation Governing Council of the United Nations Environment Program (UNEP), representing the world's governments and with 103 in attendance, approved an agenda of eight priority areas in which UNEP plans to allocate at least 70 percent of its resources. The areas are: climate and atmospheric pollution, pollution and shortage of fresh water resources, deterioration of oceans and coastal areas, land degradation, biological impoverishment, hazardous wastes and toxic chemicals, management of biotechnology, and protection of health and quality of life.[11] The first six of these priority areas correspond closely to six of the priorities listed in Table 16.1.

Priorities for the United States

Coincident with the beginning of a new United States presidential term of office in 1989, many U.S. organizations

Table 16.1
International Priorities

Priority	1 WCED, Our Common Future	2 WRI, The Global Possible	3 CSM, Agenda 2000	4 Time, Planet of the Year	5 Gaia Peace Atlas	6 Benedick, Environment: An Agenda	7 WI, State of the World 1988
Slow Population Growth	X	X	X	X	X	X	X
Reduce Poverty, Inequality, and Third World Debt	X	X	X	X	X	X	X
Make Agriculture Sustainable	X	X	X			X	X
Protect Forests and Habitats, Curb Loss of Species	X	X	X	X	X	X	X
Protect Ocean and Coastal Resources	X	X		X			
Protect Fresh Water Quality, Improve Water Efficiency	X	X	X			X	
Increase Energy Efficiency	X	X	X	X	X	X	X
Develop Renewable Energy Sources	X	X	X	X	X	X	X
Limit Greenhouse Gases and Other Air Pollutants	X	X	X	X	X	X	X
Protect the Stratospheric Ozone Layer	X	X		X	X	X	
Reduce Waste Generation, Recycle Wastes	X	X	X	X			
Shift Military Spending to Sustainable Development	X		X	X	X		X

1. World Commission on Environment and Development (WCED), *Our Common Future* (New York: Oxford University Press, 1987), 383 pp. Report of a three-year program established by the United Nations. The Commission included 22 members from 21 nations.

2. Robert Repetto (ed.), "Agenda for Action," *The Global Possible: Resources, Development, and the New Century* (New Haven: Yale University Press, 1985), pp. 496-519. Recommendations from a 1984 World Resources Institute (WRI) conference of 75 leaders from 20 nations.

3. Rushworth M. Kidder, "Agenda 2000," *The Christian Science Monitor* (CSM), July 25, 1988, pp.B1-B12. Recommendations from a conference of 35 participants from 12 nations, covering environmental degradation, North-South and East-West relations, and public and private morality.

4. *Time* Magazine, "Planet of the Year: Endangered Earth," January 2, 1989, pp. 24-63. Recommendations from a conference of 26 Time journalists and 33 experts from 11 nations.

5. Frank Barnaby (ed.), *The Gaia Peace Atlas: Survival into the Third Millennium* (New York: Doubleday, 1988), Chapter 8, "Short-Term Steps for Survival;" Chapter 9, "The Choices for Humanity."

6. Richard E. Benedick, "Environment: An Agenda for the Next G-7 Summit," *International Herald Tribune*, July 25, 1989.

7. Lester R. Brown and Edward C. Wolf, "Reclaiming the Future," in Brown, et al., *State of the World 1988* (New York: Norton, 1988), pp. 170-88. A publication of the Worldwatch Institute (WI).

prepared agendas for the new administration. Of 35 such agendas released during 1988 and early 1989 (most of which were reviewed in the World Future Society's publication, *Future Survey*), seven contained no environment-related priorities, 17 included between one and three environmental items, and 11 listed more than three priorities related to the environment.[12] Table 16.2 (on the following page) summarizes the major environment-related recommendations of nine of these agendas for the United States that contain seven or more priorities related to the environment and sustainable development.[13]

Table 16.2 includes U.S. agendas from World Resources Institute, the Peace and Environment Project, The Conservation Foundation, Worldwatch Institute, and *Time* magazine; and an agenda compiled by 18 major environmental organizations, *Blueprint for the Environment*. Table 16.2 also includes three additional agendas: one sponsored by U.S. Senators Timothy Wirth and John Heinz, one chaired by former U.S. Presidents Ford and Carter, and one written by former Colorado Governor Richard Lamm and Thomas Barron.

Although these agendas represent the perspectives of a wide variety of organizations, they all include three priorities: protecting forests and species, limiting greenhouse gases and other air pollutants, and increasing energy efficiency.[14] All but one agenda listed reducing waste generation, and all but two included slowing population growth,[15] protecting water quality, developing renewable energy, and protecting the ozone layer.

Only five of the nine U.S. agendas summarized in Table 16.2 included sustainable agriculture as a priority. In view of the agriculture sector's inclusion in six of the seven international agendas, its apparent importance as a contributing factor in most of the major global problems considered in this book (see Chapter 1, Table 1.1), and its potential for helping to solve many of those problems in the United States (see Table 16.3 below), its omission from four of the nine U.S. agendas seems noteworthy, and may reflect a lack of recognition in the United States of agriculture's role in environmental degradation.[16]

Problem Impacts and Problem Solutions

In Chapter 1, five factors were considered as important underlying causes of global problems: unsustainable population growth, food production, energy use, and industrial production; and extreme poverty and inequality. These five factors, along with several priorities from the issue chapters and from the agendas in Tables 16.1 and 16.2, form the basis for seven priority solutions presented in Table 16.3 below. Table 16.3 relates the same eight major problem impacts (on the left side) identified in Chapter 1 to the seven priority solutions (across the top). For each problem impact, the symbols in Table 16.3 suggest the likely importance or effectiveness of the seven priority solutions, as follows:[17]

●   very important solution

◗   moderately important solution

○   less important but significant solution

–   unimportant or insignificant solution

Any such tabular representation of complex issues is necessarily somewhat arbitrary, and reasonable, informed observers may differ as to the definitions and values assigned. But comparisons of potential impacts of alternative policies, and their respective costs and benefits, are a necessary step toward political and economic choices. An analysis of Table 16.3 leads to the following conclusions:

● *Slowing population growth* is important for alleviating all eight problem impacts.[18]

● *Reducing poverty and inequalities*—which includes such factors as improving health, longevity, and literacy; increasing employment and political participation; and broadening opportunities for women and minorities—is important for meeting basic human needs, slowing species depletion, curbing land degradation, and reducing conflict and war. It is less important but significant for curbing water pollution.[19]

● *Making agriculture sustainable*, which includes reducing soil erosion and decreasing the use of harmful chemicals and farming practices, is important for curbing land degradation, water pollution, and depletion of water, energy, and plant and animal species. It is less important but significant for meeting basic human needs and for reducing air pollution and conflict.[20]

● *Protecting forests and other habitats*, which includes reforestation and protection of other living resources, is important for slowing species depletion, curbing land degradation, and reducing air pollution and fresh water depletion. It is less important but significant for slowing depletion of nonrenewable energy and for meeting basic needs.[21]

● *Making energy use sustainable*, which includes improving energy efficiency, conserving energy through means such as economic incentives, and developing renewable energy sources, is important for reducing air pollution, land degradation, depletion of energy and minerals, and conflict. It is less important but significant for reducing species depletion and meeting basic needs.[22]

● *Making water use sustainable*, which includes improving the efficiency of water use and protecting water quality, is important for curbing water depletion and pollution. It is less important but significant for reducing energy depletion, land degradation, and conflict, and for meeting basic needs.[23]

Table 16.2
Priorities for the United States

Priority	1 WRI, The Global Possible	2 Platform for Peace and Environment	3 Blueprint for the Environment	4 Project 88 Wirth, Heinz	5 Conservation Foundation	6 Lamm and Barron	7 World-watch Institute	8 American Agenda Ford, Carter	9 Time, Planet of the Year
Slow Population Growth	X	X	X	X		X	X		X
Reduce Poverty, Inequality, and Third World Debt	X			X	X		X	X	X
Make Agriculture Sustainable	X	X	X		X		X		
Protect Forests and Habitats, Curb Loss of Species	X	X	X	X	X	X	X	X	X
Protect Ocean and Coastal Resources		X	X	X	X	X			X
Protect Fresh Water Quality, Improve Water Efficiency	X	X	X	X	X	X		X	
Increase Energy Efficiency	X	X	X	X	X	X	X	X	X
Develop Renewable Energy Sources	X	X	X	X		X	X	X	
Limit Greenhouse Gases and Other Air Pollutants	X	X	X	X	X	X	X	X	X
Protect the Stratospheric Ozone Layer	X	X	X	X	X	X		X	
Reduce Waste Generation, Recycle Wastes	X	X	X	X	X	X		X	X
Shift Military Spending to Sustainable Development	X	X					X		

1. World Resources Institute (WRI), *The Crucial Decade: The 1990s and the Global Environmental Challenge* (Washington, DC, 1989). 22 pp.

2. *Summary Platform for Peace and Common Security and for a Healthy, Just, and Sustainable Environment.* May 1988. Endorsed by 70 U.S. organizations. Available from Peace and Environment Project, 13122 Parson Lane, Fairfax, Virginia 22033.

3. *Blueprint for the Environment: Advice to the President-Elect from America's Environmental Community* (Washington, DC, 1988).. For publication details, see listing under Further Information. A compilation of recommendations from 18 U.S. environmental organizations.

4. *Project 88: Harnessing Market Forces to Protect Our Environment: Initiatives for the New President.* A Public Policy Study sponsored by Senator Timothy E. Wirth, Colorado, and Senator John Heinz, Pennsylvania (Washington, DC, 1988). 76 pp. Compiled by a 50-member bipartisan team from academia, industry, the environmental community, and government. See listing under Further Information.

5. Russell E. Train, "Bush Faces Major Environmental Challenges," *Conservation Foundation Letter*, 1989, No. 1. 8 pp.

6. Richard D. Lamm and Thomas A. Barron, "The Environmental Agenda for the Next Administration," *Environment*, May 1988. 7 pp

7. Lester R. Brown, Christopher Flavin, and Sandra Postel, "No Time to Waste: A Global Environmental Agenda for the Bush Administration," *World Watch*, January-February 1989, pp. 10-19. Prepared by the Worldwatch Institute.

8. *American Agenda: Report to the Forty-First President of the United States of America* (Camp Hill, PA: Book-of-the-Month Club, 1988). 289 pp. A bipartisan report chaired by former U.S. Presidents Gerald R. Ford and Jimmy Carter, with recommendations and commentary by 44 contributors.

9. *Time* Magazine, "What the U.S. Should Do," in "Planet of the Year: Endangered Earth," January 2, 1989, p. 65.

Table 16.3
Global Problem Impacts and Estimated Effectiveness of Priority Solutions

Problem Impacts	Priority Solutions						
	1 Slow Popu-lation Growth	2 Reduce Poverty and Ine-qualities	3 Make Agri-culture Sustain-able	4 Protect Forests and Other Habitats	5 Use Energy Sustain-ably	6 Use Water Sustain-ably	7 Reduce Waste Gener-ation
1. Unmet basic human needs for safe water, food, shelter, health care, education, employment, etc.	●	●	○	○	○	○	○
2. Species depletion (extinction of plants and animals), habitat degradation	●	●	◗	●	○	–	–
3. Land degradation: soil erosion, desertification, loss of soil fertility	●	◗	●	●	◗	○	–
4. Depletion of nonrenewable energy and minerals	◗	–	◗	○	●	○	◗
5. Depletion of fresh water (groundwater and surface water)	●	–	●	◗	–	●	◗
6. Water pollution: chemical and bac-terial contamination of groundwater and surface water	◗	○	◗	–	–	●	◗
7. Air pollution: urban air pollution, acid deposition, ozone layer depletion, greenhouse gas buildup	◗	–	○	◗	●	–	◗
8. Conflict and war: domestic and international	◗	◗	○	–	◗	○	–

●   very important solution
◗   moderately important solution
○   less important but significant solution
–   unimportant or insignificant solution

• *Reducing waste generation*, which includes improving production processes and recycling liquid and solid was-tes, is important for reducing air and water pollution and energy, mineral, and water depletion. It is less important but significant for meeting basic needs.[24]

Based on the estimates in Table 16.3, slowing population growth is the most effective single measure for alleviating the range of global problem impacts listed in Table 16.3. Making agriculture sustainable is the next most effective measure across the entire range of problems. Protecting forests and using energy sustainably rank next in effective-ness, followed by reducing poverty, using water sustainab-ly, and reducing waste generation.

For each problem impact, the estimates in Table 16.3 suggest which of the seven priority solutions are most important. For example, to meet unmet needs, slowing population growth and reducing poverty are estimated to be very important measures. To curb species depletion, slow-ing population growth, reducing poverty, and protecting forests are very important priorities. To reduce land degradation, population curbs, sustainable agriculture, and forest protection are all very important. To limit water depletion, slowing population growth and making agricul-ture sustainable are very important measures, along with achieving sustainable water use. And to curb air pollution, sustainable energy use is judged to be a very important priority.

Table 16.3 suggests that for problems typical of a number of developing countries and rural areas—including unmet human needs, species depletion, and land degradation (rows 1-3 in Table 16.3)—population, poverty, agriculture, forestry, and energy programs (columns 1-5) can be especially effective. For problems typical of many industrial nations and urban areas—including energy and water depletion and water and air pollution (rows 4-7)—energy, water, and waste reduction programs (columns 5-7) can be especially important.

Multisector Benefits

These conclusions lend support to the evidence presented in Chapter 1 that the Earth's social, political, economic, ecological, and resource problems are highly interdependent. The conclusions also suggest to what extent effective solutions can have multiple benefits across several sectors, and they point to the importance of a holistic, integrated view of global problem-solving and priorities for sustainable development. As ecologist Norman Myers has written, "We shall tend to underestimate the value of policy initiatives and funding support in many sectors as long as we view these sectors as distinct entities. The world does not work in discrete, isolated spheres, even though we try to manage it through separate institutional devices such as departments of agriculture, energy, forestry, and the like, that do not always coordinate their activities so as to match the integrative workings of the 'real world.'"[25]

Strategies for Achieving Sustainable Development

The range and intensity of global problems we face represent an acute challenge to the existing institutions and methods we now use to manage our affairs. We face a crucial question: Are our existing organizations, policies, and practices adequate to cope with emerging resource and environmental problems, or do we need to develop new structures and strategies to meet these challenges? In considering this question, Norman Myers asserts that "for the most part many current approaches are inherently incapable of doing the job." He suggests that many international institutions are basically large-scale intergovernmental committees and therefore mirror the nation-state system with its tendency toward competition rather than cooperation. Myers notes that many international organizations were established in the 1950s before global interdependence became an accepted reality, and that constitutionally they reflect an earlier era in international relations. Myers also asserts that we have a number of powerful options available to improve management of our affairs, including more effective use of international organizations, private enterprise, and nongovernmental organizations.[26]

Let us now consider some of the areas in which innovative ideas and methods can improve the management of global problems and achieve progress toward sustainable economic and social development. These include science, technology, and education, values and ethical standards, natural resource accounting, national governments,

economic incentives, management capabilities, international organizations, business and industry, and nongovernmental organizations.

Science, Technology, and Education

A major scientific challenge for the next few decades is to improve our understanding of the biosphere, the complex system of closely linked biological, chemical, and physical processes that supports all forms of life on Earth.

We urgently need to develop a science of the biosphere so we can recognize and understand the impacts of human activity on the Earth's life-support systems. As part of the effort to develop such a science, the field of global ecology is growing, crossing traditional academic disciplinary boundaries. It focuses on the interaction between life and its environment, including processes such as biogeochemical cycling, biological productivity, atmospheric and oceanic circulation, and global feedback mechanisms.[27]

To train scientists in this emerging field, and more broadly, to educate generalists able to deal with the interrelations among global issues, we need to establish more interdisciplinary programs at colleges and universities that encourage students to explore the field of global ecology and select courses that combine several fields such as the earth, life, and social sciences as well as the concept of systems analysis.[28]

Much of the new knowledge about how the biosphere functions and how human activity affects the global system comes from programs designed to monitor global processes and global change, and to develop computer models to simulate and analyze these processes. A number of international programs in these areas have already been completed, including the International Geophysical Year, the International Biological Program, and the International Hydrological Decade. Other programs currently under way include the Global Environment Monitoring System (GEMS) and the Global Resource Information Database (GRID), both coordinated by the United Nations Environment Program; UNESCO's Man and the Biosphere Program; and the International Geosphere-Biosphere Program's study of global change.[29]

The objective of the United States' component of the global change study, known as the U.S. Global Change Research Program, exemplifies the holistic, interdisciplinary nature of that study. The aim of the U.S. program is "to gain a predictive understanding of the interactive physical, geological, chemical, biological, and social processes that regulate the total Earth system."[30]

Many of these programs depend, at least in part, on remote sensing of global phenomena, usually from orbiting satellites. Global assessments of critical indicators are produced periodically—for example, yearly for stratospheric ozone, every five years for tropical forests, and every 20 years for glaciers.[31]

At present, however, the capabilities of remote sensing technology greatly exceed the ability of ecological scientists who utilize its product; there is a need for better communication between those who develop remote sensing technology and systems and those who could apply the methods to environmental issues.[32]

The process of systems engineering, which comprises a method of analyzing the links between many variables in complex systems, offers the potential of a more effective method for assessing the impact of human actions on the global environment. Systems engineering can manage the high degree of complexity and change typical of global environmental issues, especially as an adjunct to political decision-making.[33]

As we gain more information and understanding about the functioning of the Earth's ecological systems, we can better use them for human benefit while at the same time protecting them from damage. For example, as we have learned more about interactions between food crops and pests, we have developed appropriate technology in the form of biological pest controls and integrated pest management techniques that maintain crop yields while minimizing damage to soils and groundwater.[34]

Technology for development that is both sustainable and appropriate in terms of available skills, resources, and environmental conditions needs to be identified or created, and applied to all sectors of development, including areas such as agriculture, housing, health and sanitation, waste management, transportation, energy, manufacturing, education, and communication. The capacity for technological innovation in developing countries should be greatly expanded.[35]

Technology should be environmentally appropriate. Programs to develop innovative technology, generate alternative technology, adapt imported technology, and upgrade traditional technology should all be coupled with efforts to minimize environmental impacts.[36]

As nations develop large-scale, complex technologies such as electric power and other energy distribution systems, communication networks, and mass transit facilities, their vulnerability to failure or sabotage can become substantial. Such systems should be analyzed for risk, and designed to minimize the serious consequences of breakdown or sabotage.[37]

Modern technology, prudently applied, has great potential to monitor the state of the environment, reduce the impacts of human activities on the Earth's life-support systems, and improve the quality of life.[38]

Ethics and Values

In addition to the impact of unsustainable technology and the growing human population, the current deterioration of the Earth's life-support systems stems in part from inadequate understanding of our dependence on those systems, and our failure to accept responsibility for the future consequences of their deterioration. Simply put, we tend to place a relatively low value on nature and natural resources, and on the future.

In fact, the natural world provides us with many benefits that we often take for granted. In addition to other values, nature and natural resources support animal and plant life, supply genetic diversity, and provide economic, recreational, scientific, historic, aesthetic, and religious value.[39] Nature is a veritable storehouse of essential materials for our everyday lives—it provides us with food, clothing, shelter, and a wide range of products for commerce and industry. Our economic system, and ultimately our survival on Earth, depend on natural resources.[40]

Among the ethical questions relevant to natural resources are these: To what extent should we distribute the Earth's resources and environmental well-being according to principles of equality, utility, or entitlement? In protecting natural resources, what balance should there be between individual autonomy and government coercion? And to what extent should standards governing resource use and environmental protection be anthropocentric—emphasizing the fulfillment of human needs, or ecocentric—stressing the importance of preserving natural ecosystems?[41]

Environmental quality is increasingly seen not only as a public good but as a source of both economic and social value. Recent experience with "debt-for-nature" exchanges confirms that natural resources can have economic value when they are protected rather than exploited; for example, a nonprofit U.S. organization, Conservation International, recently purchased and agreed to "forgive" $650,000 of Bolivia's foreign debt in exchange for Bolivia's agreement to protect 3.7 million acres of threatened land from deforestation. And since Costa Rica's first debt-for-nature exchange in 1987, the country has purchased more than $75 million in debt titles for conservation and sustainable development projects.[42]

It is consistent with the definition of sustainable development presented earlier to argue that actions by individuals, groups, and nations must not excessively harm the present and future quality of life of others. Education should help citizens realize that their individual actions and policy choices today affect the present and future quality of life of others, and ultimately, the Earth's habitability.[43]

As Worldwatch Institute's *State of the World 1989* puts it, a person may be able to pay for a large, energy-inefficient automobile, but can the planet afford it? And a couple may be able to support several children, but can the Earth support several children per family?[44]

Robert McNamara has noted that, in the past, actions by governments, corporations, and other organizations have less often been held to moral and ethical standards than the behavior of individuals. Yet actions by governments and institutions affect all of us, and they share a moral responsibility to protect the basic rights of others. Conspicuous examples of immorality include the risk from the superpowers' nuclear arsenals to the survival of other nations, and rich nations' disproportionate share of global pollution and resource consumption.[45]

Just as individuals are affected by what happens in their communities, nations are influenced by events in neighboring regions of the world. Transnational problems are becoming increasingly severe and cannot be effectively managed or solved by single nations. Increasingly, both individuals and nations need ethical guidance in balancing their own rights against their responsibilities and obligations to others, and in weighing their own autonomy against the growing need for cooperation with others to solve problems that threaten the welfare and even the survival of all.[46]

Economic activity in particular needs ethical guidance; business must begin to evaluate the long-term impacts of its actions as well as the short-term results. For example, the automobile industry should take the lead, competing in the marketplace by making its products safer, more energy efficient, and more environmentally benign over the long run rather than be forced to do so by government regulation.[47]

To strengthen the moral dimension in human activity, futurist Theodore Gordon suggests that humanity should "inculcate a reverence for the future." And Russell Peterson sees a critical need to educate more people, and especially more leaders, who see the world as a whole, who are concerned about protecting the biosphere they have inherited, and who are committed to help ensure a decent quality of life for future generations.[48]

Natural Resource Accounting

Forests, soil, fresh water, wildlife, and other natural resources clearly have economic value and are essential for economic development and for human survival. They are economic assets because they can generate future income. Yet under current national income accounting procedures, natural resource assets are not valued as productive assets and depreciated over time as are assets such as buildings and equipment. National budgets and other annual audits rarely consider the depletion of natural resources. The standard United Nations national-income accounting framework—used by all market economies for economic analysis and planning—fails to distinguish between the creation of income and the destruction of natural-resource assets.

This gross discrepancy between how we measure economic activity and how we appraise the use of natural resources that ultimately sustain the economy often results in a distorted view of economic health, and sends misleading signals to policymakers. It is ironic that when we destroy forests, deplete the fertility of croplands, and pollute air and water, income as measured by the gross national product shows an increase. The lack of natural resource accounting supports and appears to validate the notion that rapid economic growth can be realized and sustained by exploiting the natural resource base; the result is deceptive gains in current income at the expense of permanent losses in a nation's economic wealth. In short, we fail to view our natural resources and environment as productive capital, even though we use them as such, and we are rapidly consuming our natural resource capital instead of living within the income derived from it.[49]

The problem of natural resource degradation encouraged by misleading economic accounting is widespread and endemic to economic systems in industrial and developing countries alike.

In West Germany, the estimated cost of forest damage from acid deposition is approaching $3 billion a year; in the United States, soil erosion may result in yearly costs of more than $6 billion, and water pollution is estimated to cost over $20 billion each year, yet none of these figures is currently subtracted from the gross national product.[50]

Over the past 20 years, Indonesia has rapidly cut its timber and greatly increased soil erosion by removing its trees and farming steep slopes. Between 1970 and 1984, Indonesia's GNP grew by 7 percent annually; but if a capital-consumption allowance were subtracted for the depleted value of forests, soils, and other natural resources, the GNP growth rate would fall to about 4 percent.[51]

Since 1960, the Philippines has cut 90 percent of its old-growth hardwood forests, thus losing a resource that could have produced substantial annual income indefinitely. Timber exports were the country's leading source of foreign exchange during the 1960s; they have since fallen sharply as Philippine forests have been depleted. The country will face a trade deficit in the coming decade, and now must cope with heavy runoff and flooding from deforested watersheds that damage irrigation systems, food production, and fisheries.[52]

Brazil has spent billions of dollars to encourage cattle ranching in the Amazon, despite evidence that it is neither ecologically nor economically viable; to support iron production that can break even only by consuming 70,000 acres of forest each year; and to plan dams that would flood vast areas of forest but produce power costing more than twice that of alternative energy-saving investments. Such projects are clearly environmentally destructive, but the irony and unrecognized tragedy is that they are also economically wasteful. Studies in Brazil and Peru demonstrate that net revenues from long-term harvesting of non-timber forest products such as rubber, oils, resins, nuts, and fibers are two to three times greater than from commercial logging or clearing a forest for cattle ranching.[53]

The first step toward including natural resources in national income accounts is to estimate values for these resources and for the services they provide. Using Indonesia's timber, soils, and petroleum as examples, a recent World Resources Institute (WRI) study has developed methods to integrate resource depletion into national accounting so that it more accurately reflects economic reality.[54]

Some countries are beginning to develop systems of environmental accounting. France and Norway are incorporating natural resource depletion into their economic analyses; Japan and the United States are developing quantitative measures of pollution and environmental quality; and Canada and the Netherlands are working in both areas. According to Norman Myers, the basic objective is to develop a measure of "net national product" or a "net national welfare indicator."[55]

Economist Robert Repetto of WRI has recommended that the United States take the lead in adopting national-income accounting and promoting this change at the United Nations, in the multilateral development banks, and in other nations. He observes that "no single innovation would so powerfully demonstrate that steps to protect the environment are in countries' own economic interests."[56] This innovation in accounting would also go far toward enabling individual enterprises to institutionalize growing concern for environmental quality and natural resource management.

Improving Governance

Inclusion of natural resource accounting in the governing process could be an important step in improving the ability of national governments to conserve resources, protect the quality of life, and move toward sustainable development. As noted earlier, economic and ecological systems do not work in discrete sectors such as energy, industry, and agriculture. Problems of air, water, and soil pollution involve many sectors and need to be broadly addressed. Many of these problems are made worse by fragmentation of responsibility between sectors. Ecological and economic considerations need to be integrated.[57]

Rather than having only a single agency responsible for controlling pollution, such as the U.S. Environmental Protection Agency, the goal of preventing environmental degradation should be shared by all sectors, each with its own institutional environmental protection program and responsibilities, but with an agency such as EPA exercising primary leadership.[58]

As natural resource accounting becomes part of the governing process, all government agencies will be able to develop clearer ideas of the relative cost of pollution damage and resource degradation, the expense of repairing damage after it has occurred, and the savings from preventing damage. Such cost estimates could be used to help develop ways to "internalize" both the high cost of repairing current and future damage—the "polluter pays" principle, and the lower cost of prevention—the "pollution prevention pays" concept used successfully by some chemical manufacturers.

Natural resource accounting by governments could be coupled with the development of improved governmental foresight capability, to allow the assessment of long-term costs and benefits, for all economic and social sectors, of policies to protect natural resources and environmental quality. Those assessments would be linked directly to current decision-making. (See Chapter 2.)

Finally, environmental objectives should become part of tax policies, policies affecting investment in research and development, and foreign trade incentives. It is especially important that environmental goals become part of both government and private sector policies that affect the pricing of goods and services. Economic activity must account for the environmental costs of production, and environmental goals should be an integral part of economic policy.[59]

Economic Incentives

In its 1989 agenda of proposals to the U.S. President and Congress for action to meet the global environmental challenge, the World Resources Institute listed several guiding precepts, including reliance on the market mechanism whenever possible, in part by ensuring environmentally accurate prices.[60] And WRI's Vice President, Jessica Mathews, recently stated that in order to achieve sustainable economic growth, all sectors will have to price goods and services to reflect the environmental costs of their provision.[61]

When resources are priced to reflect the cost of increasing their supply, they are used more efficiently and resource management is improved. And when resources are priced to include the costs of minimizing environmental harm from their production and processing, the expense of environmental protection can be covered and environmental management can be improved. One of the clearest examples of how pricing can affect resource use is the effect of OPEC oil price increases in the 1970s on energy consumption. In the United States, total energy use dropped by 2 percent between 1972 and 1983 even though U.S. GNP increased more than 30 percent; the efficiency of U.S. energy use during the period increased by one-third.[62]

In the area of environmental protection, market-based economic incentives have been implemented to a limited degree in several European nations and the United States, and have the potential for creating both economic and environmental benefits. Greater integration of market incentives into environmental policy could promote economic growth, improve environmental quality, encourage recycling and energy efficiency, and lower pollution control costs. In the area of reducing environmental pollution, there is evidence that control can be made more efficient by shifting from fixed control standards to approaches using market-based incentives.[63]

West Germany, France, and the Netherlands have had substantial experience with the use of emission fees to control point-source water pollution. In West Germany, a combination of government standards and market-based incentives is used to control pollution from liquid effluents; a charge based on the quantity of waste released supplements fixed limits on discharges set by regional authorities.[64]

In the United States, a transferable discharge permit system has been developed in which emissions of air or water pollutants are controlled by trading: an emission standard need not be met by each individual source of pollution but by a group of sources within a specified region; or a new source of pollution may begin operation if its pollution is more than offset by reduced emissions of other sources already in the area.[65]

The U.S. Environmental Protection Agency (EPA) began to experiment with "emissions trading" in 1974 as part of its air quality improvement program. Although not yet widely used, several large companies have traded emissions credits, and EPA estimates that the program has resulted in savings of over $4 billion in control costs, without adverse effect on air quality. Emissions trading could be expanded by incorporating it in the reauthorization of the Clean Air Act, and could become an affirmative tool to achieve a wider range of environmental objectives.[66]

Economic incentives can also be used to help conserve natural resources. One approach to water supply problems is to allow the voluntary exchange of water rights between users to improve the efficiency of water use. In Southern California, the two major water users are the Imperial Irrigation District (IID) and the Los Angeles Metropolitan Water District (MWD). After five years of negotiation, in 1988 IID and MWD reached a $230 million water conservation and transfer agreement from which both parties can benefit: farmers have a financial incentive to save water, urban needs can be met without building new dams and reservoirs, and environmental protection is improved.[67]

A recent study by the International Union for the Conservation of Nature and Natural Resources (IUCN) reviewed ways in which economic incentives can be used to conserve biological resources. The IUCN report stressed that resources are often threatened when the responsibility for their management is removed from those who would directly benefit from their sustainable use. Drawing on 25 case studies, the report concluded that economic induce-

ments may be the most effective measures for achieving sustainable use of biological resources.[68]

Other examples of incentive-based protection of resources and the environment include EPA's tradable permit system for meeting the Montreal Protocol's restrictions on emissions that deplete the stratospheric ozone layer, and experimental use of least-cost bidding by electrical utilities to meet power needs. A number of states have "bottle bills" that encourage recycling of containers, and a recent report has recommended a deposit-refund system for the management of containerized hazardous waste.[69]

To conserve fossil fuel supplies and reduce air pollution from their use, a variety of economic incentives has been proposed. Higher fuel-economy standards for motor vehicles, and incentives such as a gasoline tax would encourage the use of more efficient vehicles, as would graduated taxes on new vehicles and variable annual registration fees that favored efficient vehicles. Fossil fuels could also be taxed in proportion to their carbon emissions as a means of limiting greenhouse gas emissions.[70] (See Chapter 12.)

Management Capability

In addition to the benefits from the use of economic incentives, important gains can be realized by improving management of resource use and environmental protection, especially in the Third World. Improvements can include employment of appropriate technical personnel, information systems, and legal and administrative means to plan and guide resource use.[71]

To be effective, any such resource management program must involve, in both planning and implementation, the people and groups that are directly affected. Central governments and large organizations are typically unaware of local needs, preferences, opportunities, and managerial skills; yet these factors are often the key to successful resource management and environmental protection.[72] According to the United Nations Population Fund, a characteristic of many successful projects around the world dealing with population and resource-related problems is that "local communities or interest groups were given some power to control crucial decisions affecting their livelihoods."[73]

Outside specialists who are called upon to provide advice on natural resource management to a community or region should carefully examine existing local practices before recommending any changes. Local structures and practices usually have evolved slowly in response to local needs, and are often sustainable in ecological, economic, and social terms. Sudden changes can lead to unanticipated and undesirable results.[74]

International Institutions and Initiatives

In its report, *Our Common Future*, the World Commission on Environment and Development devoted a major section to proposals for needed institutional and legal changes at the national and international level, including recommendations for strengthening bilateral and multilateral development assistance agencies, United Nations agencies, and international nongovernmental organizations.[75]

There are encouraging signs that some international organizations are taking effective action on global environmental and resource concerns. Although the multilateral development lending institutions generally have not integrated environmental protection into their development programs, the World Bank has increased environmental staffing sevenfold and has strengthened the link between lending policies and environmental improvement.[76] Some United Nations agencies have been effective on resource and environmental issues, notably the Food and Agriculture Organization, UNESCO's Man and the Biosphere Program, the U.N. Population Fund, the U.N. Development Program, and especially the U.N. Environment Programme (UNEP). In July 1988, the directors of 22 U.N. agencies met in Oslo with WCED Chairman Gro Harlem Brundtland for the first time to discuss the global environment and to plan and coordinate their plans for sustainable development.[77]

Serving as the coordinating mechanism on environmental issues for the United Nations system and among nations, UNEP has helped initiate action to protect the atmosphere, forests, ocean and coastal areas, fresh water, soil, and biological diversity. As described earlier, UNEP occupies a key position in the Global Environment Monitoring System. Dr. Mostafa Tolba, Under Secretary General of the United Nations and Executive Director of UNEP, played the leading role in achieving the 1987 Montreal protocol, a 24-nation accord to protect the Earth's stratospheric ozone layer. Considering the scope and effectiveness of its efforts, UNEP merits increased support beyond its very modest current annual budget of less than $40 million.[78]

Jessica Mathews of World Resources Institute sees merit in the recent proposal by the Soviet Union to transform the U.N. Trusteeship Council, which has outlived the colonies it supervised, into a trusteeship to manage the global commons (including the oceans, biological diversity, and the atmosphere). She suggests that, properly structured, it could serve as a forum for reaching global environmental decisions at a higher political level than anything now available.[79]

A number of observers have addressed the need for new institutions and initiatives at the international level. For example, ecologist and policy analyst William Clark, noting the existence of international economic forums, has suggested the creation of analogous forums of world leaders that would address the issue of how to increase global security.[80]

Recently, there have been several important international initiatives to protect the Earth's natural resources. At the 1987 World Wilderness Congress in Colorado, some 1,600 delegates from 60 countries approved an initiative to create a World Conservation Bank that would raise money to protect critical biological resources.[81] (See Chapter 6.) In March 1989, 116 governments and the European Communities adopted the Basel Convention on the Control of Transboundary Movements of Hazardous Wastes. And in May 1989, 81 governments and the European Communities adopted a strong declaration committing them to phase out production of ozone-depleting chlorofluorocarbon gases (CFCs) by the year 2000.[82]

In 1988, *The Christian Science Monitor* and the Johnson Foundation organized a conference of 35 participants from 12 nations on issues of environmental degradation and international relations. The conference proceedings, titled "Agenda 2000," included three innovative proposals in the area of international environmental accounting that represent an extension of the recommendations on national resource accounting summarized earlier in this chapter. These proposals are:

- Establish an international system of environmental accounts designed to assess such transboundary issues as crossnational pollution, manufacturing, heating, and power-generating activities; and operations of multinational corporations.

- Strengthen those international environmental institutions that can perform environmental accounting tasks, provide mixed-nation research teams for basic fact-finding and analysis, and establish computer models for sound accounting.

- Consider creating a system of global rents for conservation of natural resources, with payments scaled to the GNP of contributing nations and to their historic use of nonrenewable resources.[83]

Because many environmental problems involve ecosystems, watersheds, or atmospheric regions that cross national borders, the solutions to these problems require regional analysis and action. The impacts, risks, and costs of such problems, the potential benefits and costs of corrective programs, and the responsibility for bearing those costs often can be estimated only by considering the affected region as a whole rather than individual nations.[84]

To meet the need for evaluation of critical threats to the world community and the biosphere, the World Commission on Environment and Development recommended establishment of a Global Risks Assessment Program that would:

- identify such global or regional threats;

- assess their causes and likely consequences;

- provide advice and proposals on what should be done to avoid, reduce, or adapt to the threats.

The Commission noted that the program would not require a new institution, but would serve to promote cooperation among nongovernmental organizations, scientific bodies, and industry groups.[85] As new technologies and burgeoning populations place increasing stress on global life-support systems, effective risk management is becoming a planetary necessity.[86]

To keep pace with the growing number and severity of global environmental problems, some observers feel that traditional patterns of international negotiation and agreement, which can take years to complete and implement, must be modified. Agreements such as the Law of the Sea Treaty typically require up to 15 years to negotiate and bring into force. Agreements to protect the environment, such as the Montreal protocol to protect the ozone layer, must be negotiated in the face of scientific uncertainty. Jessica Mathews suggests that a more flexible model is needed, one that allows a process of intermediate or self-adjusting agreements that can respond to growing scientific knowledge and understanding, that uses new economic methods for assessing risk, and that allows a more active role for scientists, for policymakers with technical competence, and for the private sector.[87]

The World Commission on Environment and Development called for a universal declaration and a subsequent convention on environmental protection and sustainable development, and for strengthened procedures to avoid or resolve disputes on issues of environmental and resource management.[88]

More generally, the Commission noted that the nation-state is inadequate for dealing with threats to global ecosystems, and concluded that the global commons can only be managed and protected from environmental threats through joint international efforts and multilateral procedures.[89]

Business and Industry

Private enterprise—especially the multinational corporation—has critical roles to play in protecting the global environment and moving toward sustainable development. Norman Myers has suggested that if multinational corporations (MNCs) could recognize their own and humanity's long-term interests in protecting global resources and the environment, it could open access to their vast financial resources, technological talent, and organizational skills.[90]

Business and industry can be especially effective in implementing environmentally sound practices in agriculture, forestry, ocean fishing, industrial manufacturing, energy production, and waste management. In agriculture, business could play a greater role in protecting cropland and promoting alternatives to the heavy use of agricultural chemicals. In forestry, corporations could make a greater effort to institute policies of sustainable forest management and reforestation.[91]

In the energy sector, private enterprise has already made impressive progress in conservation. Since the first oil price rise in 1973, annual global energy savings have averaged some $300 billion, mostly through private efforts. Notable progress in conserving energy and using it more efficiently has been made in areas such as burning waste to power industrial operations, developing cleaner and more efficient motor vehicles, and recycling of aluminum, steel, and other materials.[92]

The private sector has an especially important role to play in controlling air and water pollution and in reducing and recycling industrial wastes. Impressive accomplishments have already been made by Minnesota Mining and Manufacturing, a large multinational corporation with annual sales of over $8 billion and operations in more than 50 nations. Over a 12-year period, the firm reported saving nearly $400 million by recovering waste and recycling materials, including some 280,000 tons of solid waste and 4 billion liters of wastewater. In Europe, Monsanto Corporation, a large chemical manufacturing firm, has reported a 50 percent reduction of waste output per ton of product.[93]

Participation by business and industry in the kinds of environmentally-beneficial programs cited above could be expanded through the use of some of the economic and market incentives discussed earlier.

Norman Myers observes that although relatively few businesses have institutionalized environmental protection in their organizational structures, the multinational corporation, despite many defects, is an innovative entity for international collaboration with exceptional scope for operating across a broad range of jurisdictions and issues. Myers suggests that MNCs could offer an institutional pattern for certain kinds of international environmental management.[94]

Nongovernmental Organizations

Environmental groups, policy study centers, and other nongovernmental organizations (NGOs) have important roles to play in creating public awareness of environmental issues, developing policy recommendations, and generating political support for action on global issues. By their nature, NGOs tend to reflect the views of the public at large. Moeen Qureshi, World Bank Senior Vice President for Operations, has observed that "where bureaucratic eyes are astigmatic, NGOs provide vivid images of what is really happening at the grassroots." With their flexibility and their regional and global networks, NGOs can avoid bureaucratic inertia, adapt to new opportunities, and reach across national boundaries.[95]

Until about 1970, NGOs were relatively uncommon in the Third World, but in recent years their numbers in developing countries have grown rapidly. The Environment Liaison Center International in Nairobi maintains a registry of some 7,000 NGOs worldwide.[96]

The effectiveness of NGOs in the United States on issues of the environment, civil rights, women's rights, peace and disarmament, consumer protection, and other areas is well known. In developing countries, there are many examples of successful environmental action by NGOs, including initiatives to stem deforestation and plant trees in Colombia, Kenya, and India; a village-based campaign in Malaysia to protect fisheries from pollution and habitat destruction; a squatter upgrading program in Zambia, and campaigns to protect tropical forests and combat hazardous chemicals in Brazil, to cite only a few examples.[97]

In his survey of grassroots organizations in developing countries, Worldwatch Institute's Alan Durning reports that there are some 250 independent development organizations in Mexico, some 300 in Peru, and more than 1,000 in Brazil. In Indonesia, Durning reports 600 such groups working on environmental protection; in Bangladesh, he found that 1,200 such groups have been formed since 1971, and in India, he estimates the number of independent development groups at 12,000. Durning found that these organizations focused on a variety of environmental, economic, and social concerns, including tree planting, farming, developing appropriate technology, health care, and social welfare.[98]

Other studies have shown that partnerships between government and community organizations have led to progress in many sectors, including community forestry, soil conservation, watershed protection, the urban environment, and health and family planning.[99]

Norman Myers believes that many types of NGOs have potential for effective action in resource conservation and environmental protection, including academic and professional associations, service clubs, charity organizations, labor organizations, and private entrepreneurs. He suggests that they all can play an important part in meeting the global environmental challenge.[100]

Beginning in 1985, U.S. NGOs in the fields of development, environment, and population began meeting to develop ways to increase cooperation between groups in these fields. The effort produced a plan of action, *Making Common Cause*, that was endorsed by more than 100 U.S. organizations. In 1987, the International Council of Voluntary Agencies began work on an international version of the document, which was completed and endorsed by the ICVA Governing Board in 1988.[101]

Looking at the overall potential of private organizations and institutions for improving the prospects for sustainable development, Norman Myers envisions a network of sectoral and functional organizations, operating at the local, regional, national, and international level, with appropriate linkages between levels and sectors. He suggests that such a network could have the potential to provide a greater degree of self-regulating stability than any "global super-agency."[102]

As the world becomes increasingly interdependent, the importance of relationships based on cooperation and mutual benefit—among individuals, organizations, communities, and nations—is growing; and competitive relationships based on dominance, and concepts such as isolationism, radical individualism, rigid self-reliance, and absolute sovereignty are becoming obsolete.[103] Growing global interdependence may be accompanied by transformations in how we organize our affairs and work with others to achieve our objectives.

Needed Transitions to Achieve Sustainable Development

In 1984, participants at the World Resources Institute's Global Possible Conference agreed that several critical transitions must be completed if the goal of sustainable development is to be achieved:

- a *demographic* transition to a stable world population of low birth and death rates;

- an *energy* transition to high efficiency in production and use, and increasing reliance on renewable sources;

- a *resource* transition to reliance on nature's "income" without depletion of its "capital;"

- an *economic* transition to sustainable development and a broader sharing of its benefits;

- a *political* transition to a global negotiation grounded in complementary interests between North and South, East and West.[104]

In his overview of the Global Possible Conference, Robert Repetto noted that the conference results suggested that the prosperous and sustainable future that would be created by these transitions is attainable, and that there are ample opportunities to build such a future. Population can be stabilized, agricultural production can be expanded, economic growth can be sustained with lower energy input, forest resources can be stabilized and expanded, the loss of biological diversity can be arrested, minerals can be supplied, environmental pollution can be reduced, environmental quality can be preserved, and cities can be made healthier and more livable. Repetto concluded that these goals are not expensive to achieve, are usually less costly than the present course, and in many cases represent least-cost approaches.[105] Some of the economic benefits and costs of these objectives are discussed later in this chapter.

In its report, *Our Common Future*, the World Commission on Environment and Development concluded that the pursuit of sustainable development requires a number of developments, including:

- production that preserves the ecological basis for development;

- an economy that can sustain the generation of surpluses and new technical knowledge;

- technology that can find new solutions to problems;

- sustainable patterns of international trade and finance;

- effective citizen participation in decision-making;

- governance that is flexible and has the capacity for self-correction.[106]

To achieve these objectives will require improvement of existing institutions and management capabilities, adoption of some new structures, strategies, and methods, and concerted efforts and investments by government, private enterprise, nongovernmental organizations, and individual citizens.[107]

Current Progress toward Sustainable Development

As the issue chapters in this book have shown, there have been substantial advances in identifying effective strategies to limit the growth of human numbers, to use the Earth's natural resources more efficiently, and to prevent environmental degradation

There are many cases in which such strategies are being put into practice, and many examples of progress in slowing population growth and moving toward sustainable food production, energy use, and industrial output, both in developing and industrial countries.[108] This section summarizes the progress reported in the issue chapters.

Foresight Capability

Beginning with *The Limits to Growth* study for The Club of Rome in 1972, a number of global, regional, and national studies have analyzed demographic, economic, resource, and environmental trends and their interrelations. More than 30 nations and regions around the world have undertaken 21st century studies to explore alternative strategies for achieving sustainable economic development and security. In the United States, a number of groups have recommended that the President establish a government-wide process to ensure that all the consequences of proposed decisions including long-term, international, and cross-cutting effects are taken into account. Various United Nations agencies and other international and nongovernmental organizations analyze trends in population, food, energy, and other factors, and the International Geosphere-Biosphere Program has begun a study of global change, focusing on long-term future prospects and the continued habitability of the Earth.

Population Growth

In spite of rapid population growth in many countries, birth rates are declining for more than 90 percent of the world's population. Between 1960 and 1987, eight countries in East Asia and Latin America lowered their fertility rates by more than 50 percent. In Asia, the countries of Taiwan, South Korea, and China, along with the city-state of Singapore, have made remarkable strides. Thailand, Malaysia, Indonesia, and Turkey have also made substantial progress. In Latin America, the countries of Cuba, Chile, Colombia, and Costa Rica have achieved major reductions in population growth rates, while Mexico and Brazil have realized substantial declines.

Development and the Environment

There are many examples of progress toward sustainable development and use of natural resources in both developing and industrial nations. These include improvements in providing health care and meeting other basic human needs; the use of low-input, sustainable agricultural systems; sustainable management of tropical forests; expansion of fresh water and saltwater aquaculture; increases in irrigation efficiency and wastewater reuse; advances in energy efficiency and the use of renewable energy sources; reduction and recycling of solid and toxic wastes, and reductions in barriers to trade between nations.

Food and Agriculture

Farmers in a number of countries are beginning to use low-input, regenerative agricultural methods that reduce expensive and environmentally-harmful material inputs of pesticides and inorganic fertilizers, and that increase the use of biological pest controls, organic fertilizers, crop management, and other sustainable farming methods. In most cases, such methods result in a net economic gain for the farmer, compared with more costly high-input methods, and in greater long-term productivity of the land.

Biological Diversity

To safeguard the world's wild plants and animals, a global network of parks and reserves is being developed. Worldwide, the number of protected areas has grown from about 600 in 1950 to some 3,500 areas today, containing about 425 million hectares. A number of initiatives are under way to protect wild species and biologically-rich habitats. These include a range of laws and treaties, a number of "debt-for-nature" exchanges that provide economic incentives for conservation, and new policies by development lending agencies designed to protect biological diversity. Sustainable uses of forests and other ecosystems are also being pioneered that will allow species to survive despite commercial activity.

Tropical Forests

There is substantial evidence of progress in efforts to curb deforestation. Development agencies are providing increased funding for forestry and conservation projects. Forestry reviews are now planned for 30 tropical countries, under way in 11, and completed in three. More than a dozen countries have national forestry plans, and more than two dozen are developing projects for the sustainable production of forest products. As noted earlier, studies in Brazil and Peru show that net revenues from long-term harvesting of non-timber forest products are two to three times greater than from commercial logging or clearing the forest for cattle ranching. Worldwide, there are over 5,000 nongovernmental forestry and conservation organizations; many are working to protect the forests while meeting local needs for forest products.

Ocean and Marine Resources

Relatively few successes in the management and protection of marine resources can be cited, but there are hopeful signs. The management of ocean fisheries is improving, and a number of international and regional initiatives have been launched to control pollution and protect marine species and environments. The U.N. Environment Program's Regional Seas Program to control pollution has made progress, most notably in the Mediterranean Sea. The concept of a common marine heritage, though weaker than initially conceived, is still part of the Law of the Sea Convention, now signed by 159 nations and ratified by 39 (although not by the United States and several other major maritime nations).

Fresh Water

While global demand for water continues to grow, in the United States water use declined by 11 percent between 1980 and 1985. Because agricultural use accounts for more than 70 percent of global water consumption, new micro-irrigation techniques have major potential to slow the depletion of underground water supplies. Wastewater reuse in Israel, industrial water recycling in a number of countries, and a variety of household water-saving devices all offer the promise of slowing the growth in water demand. In the United States, industrial programs to reduce waste generation are saving water and reducing pollution, and a number of states have developed effective programs to protect drinking water quality.

Nonfuel Minerals

In the production of mineral commodities, there is potential for improved efficiency in all stages of production. In a number of industrial countries, recycling of metals, building materials, and other mineral products is saving money, creating jobs, reducing air and water pollution, and conserving energy and water as well as mineral supplies. When aluminum is recycled instead of produced from ore, energy use and air and water pollution are reduced by 90 percent or more. Although extraction of minerals from the sea poses difficult problems, the oceans are an important potential source for many minerals in the future.

Energy

Since the oil price increases of the 1970s and early 1980s, many countries have improved their energy efficiency. Between 1973 and 1985, per capita energy use in OECD nations dropped 6 percent while per capita gross domestic product rose 21 percent. Many methods have been developed for improving energy efficiency in transportation and industry, and for buildings and appliances. Technology now commercially available can reduce energy consumption in these sectors from 20 to 60 percent or more. A range of renewable energy technologies is being developed that could be competitive with fossil fuel energy sources within 20 to 30 years. These include geothermal, wind, solar thermal, photovoltaics, and alcohol fuels and other biomass sources.

Air, Atmosphere, and Climate

There has been substantial progress by industrial nations in protecting the atmosphere through improved energy efficiency, control of sulfur dioxide emissions, and reduction of chlorofluorocarbon (CFC) emissions. In 1988, 22 nations agreed to reduce sulfur emissions within 5 years by at least 30 percent of 1980 levels. In early 1989, the European Community agreed to end all use of CFCs by the year 2000. Recent international efforts suggest that a convention to regulate emissions of carbon dioxide and other greenhouse gases could be achieved in the near future.

Hazardous Substances

Several different methods are being developed to slow the proliferation of hazardous substances. These include reduction of toxic waste production in industry, the use of integrated pest management in agriculture to lessen dependence on hazardous pesticides, and methods to improve the storage of nuclear wastes and limit their future production. Genetically-engineered bacteria have been developed that can digest some toxic substances. Successful waste-reduction programs are in place in a number of countries, including Sweden, the Netherlands, Colombia, Venezuela, and the United States. Between 1975 and 1985, the Minnesota Mining and Manufacturing Company reported that its Pollution Prevention Pays program cut its waste generation in half and saved $300 million.

Solid Waste Management

Several industrial nations have effective programs to minimize the amount of municipal solid waste that goes to landfills. West Germany reuses or incinerates 45 percent of its solid waste, while Japan recycles or burns 73 percent. While the United States recycles only 11 percent of its solid waste, this represents an increase from less than 7 percent in the 1960s. State-of-the-art landfills can prevent groundwater pollution and recover methane gas as a fuel. Modern incinerators can greatly reduce waste volume and generate electricity. In the Third World, China and India have launched extensive waste recycling programs.

Global Security

Since 1946, there have been 16 international treaties or agreements relating to nuclear weapons, including agreements to prevent the spread of nuclear weapons, reduce the risk of nuclear war, limit nuclear testing, and limit stocks of nuclear weapons. In 1987, the United States and the Soviet Union agreed to eliminate an entire class of ballistic missiles, and made progress toward a 50 percent reduction in long-range nuclear weapons. Continued improvements in U.S.-Soviet relations could enhance chances for diversion of resources to sustainable development priorities. In less than a decade, China cut its military budget by 10 percent and substantially increased its investments in food production, reforestation, and family planning. These shifts helped China raise per capita food output by half and dramatically lower its birth rates.

The examples of progress cited in the issue chapters and summarized above represent only a sampling of the successful efforts around the world to move toward sustainable development. A number of organizations in the United States and elsewhere have published compilations of "success stories," and regularly publish magazines, newsletters, and other periodicals that document successful programs.[109]

In July 1989, Renew America launched a program called "Searching for Success: Meeting Community Needs Through Environmental Leadership." The program is designed to identify successful U.S. environmental programs in 22 categories ranging from air pollution to food safety to wildlife conservation, to list all successful programs in an Environmental Success Index, and to recognize successful programs with three levels of awards. The best program in each category will be given national recognition in 1990 in connection with the 20th anniversary of Earth Day.[110]

As the examples of progress summarized above demonstrate, we are improving our capability to recognize our global predicament, and to take action to reverse ecological and social deterioration and shift to a more sustainable course.

Even with these encouraging signs of progress, however, the preponderant pattern worldwide is "business as usual." In areas where change from environmentally damaging activity to sustainable practices is most needed, there are many sources of resistance to change—vested interests in the status quo, organizational inertia, bureaucratic rigidity. Many of the modest but real examples of progress documented in this book may stem directly from growing understanding of, and apprehension over, global threats such as air and water pollution, deforestation, and loss of species. We are in the midst of a vast global experiment—testing whether societies and governments can learn, adapt, and change through analysis and anticipation, rather than waiting for confrontation with crisis or catastrophe. And the outcome of the experiment is not yet clear. Those who want to assure that we can avoid crises, rather than wait and cope with them, still have an immense and urgent challenge to stimulate and guide change in policies and institutions toward a more sustainable course.

Benefits and Costs of Progress

The costs of implementing sustainable programs in the priority areas identified earlier in this chapter (population, poverty, agriculture, forestry and habitats, energy, water, and waste management) are generally small compared with the costs of allowing unsustainable practices to persist in these areas.

The short-term monetary costs of preventing pollution, resource depletion, and other environmental damage can be substantial,[111] but the economic and other costs of repairing ecological damage after it has occurred are often many times greater. And in some cases, such as loss of topsoil, extinction of species, and pollution of groundwater, the damage may not be reversible at any cost.[112]

Experience in industrial nations has shown that pollution control has avoided costly damage to health, property, and the environment, and has made many industries more resource-efficient and thus more profitable.[113]

A number of estimates of the magnitude of savings from beneficial programs can be cited. To help slow population growth, family-planning services cost about $20 per year for each user, a tiny fraction of the expense of meeting basic human needs of rapidly growing populations. The use of agroforestry techniques can eliminate the need for costly commercial fertilizers and make agriculture more sustainable. Watershed protection methods using mixed-farming systems that curb soil erosion and improve water retention can improve productivity and income. Fisheries could maintain their present yield at less than half the present cost if overfishing in certain areas were eliminated. Available options for improving the efficiency of energy use can save up to 85 percent of the cost of new energy supply, and since 1973, annual energy savings worldwide through conservation have been estimated at $300 billion. Safe drinking water can be supplied to low-income urban neighborhoods at one-tenth the cost that some slum dwellers pay water carriers and vendors. And the most cost-effective measures to reduce industrial pollution can be 40 to 60 percent cheaper than conventional methods of pollution control.[114]

Expenditures to protect natural resources are increasing. In the United States, for example, spending for environmental protection increased from less than $20 billion in 1972 to about $70 billion in 1985, a 3.5-fold increase in less than 15 years. In Europe, the pollution-control market is approaching $30 billion per year, and it is growing at a rate of more than 6 percent a year.[115]

The future quality of life on Earth will depend on whether we can recognize the true long-term benefits and costs of economic activity, and on whether we can act to minimize the costs and bring economic development and population growth into a long-term balance with the Earth's natural resource base.

Prospects for Future Progress

Concern about environmental trends appears to be growing among leaders, opinion-makers, and the public at large. In the World Future Society's review of recent literature, *Future Survey 1988-89*, editor Michael Marien judged unfavorable environmental trends to be number one in a ranking of ten top fears based on perceptions of "unfavorable trends and plausible pessimism" expressed in the literature reviewed. A year earlier, environmental fears had ranked number two, and two years earlier, they had ranked number three.[116]

Important prerequisites for effective action to deal with environmental problems include public awareness of the problems and political support for programs to deal with them. Based on the results of public opinion polls over the last two decades, there is strong evidence that support is increasing for progress on environmental protection and toward a sustainable future.

Public Opinion

There are many indications that concern about the state of the environment is growing. Polls on environmental issues in the United States conducted by *The New York Times* and CBS News, the Gallup, Harris, and Roper organizations, and other groups confirm that U.S. support for environmental protection has been increasing during the past 15 years. Membership in national environmental organizations has grown from about 4 million in 1980 to around 7 million in 1987, and there may be as many as 25 million U.S. citizens involved in local and regional environmental issues.[117]

Internationally, the environmental organization Greenpeace has grown from a membership of some 1.4 million members around the world with revenues of about $24 million in 1985, to more than 3.3 million members with about $100 million in revenues today. Greenpeace now maintains 33 offices in 20 countries.[118]

In the United States, a public opinion survey by the Roosevelt Center for American Policy Studies in October 1988 showed that energy and the environment was ranked as the third most important priority for the new president, after the federal budget deficit and education. Another poll by the Gallup organization in October found that "laws to protect the environment" were ranked as the second highest priority for the next administration, after federal budget deficit reduction. And a *New York Times*-CBS poll in November 1988 indicated that 59 percent of the respondents wanted the government to spend more on the environment.[119]

In mid-1989, a *New York Times*-CBS poll reported that the proportion of Americans supporting environmental protection *regardless of cost* had risen from 45 percent in 1981 to 80 percent in 1989. In May 1989, a national survey by Opinion Research Corporation showed that most Americans are worried about water and air pollution. Roughly eight out of 10 were very concerned about drinking water quality and air quality, two-thirds were very concerned about pesticides in food, nearly six out of 10 were very concerned about ozone depletion, and half were very concerned about the greenhouse effect.[120]

Between February 1988 and June 1989, Louis Harris and Associates conducted the first ever worldwide environmental survey, commissioned by the United Nations Environment Program. The survey covered 16 nations, including three in Asia, four in Africa, one in the Middle East, three in Europe, four in Latin America and the Caribbean, and the United States. The survey revealed that many people in industrial and developing countries alike expressed serious concern about environmental quality and about their leaders' ability to control environmental degradation. But the survey also showed that both citizens and their leaders believed that the trends could be reversed if environmental protection became a major priority.[121]

Among the survey findings are the following:

- Most people, in all countries except Saudi Arabia, rated the environments of their countries as only fair or poor; majorities or pluralities in these countries believed that their environments had become worse in the last 10 years.

- People in nearly all countries were highly concerned about the pollution of drinking water, rivers and lakes, the air, and the land. Large majorities in all countries believed that there was a direct link between environmental quality and public health.

- Sizable majorities in almost all countries also expressed grave concern about the loss of agricultural land, deforestation, desertification, radioactivity, toxic wastes, and acid rain. There was less awareness of, and less concern about climate change or depletion of the ozone layer.

- Industrial activity and governmental inaction were seen as the two most serious causes of pollution.

- In spite of their pessimism, most people did not believe that environmental degradation was inevitable. Majorities of both the public and leaders in all countries appeared to believe that the environment could be protected and improved if these goals became a major priority.

- In all countries, large majorities of the public believed that environmental protection should be a major governmental priority. With the exception of Brazil, leaders in all countries agreed.

- Most people in the survey would choose a lower standard of living with fewer health risks over a higher living standard with more health risks.

- Majorities of both the public and leaders in all countries except Argentina would be willing to pay higher taxes to protect the environment.

- Most people believed that protecting the environment should involve governments, international organizations, business, farmers, voluntary organizations, and individuals.

- In all countries except Japan, large majorities said they would be willing to work with others in their communities to improve their environment, either by working on environmental projects two hours a week or by contributing money.[122]

The Harris survey showed that nine in ten people polled want stronger national and international action to curb pollution and reverse environmental decay; people in less developed countries are even more alarmed about environmental destruction than those in more developed nations; and the public in almost every country is more concerned than the leadership, which "grossly underestimates the worldwide environmental crisis." The survey suggested that it could take "the ouster of some governments due to a failure to act on cleaning up the environment to convince leaders of just how serious this issue is."[123]

Political Leadership

The Harris survey makes it clear that people around the world want action to protect the environment, and are will-ing to help pay for it. The survey shows that the public is ahead of their leaders on environmental issues, and wants national and international leadership to reverse environmental decline.

The environmental agendas for the United States reviewed earlier in this chapter contain a number of recommendations about how the United States can exercise leadership on environmental issues, many of which are relevant to other nations. Among these recommendations are the following:

- Elevate the administration of environmental programs to cabinet-level status.

- Through high-level diplomacy, focus the attention of national leaders on the range of growing environmental threats to national and global security, and on the growing interdependence of all nations with regard to global environmental issues.

- Promote cooperation with other nations, and with international organizations such as the United Nations Environment Program, as essential for effective action on environmental problems.

- Exercise leadership in convening international conferences and in working for international treaties to protect the global environment.

- Make the environment a top agenda item at economic summit conferences and other meetings of international leaders.[124]

In their environmental agenda for the United States, former Colorado governor Richard Lamm and Thomas Barron suggest that a major leadership challenge facing the U.S. administration is to tap the U.S. public's strong support for environmental action, and maintain it even if such action means higher costs for some goods and services or spending cuts in other areas. They urge the United States to mobilize the private sector in the cause of environmental protection, and to support new treaties and programs to encourage sustainable development around the world. Lamm and Barron note that America still has the time and resources needed to regain the initiative against our growing environmental problems, but they pose the crucial question: Does America also possess the will?[125] The World Commission on Environment and Development concluded that ultimately, sustainable development depends on political will.[126]

Citizen Action

While national governments increasingly move toward programs designed to promote sustainable development, initiatives by provincial, state, and local governments, and action by nongovernmental organizations and individuals, may account for much of the actual progress in making development more sustainable. The efforts by local organizations, citizen groups, and households to create sustainable communities and adopt sustainable lifestyles are a fundamental part of efforts to achieve a sustainable world.

Each issue chapter in this book suggests specific ways in which individuals and local groups can contribute to sustainable development and use of natural resources. A separate summary chapter, "What You Can Do," offers general guidelines as to how citizens can organize activities and accomplish objectives related to the issues discussed in this book. There are many ways that each of us, acting as individuals or in concert with others, can work effectively to reverse environmental decline and move toward sustainable development. In fact, timely action by individuals and groups is the essential ingredient in achieving progress toward a sustainable future.

Further Information

Books

Berger, John J. *Restoring the Earth: How Americans Are Working to Renew Our Damaged Environment*. New York: Anchor Press, 1987. 241 pp. Paperback, $9.95. Presents case studies of how citizens are rescuing rivers, streams, lakes, forests, prairies, and wildlife.

Blueprint for the Environment: Advice to the President-Elect from America's Environmental Community. Washington, DC, 1988. Executive Summary, 32 pp., available from Natural Resources Council of America, single copies free. Book-length version, 352 pp., paperback, $15.45 including postage from Howe Brothers, P.O. Box 6394, Salt Lake City, UT 84106. A compilation of recommendations from 18 U.S. environmental organizations.

Cahn, Robert. *Footsteps on the Planet: A Search for an Environmental Ethic*. New York: Universe Books, 1978. 277 pp. Hardback, $10.95. Chronicles the struggle between public need and private gain, and examines attitudes toward the environment in business, industry, and government.

Clark, William C. and R.E. Munn (eds.). *Sustainable Development of the Biosphere*. A publication of the International Institute for Applied Systems Analysis. New York: Cambridge University Press, 1986. 491 pp. Paperback, $17.95. Chapter 1 provides an overview of current research results and future research priorities.

Cornish, Edward (ed.). *Global Solutions; Innovative Approaches to World Problems*. Bethesda, MD: World Future Society, 1984. 160 pp. Paperback, $6.95. Contains innovative proposals in such areas as population, health, energy, agriculture, and international relations.

Darrow, Ken and Mike Saxenian. *Appropriate Technology Sourcebook: A Guide to Practical Books for Village and Small Community Technology*. 800pp. Paperback, $17.95. Available from Appropriate Technology Project, Volunteers in Asia, P.O. Box 4543, Stanford, CA 94305.

Global Tomorrow Coalition. *Sustainable Development. A Guide to Our Common Future*, the Report of the World Commission on Environment and Development. Washington, DC, 1989. 77 pp. Paperback. Available from the Coalition for $2.00 to cover shipping and handling.

Kidder, Rushworth M. *Reinventing the Future: Global Goals for the 21st Century*. Cambridge, MA: The MIT Press, 1989. 185 pp. Hardback, $17.95. Results of a 1987 conference of 35 participants from 12 nations entitled "Agenda 2000: Reasonable Goals."

Marien, Michael. *Future Survey Annual 1988-89*. Volume 9. A Guide to the Recent Literature of Trends, Forecasts, and Policy Proposals. Bethesda, MD: World Future Society, 1989. 212 pp. Paperback, $25.00 Contains abstracts of recent publications on economics, energy, environment and resources, food and agriculture, science and technology, world futures, and many other topics.

Myers, Norman (ed.). *Gaia: An Atlas of Planet Management*. Garden City, NY: Anchor Books, 1984. 272 pp. Paperback, $18.95. Contains many examples of progress toward sustainable development.

Our Common Future: Healing the Planet. A Resource Guide to Individual Action. Los Angeles, CA: Physicians for Social Responsibility and Beyond War, 1989. 32 pp. Paperback, $2.50. Includes sections on diet, consumer purchasing, resource use, political and social action, and use of leisure time.

Project 88: Harnessing Market Forces to Protect Our Environment: Initiatives for the New President. A Public Policy Study sponsored by Senator Timothy E. Wirth, Colorado, and Senator John Heinz, Pennsylvania. Washington, DC, 1988. 76 pp. Available from the offices of Senator Wirth or Senator Heinz. For a summary, see Robert N. Stavins, "Harnessing Market Forces to Protect the Environment," *Environment*, January-February 1989.

The Panos Institute. *Towards Sustainable Development*. London, 1987. Paperback, $15.00. Available from The Panos Institute (see Appendix for address). Fourteen case studies prepared by African and Asian journalists for the Nordic Conference on Environment and Development, Stockholm, May 1987.

Rambler, Mitchell B. et al. (eds.). *Global Ecology: Towards a Science of the Biosphere*. San Diego, CA: Academic Press, 1989. 204 pp. Hardback, $24.95. Transcends the parochial boundaries of academic disciplines to show the various facets of the global system within a planetary perspective.

Redclift, Michael. *Sustainable Development: Exploring the Contradictions.* New York: Methuen, 1987. 221 pp. Paperback, $14.95. Argues that environmental degradation is not "natural," but an historic process linked to political and economic structures. Calls for more emphasis on indigenous knowledge and experience, and for effective political action on behalf on the environment.

Reid, Walter V. et al., *Bankrolling Successes: A Portfolio of Sustainable Development Projects.* Washington, DC: Environmental Policy Institute and National Wildlife Federation, 1988. 48 pp. Paperback, $6.00. Includes reviews of 20 projects in Asia, Africa, and Latin America.

Rolston, Holmes. *Environmental Ethics: Duties to and Values in the Natural World.* Philadelphia: Temple University Press, 1988. 391 pp. Paperback, $16.95.

Renew America. *The Environmental Success Index.* Washington, DC, forthcoming. A compilation of successful environmental programs in the United States, grouped in 22 categories. *Searching for Success: Meeting Community Needs Through Environmental Leadership.* Washington, DC, forthcoming. Will contain descriptions of the top three environmental programs in each of the 22 categories. Both publications to be released in April 1989 to coincide with the 20th anniversary of Earth Day.

Repetto, Robert et al. *Wasting Assets: Natural Resources in the National Income Accounts.* Washington, DC: World Resources Institute, 1989. 68 pp. Paperback, $10.00. Instead of treating natural resources as gifts of nature, the book calls for their treatment as productive assets whose value must be depreciated if they are used up.

Schumacher, E.F. *Small is Beautiful: Economics as if People Mattered.* New York: Harper & Row, 1973. 324 pp. Paperback, $9.95.

Seymour, John and Herbert Girardet. *Blueprint for a Green Planet: Your Practical Guide to Restoring the World's Environment.* New York: Prentice Hall, 1987. 192 pp. Paperback, $17.95.

World Commission on Environment and Development (WCED). *Energy 2000: A Global Strategy for Sustainable Development.* 1987. 76 pp. Paperback. *Food 2000: Global Policies for Sustainable Agriculture.* 1987. 131 pp. Paperback. *Industry 2000: Strategies for Sustainable Industrial Development.* 1987. 47 pp. Paperback. Available from the Centre for Our Common Future.

WCED. *Our Common Future.* New York: Oxford University Press, 1987. Paperback, $10.95. Contains a comprehensive series of recommendations for achieving sustainable development.

World Resources Institute. *The Crucial Decade: The 1990s and the Global Environmental Challenge.* Washington, DC, 1989. 22 pp. Paperback, $5.00. Contains brief descriptions of critical world environmental problems and measures to deal with them.

Articles

Bingham, Tayler H. "Social Values and Environmental Quality." In Daniel B. Botkin et al. (eds.), *Changing the Global Environment: Perspectives on Human Involvement.* San Diego, CA: Academic Press, 1989, pp. 369-82.

Brown, Lester R., Christopher Flavin, and Sandra Postel. "No Time to Waste: A Global Environmental Agenda for the Bush Administration." *World Watch*, January-February 1989, pp. 10-19.

Brown, Lester R. and Edward C. Wolf. "Reclaiming the Future." In Brown et al., *State of the World 1988.* New York: Norton, 1988, pp. 170-88.

Durning, Alan B. *Action at the Grassroots: Fighting Poverty and Environmental Decline.* Worldwatch Paper 88. Washington, DC: Worldwatch Institute, January 1989. 70 pp. $4.00.

Mathews, Jessica Tuchman. "Redefining Security." *Foreign Affairs*, Spring 1989, pp. 162-77. Analyzes the importance of natural resource and environmental determinants of security, and the need for new strategies for planetary management.

Myers, Norman. "Environmental Challenges: More Government or Better Governance?" *Ambio*, Vol. 17, No. 6, 1988, pp. 411-13.

Myers, Norman. "Writing Off the Environment." In John Elkington et al. (eds.), *Green Pages: The Business of Saving the World* (London: Routledge, 1988), pp. 190-92.

Peterson, Russell W. "An Earth Ethic: Our Choices." In National Geographic Society, *Earth '88: Changing Geographic Perspectives.* Washington, DC, 1988, pp. 360-77.

Pollock, Sarah. "A Time to Mend," *Sierra*, September-October 1988, pp. 51-55. A review of environmental restoration projects, mostly in the United States.

Reid, Walter V. C. "Sustainable Development: Lessons From Success," *Environment*, Vol. 31, No. 4, May 1989. A summary of the book, *Bankrolling Successes*, listed above.

Repetto, Robert (ed.). "Agenda for Action." In Repetto (ed.), *The Global Possible: Resources, Development, and the New Century*. New Haven, CT: Yale University Press, 1985, pp. 496-519. Recommendations from a 1984 World Resources Institute conference of 75 leaders from 20 nations.

Ruckelshaus, William D. "Toward a Sustainable World." *Scientific American*, September 1989, pp. 166-74. Outlines policies that can allow development and growth to take place without exceeding ecological limits.

Scientific American. "Managing Planet Earth," September 1989. Contains authoritative articles on population growth, climate and atmospheric changes, threats to biological diversity and fresh water supplies, strategies for agriculture, energy use, industrial manufacturing, and sustainable economic development; and needed changes in values, institutions, technology, management, and leadership.

Shrader-Frechette, Kristin. "Environmental Ethics and Global Imperatives." In Repetto, *The Global Possible*, pp. 97-127.

Talbot, Lee M. "Man's Role in Managing the Global Environment." In Botkin, *Changing the Global Environment*, pp. 15-34.

Time Magazine, "Planet of the Year: Endangered Earth," January 2, 1989, pp. 24-63. Includes an agenda of recommendations from a conference of 35 participants from 12 nations.

Periodicals

Ambio: A Journal of the Human Environment. Published eight times a year by Pergamon Press, Maxwell House, Fairview Park, Elmsford, NY 10523. Subscription $43 a year.

Environment. Published 10 times a year by Heldref Publications, 4000 Albemarle Street, Washington, DC 20016. Subscription $23 a year.

Future Survey. A Monthly Abstract of Books, Articles, and Reports Concerning Forecasts, Trends, and Ideas about the Future. Published monthly by the World Future Society.* Subscription $59 a year; includes *Future Survey Annual* (see listing under books).

Regeneration Newsletter. Published quarterly by Rodale Press, 33 East Minor Street, Emmaus, PA 18049. Subscription $12 a year.

The Renew America Report. Published quarterly by Renew America.* Subscription is included in the annual membership fee of $25.

World Watch. Published six times a year by Worldwatch Institute.* Subscription $20 a year.

Worldwatch Papers. Published monthly by Worldwatch Institute.* Annual subscription $25, which includes the annual publication, *State of the World*.

Films and Other Audiovisual Materials

Our Common Future. 1988. Presents the findings of the World Commission on Environment and Development, set up by the United Nations in 1983 to re-examine the planet's critical environment and development problems and formulate realistic proposals to solve them. Available in English or Spanish for purchase ($25.00) or rental ($7.50) from Global Tomorrow Coalition.

What is the Limit? 1987. Discusses interrelations between human population growth, environmental degradation, resource depletion, habitat destruction, and ethical considerations for the future. Includes a discussion guide, "Where Do We Go from Here?" 23 min. Grade 10 to adult. Video: VHS, Beta, or 3/4 inch, $25. National Audubon Society.

Teaching Aids

Christensen, John W. *Global Science: Energy, Resources, Environment*. 2nd Edition. Dubuque, IA: Kendall-Hunt, 1984. Textbook, 355 pp.; Laboratory Manual, 265 pp.; Teacher's Guide, 365 pp. A new edition is in preparation.

For additional teaching aids, see Chapter 1.

---

* Addresses for the publishers of these periodicals are given in the Appendix.

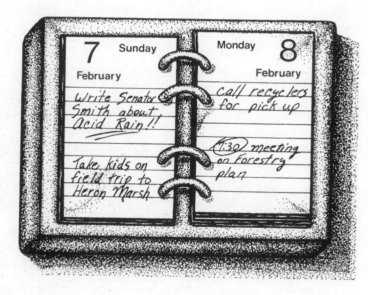

What You Can Do

Never doubt that a small group of thoughtful, committed citizens can change the world. Indeed, it's the only thing that ever has.

— Margaret Mead

When it comes to the future, there are three kinds of people: those who let it happen, those who make it happen, and those who wonder what happened.

—Carol Christensen, quoted in *Making It Happen: A Positive Guide to the Future*[1]

Global Problems, Local Actions

Previous chapters in this book have suggested specific actions that citizens can take to help solve global problems related to energy use, tropical deforestation, species loss, pollution of soils, air, and water, and many other factors. This chapter offers general guidelines for what you can do to respond to any of the issues posed in the book, including how to inform yourself, join with others, publicize your views, work with elected officials, organize for greater effectiveness, and raise money to support your projects.

Getting Started

There are a number of steps you can take to become active and to make a difference on a global issue that concerns you.

Inform Yourself

The power of information can be magnified by the numbers of people who have it and use it. Learn about the local aspects of a global issue. For instance, are there endangered species living in or near your area? In your state? How does your community dispose of its wastes? Are there industries or farms whose practices endanger the air, water, or soil? What are the projected growth patterns for your community and state? How many jobs in your state are dependent on exports? What committees do your congressional representatives influence? What are their voting records?

Where to Look

As you begin to gather information and gain experience on global issues, use your own contacts and also visit local chapters of Global Tomorrow Coalition (GTC) member groups. They have more than 10,000 chapters and affiliates, and over ten million individual members, throughout the United States. Many of them are people like yourself who, while long concerned and active on domestic issues, are now becoming increasingly concerned about global problems.

Get to know other people and organizations who recognize, appreciate, and respond to your concerns. You might start with a local chapter of a United Nations Association, Y.M.C.A., or Y.W.C.A.; a Zoo, World Affairs Council, 4H Club, or Association of Returned Peace Corps Volunteers; an international exchange organization or international program of a college, development association, or religious group; or a dedicated computer bulletin board service such as ECONET/PEACENET. Identify your shared concerns, interests, and skills. Many jobs that are too much for one individual can be easily managed when divided among several people. Besides, you'll find that your understanding of the issues and your approach will be improved by the comments, criticisms, and suggestions of others.

Become Active

Take an active role in public meetings and activities focused on global issues. Contact prominent business leaders in your area and seek their views on these concerns. Write letters to the editors of local newspapers, and talk to local officials, boards, and regulatory bodies. Work to get important issues on local and state ballots.

Join the Nongovernmental Organization (NGO) Community

Each group has its own identity and purpose. Yet the work of one organization often parallels or even overlaps that of another. There is a good chance that you already know or have worked with groups that have goals similar to yours. But it is also important to get acquainted with organizations that may differ from yours in purpose but can be very helpful in working on a particular issue.

Build Cooperative Alliances

Look for opportunities to form "cooperative alliances"— that is, identify areas of common cause with groups that you normally would not join. Even if you don't share an organization's stand on some issues, you may find that you agree sufficiently on other important issues to make collaboration valuable. For example, environmental groups can find ad hoc allies in civic, social, and cultural organizations, as well as in health, peace, disarmament, development, education, and youth groups, to name only a few.

Forming a Global Issues Committee

Organize a meeting to discuss formation of a global issues committee. Personally invite other potential members. If you are meeting under the auspices of a group with which you are already affiliated, you could advertise the meeting in their bulletin. Discuss which global issues are most pertinent to your community. Analyze those issues. Examine how local practices and policies affect those issues on a global scale. Look at the impact of global practices and policies on local and national interests, well-being, and quality of life.

Plan Your Strategy

Discuss how you and your colleagues can make a difference. Find others who may already be working on the issues within your community or state; don't duplicate their work, rather amplify their effectiveness. Set priorities in terms of time, energy, resources, and impact. Seek ways to broaden your outreach. Rather than form a new, separate organization, try to work within an existing structure whenever possible.

Benefits and Rewards

Any organization with which your global issues committee affiliates can benefit from the relationship. The benefits can include: (1) increased public exposure and membership; (2) new networks that can lead to cooperative alliances with local, national, and international groups; (3) an enlarged information base that can improve program planning; and (4) greater respect and name recognition. A successful effort can lead to the satisfaction of knowing that you have taken concrete, positive action towards solving a critical world problem.

Become a Pioneer

Working locally on global issues is a new and exciting enterprise. You will learn much in the process and contribute to a movement that, during the 1990s, will shape the future our children will inherit. Create a positive model. Your successes might be pertinent to other locales and could be transferable to other groups, through GTC's membership network.

Make Local Connections to Global Issues

Start by linking the principal global issue on which you're working to your own community. Look for some local, tangible connections to make the larger global concern of immediate importance to others. Some examples of global/local connections might be:

- Global Warming - local and state energy policies and plans.

- Endangered Species - a propagation program at your local zoo or local efforts to monitor or protect endangered migratory species.

- Acid Rain - local industrial practices, and their effects on local, state, and regional environment.

- Tropical Forests - local use of food, medicines, and other products from tropical forests; local observation of migratory birds from the tropics; seed bank programs at botanical gardens; local forestry practices.

- Foresight - local or state Year 2000 studies and trend analysis reports.

- Food Security - local poverty and hunger issues, local export/import patterns, local agricultural practices.

- Population - local growth patterns, urban planning, land-use laws, traffic congestion, open space, environmental quality, tax measures.

Organize Exciting Projects

Successful projects should be fun, educational, and carefully focused; they should achieve tangible results. You may wish to develop your ideas through activities such as meetings, reports, newsletter articles, book reviews, slide shows, films, field trips, and presentations with guest speakers. After you have gathered a core group and have gained some momentum, you may take on larger awareness-building events like a GTC Global Town Meeting, Cooperative Alliance Forum, Teacher-Training Workshop, or help to organize a Globescope Assembly. Details about these activities are available from the Global Tomorrow Coalition. There are also opportunities through other organizations for related activities including fairs, tours, exchange programs, sister-city programs, and international liaison programs with nongovernmental groups in other countries.

Changing Lifestyles

We often underestimate the importance of personal actions and consumer choices in shaping our society. Institutions, businesses, and government are geared to respond to external stimulus. You can initiate that stimulus, and be at the center of change. The world is shaped through the accumulated personal choices and daily decisions of people just like you. Take time to examine how you and your family are living and what you are buying. Check to see if the patterns of your life are consonant with the values of global sustainable development.

Consider the different ways that families in other societies meet needs that are similar, or identical, to yours. For example, think through the implications of the fact that the citizens of several West European nations enjoy standards of living at least comparable to that of the United States, with roughly half the per capita energy consumption.

Take Personal Responsibility

Talk with your family about your collective lifestyles. You don't have to make radical changes — sell your house, quit your job, or join a bean curd cult in Tibet — to make a difference. Start by making small changes or modifications and let others know why you are changing your habits.

Effective Changes

Here are some examples of simple changes in lifestyle that can make a difference:

- Minimize your consumption of resources. Lower the thermostat in your house in winter. Conserve water. Use low-wattage, energy-conserving light bulbs. Buy refrigerators and air conditioners that are ozone-friendly. Notice the energy-efficiency ratings of all appliances.

- Be aware of the momentum of global and national population growth, and its relevance to every environment and development issue, as you make choices about the size of your family.

- Properly dispose of hazardous substances used in the home. Use washable cotton diapers. Recycle aluminum cans, newspapers, glass, and other household waste products.

- Eat lower on the food chain by consuming more vegetables and fruits and less meat. Ask your grocer to stock organically-grown produce.

- Give priority to gas mileage when you buy a car. Get more exercise. Leave the car at home when you can ride a bike or walk. Use public transportation or organize a car pool.

- Avoid buying exotic or endangered animals as pets, or any product made from wild animal fur, reptile skin products, ivory, or coral.

- Maintain conditions in your yard to benefit birds, insects, mammals, and plants. Reduce or eliminate your use of chemical fertilizers and pesticides. Plant trees and bushes. Grow more of your own vegetables at home.

- Ask for paper bags at the grocery store, or better yet, bring reusable shopping bags with you. Use ceramic and glass containers instead of styrofoam or plastic. Buy pump, rather than aerosol, sprays.

- Use recreational time to increase your own awareness of global issues and get to know others interested in the same issues.

Communicating Your Views

Become confident of your views on global issues, and then begin to make them known to leaders in key sectors.

Get Around Town

Volunteer to speak to clubs, classes, and assemblies. Discuss the coverage of global issues with local newspaper editors and program directors of radio or television stations. Work with existing voluntary and nonprofit groups, schools, and communication media. If these channels prove inadequate, collaborate with others in creating new forums and citizen coalitions on the issues.

Know Your Facts

Explain your views within a broad perspective, demonstrating the significance and long-term implications of the local actions required on each problem. Be accurate and fair; both qualities breed credibility. Base your views on the most current information available. Select relevant facts and figures to back up your concern and support your viewpoint. Be clear and concise.

Make News

The best way to communicate with the largest audience most effectively—and often most quickly—is to work with the communication media. The media can help you reach an audience much larger than you could ever hope to reach through your own means. A letter to the editor or an op-ed piece printed in a newspaper not only will be read by thousands of people with whom you could not speak personally, but also will reach elected officials whose staffs monitor both the print and electronic media for new issues about which voters back home are concerned.

Seek Advice

Talk to the public relations office or publication editor of a GTC member organization or other sympathetic group. Through their national and state offices and local chapters, they have day-to-day experience in working with the press. Ask them for samples of effective and well-written news releases. Get tips on organizing press briefings and press conferences that provide the background information in sufficient detail to assist reporters in writing their stories. Learn how to cooperate with and cultivate journalists over the long term.

Attend Events and Interview Others

Go to local events—a public speech, a symposium, a demonstration, a hearing—and examine what type of media coverage they generate. Try to understand why some topics that interest you greatly are sometimes not "news." Learn how to turn events into photo opportunities. Your stories are always enhanced by pictures.

Look for different angles for stories. Interview a visitor from a developing country, a local business executive whose firm has ties around the globe, an official of a union whose members' employment depends on the import of natural resources from overseas or the export of some manufactured product. Their perspectives and opinions on global issues will be especially valuable in communicating with others in your community because they are locally based.

Starting with Education

No major problems will be resolved in our society, or in any other, without a major, long-term educational effort. Today's school and college students are tomorrow's voters, taxpayers, and leaders.

Work with Teachers

Suggest ways in which topics such as tropical deforestation, biological diversity, sustainable development, population growth, ocean and coastal resources, food security, global warming, ozone depletion, acid precipitation, and the interrelations among local and global issues, can be infused into teachers' current lesson plans. Assist in making contacts between teachers and institutions such as zoos, research centers, and nongovernmental organizations. Arrange for speakers to address combined classes or assemblies. Donate posters for classroom use or books and video cassettes to school and college libraries.

Support Global Issue Mandates

Attend PTA meetings and let school superintendents and school board members know your concerns. Work with them and with state governments to establish new — and/or strengthen existing — global issue mandates in your state. To date, more than 34 states have passed such mandates, which authorize funds for the purchase of materials (curriculum units, books, posters, films and other audiovisual materials), and require that a certain percentage of curricula, usually at the secondary school level, be devoted to understanding global issues as credit for graduation. Most school systems are interpreting such mandates only in limited terms of teaching foreign languages, broadening cultural awareness, and increasing economic competitiveness. Few administrators and educators are actively using the concept of global sustainable development in their work, or preparing curriculum materials on global resources, population, environment, and development from a holistic viewpoint that stresses the interrelations among issues. The GTC, several of its member groups, and a growing number of other organizations have materials on these subjects (see Further Information section). Review these materials and propose their use in local schools. Help organize a teacher-training workshop or presentation to school superintendents or a school board as a starting point.

Work with Colleges and Universities

Review curricula of colleges and universities to see how they cover population, resource, environment, and development issues. Encourage colleges to develop programs of integrated studies that include these issues and are designed to train generalists to deal with the interrelations among global issues.

A number of colleges and universities now offer courses on global issues. Dr. Russell Peterson, past president of the National Audubon Society and a founder of the Global Tomorrow Coalition, has taught such courses at Dartmouth College, Carleton College, and the University of Wisconsin.[2] Miami University in Ohio is actively considering establishment of an undergraduate school of interdisciplinary

studies, and Texas A & M University is considering creation of a graduate school of integrated studies.

The first edition of this book, titled *Citizen's Guide to Global Issues*, has been used not only by citizen group leaders but in courses at some 50 colleges and universities. Worldwatch Institute's annual publication, *State of the World*, has been used in some 750 courses at more than 450 colleges and universities. The Morrison Institute for Population and Resource Studies at Stanford University recently published a collection of readings titled *Population and Resources in a Changing World*. In addition to population, the book includes selections on energy, fresh water, land, food, biological resources, and the atmosphere.[3]

To become involved in global issues, college students in the United States and Canada can join public interest research groups (PIRGs) based at colleges and universities and run by students. Such PIRGs now exist in a number of U.S. states, including California, Colorado, Massachusetts, New York, and Oregon.

Create an Exchange Program

Encourage schools to form partnerships with schools in Third World countries. Look into founding an exchange program or making use of one of the several existing channels for student exchange. Seek assistance from local college students who come from the country with which you want to establish an exchange. Foreign students in the United States are a greatly under-utilized resource. They have personal knowledge about their countries' history, customs, institutions, and problems, and may have useful contacts to help get the program started. Contact university department heads for recommendations of able foreign students in your area.

Shaping Public Policy

There remains much to be done to encourage more positive and effective public policies at every level on long-term national and global issues.

Your Opinion Can Make a Difference

On most domestic issues that preoccupy Members of the U.S. Congress, there are organized and vocal constituents back in the home states and districts. They make themselves heard, and Senators and Representatives listen, knowing that the message they receive today in a letter, phone call, or visit may very well influence how a vote will be cast on the next election day. On global issues, however, the constituency is often less visible, less organized, and much less often heard. The result, not surprisingly, is that Members of Congress are often reluctant to appropriate funds for foreign aid or to support international organizations and activities that affect future global trends.

Changing this climate of apathy and inaction is one of GTC's key functions. GTC members recognize the need to address significant global challenges ahead of us, but know that we must work together, or else the multiplicity of voices may simply overwhelm our elected representatives. Together we can pool our unique resources in ways that will capture public attention for global issues. Working cooperatively, citizen activists can tap those resources and magnify their influence at the local, state, and national level. Progress on this front must begin with the involvement of concerned individuals like yourself.

Think Globally, Act Locally

This is the essential guideline for your community activities and constituency building. Broader public awareness of global problems is essential to move our country toward greater participation in effective remedial action in our own self-interest. In the community at large, there is much that can be done by a local organization to promote awareness of the need for dealing with global issues. Because an issue may have consequences in many different areas, you should look for aspects of the problem that are likely to be of concern to local organizations.

Citizen Action

Information, organization, public outreach, and direct communication with elected officials are the keys to effective citizen activism. Start by carefully defining your objectives. Decide which global issues concern you the most and are most worthy of your devotion of time and energy. Identify local organizations and individuals that you think may have similar interests. Seek them out, verify your shared concerns, and find out what they might be willing to do together with you.

Get to Know Your Elected Officials

Know who they are, meet their staffs, learn where their offices are, and what views they have expressed.[4] When you are seeking to involve your elected representative in an issue, it is vital to communicate its importance to his or her political base as well as to the nation as a whole. Don't neglect the staffs of local constituent-service offices.

Write Letters and Make Personal Contact

Begin your letter-writing campaign. A letter to a member of the House of Representatives is addressed to:

> The Honorable _____
> U.S. House of Representatives
> Washington, D.C. 20515
>
> Dear Representative _____:

A letter to a member of the Senate is addressed to:

> The Honorable _____
> U.S. Senate
> Washington, D.C. 20510
>
> Dear Senator _____:

Persuade others to write letters, too. Letter-writing is a simple but powerful political tool for communicating with elected officials. The staffs of Members of Congress and other elected officials read their mail and *count* it, too (sometimes, when the volume is exceptionally high, they even *weigh* it). Their offices have sophisticated systems for registering the receipt of mail, formulating an official response on an issue, and communicating that response to citizens who request it.

Writing a letter on an issue to your elected representative should not be a once-in-a-lifetime event but the beginning of a steady and increasingly respectful and productive relationship. Your letter should combine all the benefits of your information search, organization, and outreach. It should be short and direct, stating the facts of your concern clearly and politely in one page, if possible. Make the local relevance of your position clear. Work with others to make sure that they, too, write personal letters. Always ask your representative for a specific response, and always follow up politely, regardless of whether the response is positive or negative. Threats and accusations are counterproductive with most elected officials. If they disagree with you or are indifferent today, don't be put off. Instead, pursue the issue further, addressing or soliciting the particular objections the official may have and showing him or her that there are authoritative people who share your point of view.

The same is true if you send a telegram, make a telephone call, or pay a visit. Minds are made to be changed, and the holders of public office are as sensitive to that imperative of political life as anyone else. Successful political careers are based on responsive positions. No matter how urgent the crisis of today, keep in mind that, more often than not, you will have another opportunity to work on it tomorrow. An elected official may be turned around the next time, if you don't burn all your bridges now. And every public official is keenly aware that there will be another election in two, four, or six years.

Some membership groups within the Coalition have detailed manuals describing the political process, and suggesting how you can help shape policy.[5] Learn from these publications and from the experience of others.

Organizing Community Activities

In every community, the number of strong leaders who can command wide respect and attention, and motivate action, are few. Learn to be one of them.

Unite the Community

Community activities mobilize, maintain, or increase public interest. Any organization or group of concerned individuals can plan community activities, which can be as diverse in form and character as fairs, contests, sporting events, shows (flower shows, fashion shows, art shows), public debates, receptions, lectures, barbecues, picnics, clam bakes, or other occasions for eating and drinking. Other possible community activities are seminars, film showings, parades, "happenings," "earth days," "days" focusing on local or regional landmarks, "days" commemorating historical events, and combinations of these and other activities.

These activities should unify a community, not divide or polarize it. They should bring people together and inform them on the issues in question, arouse or increase their concern, suggest or reaffirm solutions or objectives, and encourage further action.

Attract Attention

A major purpose of the activity may be to draw public attention to a particular issue via the communication media, to engage the interest of policymakers, or even more frequently, both. Both the media and the political leaders will be attracted to the activity according to the potential number of participants. Furthermore, their goals are reciprocal: for the politician, the more media attention the better, and for the media, the more political participation the better. Remember that policymakers and journalists are very busy and also very adept at discerning whether an event merits personal attendance — you need to convince them.

Raise Funds

Community groups such as Kiwanis, Rotary, Junior Chamber of Commerce, Junior League, League of Women Voters, 4H Club, Garden Clubs, and the American Association of University Women can help by providing volunteers, fundraising and financial assistance, public promotion, and contacts with other groups and individuals. Of course, those most capable of funding projects are businesses and industries. Some community organizations avoid seeking money from business and industry, because they fear they may be "tainted" by such support; some prefer to maintain an adversary posture. Remember, however, that an organization does not have to sell its soul—only its ideas. The real challenge is to conceive of a good project with worthwhile ends that will also sell well in the corporate marketplace. Corporations are like individuals; they want to know where their money is going, and to receive public credit for their assistance.

Fundraising has developed into a profession in its own right. However, it is not necessary to hire a professional. There is usually plenty of free, voluntary advice and assistance available if you know where to look for it. Local independent institutions like colleges, schools, and charities have staff persons, advisers, or board members who know a lot about the subject. If properly approached, they usually can give you all the help you need. Sometimes the head of a community foundation can provide invaluable guidance on local sources of support.

Learning From Success

One good way to gain visibility for your ideas within the community and to bring people together in public association is to hold a fair. Many groups could probably participate at some level in an environmental fair, for example.

Where to Start

Here are some steps in planning such an event:

1. Select the right site. It should be attractive, easily accessible to many people, and able to accommodate booths, activities, and crowds.

2. Pick an attractive theme — even community groups can focus on issues of international interest and importance.

3. Invite other organizations to participate as planners, exhibitors, activity sponsors, and fundraisers.

4. Make certain to choose a chairperson or a steering committee to which the community can bring its ideas and concerns. The chairperson or the committee should also serve as the official communications link with the media.

5. Choose the dates and activities carefully. Avoid conflicts with traditional local events. Selection of activities will depend on the location, number of participating groups, anticipated attendance, promotional and publicity value of the activities, and strength of group support, not to mention available budget. If, for example, the theme "Oceans" were chosen for the fair, here are some activities that might work well:

» "Beach-In" — a teach-in on the beach to focus on ocean problems and their solutions.

» Free University Mini-Workshops — short community outreach sessions at a shopping mall or other public site on different marine subjects. These could be led by faculty from a local college or university.

» "Swim-a-Thon" — a fundraiser staged in the school pool. Pledges are made for each lap swum, with the money allocated to some worthwhile project related to the oceans.

» Ocean Film Festival — show your favorite surfing, fishing, or beach footage, and collect a small fee for attendance.

» Surf Fishing Contest — don't let the big ones get away; release them!

» Sand Sculpture Contest — create your favorite sea creature.

» Surfing Contest — another charitable, competitive event, provided you have the waves to make it worthwhile.

» King Neptune/Pisces Parties — social events just for fun, but with an ocean theme. Costumes are limited only by imagination and local regulation.

» Seafarer's Luncheon — Menu: a fish dish, submarine sandwiches, and Old Joe Frogger whaler cookies.

» Ocean Auction — a great way to dispose of valuable and not-so-valuable sea-related items and memorabilia.

» Ocean Art Sales or Contests — make waves in your local artists' community.

» Ocean Poster Contest — a prize for the best poster/slogan combination about marine conservation.

» Concerts — anything from Handel's "Water Music" to the Beatles' "Yellow Submarine."

A Word of Caution

Don't be overly ambitious or expect too much. Fairs can be very effective in attracting and providing diversion for large numbers of people, and can incorporate many different kinds of events that provide a broad range of appeal and educational opportunity. But they usually require very strong and broad community support, plus a lot of hard work by many loyal volunteers. Something less ambitious may be more appropriate as a starter.

Don't expect an event such as a fair to solve environmental problems instantly. That is not its main purpose. But the attention generated will help establish useful relationships for future problem-solving, and will promote your interests more widely and effectively throughout the community.

The event may also raise money that the organizations can use in a follow-up project. Prior designation of such an activity as the recipient of fair proceeds could boost attendance if the project is popular in the community or perceived as worthwhile and worthy of support.

A Real World Example

Here is a real-life example of a very modest but still broadly appealing and successful community activity: a seafood feast held by an environmental organization in Southern Virginia. The Environmental Organization (EO) aimed to dramatize the importance of protecting an ecologically productive and sensitive area from pollution-prone industrial development. To sponsor, supply, prepare, and serve the feast, the EO enlisted the cooperation of the local seafood industry (fishermen, processors, wholesalers, retailers, restaurant owners, etc.), rented a large waterfront area, advertised the event, invited the public, and served a variety of delicious regional seafood dishes at modest prices. Some speeches and presentations were made, but the emphasis was on food and conviviality. The attendance proved large and enthusiastic, and the EO made a substantial net profit to help finance its other activities. Persuading public policymakers, civic leaders, and the media to attend was easy. The activity was successful for the following reasons:

1. The EO selected a sure winner for an activity in that region — one bound to be popular and attractive.

2. The activity centered around a local enterprise that united the community in appreciation and pride.

3. The issues were handled in the most positive and happy way possible; acrimony was avoided; merchants got advertising; politicians got exposure; the media got a story; people enjoyed good food at low prices; and everybody had fun.

4. The EO acquired friends, money, favorable publicity, and wider public support.

Other Activities

The kinds of community activities that may be productive and successful vary widely. Here are some examples:

Educational activities such as seminars, lectures, debates, and courses. For small groups of very interested individuals, such activities can be effective. Instructors with expertise in the subject as well as teaching experience are essential. Well-known lecturers can be a special asset, but they require careful selection and may demand fees.

Film showings are often effective, particularly when combined with some other social or educational event. Caution: wherever possible, view the film in advance. It may be different from what you anticipated from the film's title.

Public debates and speeches are most apt to be effective when two other ingredients are present: well-known speakers and a high degree of current public interest in the subject.

"Earth Days" focusing on conservation and other ideas may be effective depending on the pre-existing level of public interest and the ability of the promoters to put together a variety of attractive events.

Shows such as art shows, fashion shows, flower shows, exhibits of artifacts, or even natural phenomena may be effective depending on local circumstances. The subject matter of the show should ordinarily relate to the subject or issue or objective that is the ultimate purpose of the show. Otherwise, the two will tend to detract from each other.

Parades, demonstrations, and sit-ins may be appropriate, depending on the circumstance, but they tend to provoke adverse or extreme, as well as positive, reactions. They also may raise problems with public authorities. They probably should be resorted to only under special circumstances where a favorable climate of opinion prevails and where there are reasonable assurances or safeguards against rowdy behavior and violence.

Library exhibits can often be arranged by calling the local library and asking for exhibit space to display materials produced by the United Nations and other leading international organizations. Find out how much space is available and order your materials — posters and booklets. Ask your librarian to announce the exhibit in regional library bulletins, and mount a companion display of pertinent books.

Church sermons are highly effective means of broadening the constituency of concern. Contact local clergy and ask them to give sermons on long-term global issues. Provide them with any necessary background and materials. Work closely with church fellowship programs and adult seminar series to follow-up on specific issues.

Publicity materials are always vital. Where possible, put up posters in public places and on college campuses. Arrange free Public Service Announcements (PSAs) on radio and television. Contact the calendar section of the local newspaper.

A Reminder

A variety of different activities, carefully selected to interest as many different kinds of people and groups over as long a period of time as possible, is likely to produce better results than concentrating on a big, one-shot effort.

Entertainment value is of fundamental importance to virtually any kind of community activity. In some appropriate fashion, the activity must attract and hold the voluntary attention or participation of the people for whom it is intended. One of the best ways to do that is to make sure people have fun.

Forming International Ties

Contributions from your local group to NGOs in Third World countries can go a long way. Even $2,000 a year in direct transfers or in-kind services could *double* the capacity of many local organizations working in Latin America, Africa, Asia, and the Caribbean, where a worker's annual wage may be less than $1,500. And contributions need not always be in cash. Most Third World NGOs are starved for information and international contacts, so that books, jour-

nals, reports, and other material can be of great value. Moreover, you can develop a continuing special relationship—an NGO tie—between your committee or group and a compatible group in another country. The advantages always run both ways; the insights you will gain from their perspective and field experience will bring your group many benefits.

NGO Ties

Long-term global problems in environment, development, population, and resource management are *not* imperialistic themes of the "North." You'll find dedicated, educated, hard-working citizens in every country of the Third World who have long been seized by the urgency of these issues and are working for their resolution. Indeed, most face far greater obstacles than you do; some have even been persecuted or killed because of their strong stands on these issues. They are in every occupation, including government, business, labor unions, education, law, and the nongovernmental organization community. Their numbers and their effectiveness are growing. Groups in the United States can learn much from the approaches, successes, and knowledge of Third World groups. They, in turn, can benefit from our access to data and comparative sophistication in organization, administration, and fundraising. Not all of our information is directly transferable, of course, but much is.

A Two-Way Street

NGO ties are international partnerships, built on mutual respect, goodwill, and a strong bond of common concern. Match the global issues you are addressing with a country that is directly affected by that issue. If, for example, you are working on tropical rainforest issues, you might consider working with groups in Central or South America. Some issues, like global warming, might lend themselves more to working with groups in Europe, China, or the Soviet Union. After you have identified likely countries, be guided by the people who have knowledge about those places and may also want to become involved.

Find a Partner

Once you have selected a country, discuss possible NGO choices with the international program directors of GTC member groups. You might also consult Returned Peace Corps Volunteers, officials of the United Nations Environment Program, tour guides and travel agents, and zoos or aquaria. Check with department heads of local universities. Often they have research programs overseas or have foreign students enrolled in their programs. Parks or reserves in other countries can also be directly supported.

The World Wildlife Fund, the Nature Conservancy's International Program, and the East-West Center in Hawaii have listings of NGOs in other countries. Partners of the

Americas has published a Natural Resources Directory that includes many NGOs in Latin America and the Caribbean. The Environment Liaison Center in Nairobi, Kenya has more than a hundred NGO members in Third World countries. And as part of its child and family welfare development education project, the National Association of Social Workers' International Activities Committee has a program to promote "twinning" between its chapters and affiliated social worker organizations from developing countries.

Points to Consider

Maintaining correspondence may be difficult because of language barriers. If you don't speak the foreign language or have access to people who can translate documents, then choose an NGO that can communicate in English. Many can and are also interested in receiving data and current studies in English. In your preliminary contacts, state clearly the goals, purposes, and short- and long-term aspirations of the partnership. Exchange background information about your organizations, including by-laws, financial status, policies, boards, and administrative structures. If the partner group has an established track record, the chances for long-term program cooperation are much greater.

Direct Support

You may wish to send equipment and materials to your "ties" group. Books, reports, computers, typewriters, adding machines, binoculars, and slide projectors are all useful items that might be solicited through a newsletter or newspaper article. Ask a local shipping company to donate space for sending the materials. If you decide to contribute cash, consider raising funds by selling posters, T-shirts, and other items produced by the partner group. Many of the designs from these organizations in developing countries are unusual, artistically striking, and sell very well in the United States.

If you are organizing a tour to visit your partner's country, add a tax-deductible contribution to the price of the ticket as a contribution to their work. You can also subcontract with them to organize the land portion of your trip, or a particular 1-3 day segment, such as a visit to a park or nature reserve. It is important for U.S. tour groups to meet with their local members and establish personal relationships. Encourage reciprocal visits to the United States, with home-stay arrangements to reduce costs and foster closer friendships.

Support International Programs of U.S. Groups

Many organizations in the United States support efforts on a variety of global concerns through public information activities and political action. A number of organizations have programs that help establish nature preserves in tropical forests and fund projects to conserve endangered species; these include the African Wildlife Foundation, Conservation International, Nature Conservancy's International Program, Wildlife Conservation International, and World Wildlife Fund (see the Appendix for addresses). Financial support for these private conservation initiatives is essential. Your committee could "adopt" one of these programs, and provide support by organizing a benefit dance, a special movie preview, or other activities.

Working with the Media

The communication media are "blessed" with a great deal of independence and discretionary power. Because of this independence, you cannot capture the media's attention by bullying or coercion; such tactics simply won't work. Instead, cooperate, cultivate, and assist.

Establishing Contacts

With smaller newspapers, one contact, perhaps the city editor, may be all you need. Meet with this person to describe your group and its main concerns. Thereafter, you should refer all questions and materials to that person. For larger newspapers with many departments, you may establish several contacts, such as news and feature writers covering science, health, environment, or foreign policy. But it is best to give a single story to a single contact. You can easily lose the support of the entire newspaper if the same story appears in two different columns of the same edition.

If your community group seeks a working relationship with the broadcast media, your most likely contacts will be public service directors, program directors, and news directors.

Many local public access networks are interested in airing local community news and can be helpful in filming, editing, and producing video segments for local and state distribution.

Promotional and Publicity Activities

Once your organization has established relationships with the media, here are ways that they can help publicize the issue:

- Newspaper Articles—identify leading reporters and editors of local newspapers and ask them to write stories or a series on global issues. Offer to provide them with materials.

- Letters to the Editor—submit brief but informative letters to the editor of your local newspaper(s), either to respond to their coverage of a specific issue or event, or to call their readers' attention to global issues or events of general interest.

- Press conferences—for significant events of media interest, you may wish to hold a news conference to which you invite print and broadcast media. Examples of appropriate topics for a news conference are a visit by a prominent public figure, announcement of a new campaign or success in a major endeavor, a public event or demonstration protesting government policies, or a summary of the major actions taken at a large conference. For details on organizing a press conference, consult one of the media guidebooks listed under Further Information below.

- Calendar of Events—notify your local newspapers well in advance of local meetings, fairs, and other activities for listing in their "calendar of events" section. Your local wire service offices or press club may also maintain a listing of daily events of interest to the press.

- Newspaper Advertisements—design ads and take them to your local newspaper and request that they be used on a space-available basis.

- Opinion Polls—ask the local newspaper to undertake an opinion poll related to the U.S. role in global issues, perhaps linked to some current newsworthy event, and to publicize the results.

- Television and Radio Shows—write or call local television and radio stations requesting them to put on special programs. Offer to locate films for them or request a rerun of a documentary on global issues. Offer to locate spokespeople to participate in panel discussions or interviews.

- Television and Radio Spot Announcements—ask the local station to play spots on global themes. Local stations and/or your local university's department of communications might be willing to help in the preparation of the spot announcements.

Public Service Announcements

The broadcast media are required by law to devote a certain amount of time or space to community (nonprofit) organizations, and newspapers cover community events and issues as a means of increasing local circulation. Find out what time or space is available to your group and take advantage of it as often as possible. In broadcast media, the Public Service Announcement (PSA) is your organization's most promising approach. For print media, particularly for magazines, a camera-ready advertisement or announcement may be a good investment.

Only in exceptional cases should community groups pay for time or space. If you begin this practice, the media will assume that you can afford the expense and will no longer be willing to provide pro bono services. Exceptions include advertisements sponsored by an outside source for your organization, or supported by grants dedicated to promotion. In the latter case, make it clear to the media that only the one-time grant made payment for the advertisement possible.

Many groups compete for the relatively small amount of public service time or space available. To improve your chances of success, make sure that your material is brief, interesting, accurate, of good quality, and presented in a format to which the media is accustomed.

General Guidelines

Here are some general suggestions for working with the media:

- Know the deadline requirements for each member of the media with whom you deal, and time the presentation of your material accordingly. Missing a deadline is deadly!

- Be newsworthy. Don't cry wolf to the media; they will quickly turn you off. Your existence as an organization does not automatically guarantee success with the media. Success must be earned by making news and providing factually accurate information.

- Take time to thank the media for the coverage they give your issues. You might consider instituting a special award for the best local coverage of global issues or having a special recognition lunch or reception in their honor.

Working with the Global Tomorrow Coalition

The Coalition encourages a wide diversity of approaches to the U.S. role in dealing with global problems—including the perspectives of leaders from business, labor, civic organizations, government, church-related groups, academia, research, and communication media — in the belief that no single source has the ultimate answers on how to achieve a more sustainable future. Only through vigorous and open public dialogue can consensus be achieved as the basis for more creative and energetic policies and programs in both the private and public sectors.

Our Role

Founded in 1981, GTC provides a framework for nongovernmental organizations, institutions, and businesses — especially those active in the fields of environment, conservation, public health, resources, population, development, food and agriculture, biological diversity, education, and related areas — to share information, broaden public understanding of long-term global trends and their potential effects on the United States, and encourage active U.S. leadership, at all levels, in the search for solutions to related problems.

How We Work

The GTC builds new action constituencies for a sustainable future by:

- organizing leadership forums on sustainable development with universities and a wide spectrum of local citizen groups;

- publishing newsletters, reports, and reference documents;

- promoting the replication of community-access Global Issues Resource Centers;

- conducting teacher-training workshops and producing curriculum materials on global issues for elementary and secondary schools; and

- extending the Globescope Process, a series of major public assemblies that stresses the relevance of long-term global trends to the average United States citizen, fosters stronger U.S. leadership on global issues, and promotes direct dialogue between citizen leaders from developing countries and community leaders in the United States. A listing of proceedings and action plans from Globescope assemblies is included in the Further Information section.

GTC also serves as a clearinghouse of information about its members' legislative goals, which are summarized in periodic reports and occasional action alerts. The Coalition maintains liaison with a broad spectrum of governmental and nongovernmental activities in the international field, and encourages cooperation between groups in the United States and in developing countries.

Constituency Building

GTC can assist individuals and groups concerned with global issues by working with them to coordinate events such as Global Town Meetings or Cooperative Alliance Forums, and by providing background information, lists of publications, descriptions of audiovisual programs, and suggestions of appropriate speakers. For periodic updates on legislative initiatives on global issues, as well as names, addresses, and phone numbers of people and organizations who are working with Congress, write:

Citizen Action Network
Global Tomorrow Coalition
1325 G Street N.W., Suite 915
Washington, D.C. 20005-3104

United States Assistance to Other Countries

In many areas, concerned Americans can have the greatest impact on the resolution of global issues by influencing the policies and programs of the U.S. government. However, for those whose previous grassroots activity has been confined to domestic issues, it is necessary to become acquainted with a new set of institutions and their functions — particularly those having to do with foreign assistance. The United States provides assistance to other countries directly ("bilaterally") through several U.S. government agencies and departments, and indirectly ("multilaterally") through various international organizations and banks. The main bilateral and multilateral channels for U.S. foreign assistance are described below.

Bilateral Assistance

Among the U.S. government agencies and departments that assist other countries are the Agency for International Development, Peace Corps, Department of Agriculture, and Department of the Interior.

Agency for International Development (AID) AID provides grants and loans for projects in developing countries and helps support international assistance programs of other federal agencies. Historically, AID has served the dual functions of providing humanitarian aid and development assistance to developing nations while simultaneously promoting U.S. political, economic, and security interests. There is no escaping this duality, which is mandated by the U.S. Congress. U.S. foreign assistance is thus an expression of both American generosity and American political and economic self-interest in helping to shore up the fragile economies of friends and allies in the Third World.

AID also must respond to a great many categoric requirements embodied in the Foreign Assistance Act and related legislation. Congress has instructed AID, for example, to address energy-related problems in the developing world and use its programs to assist in the preservation of biological diversity. AID has supported projects for watershed management, reforestation, and fuelwood planting. However, some feel AID could do even more to help slow the rapid destruction of tropical moist forests. The Agency has been urged to make greater efforts to stimulate planting in forest regions; to create incentives for setting aside protected areas; to increase support for research and testing of agroforestry techniques; to announce its refusal or reluctance to support destructive projects such as cattle ranching, road building, and hydroelectric development in tropical forests; and to support development and use of sustainable forestry practices.

Through its small Biden-Pell Development Education Program, AID also is able to extend modest matching support to the efforts of selected U.S. non-profit organizations that are involved in broadening public understanding of the U.S. stake in overseas development.

Peace Corps   The Peace Corps places experienced U.S. volunteers in developing countries for periods of two years or more to work independently on projects to improve the quality of life and economic opportunity for the local people. Often Peace Corps volunteers work in cooperation with specialists from other agencies such as AID, or with the development specialists of other nations. Because the volunteers commonly live and work for a protracted length of time in direct contact with the indigenous populations of Third World countries, the Peace Corps experience is unique and intense. Returned volunteers commonly possess an unusual level of understanding of the practical, everyday problems of environment and development, and the enormous impact of U.S. policies and programs throughout the world. They are now widely dispersed throughout the public and private sectors, often in influential positions, and can be extremely valuable allies in activities related to global issues.

Department of Agriculture (USDA)   As one of its many mandated responsibilities, USDA provides training and information to tropical countries on soil and forest management. USDA's Forest Service operates an Institute of Tropical Forestry in Puerto Rico and an Institute of Pacific Islands Forestry in Hawaii, both of which could usefully serve as national demonstration and training centers for tropical forest management, if so designated by the President and given increased funding.

Department of the Interior   Interior provides training and information on the establishment and management of parks and natural areas as well as environmental interpretation through the National Park Service. It also conducts cooperative programs with tropical countries in conservation of forest ecosystems through the Fish and Wildlife Service.

Multilateral Agencies

Multilateral assistance to developing countries is provided through agencies of the United Nations and through development banks. The influence of the United States within these organizations is substantial, since it generally contributes a large proportion of their funds. Our government's International Development Cooperation Agency (IDCA) and State Department plan the budgets and policies for U.S. participation in the U.N. agencies. The Treasury Department's Office of Multilateral Development Banks performs the same functions in relation to the banks. The multilateral agencies listed below are particularly strongly involved in development projects.

Food and Agriculture Organization (FAO)   FAO helps countries improve nutrition and living standards, increase food production, and improve food distribution. FAO has the largest concentration of tropical forestry expertise in the world. It assists the multilateral development banks in appraising projects for investment, implements development projects of the United Nations Development Program (UNDP), produces technical publications on natural resources, and sponsors technical assistance projects.

United Nations Development Program (UNDP)   UNDP is the world's largest channel for multilateral technical assistance. UNDP projects are designed to help developing countries make better use of their human and natural resources, improve living standards, and expand economic productivity. There are some 5,000 UNDP-supported projects currently under way, valued at about $7.5 billion.

United Nations Environment Program (UNEP)   UNEP serves as the coordinating mechanism on environmental issues for the United Nations system and among nations. The small secretariat of some 175 professionals are engaged in environmental monitoring, and in the formation of scientific, technical, and legal working groups to initiate programs, conventions, and concerted actions to avert the destruction of the atmosphere, forests, ocean and coastal areas, fresh water, soil, and biological diversity. UNEP's strength has been shown in its capacity to bring the international community together on accords addressing shared seas and rivers, the export of hazardous waste and chemicals, and the Montreal Protocol on Ozone.

United Nations Educational, Scientific, and Cultural Organization (UNESCO)   UNESCO's primary objective is to contribute to world peace and security by promoting collaboration among nations through education, science, culture, and communication. UNESCO coordinates the Man and Biosphere (MAB) program, which supports pilot projects in ecological training and research for management of tropical forest areas. The MAB program also includes an important project for setting aside protected natural areas (biosphere reserves). MAB has been a successful, low-budget program that relies largely on participating countries to fund their own efforts independently. Nonetheless, U.S. support of the program has been diminishing.

United Nations Population Fund (UNFPA)   UNFPA is the largest internationally-funded source of assistance to population programs in developing countries. The major portion of its funds, almost all of which come from voluntary governmental contributions, is allocated to family-planning projects. After pledging to contribute $46 million to UNFPA in 1985, the United States reduced its 1985 contribution by $10 million, and has eliminated it entirely since 1986.

Multilateral Development Banks   The World Bank and other multilateral development banks (including the Inter-American, African, and Asian Development Banks) provide loans to governments for development projects. The projects they fund, such as dams, roads, and industrial forest exploitation, are motivated by the goal of economic development but can sometimes have major adverse impacts on the natural resource base. In 1980, the chief executives of the World, Inter-American, and Asian Banks and seven other multilateral development agencies pledged to institute environmental assessment procedures and to support projects to protect the environment.

The World Bank, which earlier had only a small staff to assess the environmental impacts of development projects, has recently created an Environment Department and now has about 60 people working full-time on environmental aspects of development. The Bank also has started to support projects to sustain natural resources, and its executives have begun to speak widely and positively on the issues of environment and development. But it is still unclear what impact the new Department will have on the Bank's overall ability to promote sustainable development in the Third World.

The regional banks have made less progress in considering the environmental consequences of their projects. Explicit instructions from Congress to the U.S. directors of these banks on the importance of a stronger environmental effort, and stronger expressions of public interest, are needed.

What You Can Do

Opportunities for improving U.S. foreign assistance programs arise whenever Congress considers bills to authorize the programs or to appropriate funds for agency activities. By making your views known to the appropriate Members of Congress at these times, you can help shape U.S. policies and programs. The National Audubon Society is conducting a major national public awareness-raising campaign on the opportunities to shape U.S. foreign policy through the revision and reauthorization of the Foreign Assistance Act. Other sources of information on congres-

sional activity are the GTC annual report on members' legislative priorities, and the publications and newsletters of many GTC member groups.

By becoming informed about legislation on important global issues and by timely communication of your views on the issues to your elected officials, you can help shape U.S. policies that can contribute to a more sustainable global future.

Further Information

A number of organizations can provide information about how to become active on global issues.*

Books

Anzalone, Joan (ed.). *Good Works: A Guide to Careers in Social Change*. New York: Dembner Books (80 8th Ave., New York, NY 10011), 1985. 288 pp. Paperback, $18.00 postpaid. Lists public interest organizations, gives information on job and internship opportunities.

Barone, Michael and Grant Ujifusa. *The Almanac of American Politics 1990*. Washington, DC: National Journal, 1989. 1536 pp. Paperback, $39.95. Provides complete information about members of the U.S. Congress and U.S. state governors.

Boyte, Harry C. et al. *Citizen Action and the New Populism*. Philadelphia, PA: Temple University Press, 1986. 232 pp. $19.95.

CEIP Fund. *The Complete Guide to Environmental Careers*. Washington, DC: Island Press, 1989. 350 pp. Paperback, $14.95. Presents information needed to plan a career search: job outlook, salary levels, volunteer and internship opportunities, and entry requirements. Case studies show how environmental organizations, government, and industry are working to manage and protect natural resources.

Center for Innovative Diplomacy. *Building Municipal Foreign Policies*. An Action Handbook for Citizens and Local Elected Officials. Irvine, CA, 1987. 59 pp. Paperback, $6.00. A guide to thinking globally and acting locally.

Commission on Voluntary Service and Action. *Invest Yourself: The Catalog of Volunteer Opportunities*. A Guide to Action. New York, 1986. 108 pp. Paperback, $5.00.

Council on Economic Priorities. *Shopping for a Better World*. A Quick and Easy Guide to Socially Responsible

* These include: Center for Innovative Diplomacy, Common Cause, Co-op America, Friends Committee on National Legislation, Friends of the Earth, Global Tomorrow Coalition, League of Women Voters, National Audubon Society, Parliamentarians Global Action, Public Citizen, RESULTS, Renew America, Sierra Club, and World Future Society. Addresses for these organizations are listed in the appendix.

Supermarket Shopping. New York, 1988. 124 pp. Paperback, $4.95. Rates the makers of over 1300 brand name products on 10 social issues.

CEIP Fund. *The Complete Guide to Environmental Careers.* Washington, DC: Island Press, 1989. 350 pp. Paperback, $14.95. Presents information needed to plan a career search: job outlook, salary levels, volunteer and internship opportunities, and entry requirements. Case studies show how environmental organizations, government, and industry are working to manage and protect natural resources.

Council on International Educational Exchange. *Volunteer! The Comprehensive Guide to Voluntary Service in the United States and Abroad.* New York, 1988. 155 pp. Paperback, $4.95 plus $1.00 for postage and handling.

Dunn, Thomas G. *How to Shake the New Money Tree: Creative Fund-raising for Today's Nonprofit Organizations.* New York: Penguin Books, 1988. 185 pp. Paperback, $7.95.

Global Action Network. *The Action Guide.* A Guide for Citizen Group Action. Ketchum, ID, 1989. 8 pp. Copies of the Action Guide, a sample Action Manual, a Sample Action Group Charter, and an Organizer Kit are available from the Global Action Network at cost.

Hamilton, John Maxwell. *Mainstreet America and the Third World.* Cabin John, MD: Seven Locks Press, 1986. 220 pp. Paperback, $10.95.

League of Women Voters. *Action Handbook.* Publication No. 161. Washington, DC, 1978. $1.50.

League of Women Voters. *Impact on Congress: The Grassroots Lobbying Handbook.* Publication No. 835. Washington, DC, 1987. $2.25.

Lipnack, Jessica and Jeffrey Stamps. *The Networking Book: People Connecting with People.* New York: Routledge and Kegan Paul, 1986. 192 pp. Paperback, $12.95. See especially Chapter 4, "Caretakers of the Planet."

Making Common Cause: A Statement and Action Plan by U.S.-Based International Development, Environment, and Population NGOs. Washington, DC: World Resources Institute, undated. 23 pp. Includes specific recommendations for public education, public policy, and improving collaboration in field activities.

National Audubon Society. *Foreign Assistance Action Project: Creating a Better World Through Political and Community Action.* Washington, DC, 1989. 12 pp. Paperback, free.

National Audubon Society. *The Guide for Citizen Action.* Washington, DC, 1981. 40 pp. Paperback, $2.50. Topics covered include the U.S. Congress, letters and telephone calls, organizing for action, using the media, and lobbying.

National Wildlife Federation. *Conservation Directory 1989.* 34th Edition. Washington, DC, 1989. 331 pp. Paperback, $15.00 plus $3.25 for shipping. A comprehensive listing of governmental and nongovernmental organizations and personnel engaged in conservation work at state, national, and international levels.

Richardson, John M. Jr. (ed.). *Making It Happen: A Positive Guide to the Future.* Washington, DC: U.S. Association for The Club of Rome, 1982. 229 pp. Paperback, $9.95.

Ross, Donald K. *A Public Citizen's Action Manual.* New York: Grossman Publishers, 1973. 328 pp. Available in limited quantity from Public Citizen, P.O. Box 19404, Washington, D.C. 20036. $2.00.

Seymour, John and Herbert Girardet. *Blueprint for a Green Planet: Your Practical Guide to Restoring the World's Environment.* New York: Prentice Hall, 1987. 191 pp. Paperback, $17.95. Describes hundreds of simple and sensible alternatives to help restore the planet to health.

Sierra Club. *Conservation Action Handbook.* San Francisco, CA, 1987. 130 pp. $5.00 postpaid. Contains guidelines for lobbying, planning and implementing conservation campaigns, and preparing testimony and action alerts.

Voyach, Robert. *Making Decisions: Our Global Connection.* Columbus, OH: The Mershon Center, Ohio State University. $5.00.

Winston, Martin Bradley. *Getting Publicity.* New York: John Wiley and Sons, 1982. 193 pp. $8.50.

Articles, Pamphlets, and Brochures

Durning, Alan B. "Mobilizing at the Grassroots." In Lester R. Brown et al., *State of the World 1989.* New York: Norton, 1989, pp. 154-73.

Sierra Club. *Gaining a Voice.* 1983. 8 pp. 50 cents.

Sierra Club. *How to Become an Environmental Activist.* 1984. 30 cents.

Sierra Club. *The Right to Write.* 1967. 10 cents.

"Why Cities Are Entering International Affairs," *Bulletin of Municipal Foreign Policy,* Vol. 1, No. 1, Winter 1986-87 (see periodicals list).

Globescope Assembly Action Plans

● Action Plans are available from the Global Tomorrow Coalition for the following Globescope Assemblies:

   » Globescope National Assembly, Portland, Oregon, April 1985;

» Globescope Idaho, Sun Valley, Idaho, October 1987;

» Globescope Wisconsin 88, Oshkosh, Wisconsin, March 1988; and

» Globescope Pacific Assembly, Los Angeles, California, November 1988.

Globescope Assembly Proceedings

Proceedings, Globescope II, An International Forum. Tufts University, Medford, Massachusetts, October 24-26, 1986. 156 pp. $8.00. Available from the Lincoln Filene Center for Citizenship and Public Affairs, Medford, MA 02155.

David Barnhizer (ed.). *Strategies for Sustainable Societies.* Proceedings and Materials from the Globescope 87 International Forum, Cleveland State University, Cleveland, Ohio, April 29-May 2, 1987. 620 pp. $30.00. Available from Professor David Barnhizer, Cleveland State University College of Law, Cleveland, OH 44115.

Periodicals

Building Economic Alternatives. Published quarterly by Co-op America, 2100 M Street, N.W., Suite 310, Washington, DC 20063. Single issue $1.00; mailed free to members with $25 annual membership payment, which also includes two issues of Co-op America's *Alternative Catalog.*

Bulletin of Municipal Foreign Policy. Published quarterly by the Local Elected Officials Project, Center for Innovative Diplomacy, 17931 Sky Park Circle, Suite F, Irvine, CA 92714. Annual subscription $35.

The Futurist. A journal of forecasts, ideas, and trends about the future, published bimonthly by the World Future Society, 4916 St. Elmo Avenue, Bethesda, MD 20814. Annual subscription $25.

Interaction. Newsletter of the Global Tomorrow Coalition, 1325 G Street, N.W., Washington, DC 20005-3104.

The Seventh Generation Catalog. Products for a Healthy Planet. Available from 10 Farrell Street, South Burlington, VT 05403. $2.00.

Transitions Abroad: A Guide to Learning, Living, and Working Overseas. Five issues, $15, from 18 Hulst Road, Box 344, Amherst, MA 01004.

World Watch. Published bimonthly by the Worldwatch Institute, 1776 Massachusetts Avenue N.W., Washington, DC 20036. Annual subscription $20. The magazine's goal is to help reverse the environmental trends that are undermining the human prospect.

Films and Other Audiovisual Materials

An Act of Congress. A record of lawmakers at work, the film traces the progress of a bill—The Clean Air Act of 1977—through committee, the House of Representatives, and finally to enactment. Shows how conflicting environmental and industrial interests were resolved. 58 min. Film, $795; video, $650; rental, $150. Coronet/MTI Film and Video.

Branches of Government: The Legislative Branch. 1982. Follows a congressman through several hectic weeks of work, and shows how he serves many interests: his local constituency, his state, his political party, the national interest, special interest groups, and his own conscience. 23 min. Grade 7 to adult. Color video $69.95, color film $279.50. National Geographic.

Downwind/Downstream: Threats to the Mountains and Waters of the American West. Documents the threat to water quality and public health in the Colorado Rockies from mining operations, acid rain, and urbanization, and highlights the need for action. 58 min. Grade 9 to adult. Color film $850, video $450, rental $85. Bullfrog Films.

Environmental Enrichment-What You Can Do About It. 21 min. 16mm color. Michigan Media.

From Sea to Shining Sea. 1986. Documents a small town's grassroots movement against a multinational corporation whose pipeline is discharging chemical waste into the ocean. 20 min. Grade 9 to adult. Video $95, rental $35. Bullfrog Films.

People Who Fight Pollution. 1971. 18 min. 16mm color. Michigan Media.

The Power to Change. 1979. 28 min. 16mm color. Michigan Media.

What is the Limit? 1987. Discusses interrelations between population growth, environmental degradation, resource depletion, habitat destruction, and ethical considerations for the future. Includes a discussion guide, "Where Do We Go From Here?" 23 min. Grade 10 to adult. Video: VHS, Beta, or 3/4 inch, $25. National Audubon Society.

Teaching Aids

Engleson, David C. *A Guide to Curriculum Planning in Environmental Education.* Madison, WI: Wisconsin Department of Public Instruction (125 South Webster St., P.O. Box 7841, Madison, WI 53707), 1985. 103 pp. $7.00.

Global 2000 Countdown Kit. 1982. Contains 14 units covering topics such as population, food, forests, water, energy, agriculture, climate, and species extinction. A final section titled "Making a Difference" describes community actions students can take to help improve the situations noted in the units. Encourages students to become active and effective citizens. Designed for independent study with minimal teacher guidance, grades 9-12. Zero Population Growth.

Notes

Chapter 1: A Global Awakening

1. *Making Common Cause: A Statement and Action Plan by U.S.-Based International Development, Environment, and Population NGOs* (Washington, DC, World Resources Institute, undated [1986]). The document, which has been endorsed by more than 100 U.S.-based nongovernmental organizations (NGOs), was presented at a public hearing before the World Commission on Environment and Development at Ottawa, Canada, May 26-27, 1986.

2. Jessica Tuchman Mathews, "Redefining Security," *Foreign Affairs*, Spring 1989, p. 163.

3. Nafis Sadik, *The State of World Population 1988* (New York: United Nations Population Fund, 1988), p. 1.

4. Flora Lewis, "The Next Big Crisis," *The New York Times*, July 27, 1988, p. A25.

5. Quoted in *Science*, June 17, 1988, p. 1611.

6. See Peter Westbroek, "The Impact of Life on the Planet Earth: Some General Considerations," in Daniel B. Botkin et al., *Changing the Global Environment: Perspectives on Human Involvement* (San Diego, CA: Academic Press, 1989), pp. 37-48; and Lester R. Brown and Sandra Postel, "Thresholds of Change," in Brown et al., *State of the World 1987* (New York: Norton, 1987), p. 17. For a review of the history of human impacts on the environment and the evolution of attitudes toward the environment, see Donald Worster (ed.), *The Ends of the Earth: Perspectives on Modern Environmental History* (New York: Cambridge University Press, 1988).

7. Mathews, "Redefining Security," p. 177; George M. Woodwell, Director, Woods Hole Research Center, Woods Hole, MA, quoted in Rushworth M. Kidder, "Agenda 2000," *Christian Science Monitor*, July 25, 1988, p. B4; see also World Commission on Environment and Development (WCED), *Our Common Future* (New York, Oxford University Press, 1987), p. 44. Human activity, especially the burning of fossil fuels, the clearing of forests, and the use of agricultural fertilizers, has already significantly altered some of the Earth's biogeochemical cycles, including the global cycling of carbon, nitrogen, and sulfur; see Berrien Moore et al., "Biogeochemical Cycles," in Mitchell B. Rambler et al., *Global Ecology: Towards a Science of the Biosphere* (San Diego, CA: 1989), pp. 115-24.

8. WCED, *Our Common Future*, p. 323. See also Anders Wijkman and Lloyd Timberlake, *Natural Disasters: Acts of God or Acts of Man?*, International Institute for Environment and Development (London, Earthscan, 1984); and Jessica Tuchman Mathews, "Is There More Risk in the World?, *The Washington Post*, March 29, 1989, p. A25.

9. *Time Magazine*, "Planet of the Year: Endangered Earth," January 2, 1989, pp. 26-27; Lester R. Brown, Christopher Flavin, and Sandra Postel, "A World at Risk," in Brown et al., *State of the World 1989* (New York: Norton, 1989), p. 1. For additional articles appearing in major publications during 1988 on the deterioration of the global environment, see *Time*, August 1, pp. 44-50, and August 15, p. 20; *Newsweek*, July 11, pp. 23-24, and August 1, pp. 42-48; *U.S. News and World Report*, June 12, pp. 48-54, and October 31, pp. 56-68; *National Geographic*, December; and *The New York Times*, December 25, p. E3.

10. An international survey commissioned by the United Nations Environment Program and carried out by Louis Harris and Associates during 1988 and early 1989 interviewed more than 9,000 citizens and leaders in 16 industrial and developing nations. The survey found that most people in all nations expressed serious concern about the quality of their environment, and 9 out of 10 favored stronger action nationally and internationally to curb pollution and reverse environmental decay. Louis Harris and Associates, *Public and Leadership Attitudes to the Environment in Four Continents*, revised draft, August 1989. Highlights of the survey are given in Chapter 16.

11. As evidence of a growing international commitment to environmental protection, at their economic summit meeting in Paris, July 15-16, 1989, leaders of the seven major democracies released a communique covering virtually every environmental issue and calling for decisive action to understand and protect the Earth's ecological balance. See United Nations Environment Program, *UNEP North American News*, Vol. 4, No. 4, August 1989, p. 1.

12. See Lester R. Brown, "A False Sense of Security," in Brown et al., *State of the World 1985* (New York: Norton, 1985), pp. 8-9. Beginning in the late 1960s, a variety of economic, ecological, and other problems began to emerge around the world, including the global energy crisis precipitated by the tripling of oil prices by the Organization of Petroleum Exporting Countries (OPEC) in 1973. During

the 1970s, more than a dozen world conferences were held in attempts to deal with these emerging problems, including major international meetings on population, food, water, desertification, and the human environment. See R.J. Johnston and P.J. Taylor (eds.), *A World in Crisis?: Geographical Perspectives*, Second Edition (Cambridge, MA: Basil Blackwell, 1989), pp. 1-2.

13. See World Resources Institute and International Institute for Environment and Development, *World Resources 1987* (New York: Basic Books, 1987), p. 1; Mathews, "Redefining Security," p. 174; WCED, *Our Common Future*, pp. 37-38; and W.C. Clark and R.E. Munn (eds.), *Sustainable Development of the Biosphere* (New York: Oxford University Press, 1986), pp. 5-6. Recent research indicates that environmental interdependence has existed since long before humans began to alter the environment, and perhaps since life began some 3.5 billion years ago. There is extensive evidence that life and geological processes are closely linked, and that Earth's lower atmosphere is maintained by life on the surface. See Westbroek, "The Impact of Life on the Planet Earth," p. 46; and Lynn Margulis and Ricardo Guerrero, "From Planetary Atmospheres to Microbial Communities," in Botkin, *Changing the Global Environment*, p. 65. For an analysis of how the various parts of the Earth's biosphere (land, water, atmosphere, and living matter) are interrelated, see John F. Stolz et al., "The Integral Biosphere," in Rambler, *Global Ecology*, pp. 21-49. Perhaps the ultimate indication of global interdependence is the evidence that a major nuclear war between the superpowers could kill a billion people and put a majority of the world's population at risk of starvation; see Frederick Warner, "The Environmental Consequences of Nuclear War," *Environment*, June 1988, pp 2-7.

14. See Joan Martin-Brown, *Consider the Connections* (Washington, DC: Office of Public Awareness, U.S. Environmental Protection Agency, December 1980).

15. See Richard Elliot Benedick, "Population-Environment Linkages and Sustainable Development," *Populi*, Journal of the United Nations Population Fund, Vol. 15, No. 3, 1988, pp. 14, 19; WCED, *Our Common Future*, pp. 4-7, 38, 46; World Resources Institute, *The Crucial Decade: The 1990s and the Global Environmental Challenge* (Washington, DC, 1989), esp. pp. 1-2; *Making Common Cause*, p. 1; and Joan Martin-Brown, "Converging Worlds: The Implications of Environmental Events for the Free Market and Foreign Policy Developments," *The Environmentalist*, Vol. 4, No. 2, 1984, pp. 139-42. For a collection of essays on the links among population, resources, and environment, see Kingsley Davis et al. (eds.), *Population and Resources in a Changing World: Current Readings* (Stanford, CA: Morrison Institute for Population and Resource Studies, Stanford University, 1989).

16. The security implications of increasing environmental problems and growing global interdependence are discussed in Chapter 15. See also Lester R. Brown et al., *State of the World 1988* (New York: Norton, 1988), Chapters 1 and 10; and Brown, *State of the World 1989*, Chapters 1, 8, and 10.

17. WCED, *Our Common Future*, esp. p. 2. While the world's total food output per person may still be growing, as noted in Chapter 5 total grain output per person has been declining since 1984. See Lester R. Brown, *The Changing World Food Prospect: The Nineties and Beyond*, Worldwatch Paper 85 (Washington, DC: Worldwatch Institute, October 1988), p. 42.

18. WCED, *Our Common Future*, esp. pp. 2-3.

19. Ibid., p. 22.

20. Ibid., p. 9.

21. Brown and Postel, "Thresholds of Change."

22. Lester R. Brown and Christopher Flavin, "The Earth's Vital Signs," in Brown, *State of the World 1988*, p. 6.

23. Population data is from Population Reference Bureau, *World Population Data Sheet*, various years, and other sources; grain data is from Brown, *The Changing World Food Prospect*, pp. 9, 43; energy data is from United Nations, *United Nations Statistical Yearbook* (New York: United Nations, various years); economic output is from Mathews, "Redefining Security," p. 162; manufacturing output is from WCED, *Our Common Future*, p. 15; U.S. organic chemical production is from Sandra Postel, "Controlling Toxic Chemicals," in Brown, *State of the World 1988*, p. 119; and human consumption of food energy is from Mathews, "Redefining Security," p. 171. The concept of food and energy production as measures of the load on global carrying capacity is developed in Walter H. Corson, "Food, Energy, and Global Carrying Capacity," Washington, DC, World Population Society, unpublished manuscript, December 1979.

24. WCED, *Our Common Future*, p. 45.

25. Sources for the information in this section are given in the issue chapters.

26. Table 1.1 assumes that two interrelated kinds of problems can be distinguished: (1) *problem impacts* or visible effects that appear to be caused by or result from other factors (e.g., deforestation), and (2) underlying *problem causes* that may be less visible, but seem to play a role in creating or causing the more visible impacts (e.g., population growth). The estimates of the relative importance of the underlying causes are very tentative, subjective appraisals based on the material in the relevant issue chapters. For each cause, the most relevant material supporting the estimates is given in the next five endnotes.

Theodore J. Gordon, Chairman of The Futures Group, has suggested that the problems that matter are those that are irreversible, have severe and immediate impacts, affect many people, and are potentially solvable; Gordon, quoted in Kidder, "Agenda 2000," p. B3. Of the problem impacts listed in Table 1.1, many are at least partly irreversible and have severe impacts, most have immediate impacts, and nearly all affect many people and are potentially solvable. The effectiveness of a number of promising solutions to these problem impacts is evaluated in Chapter 16.

27. See WCED, *Our Common Future*, pp. 43-45. The WCED report linked a range of global environmental problems—including erosion, desertification, water depletion, air and water pollution, and hazardous waste—to agricultural, forestry, energy, transportation, and industrial policies and practices; *Our Common Future*, p. 10.

28. Jessica Tuchman Mathews maintains that "population growth lies at the core of most environmental trends;" Mathews, "Redefining Security," p. 163. In Africa, rapid population growth appears to be an underlying cause of most if not all of the problem impacts in Table 1, especially unmet needs (including food and water), habitat degradation (deforestation), land degradation (soil erosion and desertification), energy depletion (fuelwood shortages), fresh water depletion, and conflict; see Brown, "A False Sense of Security," pp. 1-17. See also National Research Council, *Population Growth and Economic Development: Policy Questions* (Washington, DC: National Academy Press, 1986), pp. 85-93.

29. See Norman Myers, "Environmental Challenges: More Government or Better Governance?," *Ambio*, Vol. 17, No. 6, 1988, pp. 411-12; and Benedick, "Population-Environment Linkages." The WCED report, *Our Common Future*, noted that poverty causes environmental stresses, and that many problems of resource depletion and environmental stress arise from inequalities in access to resources and from disparities in economic and political power; see pp. 28, 46, 48.

30. See David Pimentel and Carl W. Hall (eds.), *Food and Natural Resources* (San Diego, CA: Academic Press, 1989), esp. pp. xv-xviii and Chapters 1 and 2; Organization for Economic Cooperation and Development (OECD), *The State of the Environment 1985* (Paris, 1985), pp. 187-99; and Essam El-Hinnawi and Manzur H. Hashmi, *The State of the Environment* (London: Butterworth Scientific, 1987), p. 49. In WCED, *Our Common Future*, pp. 125-28, the authors note that agricultural policies stressing increased production at the expense of environmental protection have contributed greatly to environmental deterioration, including soil degradation and loss, water pollution, deforestation, and desertification.

31. See James J. MacKenzie, *Breathing Easier: Taking Action on Climate Change, Air Pollution, and Energy Insecurity* (Washington, DC: World Resources Institute, undated [1988]); OECD, *The State of the Environment 1985*, pp. 201-15; El-Hinnawi and Hashmi, *The State of the Environment*, pp. 73-77, 88-89; Amulya K. N. Reddy, "Energy Issues and Opportunities," in Robert Repetto (ed.), *The Global Possible: Resources, Development, and the New Century* (New Haven, CT: Yale University Press, 1985), esp. pp. 365-71; and WCED, *Our Common Future*, pp. 168-81.

32. See OECD, *The State of the Environment 1985*, pp. 217-27; El-Hinnawi and Hashmi, *The State of the Environment*, pp. 93-121; Sandra Postel, "Controlling Toxic Chemicals," in Brown, *State of the World 1988*, pp. 118-36; and WCED, *Our Common Future*, pp. 208-13.

Chapter 2: Foresight Capability

1. Council on Environmental Quality (CEQ) and U.S. Department of State, *The Global 2000 Report to the President*, Vol. 1 (Washington, DC: U.S. Government Printing Office, 1980), p. 4.

2. *Blueprint for the Environment*. Advice to the President-Elect from America's Environmental Community. Executive Summary (Washington, DC, 1988), p. 32. Suppliers for this publication are listed under Further Information.

3. Lindsey Grant, *Foresight and National Decisions* (Lanham, MD: University Press of America, 1988), pp. 3, 10-13, 173, 226.

4. Ibid., pp. x, 33.

5. CEQ and Dept. of State, *Global 2000 Report*, Letter of Transmittal, Volume 1, p. iii.

6. CEQ and Dept. of State, *Global 2000 Report*, Volume 2, p. 454; see also p. 5.

7. Ibid., p. 229.

8. Testimony of Donald R. Lesh, Executive Director, Global Tomorrow Coalition, on the Global Resources, Environment, and Population Act, before the Subcommittee on Census and Population, Committee on Post Office and Civil Service, U.S. House of Representatives, April 12, 1988.

9. CEQ and Dept. of State, *Global Future: Time to Act*, Report to the President on Global Resources, Environment and Population (Washington, DC: U.S. Government Printing Office, 1981).

10. These are reviewed in Donella Meadows et al., *Groping in the Dark: The First Decade of Global Modelling* (New York: Wiley, 1982).

11. Excerpted and paraphrased from Meadows et al., *Groping in the Dark*, pp. 15-16. For a recent collection of computer simulations covering demographic, economic and political processes and including 26 national models, see Stuart A. Bremer, *The Globus Model: Computer Simulation of Worldwide Political and Economic Processes* (Frankfurt: Campus Verlag and Boulder, CO: Westview Press, 1987).

12. Gerald O. Barney, Institute for 21st Century Studies, Arlington, VA, private communication.

13. Thomas F. Malone and Robert Corell, "Mission to Planet Earth Revisited: An Update on Studies of Global Change," *Environment*, April 1989; and Daniel B. Botkin et al., *Changing the Global Environment: Perspectives on Human Involvement* (San Diego, CA: Academic Press, 1989), esp. Chapters 18 and 19.

14. See Clement Bezold, "Lessons from State and Local Government," in Grant, *Foresight and National Decisions*, pp. 83-98, 259-64.

15. Grant, *Foresight and National Decisions*, Chapter 6 and Appendix E.

16. Ned Dearborn, who participated in preparing *The Global 2000 Report*, notes that nonstandardized data may be useful in the foresight process if the data have adequate documentation and underlying assumptions are consistent. If they are not, differences should be documented. In other words, clarity may be more important than uniformity. Dearborn, private communication, April, 1988.

17. Grant, *Foresight and National Decisions*, Chapter 10.

18. This bill, the "Global Resources, Environment, and Population Act of 1987," was introduced in the 100th Congress by Senator Robert Packwood (as S. 1171) and Representative Buddy MacKay (as H.R. 2212).

19. Obstacles to recent proposals are discussed in Grant, *Foresight and National Decisions*, Chapter 11.

20. *Blueprint for the Environment.*

21. Grant, *Foresight and National Decisions*, pp. 227-33.

22. Ibid., pp. 240-42.

23. Ibid., p. 244. In *Global Future: Time to Act*, the Council on Environmental Quality and the U.S. Department of State recommended creation of two new institutions to focus the government's attention and response to long-term global population, resource, and environment issues: a Federal Coordinating Unit located in government, and a hybrid public-private organization—a Global Population, Resources, and Environment Analysis Institute—to link government and the private sector. See CEQ and Dept. of State, *Global Future: Time to Act*, Chapter 10.

24. See Russell W. Peterson, "A College of Integrated Studies: Education for the Professional Generalist," *L&S Magazine*, University of Wisconsin, Madison, Spring 1988.

25. This discussion of the politics of foresight was suggested by Ned Dearborn in private communication, April 1988.

26. Grant, *Foresight and National Decisions*, pp. 52, 209, 214, 218, 243.

27. Clement Bezold, "Lessons from State and Local Government," in Grant, *Foresight and National Decisions*, Chapter 5 and Appendix D.

28. For a summary of recent legislation, see Grant, *Foresight and National Decisions*, Chapter 10. Among those in Congress who have worked for better foresight are Senators Mark Hatfield and Albert Gore, Jr., and Representatives Newt Gingrich and Buddy MacKay.

Chapter 3: Population Growth

1. "Statement on Population Stabilization by World Leaders," signed by the heads of state of countries with over half the world's population, presented to the United Nations in 1985. For the complete text and list of signatories, contact Population Communication, 1489 East Colorado Boulevard, Suite 202, Pasadena, CA 91106.

2. Richard Elliot Benedick, "Population-Environment Linkages and Sustainable Development," *Populi*, Vol. 15, No. 3, 1988, p. 14.

3. Thomas W. Merrick and staff, "World Population in Transition," *Population Bulletin*, Vol. 41, No. 2, April 1986, pp. 3-4; and Robert Repetto, "Population, Resources, Environment: An Uncertain Future," *Population Bulletin*, Vol. 42, No. 2, July 1987, pp. 9-10. (The *Population Bulletin* is published by the Population Reference Bureau in Washington, DC.)

4. Population Reference Bureau (PRB), *1989 World Population Data Sheet* (Washington, DC: 1988); Merrick, "World Population in Transition," pp. 3-4; Susan Okie, "World Population May Hit 10 Billion by Year 2025," *The Washington Post*, May 17, 1989, p. A3.

5. PRB, *1989 World Population Data Sheet*. As the *Data Sheet* explains, the doubling time is the number of years until the population will double assuming a *constant* rate of natural increase, and is not intended to forecast the actual doubling of a population. Doubling times can be

calculated by dividing the number 70 by the growth rate expressed as a percent.

6. This discussion is drawn from the essay, "Population Explosion," in Edward Goldsmith and Nicholas Hildyard (eds.), *The Earth Report: The Essential Guide to Global Ecological Issues* (Los Angeles: Price Stern Sloan, 1988), p. 200. See also Repetto, "Population, Resources, Environment: An Uncertain Future," especially pp. 3-9.

7. PRB, *1988 World Population Data Sheet.*

8. Ibid.

9. National Audubon Society, "What Is the Limit?" Videotape and Discussion Guide, 1987.

10. Morris David Morris, *Measuring the Condition of the World's Poor: The Physical Quality of Life Index* (New York: Pergamon Press, 1979).

11. PRB, "Population Change, Resources, and the Environment," *Population Trends and Public Policy* No. 4 (Washington, DC: Population Reference Bureau, December 1983), p. 13.

12. Sharon L. Camp and J. Joseph Speidel, *The International Human Suffering Index* (Washington, DC: Population Crisis Committee, 1987). The index was compiled by adding 10 measures of human welfare: income, inflation, demand for new jobs, urban population pressures, infant mortality, nutrition, clean water, energy use, adult literacy, and personal freedom. Using the Pearson's r, a standard statistical test for relationships, the correlation between population growth and The Human Suffering Index is $r = .83$. The authors note that "the correlation between rapid population growth and human suffering is consistent with other studies which indicate that rapid population increase restricts economic and social progress for individual families and nations." For the view that population growth and density have no apparent negative effect on life expectancy, health, or psychological and social well-being, see Julian L. Simon, *The Ultimate Resource* (Princeton, NJ: Princeton University Press, 1981), Chapters 9 and 18.

13. Papua-New Guinea is considered part of Oceania, along with Australia and New Zealand.

14. The average annual rate of population increase (expressed as a percent) for all countries within each level of suffering was computed by averaging the individual rates of increase for those countries, using the data accompanying the suffering index figures.

15. Paul R. Ehrlich has defined carrying capacity as "the maximum number of individuals that can be supported by a given habitat; it is usually related to the availability of a limiting resource, one that is in short supply in relation to the population's needs." Ehrlich, *The Machinery of Nature* (New York: Simon and Schuster, 1986), p. 51.

16. Lester R. Brown, *The Twenty-Ninth Day: Accommodating Human Needs and Numbers to the Earth's Resources* (New York: Norton, 1978), pp. 13-15.

17. See Alan S. Miller and Irving M. Mintzer, "The Sky Is the Limit: Strategies for Protecting the Ozone Layer," Research Report No. 3 (Washington, DC: World Resources Institute, November 1986); Irving M. Mintzer, "A Matter of Degrees: The Potential for Controlling the Greenhouse Effect," Research Report No. 5 (Washington, DC: World Resources Institute, April 1987).

18. In response to the possibility that growing population and technology are causing global warming, The National Energy Policy Act of 1988 seeks to establish a strategy for reducing global warming. The legislation would allocate $1.6 billion for international population and family planning assistance for fiscal years 1990-92. Hal Burdett, "Senators Seek $1.6 Billion in Population Aid," *Popline*, Vol. 10, August 1988, p. 1.

19. Lester R. Brown, *Building a Sustainable Society* (New York: Norton, 1981), pp. 3-4.

20. Brown, *The Twenty-Ninth Day*, p. 276.

21. *FAO Production Yearbook*, Vol. 40, 1986 (Rome: Food and Agriculture Organization of the United Nations, 1987), pp. 195-204.

22. See Lester R. Brown and Edward C. Wolf, "Assessing Ecological Decline," in Lester R. Brown and others, *State of the World 1986* (New York: Norton, 1986), pp. 25-26.

23. Deforestation rates are from World Resources Institute (WRI) and International Institute for Environment and Development (IIED), *World Resources 1986* (New York: Basic Books, 1986), p. 73; population growth rates are from PRB, *1988 World Population Data Sheet*.

24. United Nations Fund for Population Activities (UNFPA), "Population Images" (New York, 1987), p. 35.

25. Ibid., p. 22; WRI and IIED, *World Resources 1987* (New York: Basic Books, 1987), p. 57.

26. Norman Myers, *The Primary Source: Tropical Forests and Our Future* (New York: Norton, 1984), p. 132.

27. National Research Council, *Population Growth and Economic Development: Policy Questions* (Washington, DC: National Academy Press, 1986), pp. 35-39.

28. UNFPA, "Population Images," p. 33.

29. See Lester R. Brown and Christopher Flavin, "The Earth's Vital Signs," in Lester R. Brown et al., *State of the World 1988* (New York: Norton, 1988), p. 9; and WRI and IIED, *World Resources 1987*, (New York: Basic Books, 1987), p. 290. See also National Research Council, *Population Growth and Economic Development*, pp. 38-39.

30. Brown and Flavin, "The Earth's Vital Signs," p. 10.

31. WRI and IIED, *World Resources 1987*, p. 289.

32. For estimated soil erosion rates, see WRI and IIED, *World Resources 1988-89* (New York: Basic Books, 1988), p. 282. Estimated rates for some of the Third World regions noted exceed 100 metric tons per hectare per year, compared with an estimate of 18 tons for U.S. cropland. Population growth rates and densities are from PRB, *1988 World Population Data Sheet.*

33. Norman Myers, *Not Far Afield: U.S. Interests and the Global Environment* (Washington, DC: World Resources Institute, June 1987), p. 28.

34. Lester R. Brown and Jodi Jacobson, "Assessing the Future of Urbanization," in Brown and others, *State of the World 1987* (New York: Norton, 1987), p. 39.

35. World Commission on Environment and Development, *Our Common Future* (New York: Oxford University Press, 1987), p. 56.

36. Nafis Sadik, *The State of World Population 1988* (New York: United Nations Population Fund, 1988), visual #10.

37. Brown and Jacobson, "Assessing the Future of Urbanization," pp. 42, 54.

38. Elizabeth Raisbeck, Testimony before the House Subcommittee on Census and Population, Committee on Post Office and Civil Service, U.S. Congress, April 12, 1988, p. 7.

39. Brown and Jacobson, "Assessing the Future of Urbanization," p. 43.

40. Myers, *Not Far Afield*, pp. 28-29.

41. Brown and Jacobson, "Assessing the Future of Urbanization," pp. 38-42.

42. Essam El-Hinnawi and Manzur H. Hashmi, *The State of the Environment* (London: Butterworth Scientific, 1987), pp. 160-61.

43. Lester R. Brown and Jodi Jacobson, "The Future of Urbanization: Facing the Ecological and Economic Constraints," *Worldwatch Paper 77* (Washington, DC: Worldwatch Institute, May 1987), p. 12.

44. See Robert Repetto (ed.), *The Global Possible: Resources, Development, and the New Century* (New Haven: Yale University Press, 1985), p. 499.

45. El-Hinnawi and Hashmi, *The State of the Environment*, p. 161. See also Anders Wijkman and Lloyd Timberlake, *Natural Disasters: Acts of God or Acts of Man?* (London: International Institute for Environment and Development, 1984).

46. Sadik, *The State of World Population 1988.*

47. For a series of recommendations for action to deal with Third World urban problems, see Repetto, *The Global Possible*, pp. 499-500.

48. Sharon L. Camp (ed.), *Population Pressures—Threat to Democracy*, Demographic Factors and Their Impact on Political Stability and Constitutional Government (Washington, DC: Population Crisis Committee, 1989).

49. Lester R. Brown and Jodi L. Jacobson, *Our Demographically Divided World*, Worldwatch Paper 74 (Washington, DC: Worldwatch Institute, December 1986), pp. 25-31; Myers, *Not Far Afield*, pp. 29-37.

50. See El-Hinnawi and Hashmi, *The State of the Environment*, p. 161; Myers, *Not Far Afield*, pp. 29, 33-37, 48.

51. Civil war-related deaths calculated from data in Ruth Leger Sivard, *World Military and Social Expenditures 1987-88* (Washington, DC: World Priorities, 1987), pp. 29-31.

52. War-related deaths are from Sivard, *World Military and Social Expenditures 1987-88*, pp. 29-31; population density data are from PRB, *1988 World Population Data Sheet.* The world's average population density, 98 persons per square mile, was computed using total world land area (excluding Antarctica) from WRI and IIED, *World Resources 1987*, pp. 354-55. For the view that there is no relation between population density and war, see Simon, *The Ultimate Resource*, p. 256.

53. See, for example, Simon, *The Ultimate Resource.*

54. See El-Hinnawi and Hashmi, *The State of the Environment*, pp. 155-61; Myers, *Not Far Afield*, pp. 15-51. For evidence that rapid population growth, environmental degradation, and poverty are turning droughts, floods, and other natural hazards into major human disasters with high death tolls, see Wijkman and Timberlake, *Natural Disasters: Acts of God or Acts of Man?*

55. PRB, *1988 World Population Data Sheet*; PRB, *The United States Population Data Sheet* (Washington, DC: 1988). In the United States, the crude birth rate and death rate are 16 and 9 per 1,000 population, respectively; in Western Europe, the birth rate and death rate are 12 and 10, respectively.

56. PRB, *1988 World Population Data Sheet.*

57. Susan Weber (ed.), *USA by Numbers: A Statistical Portrait of the United States* (Washington, DC: Zero Population Growth, 1988), p. 3.

58. See, for example, Julian L. Simon, "Getting the Immigrants We Need," *Washington Post*, August 3, 1988, p. A17; Population-Environment Balance, "Reform of Legal Immigration System Underway in Congress," *Balance Report*, No. 57, November/December 1987, p. 1.

59. Estimates of the number of illegal immigrants entering the United States each year varied between 500,000 and 1.5 million before implementation of immigration reform legislation passed in 1986. Susan Weber (ed.), *USA by Numbers: A Statistical Portrait of the United States*

(Washington, DC: Zero Population Growth, 1988), pp. 35-36.

60. Karen Woodrow, U.S. Census Bureau, private communication, September 1988.

61. Weber, *USA by Numbers*, p. 41.

62. U.S. General Accounting Office, "Illegal Aliens: Limited Research Suggests Illegal Aliens May Displace Native Workers," GAO/PEMD-86-9BR, April 1986, pp. 17-18. For further information, contact the Federation for American Immigration Reform, 1666 Connecticut Avenue, N.W., Washington, DC 20009.

63. Frank Swoboda, "Immigrants and the Job Market: U.S. Work Force Unshaken Despite Massive Influx, Study Shows," *Washington Post*, January 25, 1988, p. A3.

64. See, for example, Ben J. Wattenberg, *The Birth Dearth* (New York: Pharos Books, 1987).

65. Zero Population Growth, "Fact Versus Fiction: Experts Counter 'The Birth Dearth'" (information sheet prepared in response to Wattenberg's *The Birth Dearth*).

66. Computed from data in WRI and IIED, *World Resources 1987*, p. 300.

67. Ibid., p. 96.

68. Weber, *USA by Numbers*, p. 155. The annual loss of 1.2 million hectares of cropland represents about 0.6 percent of the total U.S. cropland of 200.5 hectares reported in WRI and IIED, *World Resources 1986*, p. 47. For the view that the United States is in no danger of running out of farmland, see Simon, *The Ultimate Resource*, p. 232.

69. Weber, *USA by Numbers*, p. 117.

70. See The Conservation Foundation, *State of the Environment: A View toward the Nineties* (Washington, DC, 1987), pp. 10, 223-33.

71. Zero Population Growth, "ZPG's Urban Stress Test Targets Pressures on U.S. Cities," *ZPG Reporter*, Vol. 17, No. 4, Fall 1985.

72. See Weber, *USA by Numbers*, Chapters 8-12.

73. Zero Population Growth, "National Population Policy Moves Forward," *ZPG Reporter*, May-June 1988, p. 3.

74. Robert Gillespie, Population Communication, Pasadena, California, private communication. See also "Colombia Signs World Population Statement," *Popline*, Vol. 10, June, 1988 (Washington, DC: The Population Institute).

75. Werner Fornos, *Gaining People, Losing Ground: A Blueprint for World Population Stabilization* (Washington, DC: The Population Institute, 1987), p. 67.

76. Jodi L. Jacobson, "Anti-Abortion Policy Leads to...More Abortions," *World Watch*, May-June 1988, p. 9. See also Constance Holden, "U.S. Antiabortion Policy May Increase Abortions," *Science*, November 27, 1987, p.1222.

Legislation recently introduced in the U.S. Senate would raise funding for international population and family planning assistance to $500 million for fiscal year 1990 as part of an overall strategy for reducing global warming; Burdett, "Senators Seek $1.6 Billion in Population Aid," p. 1.

77. Sadik, *The State of the World Population 1988*, p. 18.

78. Jodi Jacobson, "Planning the Global Family," in Brown et al., *State of the World 1988*, pp. 157-58.

79. For example, studies comparing single children to those with siblings show no significant differences on a range of personality traits, and for two traits—self-esteem and motivation to achieve—the single children rated higher than those with siblings. See Martin Lasden, "Two Parents, One Child," *Washington Post-Health*, September 6, 1988, p. 18.

80. National Research Council, *Population Growth and Economic Development*, p. 86.

81. See Lester R. Brown and Edward C. Wolf, "Reclaiming the Future," in Brown et al., *State of the World 1988*, especially pp. 176-78, 188.

82. National Research Council, *Population Growth and Economic Development*, p. 90.

83. Brown and Wolf, "Reclaiming the Future," p. 176-77.

84. PCC, "Access to Birth Control: A World Assessment," *Population Briefing Paper* No. 19, October 1987, p. 1; Okie, "World Population May Hit 10 Billion by Year 2025."

85. See Repetto, *The Global Possible*, p. 498.

86. PCC, "Country Rankings of the Status of Women: Poor, Powerless and Pregnant," *Population Briefing Paper* No. 20, June 1988.

87. Fornos, *Gaining People, Losing Ground*, p. 31; Brown and Wolf, "Reclaiming the Future," p. 177.

88. For further information, contact The Futures Group, 1101 14th Street, N.W., 3rd Floor, Washington, DC 20005.

89. Jacobson, "Planning the Global Family," p. 153.

90. Ibid., pp. 153-54. For ratings of family planning program effort for 100 countries in 1982, see WRI and IIED, *World Resources 1988-89*, p. 23.

91. PCC, "Access to Birth Control."

92. See Repetto, *The Global Possible*, p. 497; Jacobson, "Planning the Global Family," pp. 151-52. For the conclusion that successful family planning programs have had a wider focus than contraception, and have included health care and programs for economic development, see Pranay Gupte, *The Crowded Earth: People and the Politics of Population* (New York: Norton, 1984).

93. Sadik, *The State of World Population 1988*, p. 4. In nine developing countries where official government population planning policies had been in effect for periods ranging from 8 to 30 years, birth rates decreased by 19 percent, on average, during the periods the policies were in effect. WRI and IIED, *World Resources 1987*, p. 10. The nine countries cited are Indonesia, Thailand, India, Pakistan, Bangladesh, Vietnam, Sudan, Mexico, and Brazil.

94. In 1979, President Nixon declared that the United States should "establish as a national goal the provision of adequate family planning services within the next five years to all those who want them but cannot afford them."

Richard M. Nixon, "Special Message to the Congress on Problems of Population Growth," July 18, 1969, cited in Lindsey Grant, *Foresight and National Decisions* (Lanham, MD: University Press of America, 1988), pp. 44, 55.

95. For policy proposals regarding future U.S. assistance to international population programs, see John Stover, "International Cooperation in Population Programs in the 1990s: The U.S. Role," presented at the Michigan State University Conference on Cooperation for International Development, East Lansing, MI, May 1988. The paper is available from The Futures Group, 76 Eastern Boulevard, Glastonbury, CT 06033.

Chapter 4: Development and the Environment

1. Walter Reid et al., *Bankrolling Successes* (Washington, DC: Environmental Policy Institute and National Wildlife Federation, 1988), p. 1.

2. From U.S. Agency for International Development (USAID), *Development and the National Interest* (Washington, DC, 1989). p. 1.

3. Lloyd Timberlake, *Africa in Crisis* (London: Earthscan, 1985), pp. 19-21.

4. United Nations Children's Fund (UNICEF), *The State of the World's Children 1989* (Oxford: Oxford University Press, 1989), pp. 2-3.

5. Gro Harlem Bruntland, Prime Minister of Norway and Chair of the World Commission on Environment and Development, Speech at the National Press Club, Washington DC, May 5, 1989.

6. Peter Zuckerman, *A New Covenant* (Washington, DC: American Peace Network, 1988), p. 128; UNICEF, *The State of the World's Children 1989*, p. 1.

7. USAID, *Development and the National Interest*, p. 4

8. Ibid., p. 3.

9. See, for example, UNICEF, *The State of the World's Children*, especially p.2; and Lloyd Timberlake, *Africa in Crisis*, especially p. 8; and Reid, *Bankrolling Successes*, especially pp. 1-3.

10. See Lester R. Brown, *Building a Sustainable Society* (New York: Norton, 1981), pp. 3-5.

11. USAID, *Development and the National Interest*, p. 25.

12. UNICEF, *The State of the World's Children 1989*, p. 32.

13. World Bank, *World Development Report 1988* (New York: Oxford University Press, 1988).

14. International Education Coalition (IEC), *Dealing with Interdependence: the U.S. and the Third World* (Washington, DC: IEC, 1987). p. 3.

15. See USAID, *Development and the National Interest*, p. 7.

16. World Bank, *World Development Report 1988*.

17. UNICEF, *The State of the World's Children*, pp. 75-80.

18. Ibid., pp. 75-77.

19. USAID, *Development and the National Interest*, p. 28.

20. Ibid.

21. UNICEF, *The State of the World's Children*, p. 1. Estimates of the number of people living in poverty range from somewhat under to somewhat over one billion; see, for example, Robert Repetto, "Population, Resources, Environment: An Uncertain Future," *Population Bulletin*, Vol. 42, No. 2, July 1987, p. 12; and Norman Myers (ed.), *Gaia: An Atlas of Planet Management* (Garden City, NY: Anchor Press, 1984), p. 220.

22. Population Reference Bureau, *World Population Data Sheet* (Washington, DC, various years).

23. Nafis Sadik, Executive Director, United Nations Population Fund, *The State of World Population 1988* (New York, 1988), press packet.

24. World Resources Institute (WRI) and International Institute for Environment & Development (IIED), *World Resources 1987* (New York: Basic Books, 1987).

25. Susan George, "Third World Debt: The Moral and Physical Equivalent of War," *Who Owes Whom?*, Newsletter of Project Abraco: North Americans in Solidarity with the People of Brazil, Spring, 1988.

26. WRI and IIED, *World Resources 1988-89* (New York: Basic Books, 1988), p. 242.

27. UNICEF, *The State of the World's Children 1989*, p. 2.

28. George, "Third World Debt."

29. Timothy Weiskel, Henry Luce Fellow, Harvard University, "The Anthropology of Environmental Decline," Summary Statement for Hearings of the U.S. Senate Committee on Environment and Public Works, September 14, 1988, p. 18.

30. Ibid.

31. WRI and IIED, *World Resources 1988-89*, p. 204.

32. World Commission on Environment and Development (WCED), *Our Common Future* (London: Oxford University Press, 1987), p. 123.

33. WRI and IIED, *World Resources, 1988-89*, p. 212.

34. Debt Crisis Network, *A Journey Through the Global Debt Crisis* (Washington, DC, 1988), p. 6.

35. Sarah Bartlett, "A Vicious Circle Keeps Latin America in Debt," *The New York Times*, January 15, 1989, Section 4, p. 5.

36. George, "Third World Debt."

37. Ibid.

38. Ibid.

39. Inter-American Development Bank, *Annual Report 1988* (Washington, DC, 1988), p. 118.

40. George, "Third World Debt."

41. UNICEF, *The State of the World's Children 1989*, p. 25.

42. Peter Stone, "The Debt Crisis," *The Baltimore Sun*, March 12, 1989, p. 1M.

43. World Bank, *World Development Report 1988*, p. 30; *Development Forum*, United Nations Department of Public Information, Vol. 17, No. 4, July-August 1989, p. 8.

44. Bartlett, "A Vicious Circle."

45. Debt Crisis Network, *From Debt to Development* (Washington, DC: Institute for Policy Studies, 1985), p. 7.

46. Stone, "The Debt Crisis."

47. Ibid., p. 1.

48. Debt Crisis Network, *From Debt to Development*, p. 13.

49. *The Washington Post*, March 12, 1989, p. 3M.

50. Stone, "The Debt Crisis."

51. Christine Bogdanowicz-Bindert and Richard Feinberg, "Third World Debt: An Analysis," *Debt For Nature: An Opportunity* (Washington, DC: National Wildlife Foundation and World Wildlife Fund, 1988), Insert #1.

52. Ibid.

53. Ibid.

54. Hobart Rowen, "Debt Reduction Agreement is Reached with Mexico," *The Washington Post*, July 24, 1989, p. A1.

55. Weiskel, "The Anthropology of Environmental Decline,"

56. WRI and IIED, *World Resources 1988-89*, p. 282.

57. WCED, *Our Common Future*, pp. 125, 128.

58. Andrew Hultkrans, "Greenbacks for Greenery," *Sierra*, November-December, 1988, p. 43.

59. Smithsonian Institution Traveling Exhibition Service (SITES), "Tropical Rainforests: A Disappearing Treasure," (Washington, DC, 1988) p. 22.

60. Norman Myers, "Writing Off The Environment," in John Elkington et al. (eds.), *Green Pages: The Business of Saving the World* (London: Routledge, 1988), p. 190.

61. SITES, "Tropical Rainforests," p. 20.

62. WRI and IIED, *World Resources 1986* (New York: Basic Books, 1986), p. 70.

63. *World Bank News*, February 9, 1989, pp. 8-9.

64. Mac Margolis, "Thousands of Amazon Acres Burning," *The Washington Post*, September 8, 1988, p. A31.

65. Peri Batliwala, "Preventable Tragedy, or Lost Cause?" *Panoscope*, November 1988, p. 22.

66. Ibid.

67. See WCED, *Our Common Future*, p. 136.

68. United Nations Environment Program, "The State of the Marine Environment 1988," *UNEP News*, April 1988, pp. 10-12.

69. Ibid.

70. Ibid.

71. *The Baltimore Sun*, November 27, 1988, p. 4P.

72. Sabine Rosenbladt, "Is Poland Lost? Pollution and Politics in Eastern Europe," *Greenpeace*, November-December 1988, p. 14.

73. WCED, *Our Common Future*, pp. 263-264.

74. Ibid., p. 262.

75. Jacques-Yves Cousteau et al., *The Cousteau Almanac* (New York: Dolphin Books, 1981), p. 411.

76. WCED, *Our Common Future*, p. 262.

77. See Ibid., Chapter 9.

78. Ibid., pp. 236-37.

79. John McCormick, *Acid Earth* (London: Earthscan, 1985), pp. 5-6.

80. Donella Meadows, "The New Alchemist," *Harrowsmith Magazine*, November-December 1988, pp. 38-39.

81. WCED, *Our Common Future*, p. 169.

82. McCormick, *Acid Earth*, p. 5.

83. Ibid., p. 27.

84. Ibid., pp. 32-33, 37; *The Clean Air Act: A Briefing Book for Members of Congress* (Washington, DC: National Clean Air Coalition, 1983).

85. WCED, *Our Common Future*, p. 206; see also Chapter 8.

86. Ibid., p. 220.

87. Ibid., see Chapter 8.

88. Ibid., p. 126.

89. Ibid.

90. World Bank, *Population Growth and Policies in Sub-Saharan Africa: A World Bank Policy Study* (Washington, DC, 1986), p. 35.

91. Hernando de Soto, *The Other Path: The Invisible Revolution in the Third World* (New York: Harper & Row, 1989).

92. United Nations Population Fund, *The State of the World's Population 1988*. Press packet.

93. Tad Szulc, "Brazil's Amazonian Frontier," *Bordering on Trouble* (Bethesda, MD: Adler & Adler, 1986), pp. 194-96.

94. Irene Dankelman and Joan Davidson, *Women and Environment in the Third World* (London: Earthscan, 1988), p. 9.

95. Maggie Black, "Earthwatch: Mothers of the Earth," *People*, International Planned Parenthood's Review of Population and Development, Vol. 5, No. 3, 1988, pp. 5-8.

96. Fred Sai, Senior Population Advisor, World Bank, Briefing to the U.S. House of Representatives Subcommittee on Natural Resources, Agriculture, and Environment.

97. WCED, *Our Common Future*, pp. 96, 106; World Bank, *Population Growth and Policies in Sub-Saharan Africa*, pp. 39-40.

98. Halfdan Mahler, Former Director-General of the World Health Organization, "On International Women's Day: A Case in Point," *International Herald Tribune*, March 8, 1989.

99. World Bank, *Population Growth and Policies in Sub-Saharan Africa*, p. 38.

100. Sai, Briefing.

101. Timberlake, *Africa in Crisis*, pp. 171-72.

102. International Fund for Agricultural Development, *Africa: Sowing the Seeds of Self-Sufficiency* (Rome, undated), p. 10.

103. Timberlake, *Africa in Crisis*, p. 83.

104. World Media Institute, *Tribute to Our Common Future* (Ottawa, 1987-88).

105. WCED, *Our Common Future*, p. 109.

106. Ibid.

107. WRI and IIED, *World Resources 1988-89*, pp. 29-30.

108. WCED, *Our Common Future*, Chapter 4.

109. WRI and IIED, *World Resources 1988-89*, p. 25.

110. Direction and material for much of this section is from WCED, *Our Common Future*.

111. James Lovelock, "Man and Gaia," in Edward Goldsmith and Nicholas Hildyard (eds.), *The Earth Report* (Los Angeles: Price Stern Sloan, 1988), p. 51.

112. See WCED, *Our Common Future*, pp. 8-9.

113. Reid, *Bankrolling Successes*, p. 38.

114. Institute for Alternative Agriculture, *Alternative Agricultural News*, April, May, and September 1988.

115. Rodale Institute, *Enough Food* (Emmaus, PA, 1985), p. 3.

116. WCED, *Our Common Future*, pp. 229-230.

117. Cynthia Pollock, *Mining Urban Wastes: The Potential for Recycling,*" Worldwatch Paper 76 (Washington, DC: Worldwatch Institute, 1987), p. 34.

118. National Wildlife Federation, *Conservation Exchange*, Spring 1988, p. 6.

119. Goldsmith and Hildyard, *The Earth Report*, p. 204.

120. Ibid., p. 229.

121. *Time*, "The Dirty Seas," August 1, 1988, p. 50.

122. John Berger, *Restoring The Earth* (New York: Doubleday, 1987), pp. 175-83.

123. IEC, *Dealing with Interdependence*, p. 16.

124. Ibid.

125. Stephen Hellinger et al., *Aid for Just Development*, The Development Group for Alternative Policies (Boulder, CO: Lynne Rienner Publishers, 1988), p. 13.

Chapter 5: Food and Agriculture

1. World Resources Institute (WRI) and International Institute for Environment and Development (IIED), *World Resources 1988-89* (New York: Basic Books, 1988), p. 3.

2. John W. Mellor, "Agricultural Development: Opportunities for the 1990s," *International Food Policy Research Institute 1988 Report* (Washington, DC, 1989), p. 9.

3. Norman Myers, "You Can't See the Future for Dust," *The Guardian*, January 8, 1988.

4. Lester R. Brown, *The Changing World Food Prospect*, Worldwatch Paper 85 (Washington, DC: Worldwatch Institute, October, 1988), p. 51.

5. WRI and IIED, *World Resources 1988-89*, p. 2.

6. Brown, *The Changing World Food Prospect*, pp. 5, 15, 45.

7. Brown, *The Changing World Food Prospect*, pp. 8, 11, 16; Lester R. Brown, "Sustaining World Agriculture," in Brown et al., *State of the World 1987* (New York: Norton, 1987), p. 135; Lester R. Brown and Christopher Flavin, "The Earth's Vital Signs," in Brown et al., *State of the World 1988* (New York: Norton, 1988), p. 11.

8. WRI and IIED, *World Resources 1987* (New York: Basic Books, 1987), pp. 40-41, 44-45; the sub-Saharan drought is documented in Peter J. Lamb and Randy A. Peppler, "North Atlantic Oscillation: Concept and an Application," *Bulletin of the American Meteorological Society*, Vol. 68, No. 10, October 1987, pp. 1218-25.

9. Brown, *The Changing World Food Prospect*, p. 8.

10. Brown, "Sustaining World Agriculture," pp. 123-24.

11. Ibid., pp. 7, 17-18.

12. David Pimentel et al., "World Agriculture and Soil Erosion," *BioScience*, Vol. 37, No. 4, April 1987, p. 277.

13. Leonardo Paulino, "Food in the Third World: Past Trends and Projections to 2000," *IFPRI Abstracts*, No. 52, (Washington, DC: International Food Policy Research Institute, June 1986).

14. Norman Myers (ed.), *Gaia: An Atlas of Planet Management* (Garden City, NY: Anchor Press, 1984), pp. 48-49.

15. Susan Okie, "Health Crisis Confronts 1.3 Billion," *The Washington Post*, September 25, 1989, p. A1.

16. Ibid., p. 48; John Robbins, *Diet for a New America* (Walpole, NH: Stillpoint Publishing, 1987), p. 352.

17. WRI and IIED, *World Resources 1988-89*, p. 245; see also James P. Grant, *The State of the World's Children 1988* (New York: United Nations Children's Fund (UNICEF)), pp. 2-3; and Edward Goldsmith and Nicholas Hildyard (eds.), *The Earth Report* (Los Angeles: Price Stern Sloan, 1988), pp. 176-77.

18. Goldsmith and Hildyard, *The Earth Report*, p. 176; see also David Pimentel, "Waste in Agriculture and Food Sectors," Cornell University, College of Agriculture and Life Sciences, unpublished manuscript, April 1989, p. 1.

19. Goldsmith and Hildyard, *The Earth Report*, p. 176; WRI and IIED, *World Resources 1988-89*, p. 54.

20. WRI and IIED, *World Resources 1988-89*, p. 53; World Bank, *World Development Report 1988* (New York: Oxford University Press, 1988), p. 278; Anastasia Toufexis, "Too Many Mouths," *Time* Magazine, January 2, 1989, p. 48.

21. WRI and IIED, *World Resources 1988-89*, p. 54; Myers, *Gaia: An Atlas of Planet Management*, p. 49.

22. Lamb and Peppler, "North Atlantic Oscillation: Concept and an Application;" Lamb, private communication, December 1988.

23. Brown, "Sustaining World Agriculture," p. 135; John Mellor and Sarah Gavian, "Famine: Causes, Prevention, and Relief," IFPRI *Articles*, (Washington, DC: International Food Policy Research Institute), January 30, 1987, pp. 539-45.

24. "250 Million People Affected by India's Continuing Drought," *Popline*, (World Population News Service, published by the Population Institute, Washington, DC), February 1988.

25. Myers, *Gaia: An Atlas of Planet Management*, p. 48.

26. U.S. Department of Health and Human Services (DHHS), *The Surgeon General's Report on Nutrition and Health, Summary and Recommendations* (Washington, DC, 1988), pp. 1-20; Robbins, *Diet for a New America*, pp. 189-349; Spencer Rich, "Nation's Health Spending Swells Nearly 10 Percent," *The Washington Post*, November 19, 1988, p. A8.

27. Robin Herman, "Diseases of Affluence," *Washington Post Health*, January 3, 1989, pp. 12-15. See also Frances Moore Lappé, *Diet for a Small Planet*, Revised Edition (New York: Ballantine Books, 1982), p. 120.

28. Herman, "Diseases of Affluence," pp. 12-15.

29. DHHS, *The Surgeon General's Report on Nutrition and Health* (Washington, DC, 1988), pp. 54, 58-59.

30. National Research Council, *Recommended Dietary Allowances* (Washington, DC; National Academy Press, 1980), p. 23; DHHS, *The Surgeon General's Report*, p. 69.

31. For example, David Pimentel cites daily U.S. protein consumption as 102 grams per person, including 70 grams of animal protein and 32 grams from plant sources, in Pimentel, "Waste in Agriculture and Food Sectors," p. 12. *The Surgeon General's Report on Nutrition and Health* estimates that in 1985, Americans received much more than the recommended dietary protein allowance: women received 144 percent; men, 175 percent; and children 222 percent of the recommended levels (pp. 68-69). Lappé has reported estimates that Americans consume about twice as much protein as their bodies can utilize, in *Diet for a Small Planet*, pp. 9, 20, 121.

32. DHHS, *The Surgeon General's Report*, pp. 329, 381-82; Robbins, *Diet for a New America*, pp. 189-201; Myron E. Winick, MD, "The Role of Protein: Most People Need Less," *Washington Post Health*, November 22, 1988; p. 20.

33. Robbins, *Diet for A New America*, p. 172; DHHS, *The Surgeon General's Report*, pp. 68-69.

34. Lappé, *Diet for a Small Planet*, pp. 162-65; Robbins, *Diet for a New America*, pp. 181-83.

35. Lappé, *Diet for a Small Planet*, pp. 158-61.

36. Robbins, *Diet for a New America*, p. 300.

37. Ibid., pp. 121-22; Alan B. Durning, "U.S. Poultry Consumption Overtakes Beef," *World Watch*, January-February 1988, pp. 11-12.

38. Myers, *Gaia: An Atlas of Planet Management*, p. 48.

39. Lappé, *Diet for a Small Planet*, p. 70.

40. Ibid., pp. 69-71; Robbins, *Diet for a New America*, pp. 350-51; Pimentel, "Waste in Agriculture and Food."

41. WRI and IIED, *World Resources 1988-89*, p. 53.

42. Food and Agriculture Organization (FAO), *Potential Population Supporting Capacities of Lands in the Developing World* (Rome, Italy: FAO, International Institute for Applied Systems Analysis, and U.N. Fund for Population Activities, 1984); Paul Harrison, *Land, Food and People* (Rome: FAO, 1984); Paul Harrison, "Land and People, the Growing Pressure," *Earthwatch* (a section of *People*, published by International Planned Parenthood Federation, London), No. 13, 1983, p. 2. For a summary of the FAO study, see Peter Hendry, "Food and Population: Beyond Five Million," *Population Bulletin*, Vol. 43, No. 2, April 1988, pp. 17-18.

43. Essam El-Hinnawi and Mansur Hashmi, *The State of the Environment*, a publication of the United Nations Environment Program (London: Butterworths, 1987), p. 36.

44. Harrison, "Land and People," p. 2.

45. Harrison, "Land and People," p. 4.

46. River blindness (onchocerciasis) and human and animal trypanosomiasis are among the diseases that limit agricultural expansion in Africa; see El Hinnawi and Hashmi, *State of the Environment*, p. 49.

47. El-Hinnawi and Hashmi, *The State of the Environment*, p. 36.

48. Ibid., p. 51; "Post Harvest Hunger," United Nations Environment Program fact sheet, undated.

49. El-Hinnawi and Hashmi, *The State of the Environment*, p. 63.

50. Pimentel, "Waste in Agriculture and Food Sectors," pp. 15-16.

51. Ibid., pp. 36-38.

52. *World Bank News*, June 8, 1989, p. 4.

53. Pimentel, "World Agriculture and Soil Erosion," p. 277.

54. Lester R. Brown and Edward C. Wolf, *Soil Erosion: Quiet Crisis in the World Economy*, Worldwatch Paper 60 (Washington DC: Worldwatch Institute, September 1984),

pp. 21-22. The global estimate was based on soil loss estimates for the four major food-producing countries, and on the assumption that soil erosion estimates for the rest of the world are similar to those of the "big four," which the authors noted "is a conservative assumption given the pressures on land in the Third World."

55. Dusan Zachar, *Soil Erosion*, Developments in Soil Science No. 10 (New York: Elsevier Scientific Publishing, 1982), p. 467. For estimated soil erosion rates in selected countries, see *World Resources 1988-89*, p. 282; and David Pimentel et al., "World Agriculture and Soil Erosion," *BioScience*, Vol. 37, No. 4, April 1987, p. 278. See also Goldsmith and Hildyard, *The Earth Report*, pp. 142-43.

56. Estimates of annual soil loss in the United States from agricultural land vary from 1.7 to 7 billion tons a year; some of the estimates include range, pasture, and forest land in addition to cropland; and some estimates include soil loss due to wind erosion as well as water erosion. See Brown and Wolf, *Soil Erosion*, p. 17; Susan Weber (ed.), *USA by Numbers* (Washington, DC: Zero Population Growth, 1988), p. 155; and Lappé, *Diet for a Small Planet*, p. 80. The estimated cost of U.S. soil erosion is from Pimentel, "Waste in Agriculture and Food Sectors," p. 8.

57. Lester R. Brown, "Reexamining the World Food Prospect," in Brown et al., *State of the World 1989* (New York: Norton, 1989), p. 57.

58. WRI and IIED, *World Resources 1988-89*, p. 111; WRI and IIED, *World Resources 1986* (New York: Basic Books, 1986), p. 111.

59. Myers, *Gaia: An Atlas of Planet Management*, p.114.

60. WRI and IIED, *World Resources 1987*, p. 71; El-Hinnawi and Hashmi, *The State of the Environment*, pp. 38-39; Brown and Flavin, "The Earth's Vital Signs," p. 6.

61. El-Hinnawi and Hashmi, *The State of the Environment*, p. 39; WRI and IIED, *World Resources 1987*, pp. 71-72.

62. WRI and IIED, *World Resources 1987*, p. 72.

63. El-Hinnawi and Hashmi, *The State of the Environment*, pp. 52-53.

64. Pimentel, "Waste in Agriculture and Food Sectors;" Pimentel, private communication.

65. El-Hinnawi and Hashmi, *State of the Environment*, pp. 49-50; WRI and IIED, *World Resources 1988-89*, p. 6.

66. Brown, "Sustaining World Agriculture," p. 125; Nafis Sadik, *The State of World Population 1988* (New York: United Nations Population Fund, 1988), p. 9; Robert Repetto, "Population, Resources, Environment: An Uncertain Future," *Population Bulletin*, Vol. 42, No. 2, July 1987, p. 21; Irene Dankelman and Joan Davidson, *Women and Environment in the Third World: Alliance for the*

Future (London: Earthscan and International Union for Conservation of Nature and Natural Resources, 1988), p. 30.

67. Pimentel, "Waste in Agriculture and Food Sectors," pp. 8-9.

68. El-Hinnawi and Hashmi, *The State of the Environment*, p. 50; WRI and IIED, *World Resources 1986*, p. 48.

69. El-Hinnawi and Hashmi, *The State of the Environment*, p. 38; WRI and IIED, *World Resources 1987*, pp. 116; Repetto, "Population, Resources, Environment," pp. 21-22.

70. Repetto, "Population, Resources, Environment," p. 21.

71. Brown, "Sustaining World Agriculture," pp. 124-27.

72. WRI and IIED, *World Resources 1988-89*, p. 78; Edward C. Wolf, "Maintaining Rangelands," in Lester R. Brown et al., *The State of the World 1986* (New York: Norton, 1986), p. 64.

73. Lappé, *Diet for a Small Planet*, p. 20.

74. FAO, *FAO Production Yearbook 1986* (Rome, 1987), pp. 195-204.

75. Robbins, *Diet for a New America*, p. 353.

76. Wolf, "Managing Rangelands," p. 65.

77. WRI and IIED, *World Resources 1986*, p. 72; WRI and IIED, *World Resources 1988-89*, pp. 263, 280.

78. Norman Myers, *The Primary Source* (New York: Norton, 1984), p. 132; United Nations Fund for Population Activities (UNFPA), *Population Images*, 2nd Edition (New York, 1987), p. 22.

79. WRI and IIED, *World Resources 1988-89*, p. 264; Robbins, *Diet for a New America*, p. 361.

80. Erik P. Eckholm, *Losing Ground: Environmental Stress and World Food Prospects* (New York: Norton, 1976), pp. 61-71.

81. WRI and IIED, *World Resources 1987*, pp. 71-72; Myers, *Gaia: An Atlas of Planet Management*, pp. 40-41; Lynn Jacobs, "How the Livestock Industry is Ruining the American West," *The Animals' Agenda*, January-February, 1988, p. 13.

82. Sierra Club, *Bankrolling Disasters: International Development Banks and the Global Environment* (San Francisco, 1986), p. 8; *Time*, January 2, 1989, p. 54.

83. Lappé, *Diet for a Small Planet*, p. 80.

84. Lappé, *Diet for a Small Planet*, pp. 74-76; Robbins, *Diet for a New America*, pp. 352, 367-69; Pimentel, "Waste in Agriculture and Food Sectors," p. 12; and Pimentel, private communication.

85. Pimentel, "Waste in Agriculture and Food Sectors," pp. 12-13; Lappé, *Diet for a Small Planet*, p. 84.

86. Sylvan Wittwer, Yu Youtai, Sun Han, and Wang Lianzheng, *Feeding a Billion: Frontiers in Chinese Agriculture* (East Lansing, MI: Michigan University Press, 1987); Brown and Flavin, "The Earth's Vital Signs," p. 11.

87. Brown, "Sustaining World Agriculture," p. 128; El-Hinnawi and Hashmi, *The State of the Environment*, p. 50.

88. El-Hinnawi and Hashmi, *State of The Environment*, p. 50.

89. Ibid., p. 96; U.S. Environmental Protection Agency, *Pesticide Industry Sales and Usage: 1987 Market Estimates* (Washington, DC, 1988), Table 1; Pimentel, private communication.

90. Sandra Postel, "Controlling Toxic Chemicals," in Brown, *State of the World 1988*, p. 120; cropland percentages were computed with data from WRI and IIED, *World Resources 1986*, p. 265.

91. "The Economics of Pesticides," *EPA Journal*, May 1987, p. 11; EPA, *Pesticide Industry Sales and Usage*, Tables 1 and 8. According to EPA, the decline in agricultural use of pesticides since 1982 is due to a number of factors, including "more potent pesticides leading to lower application rates, more efficient use of pesticides, and lower farm commodity prices:" "Summary," *Pesticide Industry Sales and Usage*.

92. Postel, "Controlling Toxic Chemicals," p. 119; U.S. Environmental Protection Agency, *Pesticide Industry Sales and Usage, 1987* (Washington, DC, 1988), Table 8; Pimentel, "Waste in Agriculture and Food Sectors," p. 10.

93. El-Hinnawi and Hashmi, *The State of the Environment*, pp. 96-97; Pimentel, "Waste in Agriculture and Food Sectors," p. 11.

94. Postel, "Controlling Toxic Chemicals," p. 121; El-Hinnawi and Hashmi, *The State of the Environment*, p. 97; *World Bank News*, November 23, 1988, p. 4; Pimentel, "Waste in Agriculture and Food Sectors," p. 11.

95. David Pimentel and Lois Levitan, "Pesticides: Amounts Applied and Amounts Reaching Pests," *Bioscience*, Vol. 36, No. 2, February 1986, pp. 86-91.

96. El-Hinnawi and Hashmi, *The State of the Environment*, p. 101; Repetto, "Population, Resources, Environment," p. 22.

97. Edward Flattau, "Loss of Farmland in China," *Chicago Tribune*, November 25, 1986; see also Brown, *The Changing World Food Prospect*, p. 19.

98. El-Hinnawi and Hashmi, *The State of the Environment*, pp. 40-41.

99. Pimentel, "World Agriculture and Soil Erosion," p. 277; El-Hinnawi and Hashmi, *The State of the Environment*, p. 40.

100. See Norman Myers, *A Wealth of Wild Species: Storehouse for Human Welfare* (Boulder, CO: Westview Press, 1983), Chapter 4; Malcolm Gladwell, "Monsanto Experiment Seeks Herbicide-Resistant Plant," *The Washington Post*, May 17, 1988, p. C1.

101. See Hendry, "Food and Population," pp. 17-18.

102. Ibid., pp. 19-24.

103. Dankelman and Davidson, *Women and Environment*, p. 9.

104. WRI and IIED, *World Resources 1986*, p. 58.

105. Smithsonian Institution Traveling Exhibition Service, *Tropical Rainforests: A Disappearing Treasure* (Washington, DC, 1988), p. 22; Brown, *The Changing World Food Prospect*, p. 50.

106. WRI and IIED, *World Resources 1987*, pp. 41-42.

107. El-Hinnawi and Hashmi, *The State of the Environment*, p. 49.

108. See El-Hinnawi and Hashmi, *The State of the Environment*, pp. 49-54; and Judith Gradwohl and Russell Greenberg, *Saving the Tropical Forests* (London: Earthscan, 1988 or Washington, DC: Island Press, 1988), p. 107.

109. Brown, *The Changing World Food Prospect*, pp. 49-50; Robert Repetto (ed.), *The Global Possible: Resources, Development, and the New Century* (New Haven: Yale University Press, 1985), p. 506; E.T. York, Jr., "Improving Sustainability with Agricultural Research," *Environment*, November 1988; El-Hinnawi and Hashmi, *The State of the Environment*, pp. 49-54; WRI and IIED, *World Resources 1987*, p. 116; Sandra Postel, "Increasing Water Efficiency," in Brown, *State of the World 1986*, pp. 41-45; *Time*, January 2, 1989, p. 71.

110. Mellor, "Agricultural Development," p. 11; Repetto, *The Global Possible*, pp. 505-06.

111. El-Hinnawi and Hashmi, *The State of the Environment*, p. 54.

112. York, "Improving Sustainability;" Repetto, *The Global Possible*, p. 506; El-Hinnawi and Hashmi, *The State of the Environment*, p. 54; Gradwohl and Greenberg, *Saving the Tropical Forests*, p. 105.

113. Pedro A. Sanchez and Jose R. Benites, "Low-Input Cropping for Acid Soils of the Humid Tropics," *Science*, Vol. 238, December 11, 1987, pp. 1521-27.

114. See Gradwohl and Greenberg, *Saving the Tropical Forests*, p. 107; and Myers, *The Primary Source*.

115. Repetto, *The Global Possible*, p. 506; York, "Improving Sustainability;" Gradwohl and Greenberg, *Saving the Tropical Forests*, p. 109.

116. Brown, *The Changing World Food Prospect*, p. 49; El-Hinnawi and Hashmi, *The State of the Environment*, pp. 54, 57; Repetto, *The Global Possible*, p. 506.

117. *World Bank News*, June 8, 1989, p. 4.

118. WRI and IIED, *World Resources 1987*, pp. 71-72; Eckholm, *Losing Ground*, p. 71.

119. *Time*, January 2, 1989, p. 71.

120. Brown, *The Changing World Food Prospect*, p. 50; Myers, *Gaia: An Atlas of Planet Management*, pp. 62-63; Repetto, *The Global Possible*, p. 506.

121. Pimentel, private communication.

122. *Time*, January 2, 1989, p. 71.

123. Postel, "Controlling Toxic Chemicals," pp. 124-25.

124. Ibid., pp. 127-28.

125. Ibid., pp. 125-26.

126. Pimentel, "Waste in Agriculture and Food Sectors," p. 11; see also WRI and IIED, *World Resources 1988-89*, p. 2.

127. El-Hinnawi and Hashmi, *The State of the Environment*, p. 51; Postel, "Controlling Toxic Chemicals," pp. 119-29; York, "Improving Sustainability;" Malcolm H. Fleming, "Agricultural Chemicals in Groundwater, *American Journal of Alternative Agriculture*, Vol. 2, No. 3, pp. 124-130.

128. Brown, *The Changing World Food Prospect*, p. 51; El-Hinnawi and Hashmi, *The State of the Environment*, p. 53.

129. Pimentel, "Waste in Agriculture and Food Sectors," p. 14.

130. See Benjamin R. Stinner and Garfield J. House, "Role of Ecology in Lower-Input, Sustainable Agriculture: An Introduction," *American Journal of Alternative Agriculture*, Vol. II, No. 4, Fall 1987; see also "High-Input vs. Low-Input Agriculture," in WRI and IIED, *World Resources 1986*, p. 57.

131. I. Garth Youngberg, Executive Director, Institute for Alternative Agriculture, private communication, 1986.

132. See Postel, "Controlling Toxic Chemicals," pp. 124-29, and El-Hinnawi and Hashmi, *The State of the Environment*, pp. 51, 102-03.

133. El-Hinnawi and Hashmi, *The State of the Environment*, p. 63.

134. See David A. Andow and David P. Davis, "Agricultural Chemicals: Food and Environment," in David Pimentel and Carl W. Hall (eds.), *Food and Natural Resources* (San Diego, CA: Academic Press, 1989), Chapter 8, esp. p. 199.

135. Lappé, *Diet for a Small Planet*; Robbins, *Diet for a New America*.

136. See Alexander King (President, The Club of Rome), "The World Food Situation: Glut and Starvation," *The World and I*, March 1988; and Susan George, "Food

Strategies for Tomorrow," *The Hunger Project Papers*, No. 6, (San Francisco: The Hunger Project, December 1987).

137. WRI and IIED, *World Resources 1987*, p. 52.

138. "EPA Gives Go Ahead", *The Washington Post*, March 30, 1988, p. A18.

139. *Time*, May 2, 1988, p. 57.

140. Brown, *The Changing World Food Prospect*, p. 51.

141. El-Hinnawi and Hashmi, *The State of the Environment*, pp. 54-57; Brown, *The Changing World Food Prospect*, p. 50; York, "Improving Sustainability;" *Time*, January 2, 1989, p. 71.

142. Pimentel, private communication.

143. El-Hinnawi and Hashmi, *The State of the Environment*, p. 56. While hydroponics and other newer technologies can be useful, they are often energy-intensive and therefore costly.

144. Ibid., pp. 57-63; Pimentel, private communication; Roger Revelle, "Present and Future State of Living Marine and Freshwater Resources," in Repetto, *The Global Possible*, pp. 439-42; World Commission on Environment and Development, *Our Common Future* (New York: Oxford, 1987), pp. 137-38.

145. Repetto, *The Global Possible*, p. 506; Mellor, "Agricultural Development," p. 10.

146. Repetto, *The Global Possible*, p. 506.

147. Sara Ebenreck, Institute for Alternative Agriculture, Greenbelt, MD, private communication, July 14, 1989.

148. Ibid., p. 506; Brown, *The Changing World Food Prospect*, p. 49.

149. Gradwohl and Greenberg, *Saving the Tropical Forests*, p. 103.

150. Repetto, *The Global Possible*, p. 506; Brown, "The Changing World Food Prospect," p. 50.

151. Mellor, "Agricultural Development," p. 13; Repetto, *The Global Possible*, p. 506.

152. Brown, *The Changing World Food Prospect*, p. 50; Repetto, *The Global Possible*, p. 506; Mellor, "Agricultural Development," p. 12; King, "The World Food Situation."

153. WRI and IIED, *World Resources 1986*, p. 58.

154. FAO, *Potential Population Supporting Capacities of Lands in the Developing World*; Harrison, *Land, Food, and People*; Brown, *The Changing World Food Prospect*, p. 51.

155. Gradwohl and Greenberg, *Saving the Tropical Forests*, pp. 102-37. For a description of a low-input cropping system for tropical agriculture, see Sanchez and Benites, "Low-Input Cropping for Acid Soils of the Humid Tropics."

156. Walter V. Reid, James N. Barnes, and Brent Blackwelder, *Bankrolling Successes: A Portfolio of Sustainable Development Projects* (Washington, DC: Environmental Policy Institute and National Wildlife Federation, 1988), pp. 6-7.

157. Ibid., pp. 8-9.

158. Ibid., pp. 26-27. A videotape, *The Miracle of Guinope*, is described under Further Information.

159. Ibid., pp. 41-42.

160. Myers, *Gaia: Atlas of Planet Management*, pp. 32-33, 62-64; Postel, "Controlling Toxic Chemicals," p. 125. See also Wittwer, *Feeding a Billion: Frontiers in Chinese Agriculture*, and Glenn Garelick's review of *Feeding a Billion* in *Time*, October 12, 1987, p. 55.

161. Orville W. Bidwell, review of Wittwer, *Feeding a Billion: Frontiers in Chinese Agriculture*, in *American Journal of Alternative Agriculture*, Vol. 3, No. 1, Winter 1988, pp. 41-42, 47. For a review of successful agricultural development in China, India, and other Asian countries, see Janos P. Hrabovszky, "Agriculture: The Land Base," in Repetto, *The Global Possible*, pp. 244-46.

162. J. Patrick Madden and Paul F. O'Connell, "Early Results of the LISA Program," (Washington, DC: U.S. Department of Agriculture, April 20, 1989), pp. 1-4; see also "USDA Establishes Low-Input Program," *Alternative Agriculture News*, Vol. 6, No. 2, February 1988, p. 1.

163. Ibid., pp. 8-11.

164. Ibid., pp. 15-16.

165. Renew America, *The State of the States 1989* (Washington, DC, 1989), pp. 26, 31; and *Alternative Agriculture News*, Vol. 7, No. 5, May 1989.

166. National Research Council, *Alternative Agriculture* (Washington, DC, 1989); and Arthur S. Brisbane, "Panel Backs 'Benign' Farming to Save Soil, Cut Chemicals," *The Washington Post*, September 9, 1989, p. A1.

167. Jan Hartke, "Food and the Environment: The Noble Pledge," *Interaction*, newsletter of the Global Tomorrow Coalition, Vol. 9, No. 1, Summer 1989.

168. Lappé, *Diet for a Small Planet*, pp. 237-39.

Chapter 6: Biological Diversity

1. Norman Myers, *A Wealth of Wild Species: Storehouse for Human Welfare* (Boulder, CO: Westview Press, 1983), p. 3.

2. Peter H. Raven, "Our Diminishing Tropical Forests," in E.O. Wilson (ed.), *Biodiversity* (Washington, DC: National Academy Press, 1988), p. 121.

3. E.O. Wilson, "The Current State of Biological Diversity," in Wilson, *Biodiversity*.

4. Reported by Boyce Rensberger, "Scientists See Signs of Mass Extinction," *Washington Post*, September 29, 1986, p. A1. The Club of Earth consists of leading biologists including Edward O. Wilson, Ernst Mayr, Thomas Eisner, and Paul Ehrlich.

5. See Mycrs, *A Wealth of Wild Species*.

6. Peter H. Raven, private communication.

7. Rensberger, "Scientists See Signs of Mass Extinction," Wilson, "The Current State of Biological Diversity," pp. 8-9.

8. Raven, private communication; World Commission on Environment and Development (WCED), *Our Common Future* (New York: Oxford University Press, 1987), p. 46.

9. See Norman Myers (ed.), *Gaia: An Atlas of Planet Management* (Garden City, NY: Anchor Books, 1984), pp. 146, 154-59.

10. Wilson, "The Current State of Biological Diversity," p. 3.

11. Wilson, "The Current State of Biological Diversity," p. 5.

12. Edward C. Wolf, "Avoiding a Mass Extinction of Species," in Lester R. Brown et al., *State of the World 1988* (New York: Norton, 1988), p. 105. For a detailed breakdown of known plant and animal species, see E. O. Wilson, "The Current State of Biological Diversity," pp. 3-5.

13. Myers, *Gaia: An Atlas of Planet Management*, pp. 154-55. See also the book's foreword by Gerald Durrell.

14. Wilson, "The Current State of Biological Diversity," p. 13; Myers, *Gaia: An Atlas of Planet Management*, p. 154. Several experts recently estimated that every day 100 species are lost completely: *UNEP North American News*, Vol. 4, No. 4, August 1989.

15. Raven, "Our Diminishing Tropical Forests," in Wilson, *Biodiversity*, p. 121.

16. Raven, private communication.

17. Projection by Daniel Simberloff; details summarized in Wolf, "Avoiding a Mass Extinction of Species," p. 103.

18. Based on data compiled by the International Union for Conservation of Nature and Natural Resources (IUCN), reported in Lee Durrell, *State of the Ark* (Garden City, NY: Doubleday, 1986), p. 95.

19. Organization for Economic Cooperation and Development, *The State of the Environment 1985* (Paris: OECD, 1985), p. 142.

20. Russell Train, "Bush Faces Major Environmental Challenges," *Conservation Foundation Letter*, 1989, No. 1, p. 8.

21. Peter H. Raven, "We're Killing Our World: Preservation of Biological Diversity." Keynote address, annual meeting, American Association for the Advancement of Science, Chicago, IL, February 1987; reprinted in *Vital Speeches of the Day*, 53:15, May 15, 1987, pp. 472-78.

22. See Myers, *Gaia: An Atlas of Planet Management*, p. 155.

23. Norman Myers, "Tropical Forests and Their Species: Going, Going...?," in E. O. Wilson (ed.), *Biodiversity*, pp. 28-29.

24. World Resources Institute (WRI) and International Institute for Environment and Development (IIED), *World Resources 1987* (New York: Basic Books, 1987), pp. 86-88.

25. Wolf, "Avoiding a Mass Extinction of Species," p. 102.

26. World Wildlife Fund, *1987 Annual Report* (Washington, DC: 1988), p.40.

27. George R. Robinson and James F. Quinn, "Extinction, Turnover and Species Diversity in an Experimentally Fragmented California Annual Grassland," *Oecologia*, Vol.76, 1988, pp.71-82; James F. Quinn and Alan Hastings, "Extinction in Subdivided Habitats," *Conservation Biology* 1:3, October 1987, pp.198-208.

28. Faith T. Campbell, "Conserving Biological Diversity," Natural Resources Defense Council, Washington, DC, unpublished paper, June 1982.

29. A. Kent MacDougall, "Worldwide Costs Mount as Trees Fall," *Los Angeles Times*, June 14, 1987, p. 16. At least two-thirds of the islands' land bird species have been lost since the Polynesian settlement around 500 A.D., according to Roger Lewin, "Hand of Man Seen in Birds," *Science*, June 19, 1987, p. 1522.

30. Campbell, "Conserving Biological Diversity."

31. WRI and IIED, *World Resources 1986* (New York: Basic Books, 1986), pp. 90-91; Smith Hempstone, "Trying to Rescue the Elephant," *The Washington Times*, April 5, 1989, p. F1.

32. Myers, *Gaia, An Atlas of Planet Management*, p. 151.

33. Bat Conservation International, "Why Save Bats" (Austin, TX: brochure, undated); Norman Myers, *The Primary Source* (New York: Norton, 1984), pp. 82-85, 213.

34. Campbell, "Conserving Biological Diversity"; Jane Netting Huff, "Ecological Aspects of Biological Diversity," unpublished paper, June 1982.

35. Myers, *Gaia: An Atlas of Planet Management*, pp. 156-57; Myers, *A Wealth of Wild Species*, pp. 13-86; Faith T. Campbell, "Food Supplies," Natural Resources Defense Council, Washington, DC, unpublished paper, June 1982; Suzanne Eaton, "Third World Food Resources and Their Dependence on Wild Species," unpublished paper, June 1982.

36. Raven, private communication.

37. *International Wildlife*, May-June, 1984, p. 35; see also WRI and IIED, *World Resources 1987*, p. 83.

38. Myers, *Gaia: An Atlas of Planet Management*, p. 147.

39. Faith T. Campbell, "Medical Progress," Natural Resources Defense Council, Washington, DC, unpublished paper, June 1982; Myers, *Gaia, An Atlas of Planet Management*, p. 147; Myers, *A Wealth of Wild Species*, pp. 89-141. Despite achievements in laboratory synthesis, some experts feel that genetic engineering will increase rather than eliminate dependence on wild species. Genetic engineering allows extraction of individual genes from species otherwise more difficult to work with. To be studied and analyzed for their usefulness, however, these genes must exist in living organisms. Thus scientists are concerned that rapid species extinction, especially in the tropics, will deprive them of useful organisms before their potential benefit to man can be assessed. Campbell, "Medical Progress."

40. Margery L. Oldfield, *The Value of Conserving Genetic Resources* (Washington, DC: National Park Service. U.S. Department of the Interior, 1984). See also Myers, *A Wealth of Wild Species*, pp. 145-92.

41. Myers, *A Wealth of Wild Species*, pp. 150-51.

42. See Bryan Norton, "The Aesthetic Value of Natural Diversity," unpublished paper, June 1982; Bryan Norton, "Commodity, Amenity, and Morality: The Limits of Quantification in Valuing Biodiversity," in Wilson, *Biodiversity*, pp. 200-05; and Bryan Norton (ed.), *The Preservation of Species: The Value of Biological Diversity* (Princeton, NJ: Princeton University Press, 1986). See also Myers, *The Primary Source*, pp. 354-55; and Holmes Rolston, *Environmental Ethics: Duties to and Values in the Natural World* (Philadelphia: Temple University Press, 1988), esp. Chapter 4.

43. Michael H. Robinson, "Beyond Destruction, Success," in Judith Gradwohl and Russell Greenberg, *Saving the Tropical Forests* (London: Earthscan, 1988), p. 16.

44. Myers, *Gaia: An Atlas of Planet Management*, pp. 164-65.

45. WRI and IIED, *World Resources 1987*, p. 78; and Wolf, "Avoiding a Mass Extinction of Species," p. 102.

46. Myers, *Gaia: An Atlas of Planet Management*, pp. 160-61.

47. Man and the Biosphere Secretariat, *Biosphere Reserves*, a 20 x 32 inch wall chart (Paris: UNESCO, 1988).

48. World Wildlife Fund, *Future in the Wild: A Conservation Handbook*, (Washington, DC: no date), p. 15.

49. Raven, private communication.

50. WRI and IIED, *World Resources 1987*, p. 78; Wolf, "Avoiding a Mass Extinction of Species," p. 102.

51. Wolf, "Avoiding a Mass Extinction of Species," p. 110. See also "Restoration Ecology: Can We Recover Lost Ground," in Wilson, *Biodiversity*, pp. 311-52.

52. Christopher Uhl, "Restoration of Degraded Lands in the Amazon Basin," in Wilson, *Biodiversity*, pp. 326-32; Daniel H. Janzen, "Tropical Dry Forests," in Wilson, *Biodiversity*, pp. 130-37; Wolf, "Avoiding a Mass Extinction of Species," pp. 111-14.

For a series of case studies of ecological restoration in the United States, see John J. Berger, *Restoring the Earth: How Americans Are Working to Renew Our Damaged Environment* (New York: Anchor Press, 1987). The case studies include restoration of strip-mined land, prairies, forests, lakes, streams, rivers, and saltwater wetlands.

53. Mark J. Plotkin, "The Outlook for New Agricultural and Industrial Products from the Tropics," in Wilson, *Biodiversity*, p. 114.

54. Myers, *Gaia: An Atlas of Planet Management*, p. 148.

55. Plotkin, "The Outlook for New Agricultural and Industrial Products from the Tropics."

56. Myers, *Gaia: An Atlas of Planet Management*, p. 149.

57. Wolf, "Avoiding a Mass Extinction of Species," pp. 112-13.

58. National Research Council, *Research Priorities in Tropical Biology* (Washington, DC: National Academy of Sciences, 1980), p.47.

59. Wilson, "The Current State of Biological Diversity," p. 14.

60. Catherine Caufield, *Tropical Moist Forests: the Resource, the People, the Threat* (London: International Institute for Environment and Development, 1982), p. 39.

61. Robinson, "Beyond Destruction, Success," p. 14.

62. Wilson, "The Current State of Biological Diversity," pp. 14-15.

63. Wolf, "Avoiding a Mass Extinction of Species," p. 108.

64. Ibid., p. 116; World Wildlife Fund, "Linking Conservation and Development: The Program in Wildlands and Human Needs," Washington, DC, December 1986.

65. Wolf, "Avoiding a Mass Extinction of Species," p. 115; WRI and IIED, *World Resources 1987*, p. 90.

66. Wolf, "Avoiding a Mass Extinction of Species," p. 115.

67. Norman Cohen, U.S. Agency for International Development, Washington, DC, private communication, March 1988.

68. Wilson, "The Current State of Biological Diversity," p. 15.

69. Wolf, "Avoiding a Mass Extinction of Species," pp. 115-16.

70. Sheldon Annis, Overseas Development Council, Testimony before the U.S. Senate Subcommittee on International Economic Policy, Trade, Oceans, and the Environment, April 27, 1988.

As of mid 1989, Costa Rica had purchased over $75 million in debt titles for conservation and sustainable development proojects, making the country's debt-exchange program the largest in the Third World: "The Impact of Debt-for Nature Swaps on Global Conservation," Environmental and Energy Study Institute, *Notice*, July 25, 1989.

71. J.A. McNeely, *Economics and Biological Diversity*, pp. vii-ix.

72. *UNEP North American News*, October 1987, p. 3. For further details, see I. Michael Sweatman, "International Conservation Finance Programme," in Vance Martin (ed.), *For the Conservation of Earth* (Golden, CO: International Wilderness Leadership Foundation, 1988), pp. 361-63; and WCED, *Our Common Future*, pp. 338-39.

73. C. MacFarland and others, "Establishment, Planning, and Management of a National Wildlands System

in Costa Rica: A Case Study;" in J.A. McNeely and K.R. Miller, *National Parks, Conservation and Development: The Role of Protected Areas in Sustaining Society* (Washington, DC: Smithsonian Institution Press, 1984), pp. 592-600.

74. This discussion is based on suggestions by Peter H. Raven in private communication, October 25, 1988.

75. Myers, *Gaia: An Atlas of Planet Management*, p. 166.

76. *Ibid.*, pp. 166-67. The World Heritage Convention designates cultural and natural properties of "outstanding universal value to mankind," and provides for positive action to preserve them.

77. WCED, *Our Common Future*, p. 162.

78. Myers, *Gaia: An Atlas of Planet Management*, p. 166.

79. *World Conservation Strategy; Living Resource Conservation for Sustainable Development* (Gland, Switzerland: International Union for Conservation of Nature and Natural Resources, 1980); Myers, *Gaia: An Atlas of Planet Management*, p. 168.

80. WRI and IIED, *World Resources 1986*, p. 100.

81. Wolf, "Avoiding a Mass Extinction of Species," p. 116.

82. Thomas B. Stoel, Jr., Natural Resources Defense Council, Washington, DC, private communication, February 1988.

83. *UNEP North American News*, Vol. 4, No. 4, August 1989.

84. "Towns Embracing 'Land Bank' Idea," *Washington Post*, October 31, 1987, p. E1. For further information, obtain a copy of *Land Banking*, published in 1986 by the Massachusetts Association of Conservation Commissions, Lincoln Filene Center, Tufts University, Medford, MA 02155. 41 pp., $6.00.

85. Wolf, "Avoiding a Mass Extinction of Species," p. 109.

86. Ibid., pp. 108, 114.

Chapter 7: Tropical Forests

1. Norman Myers, *The Primary Source* (New York: Norton, 1984), p. 333.

2. Michael Robinson, "Beyond Destruction, Success," in Judith Gradwohl and Russell Greenberg, *Saving the Tropical Forests* (London: Earthscan, 1988), p. 11.

3. Judith Gradwohl, *Tropical Rainforests: A Disappearing Treasure* (Washington, DC: Smithsonian Institution Traveling Exhibition Service, 1988), p. 1.

4. Ibid., pp. 1, 20-22.

5. Norman Myers (ed.), *Gaia: An Atlas of Planet Management* (Garden City, NY: Anchor Books, 1984), p. 30.

6. Peter H. Raven, "Our Diminishing Tropical Forests," in E.O. Wilson (ed.), *Biodiversity* (Washington, DC: National Academy Press, 1988), p. 119.

7. World Resources Institute (WRI) and International Institute for Environment and Development (IIED), *World Resources 1987* (New York: Basic Books, 1987), p. 61.

8. Hugh Itlis, "Serendipity in the Exploration of Biodiversity," in Wilson, *Biodiversity*, p. 100.

9. R. Meyer De Schauensee, *The Birds of Colombia* (Narberth, PA: Livingston Publishing Company, 1964), p. ix.

10. World Commission on Environment and Development (WCED), *Our Common Future* (New York: Oxford University Press, 1987), pp. 147-67.

11. William Booth, "Saving Rain Forests by Using Them," *The Washington Post*, June 29, 1989, p. A1. The studies were carried out by scientists from the New York Botanical Garden, the Missouri Botanical Garden, and Yale University.

12. World Resources Institute (WRI), *Tropical Forests: A Call for Action*, Part 1, *The Plan* (Washington, DC: WRI, 1985), p. 10.

13. Mark J. Plotkin, "The Outlook for New Agricultural and Industrial Products from the Tropics," in Wilson, *Biodiversity*, pp. 107-10.

14. Ibid.

15. Ibid.

16. Norman Myers, *A Wealth of Wild Species* (Boulder, CO: Westview Press, 1983), pp. 60-61.

17. Gradwohl, *Tropical Rainforests*; World Resources Institute, *Keep Tropical Forests Alive* (Washington, DC: WRI, undated), pp. 2-3.

18. Myers, *A Wealth of Wild Species*, pp. 145-48.

19. Ibid., pp. 155-56, 167-68.

20. Myers, *The Primary Source*, pp. 260-93.

21. Catherine Caufield, *Tropical Moist Forests* (London: Earthscan, 1982), pp 11-13.

22. WCED, *Our Common Future*, p. 151.

23. Smithsonian Institution Traveling Exhibition Service (SITES), *Tropical Rainforests: A Disappearing Treasure* (Washington, DC, 1988), exhibition script.

24. Lester R. Brown, Christopher Flavin, and Sandra Postel, "Outlining a Global Action Plan," in Brown et al., *State of the World 1989* (New York: Norton, 1989), p. 180.

25. WRI and IIED, *World Resources 1987*, p. 57.

26. Myers, *Gaia*, p. 44; E.O. Wilson, "The Current State of Biological Diversity," in Wilson, *Biodiversity*, p. 10.

27. Lester R. Brown and Edward C. Wolf, "Reversing Africa's Decline," in Brown et al., *State of the World 1986* (New York: Norton 1986), p. 181.

28. Wilson, "The Current State of Biological Diversity," p. 10; Edward C. Wolf, "Avoiding a Mass Extinction of Species," in Lester R. Brown et al., *State of the World 1988* (New York: Norton, 1988), p. 109.

29. WRI and IIED, *World Resources 1986* (New York: Basic Books, 1986), p. 64.

30. Gradwohl, *Tropical Rainforests*, pp. 20-23.

31. Myers, *The Primary Source*, p. 300.

32. Robert Repetto, *The Forest for the Trees? Government Policies and Misuse of Forest Resources* (Washington, DC: World Resources Institute, 1988).

33. Gradwohl, *Tropical Rainforests*, p. 20.

34. Myers, *Gaia*, pp. 42-43.

35. Sandra Postel and Lori Heise, "Reforesting the Earth," in Brown et al., *State of the World 1988*, p. 86.

36. Gradwohl and Greenberg, *Saving the Tropical Forests*, p. 41; John Spears, "Preserving Biological Diversity in the Asian Region," in Wilson, *Biodiversity*, p. 394.

37. Catherine Caufield, *In the Rainforest* (Chicago: University of Chicago Press, 1984), p. 148; Mark Mardon, "The Big Push," *Sierra*, November-December 1988, pp. 67-75.

38. Caufield, *In the Rainforest*, pp. 147-49; Patricia Castano and Adelaida Trujillo, "Colombian Peasants: The Road to Coca," *Panoscope*, September 1988, pp. 12-15.

39. Gradwohl, *Tropical Rainforests*, pp. 20-22.

40. *UNEP News*, February 1988.

41. *The Washington Post*, July 23, 1988, p. A19.

42. See WRI and IIED, *World Resources 1988-89* (New York: Basic Books, 1988), p. 286; and WRI and IIED, *World Resources 1986*, p. 73.

43. Caufield, *Tropical Moist Forests*, p. 29.

44. A. Kent MacDougall, "Worldwide Costs Mount as Trees Fall," *Los Angeles Times*, June 14, 1987, p. 12.

45. John Magrath, "Brazil to Burn Amazon Forest for Iron," *Panoscope*, November 1988, p. 24.

46. Myers, *Gaia*, p. 103.

47. WRI and IIED, *World Resources 1988-89*, pp 286-89.

48. Gradwohl, *Tropical Rainforests*, p. 22; Myers, *Gaia*, p. 43.

49. WRI and IIED, *World Resources 1986*, p. 72.

50. Alan Durning, "U.S. Poultry Consumption Overtakes Beef," *World Watch*, January-February 1988, p. 11.

51. Myers, *The Primary Source*, pp. 131-36; SITES, *Tropical Rainforests*.

52. Durning, "U.S. Poultry Consumption Overtakes Beef," p. 11.

53. Gradwohl, *Tropical Rainforests*, p. 22.

54. Myers, *Gaia*, p. 43.

55. Gradwohl, *Tropical Rainforests*, pp. 22-23.

56. Caufield, *Tropical Moist Forests*, pp. 27, 33.

57. Jean Paul Malingreau and Compton J. Tucker, "Large-scale Deforestation in the Southeastern Amazon," *Ambio*, Vol. 17, No. 1, 1988, pp. 49-55.

58. M. Margolis, "Thousands of Amazon Acres Burning," *The Washington Post*, September 8, 1988, p. A31.

59. "Brazil Announces Plans to Save Forest," *The Washington Post*, October 13, 1988, p. A34.

60. National Research Council, *Population Growth and Economic Development: Policy Questions* (Washington, DC: National Academy Press, 1986), pp. 35, 39.

61. *The Washington Post*, October 13, 1988, p. A34.

62. Caufield, *Tropical Moist Forests*, p. 24; Gradwohl, *Tropical Rainforests*, p. 23.

63. Deforestation rates from WRI and IIED, *World Resources 1986*, p. 73; population growth rates from Population Reference Bureau (PRB), *1988 World Population Data Sheet* (Washington, DC: PRB, 1988).

64. William S. Ellis, "Brazil's Imperiled Rainforest," *National Geographic*, December 1988, pp. 778-80; Caufield, *Tropical Moist Forests*, p. 24.

65. Sierra Club, *Bankrolling Disasters* (San Francisco, CA: Sierra Club, 1986), p. 7; Caufield, *Tropical Moist Forests*, pp. 24-28.

66. Gradwohl, *Tropical Rainforests*, p. 23; John Sewell et al., *Growth and Jobs in a Changing World Economy: Agenda 1988* (New Brunswick, NJ: Transaction Books, 1988), p. 227.

67. MacDougall, "Worldwide Costs Mount as Trees Fall," p. 12.

68. Myers, *The Primary Source*, p. 335.

69. United Nations Fund for Population Activities (UNFPA), *Population Images* (New York: UNFPA, 1987), p. 20.

70. Gradwohl, *Tropical Rainforests*, pp. 24-27.

71. Wilson, *Biodiversity*, pp. 29-31, 59-69, 119-22; WCED, *Our Common Future*, p. 151.

72. Eugene Linden, "The Death of Birth," *Time*, January 2, 1989, p. 34.

73. John H. Rappole et al., *Nearctic Avian Migrants in the Neotropics* (Washington, DC: U.S. Department of the Interior, Fish and Wildlife Service, July 1983), pp. 59-68; Paul R. Ehrlich, "The Loss of Diversity," in Wilson, *Biodiversity*, p. 22; Paul R. Ehrlich, "Winged Warning," *Sierra*, September-October 1988, p. 58; Shirley F. Briggs, "Results of Recent Winter and Breeding Bird Censuses in Long-term Study Area," *Atlantic Naturalist*, Vol. 36, 1986; Chandler Robbins, John Sauer, Russell Greenberg, and Samuel Droege, "Population Declines in North American Birds that Migrate to the Neotropics," *Proceedings of the National Academy of Sciences (USA)*, in press; and William Booth, "Tropical Forest Loss May Be Killing Off Songbirds," *The Washington Post*, July 26, 1989, p. A1.

74. Caufield, *Tropical Moist Forests*, pp. 12-13.

75. Myers, *Gaia*, p. 148.

76. Caufield, *Tropical Moist Forests*, pp. 11-13, 19-24; Myers, *Gaia*, p. 49.

77. Gilbert M. Grosvenor, "Will We Mend Our Earth," *National Geographic*, December 1988, p. 770.

78. WRI and IIED, *World Resources 1988-89*, p. 192.

79. Ibid., p. 282.

80. Myers, *The Primary Source*, pp. 9-10; Brown, "Conserving Soils," in Lester R. Brown et al., *State of the World 1984* (New York: Norton, 1984), pp. 65-66.

81. Linden, "The Death of Birth," p. 34.

82. WRI and IIED, *World Resources 1988-89*, pp. 192-93.

83. Brown, "Conserving Soils," p. 67.

84. Nafis Sadik, *The State of World Population 1988* (New York: United Nations Population Fund, 1988), p. 1.

85. Myers, *The Primary Source*, pp. 262-72; Myers, *Gaia*, p. 44; The World Bank, *Annual Report 1985* (Washington, DC, 1985), p. 72.

86. WRI and IIED, *World Resources 1988-89*, p. 192.

87. Norman Myers, "Tropical Forests and Their Species: Going, Going...?," in Wilson, *Biodiversity*, p. 32.

88. Caufield, *Tropical Moist Forests*, p. 17.

89. Myers, *Gaia*, p. 44.

90. Postel and Heise, "Reforesting the Earth," p. 95.

91. WCED, *Our Common Future*, pp. 147-67.

92. WRI, *Tropical Forests*, p. 10.

93. Sandra Postel and Lori Heise, *Reforesting the Earth*, Worldwatch Paper 83, (Washington, DC: Worldwatch Institute, April 1988), p. 25.

94. Robert Repetto (ed.), *The Global Possible: Resources, Development and the New Century* (New Haven: Yale University Press, 1985), p. 503.

95. Norman Myers, *Not Far Afield: U.S. Interests and the Global Environment* (Washington, DC: WRI, 1987), pp. 29-30; WRI and IIED, *World Resources 1987*, p. 284; Gradwohl, *Tropical Rainforests*, p. 26.

96. Gradwohl and Greenberg, *Saving the Tropical Forests*, pp. 18, 57.

97. F. William Burley, "The Tropical Forest Action Plan," in Wilson, *Biodiversity*, p. 405; C. MacFarland et al., "Establishment, Planning and Management of a National Wildlife System in Costa Rica: A Case Study," in J. A. McNeely and K. R. Miller (eds.), *National Parks, Conservation and Development: The Role of Protected Areas in Sustaining Society* (Washington, DC: Smithsonian Institution Press, 1984), pp. 592-600.

98. Gradwohl and Greenberg, *Saving the Tropical Forests*, pp. 60-63; The Man in the Biosphere Secretariat, *Biosphere Reserves*, a wall chart, (Paris: UNESCO, 1988).

99. Gradwohl, *Tropical Rainforests*, p. 30; Myers, *The Primary Source*, pp. 107, 110, 297; Gradwohl and Greenberg, *Saving the Tropical Rainforests*, pp. 141-59.

100. Edward Goldsmith and Nicholas Hildyard (eds.), *The Earth Report: The Essential Guide to Global Ecological Issues* (Los Angeles: Price Stern Sloan, 1988), p. 89.

101. Peter Hazlewood, World Resources Institute, private communication, March 1989.

102. Myers, *The Primary Source*, pp. 297, 325-26, 356. Although local processing of hardwoods can have advantages for tropical countries, local sawmills can be highly inefficient and wasteful, resulting in the cutting of more trees to meet a given demand level. See Robert Repetto, *The Forest for the Trees?: Government Policies and the Misuse of Forest Resources* (Washington, DC: World Resources Institute, 1988), p. 25.

103. For case studies of sustainable harvesting in Brazil, Colombia, and Cameroon, see Gradwohl and Greenberg, *Saving the Tropical Forests*, pp. 142-62.

104. Miriam Parel, "The Death of Chico Mendes," *The Washington Post*, January 19, 1989, p. A27.

105. Myers, *The Primary Source*, p. 350; Gradwohl and Greenberg, *Saving the Tropical Forests*, p. 140.

106. Gradwohl and Greenberg, *Saving the Tropical Forests*, pp. 94-96, 155-57; Burley, "The Tropical Forestry Action Plan," p. 405.

107. Wilson, "The Current State of Biological Diversity," pp. 15-16.

108. Postel and Heise, "Reforesting the Earth," p. 85.

109. These include Haiti, Costa Rica, Brazil, Ecuador, Uganda, China, Thailand, and the Philippines. See Gradwohl and Greenberg, "*Saving the Tropical Rainforests*," pp. 163-189; and Christopher Uhl, "Restoration of Degraded Lands in the Amazon," in Wilson, *Biodiversity*, pp. 326-32.

110. See Gradwohl and Greenberg, *Saving the Tropical Rainforests*, pp. 165-67; Myers, *The Primary Source*, pp. 162-65; and Robert Winterbottom and Peter T. Hazlewood, "Agroforestry and Sustainable Development: Making the Connection," *Ambio*, Vol. 16, No. 2-3, 1987, pp. 100-10.

111. Postel and Heise, "Reforesting the Earth," pp. 99-100; Burley, "The Tropical Forestry Action Plan," p. 406.

112. Gradwohl and Greenberg, *Saving the Tropical Forests*, pp. 104-09.

113. Myers, *The Primary Source*, pp. 165-69.

114. Gradwohl, *Tropical Rainforests*, p. 30.

115. Myers, *Gaia*, p. 43; Myers, *The Primary Source*, pp. 297, 328.

116. Myers, *The Primary Source*, p. 148.

117. *The New York Times*, October 13, 1987.

118. Myers, *The Primary Source*, pp. 319-28.

119. World Resources Institute, "Summary of the Tropical Forestry Action Plan," (Washington, DC, undated), and Burley, "The Tropical Forestry Action Plan," p. 404.

120. WRI, *Tropical Forests*, pp. 1-2; Hazlewood, private communication, March 1989.

121. Lester R. Brown and Edward C. Wolf, "Reclaiming the Future," in Brown et al., *State of the World 1988*, pp. 182-83.

122. Hazlewood, private communication, March 1989.

123. Ibid., p. 405; Wolf, "Avoiding a Mass Extinction of Species," p. 116.

124. Gradwohl and Greenberg, *Saving the Tropical Forests*, p. 17; Peter T. Hazlewood, *Expanding the Role of Non-Governmental Organizations (NGOs) in National Forestry Programs*, The Report of Three Regional Workshops in Africa, Asia, and Latin America (Washington, DC: World Resources Institute, 1988).

125. Linden, "The Death of Birth," p. 32.

126. Burley, "The Tropical Forestry Action Plan," p. 405; Hazlewood, private communication.

Chapter 8: Ocean and Coastal Resources

1. G. Carleton Ray, "Ecological Diversity in Coastal Zones and Oceans," in E.O Wilson (ed.), *Biodiversity* (Washington, DC: National Academy Press, 1988), p. 48.

2. Quoted in "The State of the Marine Environment 1988," *UNEP News*, April 1988.

3. Michael Weber and Richard Tinney, *A Nation of Oceans* (Washington, DC: Center for Environmental Education, 1986), p.11.

4. Ray, "Ecological Diversity," pp. 38-39.

5. World Resources Institute (WRI) and International Institute for Environment and Development (IIED), *World Resources 1986* (New York: Basic Books, 1986), p. 141.

6. Norman Myers (ed.), "Ocean," *Gaia: An Atlas of Planet Management* (Garden City, NY: Anchor Books, 1984), pp. 70, 78; and World Commission on Environment and Development (WCED), *Our Common Future* (New York: Oxford University Press, 1987), p. 262.

7. WRI and IIED, *World Resources 1986*, p. 141.

8. Myers, "Ocean," pp. 74-75.

9. Anthony Calio, "Defining the Estuary," *EPA Journal*, July-August 1987, pp. 9-11.

10. Myers, "Ocean," p. 75.

11. Clifton Curtis, "An International Perspective," *EPA Journal*, July-August 1987, p. 30.

12. Myers, "Ocean," p. 75.

13. Weber and Tinney, *A Nation of Oceans*, pp. 54-56.

14. Myers, "Ocean," p. 74.

15. Weber and Tinney, *A Nation of Oceans*, p. 9.

16. WRI and IIED, *World Resources 1986*, p. 146.

17. Myers, "Ocean," pp. 86-87; "The Dirty Seas," *Time*, August 1, 1988, p. 46.

18. *EDF Letter*, September 1988, p. 3; "Don't Go Near the Water," *Newsweek*, August 1, 1988, p. 45; *Time*, August 1, 1988, p. 48.

19. WRI and IIED, *World Resources 1986*, p. 151; Manel Tampoe, "Economic Development and Coastal Erosion in Sri Lanka," *The Ecologist*, November-December 1988, pp. 225-30.

20. "The State of the Marine Environment 1988," *UNEP News*, April 1988; WRI and IIED, *World Resources 1986*, pp. 146-56.

21. Myers, "Ocean," p. 84.

22. Beth Millemann, *And Two If By Sea: Fighting the Attack on America's Coast* (Washington, DC: Coast Alliance, 1986), pp. 30-31; *Newsweek*, August 1, 1988, p. 45.

23. WRI and IIED, *World Resources 1986*, pp. 87, 126-27; "The State of the Marine Environment 1988," *UNEP News*, April 1988.

24. Millemann, *And Two If By Sea*, p. 28.

25. *Newsweek*, August 1, 1988, pp. 45-47.

26. Natural Resources Defense Council, *Ebb Tide for Pollution: Actions for Cleaning Coastal Waters* (New York, 1989); D'Vera Cohn, "Beach Pollution Problem is Worsening, Group Says," *The Washington Post*, August 10, 1989, p. A14.

27. Millemann, *And Two If By Sea*, p. 28.

28. *The Washington Post*, December 15, 1986, p. A6.

29. Joseph E. Cummins, "Extinction: The PCB Threat to Marine Mammals," *The Ecologist*, November-December 1988, pp. 193-95.

30. Myers, "Ocean," p. 84; Millemann, *And Two If By Sea*, p. 30.

31. Millemann, *And Two If By Sea*, p. 29.

32. The Conservation Foundation, *State of the Environment: A View toward the Nineties* (Washington, DC: 1987), p. 146; Millemann, *And Two If By Sea*, p. 30.

33. *The Washington Post*, August 21, 1987, p. A9; *Newsweek*, August 1, 1988, p. 44.

34. *EDF Letter*, September 1988.

35. Myers, "Ocean," p. 86; *Time*, August 1, 1988, p. 46; *Newsweek*, August 1, 1988; *The New York Times*, December 25, 1988, p. E3.

36. *Time*, August 1, 1988, p. 49; Philip J. Hilts and Lisa Leff, "Toxic Algae Killed Dolphins," *The Washington Post*, February 2, 1989, p. D1.

37. *Time*, August 1, 1988, pp. 46-47.

38. *EDF Letter*, September 1988.

39. WRI and IIED, *World Resources 1986*, p. 146; *Time*, August 1, 1988, p. 44.

40. Millemann, *And Two If By Sea*, pp. 31-32.

41. Ibid., p. 64.

42. *The Washington Post*, July 15, 1988, p. A8.

43. WRI and IIED, *World Resources 1987* (New York: Basic Books, 1987), p. 130.

44. Millemann, *And Two If By Sea*, p. 67.

45. WRI and IIED, *World Resources 1987*, p. 131.

46. *Newsweek*, August 1, 1988, p. 46; Millemann, *And Two If By Sea*, p. 65.

47. Millemann, *And Two If By Sea*, p. 64.

48. WRI and IIED, *World Resources 1987*, p. 131; "The State of the Marine Environment," *UNEP News*, April 1988, p. 88.

49. Edward Goldsmith and Nicholas Hildyard (eds.), *The Earth Report* (Los Angeles, CA: Price Stern Sloan, 1988) p. 185; "What On Earth Are We Doing?" *Time*, January 2, 1988, p. 79.

50. WRI and IIED, *World Resources 1986*, pp. 152-53.

51. Myers, "Ocean," p. 84.

52. Millemann, *And Two If By Sea*, p. 70.

53. WRI and IIED, *World Resources 1987*, pp.128-29; "The State of the Marine Environment," *UNEP News*, April 1988, p. 11.

54. Organization for Economic Cooperation and Development (OECD), *The State of the Environment 1985* (Paris: OECD, 1985) p. 74.

55. WRI and IIED, *World Resources 1987*, p. 129.

56. Ibid., pp. 129-30.

57. "Wildlife Toll Still Rising in Alaska," *The Washington Post*, June 1, 1989, p. A3.

58. "The State of the Marine Environment," *UNEP News*, April 1988, pp. 11-12.

59. WRI and IIED, *World Resources 1987*, p. 128.

60. *Time*, August 1, 1988, p. 47.

61. Ibid., p. 47; *EPA Journal*, January-February 1988, p. 41.

62. WRI and IIED, *World Resources 1987*, p. 128.

63. WRI and IIED, *World Resources 1988-89* (New York: Basic Books, 1988), p. 159.

64. WRI and IIED, *World Resources 1987*, p. 128.

65. Ibid., p. 135; OECD, "The State of the Environment 1985," p. 75.

66. Roger Revelle, "Present and Future State of Living Marine and Freshwater Resources," in Robert Repetto (ed.), *The Global Possible: Resources, Development, and the New Century* (New Haven, CT: Yale University Press, 1985), p. 431.

67. WRI and IIED, *World Resources 1988-89*, p. 145.

68. Food and Agriculture Organization (FAO), *FAO at Work*, May-June 1987.

69. WRI and IIED, *World Resources 1988-89*, p. 146.

70. Myers, "Ocean," p. 82.

71. Lester R. Brown, "Maintaining World Fisheries," in Brown et al., *State of the World 1985* (New York: Norton, 1985), p. 76.

72. *UNEP News*, October 1985.

73. Brown, "Maintaining World Fisheries," p. 76; FAO, *Yearbook of Fishery Statistics 1986* (Rome: FAO, 1988).

74. Revelle, "Marine and Freshwater Resources," pp. 431-32.

75. The Conservation Foundation, *State of the Environment*, p. 331; U.S. Department of Commerce, National Marine Fisheries Service, *Fisheries of the United States, 1987* (Washington, DC: National Oceanic and Atmospheric Administration, 1988).

76. Sam Hall, "Whaling: the Slaughter Continues," *The Ecologist*, pp. 207-12.

77. See WCED, *Our Common Future*, pp. 18-19, 264-69.

78. Myers, "Ocean," p. 97.

79. WRI and IIED, *World Resources 1987*, p. 193.

80. Myers, "Ocean," p. 96.

81. Council on Ocean Law, *The United States and the 1982 Law of the Sea Convention*, October 1987.

82. Ibid.

83. Myers, "Ocean," p. 97.

84. Ibid.

85. Council on Ocean Law, *The United States and the 1982 Law of the Sea Convention*.

86. Myers, "Ocean," p. 92.

87. Michael Weisskopf, "Plastic Reaps a Grim Harvest in the Oceans of the World," *Smithsonian*, March 1988, pp. 64-65.

88. Myers, "Ocean," p. 92.

89. Goldsmith and Hildyard, *The Earth Report*, p. 174; WRI and IIED, *World Resources 1988-89*, p. 325.

90. WRI and IIED, *World Resources 1987*, p. 193; Council on Ocean Law, *International Ocean Law and Policy - Section G*.

91. WRI and IIED, *World Resources 1987*, p. 137; *UNEP News*, April 1988, p. 11.

92. Myers, "Ocean," p. 92.

93. Goldsmith and Hildyard, *The Earth Report*, pp. 193-94.

94. *UNEP News*, April 1988, p. 10

95. Michael Weber et al., *The 1985 Citizen's Guide to the Ocean* (Washington DC: Center for Environmental Education, 1985), pp. 57-58.

96. Ibid., pp. 61, 65.

97. *Ocean Watch*, Newsletter of the Oceanic Society, October 1988.

98. Weber, *The 1985 Citizen's Guide to the Ocean*, pp. 41-45.

99. Weber and Tinney, *A Nation of Oceans*, pp. 14, 91.

100. WRI and IIED, *World Resources 1988-89*, p. 150.

101. *The Sun* (Baltimore, MD), December 15, 1987, p. 13A.

102. *The Washington Post*, November 19, 1987, p.A3.

103. *Time*, August 1, 1988, p. 50.

104. Ibid.

105. Myers, "Ocean," pp. 91, 99.

106. Ibid., p. 91.

107. Revelle, "Marine and Freshwater Resources," pp. 445, 452-53.

108. Ibid., pp. 451-452; Myers, "Oceans," pp. 82-83.

109. Marcia Lowe, "Salmon Ranchers and Farmers Net Growing Harvest," *World Watch*, January-February 1988.

110. Revelle, "Marine and Freshwater Resources," pp. 439-42.

111. Oliver A. Houck, "America's Mad Dash to the Sea," *The Amicus Journal*, Summer 1988, p. 36.

112. Ibid.

113. See "NRDC Takes Action Against Ocean Dumping and Beach Pollution," *NRDC Newsline* (Washington, DC: Natural Resources Defense Council), September-October 1988.

114. Houck, "America's Mad Dash to the Sea," pp. 21-36.

Chapter 9: Fresh Water

1. Philip W. Quigg, *Water: The Essential Resource*, International Series No. 2 (New York: National Audubon Society, 1976).

2. United Nations Environment Program (UNEP), "What Water Shortage?," undated.

3. Armin Maywald et al., "Water Fit to Drink?," in Edward Goldsmith and Nicholas Hildyard (eds.), *The Earth Report: The Essential Guide to Global Ecological Issues* (Los Angeles: Price Stern Sloan), 1988, p. 79.

4. Robert Repetto (ed.), *The Global Possible: Resources, Development, and the New Century* (New Haven: Yale University Press), p. 500.

5. Sandra Postel, "Managing Freshwater Supplies," in Lester R. Brown et al., *State of the World 1985* (New York: Norton, 1985), p. 46.

6. Essam El-Hinnawi and Mansur Hashmi, *The State of the Environment* (London: Butterworths, 1987), p. 41; World Resources Institute (WRI) and International Institute for Environment and Development (IIED), *World Resources 1988-89* (New York: Basic Books, 1988), p. 128.

7. See WRI and IIED, *World Resources 1988-89*, p. 4.

8. WRI and IIED, *World Resources 1986*, (New York: Basic Books, 1986), p. 122.

9. Norman Myers (ed.), *Gaia: An Atlas of Planet Management* (Garden City, NY: Anchor Books, 1984), pp. 108-09.

10. Population Reference Bureau, *1987 World Population Data Sheet* (Washington, DC, 1987), p. 3.

11. Postel, "Managing Freshwater Supplies," p. 43.

12. Susan Weber (ed.), *USA by Numbers* (Washington, DC: Zero Population Growth, 1988), p. 117.

13. WRI and IIED, *World Resources 1988-89*, p. 4.

14. World Commission on Environment and Development, *Our Common Future* (New York: Oxford, 1987), p. 16.

15. WRI and IIED, *World Resources 1987*, p. 112.

16. David Pimentel, "Waste in Agriculture and Food Sectors," Cornell University, College of Agriculture and Life Sciences, unpublished paper, April 1989, p. 9, and U.S. Bureau of the Census, *Statistical Abstract of the United States 1989* (Washington, DC, 1989), pp. 198-99. Water statistics distinguish between water *withdrawn* from surface and groundwater supplies, and "consumptive use"—water that has been evaporated, transpired, or incorporated into products, plant or animal tissue; and therefore is not available for immediate reuse. In 1985, total U.S. water withdrawals were 400 billion gallons per day, while daily consumptive use amounted to only 92 billion gallons. In the same year, U.S. agriculture accounted for about 80 percent of total U.S. consumptive water use (most water consumed by agriculture is lost by transpiration and evaporation), but only 34 percent of U.S. water withdrawals.

17. Sandra Postel, "Increasing Water Efficiency," *The State of the World 1986*, pp. 41-43.

18. Postel, "Managing Freshwater Supplies," p. 49. For the difference between water withdrawals and water use, see note 16.

19. Phillip P. Micklin, "Dessication of the Aral Sea: A Water Management Disaster in the Soviet Union," *Science*, Vol. 241, September 2, 1988, pp. 1170-76.

20. Postel, "Increasing Water Efficiency," p. 59.

21. WRI and IIED, *World Resources 1986*, p. 132.

22. Tristram Coffin, "Looking Ahead: The American Water Crisis," *Washington Spectator*, Vol. 11, No. 6.

23. Pimentel, "Waste in Agriculture and Food Sectors," p. 9.

24. Weber, *USA by Numbers*, p. 117.

25. Francis Moore Lappé, *Diet for a Small Planet*, Revised Edition, (New York: Ballantine, 1982), p. 76; David Pimentel, "Waste in Agriculture and Food Sectors," p. 9.

26. Pimentel, "Waste in Agriculture and Food Sectors," p. 10.

27. *The New York Times*, October 30, 1988, p. E-4.

28. Lappé, *Diet for a Small Planet*, p. 85. A 1988 study by the Bureau of Reclamation found that the U.S. government spends more than $534 million a year to provide inexpensive irrigation water to western farms, many of which produce surplus crops that yield additional federal

farm subsidy payments: Cass Peterson, "Cut-Rate Water, Surplus Crops," *The Washington Post*, March 8, 1988, p. A17.

29. WRI and IIED, *World Resources 1986*, p. 134.

30. Cass Peterson, *The Washington Post*, March 7, 1989.

31. Marc Reisner, "No Country on Earth Has Misused Water as Extravagantly as We Have," *The New York Times*, October 30, 1988, p. 4E.

32. Pimentel, "Waste in Agriculture and Food Sectors," p. 10.

33. The Conservation Foundation, *State of the Environment 1982* (Washington, DC, 1982), pp. 92-93; National Audubon Society, "Where Do We Go From Here?," a discussion guide accompanying the videotape, "What is the Limit?" (Washington, DC, 1987).

34. In 1985, electric power generation and industrial use accounted for about 54 percent of total U.S. water withdrawals: U.S.Bureau of the Census, *Statistical Abstract of the United States 1989*, p. 198.

35. The Conservation Foundation, *State of the Environment: A View Toward the Nineties* (Washington, DC, 1987), p. 225.

36. U.S. Bureau of the Census, *Statistical Abstract of the United States 1988*, (Washington, DC, 1987), p. 191.

37. Quigg, *Water: The Essential Resource*, p. 14.

38. Lester R. Brown and Edward C. Wolf, *Reversing Africa's Decline*, Worldwatch Paper 65 (Washington, DC: Worldwatch Institute, June 1985), pp. 19-29.

39. See Norman Myers, *Not Far Afield: U.S. Interests and the Global Environment* (Washington, DC: World Resources Institute, 1987), p. 24; and Michael Renner, "Enhancing Global Security," in Lester R. Brown et al., *State of the World 1989* (New York: Norton, 1989), pp. 142-43.

40. Foundation for Advancements in Science and Education (FASE), "Water Resource Conflicts," *FASE Reports*, Vol. 7, No. 1, Spring 1988.

41. Quigg, *Water: The Essential Resource*, p. 10.

42. Myers, *Gaia: An Atlas of Planet Management*, p. 132.

43. Quigg, *Water: The Essential Resource*, pp. 8-9.

44. Ibid., pp. 10-11; see also Thayer Scudder, "Conservation vs. Development: River Basin Projects in Africa," *Environment*, March 1989.

45. *Third World Network Features*, Newsletter of the Consumers' Association of Penang, Malaysia, December 1987.

46. *The New York Times*, November 6, 1988, p. 14.

47. WRI and IIED, *World Resources 1987*, p. 114.

48. Results of the World Health Organization survey summarized in Ruth Leger Sivard, *World Military and Social Expenditures 1987-88* (Washington, DC: World Priorities, 1987), pp. 46-51.

49. Myers, *Gaia: An Atlas of Planet Management*, p. 120; WRI and IIED, *World Resources 1987*, pp. 19, 111, 114.

50. El-Hinnawi and Hashmi, *The State of the Environment*, p. 128; Myers, *Gaia: An Atlas of Planet Management*, p. 120; Quigg, *Water: The Essential Resource*, p. 5; Norman Myers, *A Wealth of Wild Species: Storehouse for Human Welfare* (Boulder, CO: Westview Press, 1983), p. 99.

51. See El-Hinnawi and Hashmi, *The State of the Environment*, pp. 129-31.

52. See Quigg, *Water: the Essential Resource*, p. 3; Myers, *Gaia: An Atlas of Planet Management*, p. 120.

53. Myers, *Gaia: An Atlas of Planet Management*, p. 120; see also El-Hinnawi and Hashmi, *The State of the Environment*, p. 128.

54. See Quigg, *Water: the Essential Resource*, pp. 2-3; and Myers, *Gaia: An Atlas of Planet Management*, p. 120.

55. James P. Grant, *The State of the World's Children 1988* (New York: United Nations Children's Fund (UNICEF), 1988.

56. Perdita Huston, "Taking Time," *Decade Watch*, United Nations International Drinking Water Supply and Sanitation Decade, Vol. 4, No. 2 (New York: United Nations, June 1985).

57. United Nations Environment Program, press release, October 17, 1985.

58. Myers, *Gaia: An Atlas of Planet Management*, p. 134.

59. El-Hinnawi and Hashmi, *The State of the Environment*, pp. 42, 43.

60. Meyers, *Gaia: An Atlas of Planet Management*, p. 134.

61. See Eric Draper, "Groundwater Protection", *Clean Water Action News*, Fall 1987, p. 4.

62. See WRI and IIED, *World Resources 1987*, p. 121; Maywald, "Water Fit to Drink?," p. 84.

63. International Rivers Network, *World Rivers Review*, July-August 1989, p. 4.

64. Lester R. Brown and Christopher Flavin, "The Earth's Vital Signs", in Brown et al., *State of the World 1988* (New York: Norton, 1988), Table 1-1, p. 6.

65. Norman Myers, "Writing Off the Environment", in John Elkington et al. (eds.), *Green Pages: The Business of Saving the World* (London: Routledge, 1988), pp. 190-92.

66. The Conservation Foundation, *State of the Environment 1982*, p. 126.

67. *The Washington Post*, February 22, 1988, p. A3.

68. See Robert Collins, "Preventing Chemical Catastrophes", *Clean Water Action News*, Winter-Spring 1988, pp. 4-5; *The Washington Post*, February 2, 1988, p. A3.

69. Sandra Postel, *Defusing the Toxics Threat: Controlling Pesticides and Industrial Waste*, Worldwatch Paper 79, (Washington, DC: Worldwatch Institute, September 1987), p. 22.

70. Ibid.

71. John Langone, *Time Magazine*, January 2, 1989, p. 45.

72. See Concern, Inc., *Drinking Water: A Community Action Guide*, (Washington, DC: 1986), pp 1-4.

73. See Quigg, *Water: The Essential Resource*, p. 11.

74. El-Hinnawi and Hashmi, *The State of the Environment*, p. 44; Malcolm H. Fleming, "Agricultural Chemicals in Ground Water: Preventing Contamination by Removing Barriers against Low-Input Farm Management," *American Journal of Alternative Agriculture*, Vol. 2, No. 3, Summer 1987; *The Washington Post*, February 2, 1988, p. A21.

A 1988 survey of U. S. farmers by the Agricultural Law and Policy Institute found that 52 percent of the farmers interviewed were concerned about the quality of groundwater in their counties and supported regulations to ban or limit the use of agricultural chemicals in areas with a high risk of groundwater contamination: Environmental and Energy Study Institute *Notice*, June 2, 1989.

75. See George R. Hallberg, "From Hoes to Herbicides: Agriculture and Groundwater Quality," *Journal of Soil and Water Conservation*, November-December 1986, pp. 358-59.

76. Lappé, *Diet for a Small Planet*, p. 84.

77. Pimentel, "Waste in Agriculture and Food Sectors", p. 10.

78. El-Hinnawi and Hashmi, *The State of the Environment*, p. 26.

79. *The New York Times*, May 22, 1988, p. A16.

80. Environmental Defense Fund, *EDF Letter*, September 1988.

81. *The Washington Post*, March 16, 1988, p. A3.

82. See Draper, "Groundwater Protection," p. 3.

83. The Conservation Foundation, *State of the Environment 1982*, p. 110.

84. Statement by Associate Director, U.S. Fish and Wildlife Service, July 17, 1985.

85. Goldsmith and Hildyard, *The Earth Report*, p. 231.

86. See The Conservation Foundation, *State of the Environment: A View toward the Nineties*, p. 226.

87. The Conservation Foundation, *State of the Environment 1982*, p. 110; Concern, Inc., *Drinking Water*, p. 5.

88. Quigg, *Water: The Essential Resource*, p. 5.

89. *The Washington Post*, February 15, 1989, p. A16.

90. Environmental Policy Institute, *Citizen's Groundwater and Drinking Water Bulletin*, Vol. 1, No. 1, January 1988, p. 8.

91. Hallberg, "From Hoes to Herbicides," p. 361.

92. See The Conservation Foundation, *The State of the Environment: A View toward the Nineties*, p. 370; and Alliance for the Chesapeake Bay, *Baybook: A Guide to Reducing Water Pollution at Home*, (Baltimore, 1986), p. 11.

93. *The Washington Post*, April 4, 1987.

94. The Conservation Foundation, *State of the Environment 1982*, pp. 97-99; see also Organization for Economic Cooperation and Development, *State of the Environment 1985*, (Paris, 1985), pp. 52-53.

95. WRI and IIED, *World Resources 1988-89*, p. 6.

96. WRI and IIED, *World Resources 1987*, pp. 112, 116; Postel, "Increasing Water Efficiency", pp. 41-43; Myers, *Gaia: An Atlas of Planet Management*, p. 132.

97. Cass Peterson, "Lujan Pledges Water Subsidy Policy Review, *The Washington Post*, February 2, 1989, p. A29.

98. WRI and IIED, *World Resources 1987*, pp. 114-115.

99. Postel, "Increasing Water Efficiency," p. 50.

100. Ibid.," p. 52.

101. Ibid., pp. 54-55; WRI and IIED, *World Resources 1987*, p. 22.

102. *The Washington Post*, November 27, 1986, p. G1.

103. WRI and IIED, *World Resources 1987*, p. 118.

104. Renew America, *The State of the States 1989*, (Washington, DC, 1989), p. 21.

105. Ibid., p. 21.

106. Ibid., p. 22.

107. Ibid., p. 22.

108. Ibid., pp. 21-23.

109. Ibid., p. 23.

110. Ibid., pp. 18-23.

111. Sandra Postel, "Controlling Toxic Chemicals," in Brown, *State of the World 1988*, p. 131.

112. Donella H. Meadows, "The New Alchemist," *Harrowsmith*, November-December 1988, pp. 38-47.

113. Renew America, *The State of the States 1989*, p. 23.

114. WRI and IIED, *World Resources 1986*, p. 137.

115. Myers, *Gaia: An Atlas of Planet Management*, p. 108.

116. Maywald, "Water Fit to Drink?," pp. 81-83.

117. Repetto, *The Global Possible*, p. 501.

Chapter 10: Nonfuel Minerals

1. Ann Dorr, *Minerals—Foundations of Society* (Alexandria, VA: American Geological Institute, 1987), p. 1.

2. Eugene N. Cameron, *At the Crossroads—The Mineral Problems of the United States* (New York: Wiley-Interscience, 1986), p. 295.

3. Dorr, *Minerals*, p. 41.

4. U.S. Bureau of Mines (USBM), Minerals Information Office, 1989.

5. Dorr, *Minerals*, p. 6.

6. Ibid., p. 8.

7. Ibid., p. 9.

8. The average annual consumption growth rates for these metals were computed from data in World Resources Institute and International Institute for Environment and Development, *World Resources 1988-89* (New York: Basic Books, 1988), pp. 311-12; see also World Commission on Environment and Development (WCED), *Our Common Future* (New York: Oxford University Press, 1987), p. 59. The slower growth in metal consumption since 1980 reflects the slower growth of the world economy during the 1980s.

9. Dorr, *Minerals*, p. 31.

10. Ibid., p. 9.

11. USBM, *Mineral Commodity Summaries, 1989* (Washington, DC, 1989).

12. USBM, Minerals Information Office, 1989; see also Dorr, Minerals, p. 69.

13. USBM, Minerals Information Office; see also Dorr, *Minerals*, p. 35.

14. Dorr, *Minerals*, pp. 2, 5.

15. H. E. Goeller and A. Zucker, "Infinite Resources: The Ultimate Strategy," *Science*, Vol. 223, February 3, 1984, p. 457.

16. Economist Julian Simon believes that the world's natural resources are not finite in any economic sense. He takes what may be a minority position among economists in maintaining that "...if the past is any guide, natural resources will progressively become less scarce, and less costly...." Julian L. Simon, *The Ultimate Resource* (Princeton, NJ: Princeton University Press, 1981, p. 5; see also Chapter 3.

17. Cameron, *At the Crossroads*, Table 6-4, pp. 158-59. The recent report of the World Commission on Environment and Development (WECD) took a generally optimistic view of prospects for future mineral supplies, noting that "although non-renewable resources are by definition exhaustible, recent assessments suggest that few minerals are likely to run out in the near future." WCED, *Our Common Future, p. 209.*

18. In its 1987 report, the World Commission on Environment and Development (WCED) recommended that the rate of mineral depletion and the emphasis on recycling and economy of use should be calibrated to ensure that the mineral resource does not run out before adequate substitutes are available: WCED, *Our Common Future*, p. 46.

19. Dorr, *Minerals*, pp. 41-48.

20. U.S. Congress, Office of Technology Assessment, *Copper: Technology and Competitiveness*, Summary (Washington, DC, 1988), pp. 16-17.

21. *The New York Times*, August 23, 1985, p. 1; *San Francisco Examiner*, September 1, 1985, p. A13.

22. USBM, *Mineral Commodity Summaries, 1989*, p. 9.

23. Wilton Johnson and James Paone, *Land Utilization and Reclamation in the Mining Industry, 1930-80*, U.S. Department of the Interior, Bureau of Mines, Information Circular 8862 (Washington, DC, 1982), p. 10.

24. Dorr, *Minerals*, pp. 49-58.

25. Robert Repetto (ed.), *The Global Possible: Resources, Development, and the New Century* (New Haven: Yale University Press, 1985), p. 511.

26. USBM, *Mineral Commodity Summaries, 1989.* For a listing of recovery ratios for aluminum, copper, iron and steel, tin, zinc, paper, and glass for a number of countries, see World Resources Institute and International Institute for Environment and Development, *World Resources 1987* (New York: Basic Books, 1987), Table 22.6, pp. 308-09.

27. Cynthia Pollock, "Realizing Recycling's Potential," in Lester R. Brown et al., *State of the World 1987* (New York: Norton, 1987), pp. 114-15.

28. Ibid., pp. 105, 109.

29. See Norman Myers (ed.), "The Ocean Potential," in Myers (ed.), *Gaia: An Atlas of Planet Management* (Garden City, NY: Anchor Books, 1984), pp. 78-79.

30. Dorr, *Minerals*, pp. 79-80.

31. Ibid., p. 80.

32. Ibid., pp. 80-83.

33. USBM, *Mineral Commodity Summaries*, various years.

34. Michael Kidron and Ronald Segal, *The New State of the World Atlas* (New York: Simon and Schuster, 1987), Figure 11.

35. U.S. Department of the Interior, Bureau of Mines, *Minerals Yearbook*, Vol. III, Area Reports (Washington, DC, 1984).

36. See Dorr, *Minerals*, p. 20.

37. Repetto, *The Global Possible*, p. 511.

Chapter 11: Energy

1. *Time* Magazine, January 2, 1989, p. 65.

2. World Commission on Environment and Development (WCED), *Our Common Future* (New York: Oxford University Press, 1987), pp. 189, 195-96.

3. World Resources Institute (WRI) and International Institute for Environment and Development (IIED), *World Resources 1988-1989* (New York: Basic Books, 1988), pp. 5, 111.

4. Commercial energy includes all fuel and electricity that is bought and sold. Noncommercial energy is primarily wood, animal wastes, and crop residues used as fuels.

5. WRI and IIED, *World Resources 1988-89*, pp. 246-47, 306-07.

6. WRI and IIED, *World Resources 1987* (New York: Basic Books, 1987), p. 96.

7. WRI and IIED, World Resources 1986 (New York: Basic Books, 1986), p. 105.

8. British Petroleum, *BP Statistical Review of World Energy, 1989* (London, 1989), p. 34.

9. U.S. Energy Information Administration (EIA), *Annual Energy Review 1987* (Washington, DC: U.S. Department of Energy, 1988).

10. See José Goldemberg et al., *Energy for a Sustainable World* (Washington, DC: World Resources Institute, 1987), esp. p. v.

11. WRI and IIED, *World Resources 1986*, p. 103.

12. EIA, *Annual Energy Review 1987*, p. 13; Daniel Yergin, "Energy Security in the 1990s," *Foreign Affairs*, Fall, 1988; *Los Angeles Times*, December 21, 1988.

13. WRI and IIED, *World Resources 1988-89*, pp. 110-11.

14. Ibid., p. 110; EIA, *Annual Energy Review 1987*, p. 263; EIA, private communication, April, 1989.

15. Christopher Flavin, "Creating a Sustainable Energy Future," in Brown et al., *State of the World 1988* (New York: Norton, 1988), p. 27.

16. John Gever et al., *Beyond Oil: The Threat to Food and Fuel in the Coming Decades*, A project of Carrying Capacity, Inc. (Cambridge, MA: Ballinger, 1986), pp. 55-57.

17. EIA, *Annual Energy Review 1987*, p. 253.

18. WRI and IIED, *World Resources 1988-1989*, pp. 109-11; EIA, *Annual Energy Review 1987*, pp. 253, 265.

19. WRI and IIED, *World Resources 1988-1989*, p. 121.

20. Matthew L. Wald, "Fighting the Greenhouse Effect," *The New York Times*, August 28, 1988, Section 3, p. 2.

21. American Gas Association, "Natural Gas Can Play a Major Role in Ozone Abatement," *The Natural Resource*, Summer 1987, p. 2.

22. WRI and IIED, *World Resources 1988-89*, pp. 109-11; EIA, *Annual Energy Review 1987*, pp. 247, 249, 263.

23. WRI and IIED, *World Resources 1987*, pp. 96-100, 302-03; EIA, *Annual Energy Review 1987*, pp. 9, 215; WCED, *Our Common Future*, p. 182.

24. See Nancy Rader et al., *Power Surge: The Status and Near Term Potential of Renewable Energy Technologies* (Washington, DC: Public Citizen Critical Mass Energy Project, May 1989), esp. p. I-4.

25. Bill Keepin and Gregory Kats, Letter, *Science*, Vol. 241, August 26, 1988, p. 1027; Yergin, "Energy Security in the 1990s," p. 121.

26. Edward Goldsmith and Nicholas Hildyard (eds.), *The Earth Report* (Los Angeles: Price Stern Sloan, 1988), p. 187; Christopher Flavin, "Reassessing Nuclear Power," in Lester R. Brown et al., *State of the World 1987* (New York: Norton, 1987), pp. 57-68; Yergin, "Energy Security in the 1990s," pp. 114, 121.

27. "Deadly Meltdown," *Time*, May 12, 1986, p. 50; Cass Peterson, "A Decade After Accident, Legacy at TMI is Mistrust," *The Washington Post*, March 28, 1989, p. A1.

28. Yergin, "Energy Security in the 1990s," pp. 114, 121.

29. *The Washington Post*, December 30, 1988, p. A1.

30. Cynthia Pollock, *Decommissioning: Nuclear Power's Missing Link,* Worldwatch Paper 69 (Washington, DC: Worldwatch Institute, April 1986), pp. 5-7.

31. WCED, *Our Common Future,* pp. 14-15.

32. WRI and IIED, *World Resources 1987,* p. 302; U.S. Agency for International Development, *Power Shortages in Developing Countries: Magnitude, Impacts, Solutions, and the Role of the Public Sector* (Washington, DC: March 1988), p. iii; Wilson Clark, *Energy for Survival: The Alternative to Extinction* (Garden City, NY: Anchor Press, 1975), p. 208; Howard Kurtz, "Tottering on the Brink of Darkness," *The Washington Post,* August 17, 1988, p. A3.

33. EIA, *Annual Energy Review 1987,* p. 193.

34. Ibid., p. 4; U.S. Department of Energy (DOE), *Energy Security: A Report to the President of the United States* (Washington, DC: March 1987), p. 100.

35. WRI and IIED, *World Resources 1987,* pp. 96, 100, 302.

36. Rader, *Power Surge,* p. I-3. Baseload power plants are designed to run full-time to meet power levels needed throughout a season or a year, in contrast to intermediate load or peak load plants that are intended to meet higher demand levels for shorter periods.

37. Rader, *Power Surge,* p. I-2.

38. WRI and IIED, *World Resources 1987,* p. 100; Cynthia Pollock Shea, "Shifting to Renewable Energy," in Brown et al., *State of the World 1988,* p. 66; Zhu Xiaozhang, "Small Hydropower Stations," *China Reconstructs,* July 1983.

39. WRI and IIED, *World Resources 1987,* p. 105; WRI and IIED, *World Resources 1988-1989,* p. 114; National Research Council, *Geothermal Energy Technology: Issues, Research and Development Needs, and Cooperative Arrangements* (Washington, DC: National Academy Press, 1987).

40. Clark, *Energy for Survival,* pp. 327-29.

41. Shea, "Shifting to Renewable Energy," pp. 72-73; Jay Mathews, "Solar-Energy Complex Hailed as Beacon for Utility Innovation," *The Washington Post,* March 2, 1989, p. A25.

42. Shea, "Shifting to Renewable Energy," pp. 74-76; Environmental and Energy Study Institute, Notice, July 24, 1989; *Greenhouse Gas-ette* (Oakland, CA: Climate Protection Institute, Fall, 1988), p. 3.

43. Shea, "Shifting to Renewable Energy," pp. 74-76; Dan Arvizu, Sandia National Laboratory, "Status of Photovoltaic Solar Technology," paper presented at the Second North American Conference on Preparing for Climate Change, Washington, DC, December 6, 1988; and Rader, *Power Surge,* p. I-3.

44. Shea, "Shifting to Renewable Energy," pp. 77-78; Rader, *Power Surge,* p. I-3. The value of 1.6 billion kilowatt hours of electricity is based on 6.3 cents per KWH from EIA, *Annual Energy Review 1987,* p. 209.

45. John H. Gormley, "A New Kind of Water Power," *The Baltimore Sun,* August 7, 1988, p. 1D; "From the Sea: Power and Potable Water," *The Washington Post,* October 26, 1987, p. A7; WRI and IIED, *World Resources 1988-89,* p. 112.

46. Shea, "Shifting to Renewable Energy," pp. 67-72; WRI and IIED, *World Resources 1987,* pp. 104-05; *World Resources 1988-1989,* p. 116-17.

47. Sandra Postel and Lori Heise, *Reforesting the Earth* Worldwatch Paper 83 (Washington, DC: Worldwatch Institute, April 1988).

48. Ibid., pp. 16-17.

49. EIA, *Annual Energy Review 1987,* p. 103; John Yemma, "Fresh Slide in Oil Prices Expected to Cool Inflation," *Christian Science Monitor,* October 8, 1988, p. 15; Yergin, "Energy Security," p. 110; *The Washington Post,* January 19, 1989, p. A1.

50. *Future Survey* (a publication of the World Future Society), Vol. 9, No. 12, December 1987, p. 9; *The Washington Post,* January 19, 1989, p. A1.

51. Yergin, "Energy Security," p. 116.

52. Ibid., p. 118; Flavin, "Creating a Sustainable Energy Future," p. 27.

53. Union of Concerned Scientists, *Briefing Paper No. 10* (Cambridge, MA, September 1984).

54. Cass Peterson, "U.S. to Allow Unrestricted Transfer of Plutonium," *The Washington Post,* April 22, 1988, p. A4.

55. Ibid.

56. *The Washington Times,* February 6, 1989; DOE, *Annual Report 1987,* Chapter 11.

57. "The Vulnerability of Telecommunications and Energy Resources to Terrorism," Hearing before the U.S. Senate Committee on Governmental Affairs, February 7, 1989; Steven Ryan, Governmental Affairs Committee staff member, private communication, June 1, 1989.

58. See WCED, *Our Common Future,* pp. 60-61.

59. Organization for Economic Cooperation and Development (OECD), *The State of the Environment 1985* (Paris, 1985), p. 209.

60. Irving M. Mintzer, *A Matter of Degrees: The Potential for Controlling the Greenhouse Effect* (Washington, DC: World Resources Institute, 1987); Brown et al., "A World at Risk," in Brown et al., *State of the World 1989* (New York: Norton, 1989), p. 9.

61. Lester R. Brown and Christopher Flavin, "The Earth's Vital Signs," in Brown et al., *State of the World 1988,* p. 17.

62. See WRI and IIED, *World Resources 1987*, p. 102; and Don Hinrichsen, "Acid Rain and Forest Decline," in Goldsmith and Hildyard, *The Earth Report*, p. 10.

63. See WRI and IIED, *World Resources 1987*, pp. 147, 154; and James J. MacKenzie and Mohamed El-Ashry, Ill Winds: *Airborne Pollution's Toll on Trees and Crops* (Washington, DC: World Resources Institute, 1988), p. 10.

64. WRI and IIED, *World Resources 1987*, pp. 128-30.

65. See National Research Council, *Coal Mining and Groundwater Resources in the United States* (Washington, DC: National Academy Press, 1981).

66. *The Washington Post*, May 12, 1988, p. A25; Goldsmith and Hildyard, The Earth Report, p. 187.

67. U.S. Department of Health and Human Services, Department of Health and Human Services *Newsletter*, February 4, 1988.

68. Goldsmith and Hildyard, *The Earth Report*, p. 188.

69. WRI and IIED, *World Resources 1988-89*, p. 114.

70. Wald, "Fighting the Greenhouse Effect."

71. WRI and IIED, *World Resources 1986*, p. 113; World Resources 1987, p. 105.

72. Goldemberg, *Energy for a Sustainable World*, p. 5.

73. William U. Chandler et al., *Energy Efficiency: A New Agenda* (Washington, DC: American Council for an Energy-Efficient Economy, 1988), pp. 14, 44.

74. Jessica Tuchman Mathews, "Redefining Security," *Foreign Affairs*, Spring 1989, p. 172.

75. Flavin, "Creating a Sustainable Energy Future," p. 25; Goldemberg, *Energy for a Sustainable World*, p. v.; Keepin and Kats, Letter.

76. Goldemberg, *Energy for a Sustainable World*, p. 55.

77. Christopher Flavin and Alan Durning, "Raising Energy Efficiency," in Brown et al., *State of the World 1988*, p. 47.

78. Flavin, "Creating a Sustainable Energy Future," pp. 33-35.

79. Goldemberg, *Energy for a Sustainable World*, pp. 52-53.

80. Flavin and Durning, "Raising Energy Efficiency," p. 48.

81. Environmental and Energy Study Institute, "New Developments in Energy Efficient Lighting," *Notice*, June 1, 1989, based on a presentation by Robert Sardinsky, Research Associate, Rocky Mountain Institute.

82. Goldemberg, *Energy for a Sustainable World*, pp. 56.

83. Craig Canine, "Generating Megawatts," *Harrowsmith*, April 1989, pp. 42-49; Time, January 2, 1989, p. 39.

84. Alliance to Save Energy, Washington, DC, press release, October 15, 1986.

85. WRI and IIED, *World Resources 1988-1989*, p. 121.

86. Marc Ross, "Improving the Efficiency of Electricity Use in Manufacturing," *Science*, Vol. 224, April 21, 1989, p. 316.

87. WRI and IIED, *World Resources 1987*, p. 99.

88. Flavin and Durning, "Raising Energy Efficiency," p. 55.

89. WRI and IIED, *World Resources 1987*, p. 99.

90. DOE, Energy Security, p. 100.

91. See Goldemberg, *Energy for a Sustainable World*, pp. 63-65; and Flavin and Durning, "Raising Energy Efficiency," pp. 50-53.

92. Chandler, *Energy Efficiency*, p. 28; Flavin and Durning, "Raising Energy Efficiency," pp. 50-51; *The Washington Post*, May 3, 1989, p. F3.

93. Chandler, *Energy Efficiency*, pp. 63-64.

94. WRI and IIED, *World Resources 1988-1989*, pp. 124-25.

95. Ibid., p. 15.

96. Goldemberg, *Energy for a Sustainable World*, pp. 56-58; J. Arnold Varney and Frederick M. Varney, "The Stirling-Cycle Energy Conversion and Utilization System," Ecological Systems Technology Association (ESTA), Los Angeles, CA, unpublished paper, February, 1989

97. John A. Adam, "Plundering the Amazon for Power," *The Washington Post*, November 27, 1988, p. D3.

98. José Goldemberg et al., *Energy for Development* (Washington, DC: World Resources Institute, 1987), p. 34.

99. Varney and Varney, "The Stirling-Cycle Energy Conversion and Utilization System."

100. WRI and IIED, *World Resources 1986*, p. 113.

101. For further information, contact Arnold Varney, 7326 Ogelsby Avenue, Los Angeles, CA 90045; or Mohebat Ahdyyih, Manager, Technology Information, VITA, 1815 Lynn Street, Suite 20, Arlington, VA 22209.

102. See Nicholas Lennsen, "Cooked By the Sun," *World Watch*, March-April 1989, pp. 41-42. For further information, contact Solar Box Cookers International, 1724 11th Street, Sacramento, CA 95814.

103. Goldemberg, *Energy for Development*, pp. v-vi.

104. Ibid., pp. 68-69.

105. Flavin and Durning, "Raising Energy Efficiency," p. 59.

106. Chandler, *Energy Efficiency*, p. 31.

107. Yergin, "Energy Security in the 1990s," p. 130; Michael Renner, *Rethinking the Role of the Automobile*,

Worldwatch Paper 84 (Washington, DC: Worldwatch Institute, June 1988), p. 17.

108.   Renner, *Rethinking the Role of the Automobile*, p. 48.

109.   Chandler, *Energy Efficiency*, p. 34.

110.   Wald, "Fighting the Greenhouse Effect."

111.   WRI and IIED, *World Resources 1988-89*, p. 122.

112.   DOE, *Renewable Energy Research and Development Outlook* (Washington, DC, October 1985), p. 33.

113.   Shea, "Shifting to Renewable Energy," p. 70.

114.   The Conservation Foundation, *State of the Environment*, p. 254; Rader, *Power Surge*, p. II-2; DOE, *Renewable Energy*, pp. vii, 34. The DOE estimates for renewable energy sources in Figure 7 exclude hydropower; the sources included add up to about 0.3 quadrillion BTUs (quads) for 1984, and 2.6 quads for 2000. The projection for 2000 is more than an eightfold increase over 1984, and is about 3.3 percent of the total energy (80 quads) used by the United States in 1988.

115.   Cynthia Pollock Shea, *Renewable Energy: Today's Contribution, Tomorrow's Promise*, Worldwatch Paper 81 (Washington, DC, January 1988), p. 15; Barry Commoner, The Poverty of Power (New York: Knopf, 1976), p. 82; Environmental and Energy Study Institute, *Energy Policy Statement: A Call to Action for the Next President* (Washington, DC, October 1988), p. 5

116.   Shea, *Renewable Energy*, pp. 41, 44; DOE, *Renewable Energy*, p. vii; Flavin, "Creating a Sustainable Energy Future," p. 37.

117.   Rader, *Power Surge*, p. I-3.

118.   WRI and IIED, *World Resources 1988-1989*, p. 114; Shea, *Renewable Energy*, p. 44.

119.   News Release, Hawaiian Electric Company, Honolulu, HI, February 6, 1989.

120.   Shea, *Renewable Energy*, pp. 37-39.

121.   Ibid., pp. 39-41.

122.   This section summarizes material presented in Rader, *Power Surge*, pp. II-39 to 41.

123.   Ibid.; see also John J. Berger, *Restoring the Earth: How Americans Are Working to Renew Our Damaged Environment* (New York: Anchor Press, 1987), p. 183; and William A. Shurcliff, *Super Solar Houses: Saunders' Low Cost, 100 Percent Solar Designs* (Andover, MA: Brick House, 1983).

124.   Mathews, "Solar-Energy Complex Hailed as Beacon for Utility Innovation;" Rader, *Power Surge*, p. I-3.

125.   Solar Energy Industries Association, *Photovoltaics: Solar Electricity in the 1990s* (Arlington, VA, November 1988); H. M. Hubbard, "Photovoltaics Today and Tomorrow," *Science*, Vol. 244, April 28, 1989, p. 297; WRI and IIED, *World Resources 1988-89*, p. 114; Nicholas Lenssen, Worldwatch Institute, private communication, September 1989.

126.   Hubbard, "Photovoltaics," p. 302.

127.   Ibid., p. 297.

128.   WRI and IIED, *World Resources 1988-89*, pp. 116-17.

129.   Flavin and Pollock, "Harnessing Renewable Energy," p. 189; Shea, "Shifting to Renewable Energy," p. 71.

130.   Shea, "Shifting to Renewable Energy," p. 71.

131.   Flavin and Pollock, "Harnessing Renewable Energy," p. 195; Dick Russell, "L.A. Air," *The Amicus Journal*, a publication of the Natural Resources Defense Council, Summer 1988, p. 15.

132.   Rader, *Power Surge*, pp. II-18 and 19.

133.   Paul J. Werbos, U.S. Energy Information Administration, "Oil Dependency and the Potential for Fuel Cell Vehicles," Society of Automotive Engineers Technical Paper Series, May 1987.

134.   Peter Hoffman, "Potential Energy Uses and Greenhouse Implications of Hydrogen," paper prepared for the Second North American Conference on Preparing for Climate Change, Washington, DC, December 6-8, 1988, pp. 1-2; Peter Hoffman et al., "The Fuel of the Future is Making a Comeback," *Business Week*, November 28, 1988, p. 130; Rader, *Power Surge*, p. II-57.

135.   Hoffman, "Potential Energy Uses;" Hoffman, "The Fuel of the Future."

136.   Postel and Heise, *Reforesting the Earth*, p. 15; Goldsmith and Hildyard, The Earth Report, p. 204.

137.   Walter V. Reid et al., *Bankrolling Successes: A Portfolio of Sustainable Development Projects* (Washington, DC: Environmental Policy Institute and National Wildlife Federation, 1988), pp. 36-38.

138.   Ibid., p. 20; Essam El-Hinnawi and Mansur Hashmi, *The State of the Environment* (London: Butterworths, 1987), pp. 79-81; WRI and IIED, World Resources 1987, pp. 63-64.

139.   Postel and Heise, *Reforesting the Earth*, p. 20.

140.   Extracted from "A Call for U.S. Action on Global Climate Change," a letter to President George Bush by a panel sponsored by the World Resources Institute, May 1989.

Chapter 12: Air, Atmosphere, and Climate

1. Don Hinrichsen, "Acid Rain and Forest Decline," in Edward Goldsmith and Nicholas Hildyard (eds.), *The Earth Report* (Los Angeles: Price Stern Sloan, 1988), p. 66.

2. *The Washington Post,* October 8, 1988, p. A30.

3. James Hansen, "I'm Not Being an Alarmist About the Greenhouse Effect," *The Washington Post,* February 11, 1989, p. A23.

4. Conclusions from an International Congress on "Nature Management and Sustainable Development," University of Groningen, Netherlands, December 1988, reported in *Ambio,* Vol. 18, No. 2, 1989, p. 145.

5. Lester R. Brown, Christopher Flavin, and Sandra Postel, "Outlining a Global Action Plan," in Brown et al., *State of the World 1989* (New York: Norton, 1989), p. 178.

6. World Resources Institute (WRI) and International Institute for Environment and Development (IIED), *World Resources 1988-1989* (New York: Basic Books, 1988), pp. 163, 172.

7. Michael G. Renner, "Car Sick," *Worldwatch,* November-December, 1988, p. 36.

8. Norman Myers (ed.), *Gaia: An Atlas of Planet Management* (Garden City, New York: Anchor Press, 1984), p. 118.

9. Renner, "Car Sick," p. 36.

10. Organization for Economic Cooperation and Development, *OECD Environmental Data Compendium 1989* (Paris, 1989), Table 2.1A, p. 17; Essam El-Hinnawi and Manzur H. Hashmi, *The State of the Earth* (London: Butterworth Scientific, 1987), p. 5.

11. WRI and IIED, *World Resources 1988-1989,* pp. 163-64.

12. James J. MacKenzie and Mohamed T. El-Ashry, *Ill Winds: Airborne Pollution's Toll on Trees and Crops* (Washington, DC: World Resources Institute, 1988), p. v.

13. WRI and IIED, *World Resources 1988-1989,* pp. 164-65.

14. Ibid., pp. 165-66.

15. Ibid., p. 165.

16. Ibid., p. 168.

17. Ibid., pp. 168-69.

18. Ibid., p. 147.

19. Ibid., p. 170; U.S. Bureau of the Census, *Statistical Abstract of the United States 1989* (Washington, DC, 1989), p. 200; Philip Shabecoff, "Health Risk From Smog Is Growing, Official Says," *The New York Times,* March 1, 1989, p. A16.

20. Postel, "Stabilizing Chemical Cycles, " p. 170.

21. Organization for Economic Cooperation and Development (OECD), *The State of the Environment 1985* (Paris: OECD,1985), p. 34.

22. Dick Russell, "L.A. Air," *The Amicus Journal* (A Publication of The Natural Resources Defense Council), Summer, 1988.

23. Edward C. Wolf, "Raising Agricultural Productivity," in Brown et al., *State of the World 1987*, p. 143; Michael Renner, *Rethinking the Role of the Automobile,* Worldwatch Paper 84 (Washington, DC: Worldwatch Institute, June, 1988), p. 36; MacKenzie and El-Ashry, *Ill Winds.*

24. MacKenzie and El-Ashry, *Ill Winds,* pp. 10, 16, 31.

25. "More Americans Lived in Harmful Smog in '88," *The Washington Post,* July 28, 1989, p. A14; Russell, "L.A. Air," p. 10; Myers, *Gaia: An Atlas of Planet Management,* pp. 118-19; Renner, "Car Sick," p. 36.

26. *The Sun* (Baltimore), December 12, 1988; *The Washington Post,* November 28, 1988, p. A1.

27. "New Tactics Emerge in Struggle Against Smog," *The New York Times,* February 21, 1989, p. C1, summarized in *Future Survey,* Vol. 11, No. 6, June 1989, p. 3.

28. WRI and IIED, *World Resources 1988-89,* p. 173.

29. Ibid., p. 167.

30. Sandra Postel, "Stabilizing Chemical Cycles," In Lester R. Brown et al., *State of the World 1987* (New York: Norton, 1987), p. 172.

31. Ibid., p. 168; Myers, *Gaia: An Atlas of Planet Management,* p. 18.

32. El-Hinnawi and Hashmi, *The State of the Environment,* p. 24.

33. Edward Goldsmith and Nicholas Hildyard (eds.), *The Earth Report* (Los Angeles: Price Stern Sloan, 1988), p. 69.

34. Ibid., p. 70.

35. Ibid., p. 67; Lester R. Brown and Christopher Flavin, "The Earth's Vital Signs," in Brown et al., *State of the World 1988* (New York: Norton, 1988) p. 6.

36. Hinrichsen, "Acid Rain and Forest Decline," pp. 66-67; Robert W. Haseltine, "Economics vs. Ecology: Problems with Solutions to Pollution," *USA Today* (Magazine), January 1986, p. 48.

37. Hinrichsen, "Acid Rain and Forest Decline," pp. 67, 76.

38. Ibid., pp. 70-74; MacKenzie and El-Ashry, *Ill Winds,* pp. 18-24.

39. *EDF Letter* (Published by the Environmental Defense Fund), September 1988, p. 5.

40. Hinrichsen, "Acid Rain and Forest Decline," p. 74.

41. Norman Myers, "Writing off the Environment," in John Elkington et al. (eds.), *Green Pages: The Business of Saving the World* (London: Routledge, 1988), pp. 190-192.

42. Myers, *Gaia: An Atlas of Planet Management*, p. 118.

43. John McCormick, *Acid Earth: The Global Threat of Acid Pollution* (London: Earthscan and IIED, 1985), p. 146.

44. Renner, "Car Sick," p. 38.

45. El-Hinnawi and Hashmi, *The State of the Environment*, pp. 29-30; Hinrichsen, "Acid Rain," *The Earth Report*, p. 75.

46. Hinrichsen, "Acid Rain," *The Earth Report*, pp. 76-78.

47. James J. MacKenzie, *Breathing Easier: Taking Action on Climate Change, Air Pollution, and Energy Insecurity* (Washington, DC: World Resources Institute, undated [1988]), p. 12.

48. Ibid., p. 8.

49. *The Washington Post*, September 9, 1988, p. A22.

50. MacKenzie, *Breathing Easier*, p. 12.

51. Ibid., pp. 8-25.

52. Orie L. Loucks, "Large-Scale Alteration of Biological Productivity Due to Transported Pollutants," in Daniel B. Botkin et al. (eds.), *Changing the Global Environment* (San Diego, CA: Academic Press, 1989), pp. 101-16.

53. Alan S. Miller and Irving M. Mintzer, *The Sky is the Limit: Strategies for Protecting the Ozone Layer* (Washington, DC: WRI, November 1986), p. 3; El-Hinnawi and Hashmi, *The State of the Environment*, pp. 11, 15.

54. Cynthia Pollock Shea, *Protecting Life on Earth: Steps to Save the Ozone Layer*, Worldwatch Paper 87 (Washington, DC: Worldwatch Institute, December 1988), p. 5; WRI and IIED, *World Resources 1986* (New York: Basic Books, 1986), p. 173; Goldsmith and Hildyard, *The Earth Report*, p. 195; WRI and IIED, *World Resources 1988-89*, pp. 175-78; *The Washington Post*, March 16, 1988, p. A1; April 6, 1989, p. A3; and July 27, 1989, p. A3.

55. Goldsmith and Hildyard, *The Earth Report*, pp. 120, 194-195.

56. Ibid., p. 195; WRI and IIED, *World Resources 1986*, p. 173.

57. Goldsmith and Hildyard, The Earth Report, p. 194; WRI and IIED, *World Resources 1988-89*, p. 175; Shea, *Protecting Life on Earth*, pp. 6, 14.

58. Recent research suggests that suppression of the immune system may be the most important effect of ultraviolet radiation. Alan S. Miller, University of Maryland, private communication, May 1989.

59. El-Hinnawi and Hashmi, *The State of the Environment*, p. 15; Miller and Mintzer, *The Sky is the Limit*, p. 10; WRI and IIED, World Resources 1988-89, p. 175.

60. Miller and Mintzer, *The Sky is the Limit*, pp. 10-11.

61. Hansen, "I'm Not Being an Alarmist."

62. Alan E. Strong, "Greater Global Warming Revealed by Satellite-Derived Sea-Surface-Temperature Trends," *Nature*, Vol. 338, April 20, 1989, pp. 642-45.

63. Brown, Flavin, and Postel, "Outlining a Global Action Plan," p. 175.

64. Irving M. Mintzer, *A Matter of Degrees: The Potential for Controlling the Greenhouse Effect* (Washington, DC: WRI, April 1987), p. 3; Christopher Flavin, "The Heat Is On," *World Watch*, November-December 1988, p. 12; WRI and IIED, *World Resources 1988-89*, p. 170.

65. Sandra Postel and Lori Heise, "Reforesting the Earth," in Brown et al., *State of the World 1988*, pp. 94-95; Lester R. Brown, Christopher Flavin, and Sandra Postel, "A World At Risk," in Brown et al., *State of the World 1989*, p. 16.

66. Renew America, *Reducing the Rate of Global Warming: The States' Role* (Washington, DC, 1988).

67. Mintzer, *A Matter of Degrees*, pp. 35-37.

68. WRI and IIED, *World Resources 1988-89*, p. 173; Flavin, "The Heat Is On," p. 12.

69. WRI and IIED, *World Resources 1988-89*, p. 173; Mark Trexler, World Resources Institute, private communication, October 1989; H. Craig et al., "The Isotopic Composition of Methane in Polar Ice Cores," Science, December 16, 1988, pp. 1535-38; *The Washington Post*, February 11, 1989, p. A23; James Hansen et al., "Global Climate Changes as Forecast by the Goddard Institute for Space Studies Three-Dimensional Model," *Journal of Geophysical Research*, August 20, 1988, pp. 9341-9364.

70. Paul E. Waggoner (ed.), *Climate and Water*, Report of the American Association for the Advancement of Science (AAAS) Panel on Climatic Variability, Climate Change, and the Planning and Management of U.S. Water Resources (Washington, DC: AAAS, 1988). "Equivalent doubling" is the point at which the concentration of all greenhouse gases achieves the same heat absorbing effect as a doubling of pre-industrial levels of atmospheric CO_2 alone.

71. George M. Woodwell, *Global Climate Change: Warming of the Industrial Middle Latitudes 1985-2050: Causes and Consequences* (Woods Hole, MA: The Woods Hole Research Institute, 1987), p. 15; Jill Jaeger,

"Anticipating Climate Change: Priorities for Action," *Environment*, September 1988.

72. Leslie Roberts, "Is There Life After Climate Change?," Science, November 18, 1988: WRI and IIED, *World Resources 1988-89*, p. 173.

73. J.F.B. Mitchell et al., "CO2 and Climate: A Missing Feedback?," *Nature*, Vol. 341, September 14, 1989, pp. 132-34; Tony Slingo, "Wetter Clouds Dampen Greenhouse Warming," *Nature*, Vol. 341, September 14, 1989, p. 104; see also William Booth, "Climate Study Halves Estimate of Global Warming," *The Washington Post*, September 14, 1989, p. A8.

74. *The Washington Post*, January 27, 1989, p. A3.

75. V. Ramanathan et al., "Cloud-Radiative Forcing and Climate: Results from the Earth Radiation Budget Experiment," *Science*, January 6, 1989, pp. 57-63; Richard A. Kerr, "How to Fix the Clouds in Greenhouse Models," *Science*, January 6, 1989, pp. 28-29; Larry R. Ephron, Letter to the Editor, *The New York Times*, July 25, 1988.

76. Brown and Flavin, "The Earth's Vital Signs," p. 16; Ephron, Letter.

77. Jaeger, "Anticipating Climate Change," p. 15; Flavin, "The Heat Is On," p. 3; WRI and IIED, *World Resources 1988-89*, p. 173; James J. MacKenzie, *Breathing Easier: Taking Action on Climate Change, Air Pollution, and Energy Insecurity* (Washington, DC: World Resources Institute, undated [1988]), p. 7.

78. Jaeger, "Anticipating Climate Change," p. 14; James Hansen, "Likely Climate Changes in North America and the Caribbean," paper presented at the second North American Conference on Preparing for Climate Change, Washington D.C., December 6, 1988.

79. Waggoner, *Climate and Water;* WRI and IIED, *World Resources 1988-89*, p. 174; MacKenzie, *Breathing Easier*, p. 7.

80. Brown and Flavin, "The Earth's Vital Signs," p. 17; Stephen Leatherman, "Likely Sea Level Rise," paper presented at the second North American Conference on Preparing for Climate Change, Washington, DC, December 6, 1988.

81. WRI and IIED, *World Resources 1986*, pp. 174, 177; Postel, "Stabilizing Chemical Cycles," p. 164; Flavin, "The Heat Is On," p. 15.

82. Flavin, "The Heat Is On," p. 15; WRI and IIED, *World Resources 1986*, p. 177; WRI and IIED, *World Resources 1988-89*, p. 174.

83. Flavin, "The Heat Is On," p. 15.

84. George Woodwell, "Rich Nations and Poor Nations on a Warmer Earth," Lecture at World Resources Institute International Environmental Forum, Washington, D.C., November 18, 1988; Roberts, "Is There Life After Climate Change?," p. 1011.

85. WRI and IIED, *World Resources 1986*, p. 177.

86. Roberts, "Is There Life After Climate Change?," p. 1011.

87. Ibid., pp. 1010, 1012.

88. Ibid., p. 1010.

89. Brown, Flavin, and Postel, "Outlining a Global Action Plan," pp. 175-76.

90. World Commission on Environment and Development (WCED), *Our Common Future* (New York: Oxford University Press, 1987), pp. 176-77.

91. Renner, "Rethinking the Role of the Automobile," p. 35; *Power Line*, November-December 1988, p. 6.

92. Renner, "Rethinking the Role of the Automobile;" James J. MacKenzie, *Breathing Easier: Taking Action on Climate Change, Air Pollution, and Energy Insecurity* (Washington, DC: World Resources Institute, undated [1988]), p. 18.

93. MacKenzie, *Breathing Easier*, p. 10.

94. Brown, Flavin, and Postel, "Outlining a Global Action Plan," p. 177.

95. William V. Chandler, Howard S. Geller, and Marc R. Ledbetter, *Energy Efficiency: A New Agenda* (Washington, DC: The American Council for an Energy-Efficient Economy, 1988), p. 52.

96. Brown, Flavin, and Postel, "Outlining a Global Action Plan," p. 177.

97. Bill Keepin and Gregory Kats, "The Efficient Response to Global Warming," *Rocky Mountain Institute Newsletter*, August 1989, p. 3; Science, August 26, 1988, p. 1027; *Scientific American*, April 1988, pp. 258-78.

98. MacKenzie, *Breathing Easier*, p. 10.

99. Brown, Flavin, and Postel, "Outlining a Global Action Plan," p. 178.

100. Ibid., pp. 175-80.

101. Ibid., pp. 177-78.

102. See Cynthia Pollock Shea, "Shifting to Renewable Energy," in Brown et al., *State of the World 1988*.

103. Lester R. Brown and Edward C. Wolf, "Reclaiming the Future," in Brown et al., *State of the World 1988*, p. 175; *Science*, October 7, 1988, pp. 19-20; *Science*, December 16, 1988, p. 1493; Roger A. Sedjo, "Forests: A Tool to Moderate Global Warming?," *Environment*, January-February, 1989.

104. El-Hinnawi and Hashmi, *The State of the Earth*, p. 30.

105. "Tackling Smog Ozone: Tougher Than Thought," *Science News*, Vol. 136, July 22, 1989, p. 53.

106. WRI and IIED, *World Resources 1987* (New York: Basic Books, 1987), p. 156.

107. Shea, *Protecting Life on Earth*, p. 22; WRI and IIED, *World Resources 1988-89*, p. 178.

108. *The Washington Post*, January 31, 1989, p. A6.

109. Miller, private communication, May 1989.

110. Shea, "Protecting Life on Earth," p. 24.

111. WRI and IIED, *World Resources 1988-89*, p. 271.

112. Frances Moore Lappé, *Diet for a Small Planet*, Revised edition (New York: Ballantine, 1982), pp. 10, 74; John Robbins, *Diet for a New America* (Walpole, NH: Stillpoint Publishing, 1987), pp. 368-69.

113. See Brown et al., *State of the World 1988*, Chapters 9 and 10.

114. Russell, "L.A. Air," pp. 12-15; National Wildlife Federation, *Conservation Exchange*, Winter 1988-1989, p. 2.

115. *The Washington Post*, September 25, 1988, p. A4.

116. WRI and IIED, *World Resources 1988-89*, pp. 177-78; *The Washington Post*, November 11, 1988, p. A26.

117. WCED, *Our Common Future*, p. 176.

118. *The Washington Post*, February 2, 1989, p. A34

119. *The Washington Post*, January 12, 1989, p. A3; *The Global Greenhouse Information Kit* (Washington, DC: Foundation on Economic Trends, 1988), pp. 1, 4.

120. Brown, Flavin, and Postel, "Outlining a Global Action Plan," *State of the World 1989*, p. 180. The World Environment Policy Act of 1989, introduced by Senator

121. Alvin L. Alm, "The Environment," *The American Agenda: Report to the Forty-First President of the United States* (Camp Hill, PA: Book of the Month Club, undated [1988]), p. 179; *Blueprint for the Environment* (Washington, DC: undated [1988]), pp. 11-12.

122. Lester R. Brown, Christopher Flavin, and Sandra Postel, "No Time to Waste: A Global Environmental Agenda for the Bush Administration," *World Watch*, January-February, 1989, p. 11; Time, January 2, 1989, p. 65.

123. *The Washington Post*, January 31, 1989, p. A6.

124. *The Washington Post*, March 21, 1989, p. A17.

125. *The Washington Post*, March 8, 1989, p. A30; and March 14, 1989, p. A18.

126. United Nations Environment Program, *UNEP North American News*, June 1989.

127. *The Washington Post*, March 12, 1989, p. A27.

128. Conclusions from an International Congress on "Nature Management and Sustainable Development," *Ambio*, Vol. 18, No. 2, 1989, p. 145.

129. "A Call for U.S. Action on Global Climate Change," a letter sent to U.S. President George Bush by a panel sponsored by the World Resources Institute in January 1989.

130. Senate Bill No. 343 was introduced in, but not passed by the Missouri General Assembly's first regular session in 1989.

Albert Gore, Jr. and others, was referred to the Senate Committee on Environment and Public Works.

Chapter 13: Hazardous Substances

1. Rachel Carson, *Silent Spring* (Boston, MA: Houghton Mifflin, 1962).

2. Joel Hirschhorn, "Cutting Production of Hazardous Waste," *Technology Review*, April 1988.

3. Sandra Postel, *Defusing the Toxics Threat: Controlling Pesticides and Industrial Waste, Worldwatch Paper 79* (Washington, DC: Worldwatch Institute, September 1987), pp. 36-37.

4. Michael Weisskopf, "Japanese Town Still Staggered by Legacy of Ecological Disaster," *The Washington Post*, April 18, 1987.

5. Sandra Postel, *Defusing the Toxic Threat*, p. 8; Sandra Postel, "Stabilizing Chemical Cycles," in Lester R. Brown et al., *State of the World 1987* (New York: Norton, 1987), p. 172.

6. Lewis Regenstein, *America the Poisoned* (Washington, DC: Acropolis Books, 1982), pp. 44-48.

7. Nafis Sadik, *The State of World Population 1988* (New York: United Nations Population Fund, 1988), p. 12.

8. World Resources Institute (WRI) and International Institute for Environment and Development (IIED), *World Resources 1987* (New York: Basic Books, Inc., 1987), p. 202.

9. WRI and IIED, *World Resources 1987*, p. 201.

10. Postel, *Defusing the Toxics Threat*, pp. 12-13; The Conservation Foundation, *State of the Environment: A View Toward the Nineties* (Washington, DC: 1987), p. 434.

11. Postel, *Defusing the Toxics Threat*, pp. 12-13.

12. Edward Goldsmith and Nicholas Hildyard, (eds.), *The Earth Report* (Los Angeles: Price Stern Sloan, 1988), p. 174; *The Washington Post*, February 24, 1988, p. A10.

13. "Hazardous and Toxic Wastes," *Inform Report*, Vol. 3, No. 2, March-April, 1983, p. 1.

14. Susan Weber, (ed.), *USA by Numbers* (Washington, DC: Zero Population Growth, 1988), pp. 141-42.

15. U.S. Bureau of the Census, *Statistical Abstract of the United States: 1989* (Washington, DC: 1989), p. 202; Weber, USA by Numbers, p. 141.

16. Susan Weber, *USA by Numbers*, p. 141.

17. Lawrie Mott and Karen Snyder, *Pesticide Alert* (San Francisco: Sierra Club Books, 1987), p. 6.

18. Postel, *Defusing the Toxics Threat*, p. 12.

19. Michael Weisskopf, "Did Water Kill Children In Woburn?" *The Washington Post*, April 3, 1986, p. A3.

20. Michael Weisskopf, "EPA Finds Pollution Unacceptably High," *The Washington Post*, April 13, 1989, p. A33.

21. "Getting Chemical Data," *Environmental Action*, September-October, 1988, p. 21.

22. William C. Walsh, National Association of Public Interest Research Groups, private communication, June 1989.

23. WRI and IIED, *World Resources 1987*, p. 207.

24. "A Problem that Cannot Be Buried," *Time*, October 14, 1987, p.76.

25. Sadik, *The State of World Population 1988*, p. 13.

26. John Langone, "A Stinking Mess," *Time*, January 2, 1989, p. 45.

27. Postel, *Defusing the Toxics Threat*, pp. 14-15.

28. WRI and IIED, *World Resources 1987*, p.202.

29. Hilary French, "Toxic Wastes Crossing Borders," *World Watch*, January-February, 1988, p. 8.

30. Langone, "A Stinking Mess," p. 47.

31. United Nations Environment Program (UNEP), *UNEP North American News*, April 1989.

32. *The New York Times*, July 5, 1988.

33. Kristin Helmore, "Dumping on Africa," *The Christian Science Monitor*, June 30, 1988, p.1.

34. *UNEP North American News*, April 1989; Edward Cody, "Pact Seeks to Shield Third World States," *The Washington Post*, March 23, 1989, p. A1.

35. *World Bank News*, July 14, 1988, p. 5.

36. David Maraniss and Michael Weisskopf, "Jobs and Illness in Petrochemical Corridor," *The Washington Post*, December 22, 1988, p. A1.

37. Michael Weisskopf, *The Washington Post*, "U.S. Air Pollution Exceeds Estimates," March 23, 1989, p. A1.

38. David Doniger and Deborah A. Sheiman, "To Avoid a Bhopal in the U.S.," *The Washington Post*, December 11, 1988; Peter H. Stone, "Bill Seeks to Aid Workers Exposed to Hazardous Chemicals," *The Washington Post*, July 13, 1986, p. A4.

39. Michael Weisskopf, "Toxic Air Far Exceeds EPA Limits," *The Washington Post*, June 9, 1989, p. A3.

40. Organization for Economic Cooperation and Development (OECD), *The State of the Environment 1985* (Paris: OECD, 1985), p. 38.

41. Ibid., p.39.

42. Ibid., p.41.

43. David Galvin and Sally Toteff, "Toxics on the Home Front," *Sierra*, September-October, 1986, p. 47.

44. Ibid., p. 44.

45. Ibid., p. 47.

46. Walsh, private communication, June 1989.

47. Michael Weisskopf, "EPA Acts on Leaking Storage Tanks," *The Washington Post*, April 3, 1987, p. A12.

48. Michael Weisskopf, "Pesticides and Death Amid Plenty," *The Washington Post*, August 30, 1988, p. A1.

49. Michael Weisskopf, "Pesticides Pose Higher Risk To Children, Group Concludes," *The Washington Post*, February 25, 1989, p. A2.

50. Sandra Postel, "Controlling Toxic Chemicals," in Lester R. Brown et al., *State of the World 1988* (New York: Norton, 1988), p. 121.

51. Essam El-Hinnawi and Mansur Hashmi, *The State of the Environment* (London: Butterworth Scientific, 1987), p. 97.

52. Mott and Snyder, *Pesticide Alert*, p. 6; Postel, "Controlling Toxic Chemicals," pp. 121-22.

53. Mott and Snyder, *Pesticide Alert*, pp. 1-2.

54. Goldsmith and Hildyard, *The Earth Report*, p. 197.

55. Postel, *Defusing the Toxics Threat*, p. 19.

56. Ruth Norris (ed.), *Pills, Pesticides, and Profits* (Croton-on-Hudson, NY: North River Press, 1982), p. 13.

57. "Report Urges Better Chemical Controls Abroad," *Conservation Foundation Letter*, No. 2, 1988, p. 1.

58. Postel, *Defusing the Toxics Threat*, p. 19.

59. Ibid., p. 16.

60. U.S. General Accounting Office, *Export of Unregistered Pesticides Is Not Adequately Monitored by EPA* (Washington, DC, April 1989); and Walsh, private communication. See also David Weir, "The Global Pesticide Threat," *Multinational Monitor*, Vol. 6, No. 13, September, 1985, p. 9.

61. *UNEP North American News*, April 1989.

62. Postel, *Defusing the Toxics Threat*, pp. 16-17.

63. "U.S. Restricts Sale of Australian Beef," *The Washington Post*, August 21, 1987.

64. "The Toxic Assault on the Rhine," *Greenpeace*, Vol. 12, No. 2, April-June, 1987, p. 5.

65. "Re:sources," *Environmental Action*, Vol. 20, No. 2, September-October, 1988, p. 23.

66. "U.S. Sues to Recoup Cost of Dioxin Cleanup," *The Washington Post*, August 18, 1988, p. A6.

67. Goldsmith and Hildyard, *The Earth Report*, p. 107; *The Washington Post*, February 15, 1989, p. D1.

68. Christopher Flavin, "Reassessing Nuclear Power," in Brown et al., *State of the World 1987*, p. 58.

69. Flavin, "Reassessing Nuclear Power," pp. 60-61; *The Washington Post*, May 12, 1988, p. A25.

70. *The Washington Post*, February 16, 1989, p. A36.

71. Goldsmith and Hildyard, *The Earth Report*, p. 221.

72. Cass Peterson, "DOE, NRC Contradicted on Accident," *The Washington Post*, October 23, 1987, p. A11.

73. Goldsmith and Hildyard, *The Earth Report*, pp. 186-87.

74. "High Contamination Found Around Nuclear Facilities," The Washington Post, September 26, 1986, p. A5.

75. Robert Alvarez and Arjun Makhijani, "Nuclear Waste: The $100-Billion Mess," *The Washington Post*, September 4, 1988, p. C3.

76. Flavin, "Reassessing Nuclear Power," p. 77.

77. The Conservation Foundation, *State of the Environment*, pp. 177-78.

78. Ibid., p. 176.

79. Richard Monastersky, "The 10,000-Year Test," *Science News*, February 27, 1988, p. 139.

80. The Conservation Foundation, *State of the Environment*, p. 175.

81. Cass Peterson, "Permanent Nuclear Waste Dump Delayed to 2003," *The Washington Post*, July 29, 1987, p. A3.

82. Alvarez and Makhijani, "Nuclear Waste: The $100-Billion Mess."

83. OECD, *State of the Environment 1985*, p. 167.

84. "Swiss Accelerate Evolution of Bacteria to Fight Toxicity," *The Washington Post*, February 2, 1987, p. A4; "Turning to New Technologies," *Time*, October 14, 1985.

85. Scott Ridley, "Hazardous Waste," in Ridley, *State of the States 1987* (Washington, DC: Fund for Renewable Energy and the Environment, 1987), p. 18.

86. This section summarizes material in *Toxics Use Reduction: From Pollution Control to Pollution Prevention* (Boston, MA: The National Toxics Campaign, February 1989), pp. 3-6.

87. Ibid., pp. 3-4; and U.S. Congress, Office of Technology Assessment, *Serious Reduction of Hazardous Waste: For Pollution Prevention and Industrial Efficiency* (Washington, DC: U.S. Government Printing Office, 1986).

88. Hirshhorn, "Cutting Production of Hazardous Waste," pp. 60-61.

89. WRI and IIED, *World Resources 1987*, p. 215; Postel, "Controlling Toxic Chemicals," pp. 130-31.

90. Joel Hirschhorn, "Cutting Production of Hazardous Waste," *Technology Review*, April 1988, p. 54.

91. *Toxics Use Reduction*, p. 5.

92. Hirschhorn, "Cutting Production of Hazardous Waste," p. 54; Postel, "Controlling Toxic Chemicals," p. 131; Langone, "A Stinking Mess," p. 45.

93. Hirschhorn, "Cutting Production of Hazardous Waste," p. 55.

94. Postel, *Defusing the Toxics Threat*, p. 44.

95. Hirschhorn, "Cutting Production of Hazardous Waste," p. 55.

97. Claudine Schneider, "Hazardous Waste: The Bottom Line is Prevention," *Issues in Science and Technology*, Vol. 4, No. 4, Summer, 1988, p. 78.

97. Ibid., p. 80.

98. *Toxics Use Reduction*, pp. 6-7.

99. Postel, *Defusing the Toxics Threat*, pp. 26-27, 31.

100. Ibid., pp. 34-35.

101. Flavin, "Reassessing Nuclear Power," p. 71.

Chapter 14: Solid Waste Management

1. James Cook, "Not in Anybody's Backyard," *Forbes*, November 28, 1988, pp. 172-82.

2. Quoted in Center for National Policy, "Solid Waste: A Practical Approach," *Focus On...*, Chapter Vol. 2, No. 2, 1988.

3. Ibid., p. 173; and *Time*, January 2, 1989, p. 45.

4. The 20 percent figure was calculated from *data* in U.S. Environmental Protection Agency (EPA), *The Waste System* (Washington, DC, November 1988), *p.* 2-2.

5. Cass Peterson, "Mounting Garbage Problem," *The Washington Post*, April 5, 1987; Peter Tonge, "All That Trash," *Christian Science Monitor*, July 6, 1987; Bill Richards, "Burning Issue," *The Wall Street Journal*, June 16, 1988.

6. Nevin Cohen et al., *Coming Full Circle* (New York: Environmental Defense Fund, 1988), p. 3.

7. Cynthia Pollock, *Mining Urban Wastes: The Potential for Recycling*, Worldwatch Paper 76 (Washington, DC: Worldwatch Institute, April 1987), p. 9.

8. "EPA Proposes Monitoring Municipal Dumps," *The Washington Post*, August 25, 1988, p. A7.

9. EPA, *Report to Congress: Solid Waste Disposal in the United States*, Volume II (Washington, DC, October 1988), pp. 4-7.

10. Philip R. O'Leary et al., "Managing Solid Waste," *Scientific American*, December 1988, p. 36.

11. EPA, *The Waste System*, p. 2-1.

12. EG&G Idaho, Inc., *Scrap Tires: A Resource and Technology Evaluation of Tire Pyrolysis and Other Selected Alternate Technologies*, EGG-2241 (Idaho Falls, ID: U.S. Department of Energy, 1983); *The* Washington Post, November 18, 1987, p. A3.

13. U.S. Bureau of Mines (USBM), *Mineral Commodity Summaries* (Washington, DC: 1989).

14. Keep America Beautiful, *Overview: Solid Waste Disposal Alternatives* (Stamford, CT, 1989), p. 1.

15. EPA, *The Waste System*, p. 2-3.

16. World Resources Institute (WRI) and International Institute for Environment and Development (IIED), *World Resources 1988-89* (New York: Basic Books, 1988), p. 46.

17. J. Winston Porter, Assistant Administrator for Solid Waste, EPA, "A National Perspective on Municipal Solid Waste Management," presentation to the Fourth Annual Conference on Solid Waste Management and Materials Policy, New York City, January 29, 1988.

18. EPA, *The Waste System*, p. 2-3; Allen Hershkowitz and Eugene Salerni, *Garbage Management in Japan* (New York: INFORM, 1987); and Elliot Marshall, "Confronting the Solid Waste Dilemma," *Focus on....*, (a publication of the Center for National Policy, Washington, DC), Vol. 2, No. 2, 1988, p. 3.

19. Allen Hershkowitz and Eugene Salerni, *Garbage Management in Japan: Lessons for the United States* (New York: INFORM, 1987).

20. *Time*, January 2, 1989, pp. 45-47; Hershkowitz and Salerni, *Garbage Management in Japan;* U.S. Bureau of the Census, *Statistical Abstract of the United States: 1989*, 109th edition (Washington, DC, 1989), Table 350.

21. WRI and IIED, *World Resources 1988-89*, p. 46.

22. Ibid., pp. 46-47.

23. Portions of this section on recyclable materials are based on Ann Dorr, American Geological Institute, "Recycling—How Much of the Solution?," unpublished manuscript, November 1988.

24. Paper Stock Institute of America, "Paper Stock Standards and Practices," (New York: April 1986), Circular PS-86.

25. Joan Rohlfs, "Background Paper on Old Newspaper," (Washington, DC: Metropolitan Washington Council of Governments, Feb. 24, 1989).

26. *The Washington Post*, October 2, 1989, p. D1.

27. Bill Metzfield, Gannett Supply Corporation, Gannett Newspapers, Washington, DC, "The Outlook for Recycled Newsprint in Newspaper Publishing," oral presentation, March 9, 1989.

28. Gary Stanley, Forest Products Division, U.S. Department of Commerce, Washington, DC, "Used Newspaper: Domestic and International Consumption," oral presentation, March 9, 1989; Joan Rohlfs, "Background Paper on Old Newspaper."

29. Metropolitan Washington Council of Governments, "U.S. Export Markets for Wastepaper from Baltimore Port of Embarkation," 1988.

30. USBM, *Mineral Commodity Summaries, 1989;* Patricia Plunkert, Aluminum Commodity Specialist, USBM, private communication.

31. USBM, *Mineral Commodity Summaries 1989*.

32. Jeanne Wirka, *Wrapped in Plastics* (Washington, DC: Environmental Action, 1988), pp. 38-40; Mark Crawford, "There's (Plastic) Gold in Them Thar Landfills," *Science*, Vol. 241, July 1988, p. 411.

33. Steve Alexander, Council for Solid Waste Solutions, Washington, DC, private communication, May 25, 1989.

34. Dennis Sabourin, Wellman, Inc., Shrewsbury, NJ, private communication, May 25, 1989.

35. Maryland Environmental Service, "Plastics," *Maryland Recycling Directory 1988* (Annapolis, MD, 1988), pp. 42-44.

36. Researchers have found that plastic can be made more biodegradable by adding substances such as starch to the synthetic polymers that make up the plastic. See William Booth, "Innovation Comes Full Circle: Making a Plastic Self-Destruct," *The Washington Post*, August 7, 1989, p. A3.

37. Mobil Corporation, *Quarterly Report to Stockholders*, March 1989.

38. Dennis Sabourin, private communication, May 25, 1989.

39. David Zykan, "The Vital Link," *Waste Alternatives* (Washington, DC, September, 1988), p. 17.

40. Environmental Defense Fund, *To Burn or Not to Burn*, (New York, 1986), p. 20.

41. Concern, Inc., *Waste: Choices for Communities*, p. 8.

42. Ibid.

43. Midwest Research Institute, *Final Report: Results of the Combustion and Emissions Research Project at the Vicon Incinerator Facility in Pittsfield, Massachusetts,* Volume 1 (Kansas City, MO, June 1987). The report was prepared for the New York State Energy Research and Development Authority.

44. *Proceedings, International Conference on Municipal Waste Combustion,* Hollywood, Florida, April 11-14, 1989 (Ottawa, Ontario: Environment Canada, 1989).

45. O'Leary et al., "Managing Solid Waste," pp. 39-40.

46. Jay Mathews, "Garbage In, Power Out: A Clean Solution to a Heap of Problems," *The Washington Post,* November 18, 1987, p. A3.

47. *Time,* January 2, 1989, p. 45; Concern, Inc., *Waste: Choices for Communities,* p. 2.

48. EPA, *Solid Waste Disposal in the United States* (Washington, DC, October 1988), Volume 1, p. 14; Volume 2, pp. 4-16; Concern, Inc., *Waste: Choices for Communities,* p. 10.

49. See Ronald J. Lofy, *Feasibility of Direct On-Site Conversion of Landfill Gas to Electrical Energy at Scholl Canyon Landfill, Glendale, California,* Final Report (Washington, DC: Office of Energy from Municipal Waste, U.S. Department of Energy, June 1981).

50. O'Leary et al., "Managing Solid Waste," pp. 40-42.

51. Center for National Policy, "Solid Waste: A Practical Approach."

52. Renew America, "Solid Waste Recycling," *The State of the States 1989* (Washington, DC, 1989), pp. 12-13.

53. EPA, *The Solid Waste Dilemma: An Agenda for Action* (Washington, DC, September 1988), p. 23.

54. Ibid., p. 23.

55. Ibid., pp. 26-69.

56. EPA, *Solid Waste Disposal,* Vol. 2, pp. 1-3.

57. Cohen, *Coming Full Circle,* pp. 66-68.

58. Ibid., pp. 66-67; *The Washington Post,* April 21, 1987, p. A4.

59. Cohen, *Coming Full Circle,* pp. 69-70.

60. Ibid., pp. 75-78.

61. Ibid., pp. 78-83; Kenneth Laden, District of Columbia Department of Public Works, Office of Planning, private communication, December 23, 1988.

62. William C. Clark, "Global Change Begins at Home," *Environment,* Vol. 31, No. 4, May 1989, p. 45.

63. Cohen, *Coming Full Circle,* p. 30.

64. Institute for Local Self-Reliance, "Selected Recycling Programs in the United States" (Washington, DC, undated).

65. *Time,* January 2, 1989, p. 45.

66. Ibid., p. 45.

Chapter 15: Global Security

1. Paul Kennedy, *The Rise and Fall of the Great Powers* (New York: Vintage Books, 1989), p. xxiii.

2. Robert S. McNamara, "An Address on the Population Problem," Cambridge, Massachusetts, April 28, 1977.

3. Ruth Leger Sivard, *World Military and Social Expenditures 1987-88* (Washington, DC: World Priorities, 1987).

4. Mikhail Gorbachev, Speech before the United Nations General Assembly, December 1988, quoted in *Time,* January 2, 1989, p. 68.

5. Sivard, *World Military and Social Expenditures,* p. 23.

6. Jessica Tuchman Mathews, "Redefining Security," *Foreign Affairs,* Spring 1989, p. 162.

7. World Commission on Environment and Development (WCED), *Our Common Future* (New York: Oxford University Press, 1987), p. 297.

8. Sivard, *World Military and Social Expenditures,* pp. 43-45.

9. Quoted in Hobart Rowen, "Conable Warns Poor Nations on Arms," *The Washington Post,* September 27, 1989, p. A28.

10. Sivard, *World Military and Social Expenditures,* pp. 46-50.

11. Lloyd Timberlake and John Tinker, *Environment and Conflict,* Earthscan Briefing Document No. 40 (Washington, DC: International Institute for Environment and Development, 1984), p. 13.

12. Lester R. Brown, "Redefining National Security," in Brown et al. *The State of the World 1986* (New York: Norton, 1986), p. 197.

13. Essam El-Hinnawi and Manzur H. Hashmi, "Military Activity," in El Hinnawi and Hashmi, *The State of the Environment* (London: Butterworth Scientific, 1987), pp. 144-45; Michael Renner, "Enhancing Global Security," in Lester R. Brown et al., *State of the World 1989* (New York: Norton, 1989), p. 134.

14. El-Hinnawi and Hashmi, "Military Activity," pp. 144-45.

15. Ibid., p. 144.

16. Brown, "Redefining National Security," p. 196.

17. Center for Defense Information, *The Defense Monitor*, Vol. 16, No. 7, 1987.

18. Ibid.; Brown, "Redefining National Security", p. 195; *Statistical Abstract of the United States: 1988* (Washington, DC: U.S. Department of Commerce, 1988), p. 297. For the United States, the amount spent on the military would be considerably larger than 29 percent if one includes military-related spending by the Department of Energy, the National Aeronautics and Space Administration, and the Veteran's Administration, plus foreign military aid and military-related payments on the federal debt. The military share becomes still greater when social security receipts are subtracted from the total budget. Michael Renner, Worldwatch Institute, private communication, July 1989.

19. Brown, "Redefining National Security," p. 197.

20. Ibid., p. 198.

21. Renner, private communication.

22. Brown, "Redefining Security," p. 197.

23. WCED, *Our Common Future*, p. 299.

24. El-Hinnawi and Hashmi, "Military Activity," p. 143; Norman Myers, *Not Far Afield: U.S. Interests and the Global Environment*, (Washington, DC: World Resources Institute, 1987), p. 37.

25. Norman Myers, (ed.), *Gaia: An Atlas of Planet Management* (Garden City, NY: Anchor Press, 1984), p. 244.

26. Renner, "Enhancing Global Security," pp. 135-36.

27. WCED, *Our Common Future*, p. 19.

28. Ibid, p. 293.

29. Timberlake and Tinker, *Environment and Conflict*, p. 54; Renner, "Enhancing Global Security," p. 142.

30. Timberlake and Tinker, *Environment and Conflict*, pp. 62-64.

31. Renner, "Enhancing Global Security," p. 142.

32. Renner, private communication.

33. Timberlake and Tinker, *Environment and Conflict*, p. 67.

34. World Resources Institute (WRI) and International Institute for Environment and Development (IIED), *World Resources 1988-89* (New York: Basic Books, 1988), p. 145.

35. WCED, *Our Common Future*, pp. 293-94.

36. Timberlake and Tinker, *Environment and Conflict*, pp. 71-72.

37. Ibid., pp. 72-73.

38. Christopher Flavin, "Creating a Sustainable Energy Future," in Lester R. Brown et al., *State of the World 1988* (New York: Norton, 1988), p. 27.

39. Timberlake and Tinker, *Environment and Conflict*, p. 13.

40. Myers, *Not Far Afield*, pp. 33-34.

41. El-Hinnawi and Hashmi, "Military Activity," p. 152.

42. Enrique A. Boloyra, "The Seven Plagues of El Salvador," *Current History*, December 1987, p. 413.

43. WCED, *Our Common Future*, p. 292.

44. Christopher Hitchen, "Minority Report," *The Nation*, December 19, 1987, p. 742.

45. Jodi Jacobson, *Environmental Refugees: A Yardstick of Habitability*, Worldwatch Paper 86 (Washington, DC: Worldwatch Institute, 1988).

46. Timberlake and Tinker, *Environment and Conflict*, p. 26.

47. Ibid., p. 19.

48. Ibid., pp. 47-53.

49. Federation for American Immigration Reform, "Immigration Report," October 1988, p. 1.

50. Lester R. Brown and Edward C. Wolf, "Reclaiming the Future," in Brown et al., *State of the World 1988*, p. 179.

51. Ibid., p. 184.

52. Michael Renner, *National Security: The Economic and Environmental Dimensions*, Worldwatch Paper 89 (Washington, DC: Worldwatch Institute, May 1989), p. 29.

53. Brown, "Redefining National Security," pp. 205-06.

54. Ibid., p. 206.

55. Ibid., pp. 197-99.

56. Ibid., pp. 201-02.

57. Ibid., p. 202.

58. Renner, "Enhancing Global Security," pp. 138-39.

59. Brown, "Redefining National Security," p. 200; Renner, "Enhancing Global Security," pp. 140-41.

60. Brown, "Redefining National Security," p. 200.

61. M.D. Oden, "Military Spending Erodes Real National Security," *The Bulletin of Atomic Scientists*, June 1988, pp. 40-41.

62. Remarks by Representative Claudine Schneider, Energy Efficiency Technology Exhibit Press Conference, February 2, 1988.

63. Oden, "Military Spending Erodes Real National Security," p. 41.

64. Renner, "Enhancing Global Security," p. 139.

65. Oden, "Military Spending Erodes Real National Security," p. 39.

66. Brown, "Redefining National Security," pp. 198-204.

67. El-Hinnawi and Hashmi, "Military Activity," p. 143.

68. Renner, "Enhancing Global Security," p. 135.

69. El-Hinnawi and Hashmi, "Military Activity," p. 147.

70. William Lewis and Christopher Joyner, "The Runaway Arms Race in the Third World," *The Washington Post,* October 25, 1988, p. A 27.

71. El-Hinnawi and Hashmi, "Military Activity," p. 145.

72. Ibid., p. 146.

73. Sir Frederick Warner, "The Environmental Effects of Nuclear War: Consensus and Uncertainties," *Environment,* June 1988.

74. Myers, *Gaia: An Atlas of Planet Management,* pp. 248-249.

75. El Hinnawi and Hashmi, "Military Activity," p. 150.

76. Warner, "The Environmental Effects of Nuclear War," p. 3 ; United Nations Study Group, "The Climatic and Other Global Effects of Nuclear War, " *Environment,* June 1988, p. 44; El-Hinnawi and Hashmi, "Military Activity," pp. 150-51.

77. United Nations Study Group, "The Climatic and Other Global Effects of Nuclear War," p. 44.

78. Warner, "The Environmental Effects of Nuclear War," p. 7.

79. Paul R. Ehrlich, Carl Sagan, Donald Kennedy, and Walter Orr Roberts, *The Cold And The Dark* (New York: Norton, 1984), p. 26.

80. Sivard, *World Military and Social Expenditures,* pp. 17-19.

81. Renner, private communication.

82. Although the publicly-announced goal of the negotiations is a 50 percent reduction, Michael Renner has noted (in private communication) that the real reductions are likely to be around 30 to 35 percent, according to Robert Norris et al., "START and Strategic Modernization," *Nuclear Weapons Databook Working Papers 87-2* (New York: Natural Resources Defense Council, 1987).

83. Sivard, *World Military and Social Expenditures,* p. 32. Murray Gell-Mann, Professor of Physics at California Institute of Technology, recently urged the superpowers to

redefine global security to include "the issues of population, environment, and sustainable development." *Time,* January 2, 1989, p. 63.

84. Brown and Wolf, "Reclaiming the Future."

85. Ibid., pp. 173-74.

86. Ibid., pp. 175-76.

87. Ibid., pp. 181-82.

88. Ibid., pp. 176-78.

89. Ibid., pp. 184-85.

90. Ibid., p. 186.

91. The Weiss legislation would provide for advance notification of military contract termination to allow advance conversion planning; mandatory occupational retraining; income maintenance for workers during the conversion process and relocation allowances where needed; creation of a national network for employment opportunities; governmental alternative capital investment planning; and creation of a conversion fund financed from military contracts: Michael Renner, "Swords into Consumer Goods," *World Watch,* July-August 1989, p. 24. The Weiss legislation is a refinement of a Senate bill first introduced by Senator George McGovern in 1963.

92. Renner, "Enhancing Global Security," pp. 149-50; and private communication, July 1989.

93. Ibid., pp. 148-49.

94. Lewis and Joyner, "The Runaway Arms Race."

95. Renner, "Enhancing Global Security," p. 147.

96. Brown, "Redefining National Security," pp. 207-08.

97. Brown, "Redefining Global Security," p. 207; Renner, "Enhancing Global Security," p. 151; Renner, *National Security;* and Renner, "Swords into Consumer Goods."

98. *The Washington Post,* December 8, 1988, p. A32.

99. For more details about activities of the United Nations and related agencies, see *Basic Facts About the United Nations* (New York: United Nations, 1987). For a proposal to restructure and democratize the United Nations, see Frank Barnaby (ed.), *The Gaia Peace Atlas* (New York: Doubleday, 1988), pp. 238-39.

100. Renner, private communication.

101. *The World Almanac and Book of Facts* (New York: Pharos Books, 1987), p. 640.

Chapter 16: Toward a Sustainable Future

1. Quoted in Richard E. Benedick, "Environment: An Agenda for the Next G-7 Summit," *International Herald Tribune*, July 25, 1989.

2. Hazel Henderson, "China: Key Player in a New World Game," *Futures Research Quarterly*, Fall 1987, p. 40.

3. Jessica Tuchman Mathews, "Redefining Security," *Foreign Affairs*, Spring 1989, p. 173.

4. World Commission on Environment and Development (WCED), *Our Common Future* (New York: Oxford University Press, 1987).

5. Ibid., pp. 43-46. For a further discussion of sustainable development, see Chapter 4. See also Herman E. Daly, The World Bank, *Sustainable Development: Some Basic Principles*, undated. Economist David Pearce suggests that there are two fundamental dimensions of sustainability: sustainable economic development—the sustainable growth of per capita real income over time, and sustainable use of natural resources and the environment. For a discussion of how the goal of sustainable development could be attained, see David Pearce, "Sustainable Futures: Some Economic Issues," in Daniel B. Botkin et al., *Changing the Global Environment: Perspectives on Human Involvement* (San Diego, CA: Academic Press, 1989), pp. 311-23.

6. David Brokensha and Bernard W. Riley, "Managing Natural Resources: The Local Level," in Botkin, *Changing the Global Environment*, pp. 341-66.

7. Pearce, "Sustainable Futures," pp. 316-20.

8. For an expanded version of the "Agenda 2000" report in book form, see Rushworth M. Kidder, *Reinventing the Future: Global Goals for the 21st Century* (Cambridge, MA: The MIT Press, 1989).

9. Most of the 12 priorities listed in Table 1 (international agendas) and Table 2 (agendas for the United States) are broadly defined (e.g., make agriculture sustainable), and several include two or more related parts (e.g., protect forests and habitats, curb loss of species). If a particular priority is marked in Table 1 or 2 with an "X" for a given agenda, the agenda contains a reference to at least one aspect or part of that priority, but not necessarily to other aspects or parts.

The agendas summarized in Tables 1 and 2 vary greatly in length, from book-length reports (e.g., *Our Common Future)* to short articles (e.g., Richard Benedick's newspaper article).

10. At the Ecology '89 conference in Gothenberg, Sweden, a panel of about a dozen world experts compiled a list of the most important global environmental topics; they included human population growth, biodiversity and conservation, climate change, forest decline, hazardous wastes, land degradation, and environmentally-degrading energy production: Arno Rosemarin, "Global Change, Sustainable Development and the Dangers of Information Overkill," *Ambio*, Vol. 18, No. 6, 1989, p. 307.

In a recent article, Jessica Mathews listed four steps as "most important:" prompt revision of the Montreal Treaty to protect the ozone layer, implementation of the Tropical Forestry Action Plan, sufficient support for family planning programs, and an increase in energy productivity. Mathews, "Redefining Security," p. 177. For an action agenda for the Western Hemisphere, see *The Americas in 1989: Consensus for Action*, a report of the Inter-American Dialogue (IAD) (Queenstown, MD: The Aspen Institute, 1989). The report is available from IAD, 1333 New Hampshire Avenue, N.W., Suite 1070, Washington, DC 20036. The report includes 7 of the 12 priorities listed in Table 1.1.

11. Opening statement by Dr. Mostafa K. Tolba, Executive Director, United Nations Environment Program (UNEP), at the Fifteenth Session of the UNEP Governing Council, Nairobi, May 15-26, 1989; and *UNEP North American News*, Vol. 4, No. 4, August 1989.

12. For summaries of most of the agendas included in this tally, see *Future Survey*, especially Vol. 10, Nos. 4 and 10 (April and October 1988); and Vol. 11, Nos. 1, 2, 4, and 5 (January, February, April, and May 1989). For a condensation of the *Project 88* agenda and a summary of the Worldwatch Institute agenda, see Robert N. Stavins, "Harnessing Market Forces to Protect the Environment," *Environment*, Vol. 31, No. 1, January-February 1989.

13. See note 7.

14. Energy efficiency was not explicitly included in the Conservation Foundation's agenda, but several related priorities—control of air pollution, acid rain, and global warming—were included.

15. Although The Conservation Foundation's agenda omitted population, the agenda was published together with a summary of major proposals from *Blueprint for the Environment* that included population and family planning. American Agenda apparently contains no reference to population as a priority issue.

16. An earlier U.S. agenda, *An Environmental Agenda for the Future* (Washington, DC: Island Press, 1985), presented the priority issues as seen by the leaders of the ten largest conservation organizations in the United States. It included 10 of the 12 priorities in Table 16.2, omitting only a shift of military spending to sustainable development (though it strongly stressed the need to reduce the risk of nuclear war), and protection of the ozone layer (which was not a major public issue in 1985).

17. These estimates of the importance or effectiveness of the priority solutions are very tentative, subjective appraisals based on the material in the relevant issue chapters. For each solution, the most relevant material supporting the estimates is given in the next seven endnotes that accompany the conclusions drawn from Table 16.3.

18. As Lester Brown has noted, slowing population growth makes almost all global problems more manageable. See Lester R. Brown, "Stopping Population Growth," in Brown et al., *State of the World 1985* (New York: Norton, 1985), pp. 200-21. See also Lester R. Brown and Edward C. Wolf, "Reclaiming the Future," in Brown et al., *State of the World 1988* (New York: Norton, 1988), pp. 176-78, 188; and National Research Council, *Population Growth and Economic Development: Policy Questions* (Washington, DC: National Academy Press, 1986), pp. 85-93. For a discussion of "multisector payoffs" from population planning, see Norman Myers, "The Global Possible: What Can Be Gained?," in Robert Repetto (ed.), *The Global Possible: Resources, Development, and the New Century* (New Haven: Yale University Press, 1985), pp. 478-81.

19. See Myers, "The Global Possible," pp. 479-82; and Norman Myers, "Environmental Challenges: More Government or Better Governance?," *Ambio,* Vol. 17, No. 6, 1988, pp. 411-12. Richard Benedick has observed that "protecting the global environment is inextricably linked with the elimination of poverty." Benedick, "Environment: An Agenda for the Next Summit."

20. See Janos P. Hrabovszky, "Agriculture: The Land Base," in Repetto, *The Global Possible,* pp. 246-53; Myers, "The Global Possible," pp. 488-89; and David Pimentel and Carl W. Hall (eds.), *Food and Natural Resources* (San Diego, CA: Academic Press, 1989), especially Chapter 2, "Interdependence of Food and Natural Resources."

21. See John Spears and Edward S. Ayensu, "Resources, Development, and the New Century: Forestry," in Repetto, *The Global Possible,* pp. 300, 325, 333-35; and Myers, "The Global Possible," pp. 484-87.

22. See Amulya K. N. Reddy, "Energy Issues and Opportunities," in Repetto, *The Global Possible,* esp. pp. 365-71; Myers, "The Global Possible," pp. 487-88; and Stephen H. Schneider and Starley L. Thompson, "Future Changes in the Atmosphere," in Repetto, *The Global Possible,* p. 426.

23. See Sandra Postel, *Water: Rethinking Management in an Age of Scarcity,* Worldwatch Paper 62 (Washington, DC: Worldwatch Institute, December 1984), esp. pp. 11-18; Peter P. Rogers, "Fresh Water," in Repetto, *The Global Possible,* esp. pp. 266-73, 292-97; and Myers, "The Global Possible," pp. 488-89.

24. See Cynthia Pollock, *Mining Urban Wastes: The Potential for Recycling,* Worldwatch Paper 76 (Washington, DC: Worldwatch Institute, April 1987), esp.

pp. 20-31; and Philip O'Leary et al., "Managing Solid Waste," *Scientific American,* December 1988, pp. 36-42.

25. Myers notes further that "Of all areas of public policy, there is probably none that we understand less than this of intersectoral linkage. To this extent, then, the underlying problem is not so much one of economic outlays as of integrative policy, backed by institutional design." Myers, "The Global Possible," p. 487.

26. Norman Myers, "Environmental Challenges," p. 411. Robert McNamara has suggested that existing international institutions lack sufficient authority and resources to deal with growing global interdependence and environmental problems that increasingly transcend national boundaries. McNamara, quoted in Kidder, "Agenda 2000," p. B6.

27. Mitchell B. Rambler, Lynn Margulis, and Rene Fester (eds.), *Global Ecology: Towards a Science of the Biosphere* (San Diego, CA: Academic Press, 1989), pp. ix-x; and John F. Stolz et al., "The Integral Biosphere," in Rambler et al., *Global Ecology,* pp. 31-49. See also William C. Clark, "Sustainable Development of the Biosphere: Themes for a Research Program," in W.C. Clark and R.E. Munn (eds.), *Sustainable Development of the Biosphere* (New York: Cambridge University Press, 1986), pp. 5-48; and Alexander J. Tuyahov et al., "Observing the Earth in the Next Decades," in Botkin et al., *Changing the Global Environment,* pp. 285, 302.

Recent research on Earth and exploration of the planets Venus and Mars have greatly increased knowledge about the interdependence between life on Earth (the biota) and the Earth's surface and atmosphere. This research supports what is called the Gaia hypothesis (named after the ancient Greek goddess of the Earth), that links the biota to the composition of the Earth's atmosphere and surface sediments. The hypothesis states that "the temperature and composition of the Earth's surface are actively regulated by the sum of life on the planet," and maintains that many characteristics of Earth that differentiate it from other planets in the solar system, including its oxygen-rich atmosphere and surface characteristics such as limestone and other deposits, are the result of life on Earth. Lynn Margulis and J.E. Lovelock, "Gaia and Geognosy," in Rambler et al., *Global Ecology,* p. 1; Lynn Margulis and Ricardo Guerrero, "From Planetary Atmospheres to Microbial Communities," in Botkin et al., *Changing the Global Environment,* pp. 51-67. See also Clark and Munn, "Sustainable Development of the Biosphere," p. 16.

28. See Russell W. Peterson, "A College of Integrated Studies: Education for the Professional Generalist," *L&S Magazine,* University of Wisconsin, Spring 1988; and "An Earth Ethic," in National Geographic Society, *Earth '88: Changing Geographic Perspectives* (Washington, DC, 1988), pp. 363-64.

29. M.I. Dyer and D.A. Crossley, Jr., "Linking Ecological Networks and Models to Remote Sensing Programs," in Botkin et al., *Changing the Global Environment*, p. 273; Michael D. Gwynne and D. Wayne Mooneyhan, "The Global Environment Monitoring System and the Need for a Global Resource Data Base," in Botkin et al., *Changing the Global Environment*, pp. 243-56; Richard Monastersky, "Global Change: The Scientific Challenge," *Science News*, Vol. 135, April 15, 1989, pp. 232-35; National Research Council, *Toward an Understanding of Global Change* (Washington, DC: National Academy Press, 1988). Regarding the potential contributions of environmental monitoring, see Clark and Munn, "Sustainable Development of the Biosphere," p. 14.

30. William Booth, "U.S. Details Global Warming Study," *The Washington Post*, September 1, 1989, p. A3.

31. Botkin et al., *Changing the Global Environment*, pp. 165-66; Gwynne and Mooneyhan, "The Global Environment Monitoring System," p. 249. See also WCED, *Our Common Future*, pp. 274-77.

32. Dyer and Crossley, "Linking Ecological Networks and Models;" Alexander J. Tuyahov et al., "Observing the Earth in the Next Decades," in Botkin et al., *Changing the Global Environment*, pp. 285-304.

33. Brian W. Mar, "Management of High Technology: A Cure or Cause of Global Environmental Changes?," in Botkin et al., *Changing the Global Environment*, pp. 405-18.

34. See George Bugliarello, "Technology and the Environment," in Botkin, *Changing the Global Environment*, p. 401.

35. See WCED, *Our Common Future*, p. 60; Marilyn Carr, *The A.T. Reader: Theory and Practice in Appropriate Technology* (London: Intermediate Technology Publications, 1985); E.F. Schumacher, *Small is Beautiful: Economics as if People Mattered* (New York: Harper & Row, 1973); Ken Darrow and Mike Saxenian, *Appropriate Technology Sourcebook: A Guide to Practical Books for Village and Small Community Technology* (Appropriate Technology Project, P.O. Box 4543, Stanford, CA 94305); and Eric L. Hyman, "The Identification of Appropriate Technologies for Rural Development," *Impact Assessment Bulletin*, Vol. 5, No. 3, 1987, pp. 35-55 (available from Appropriate Technology International, 1331 H Street, N.W., Washington, DC 20005).

36. WCED, *Our Common Future*, p. 60.

37. Ibid., pp. 60-61.

38. See Angelo A. Orio, "Modern Chemical Technologies for Assessment and Solution of Environmental Problems," in Botkin et al., *Changing the Global Environment*, p. 169. For a list of priorities in the area of global monitoring and data analysis, see Repetto, *The Global Possible*, pp. 514-15.

39. Holmes Rolston, III, *Environmental Ethics: Duties to and Values in the Natural World* (Philadelphia: Temple University Press, 1988), Chapter 1. Roderick Nash observes that environmental ethics has come to have two meanings: (1) the idea that it is right to protect nature and wrong to harm it from the perspective of human interest; and (2) the more radical belief that nature has intrinsic value and therefore the right to exist, apart from human interest. Roderick Frazier Nash, *The Rights of Nature: A History of Environmental Ethics* (Madison, WI: University of Wisconsin Press, 1989), as summarized in *Future Survey*, Vol. 11, No. 6, June 1989, p. 6.

40. See Norman Myers, *A Wealth of Wild Species: Storehouse for Human Welfare* (Boulder, CO: Westview Press, 1983); and Robert Hamrin, *A Renewable Resource Economy* (New York: Praeger, 1983).

41. See Kristin Shrader-Frechette, "Environmental Ethics and Global Imperatives," in Repetto, *The Global Possible*, pp. 97-127.

42. Tayler H. Bingham, "Social Values and Environmental Quality," in Botkin et al., *Changing the Global Environment*, p. 380; Environmental and Energy Study Institute, "The Impact of Debt-for-Nature Swaps on Global Conservation," July 25, 1989.

43. See Rushworth M. Kidder, "Bringing Ethics to Bear," in "Agenda 2000," *Christian Science Monitor*, July 25, 1988, p. B11.

44. Lester R. Brown, Christopher Flavin, and Sandra Postel, "Outlining a Plan of Action," in Brown et al., *State of the World 1989* (New York: Norton, 1989), p. 174.

45. Robert S. McNamara, quoted in Kidder, "Bringing Ethics to Bear," p. B10. The destruction of tropical forests is an aspect of environmental degradation involving both inequity and immorality. As Michael Robinson has observed, "the rainforests are being destroyed not out of ignorance or stupidity but largely because of poverty and greed." Robinson, "Beyond Destruction, Success," in Judith Gradwohl and Russell Greenberg, *Saving the Tropical Forests* (London: Earthscan Publications, 1988), p. 11.

46. Kidder, "Bringing Ethics to Bear," p. B10.

47. Ibid., p. B12. For an analysis of attitudes toward the environment and the environmental impacts of policies in U.S. business, industry, and government, see Robert Cahn, *Footprints on the Planet: A Search for an Environmental Ethic* (New York: Universe Books, 1978).

48. Theodore J. Gordon, The Futures Group, quoted in Kidder, "Bringing Ethics to Bear," p. B12; Peterson, "An Earth Ethic," p. 375.

49. Robert Repetto et al., *Wasting Assets: Natural Resources in the National Income Accounts* (Washington, DC: World Resources Institute, 1989); Norman Myers, "Writing Off the Environment," in John Elkington et al.,

Green Pages: The Business of Saving the Environment (London: Routledge, 1988), pp. 190-92. See also Mathews, "Redefining Security," p. 173.

50.   Myers, "Writing Off the Environment," p. 190.

51.   Robert Repetto, "No Accounting for Pollution: A New Means of Calculating Wealth Can Save the Environment," *The Washington Post,* May 28, 1989, p. B5.

52.   Ibid.

53.   Ibid.; William Booth, "Saving Rain Forests by Using Them," *The Washington Post,* June 29, 1989, p. A1.

54.   Repetto, *Wasting Assets.*

55.   Myers, "Writing Off the Environment," p. 192.

56.   Repetto, "No Accounting for Pollution." For additional observations and recommendations regarding national income accounting, see Bingham, "Social Values and Environmental Quality" (reviews promising methods for placing monetary value on natural resources); Pearce, "Sustainable Futures" (discusses the concept of social prices); Herman Daly, "Steady-State Versus Growth Economics: Issues for the Next Century," paper prepared for the Hoover Institution Conference on Population, Resources and Environment, Stanford University, February 1-3, 1989 (calls for better practical measures of sustainable income for national accounts); Frank Barnaby (ed.), *The Gaia Peace Atlas* (New York: Doubleday, 1988), pp. 205-06 (recommends a unified accounting system that includes a nation's resource base); Benedick, "Environment: An Agenda for the Next G-7 Summit," (recommends international endorsement of reforms in national accounting practices); William C. Clark, Harvard University, quoted in Kidder, "Agenda 2000" (recommends environmental accounts that recognize that resources and environment are part of a society's productive stock); Nafis Sadik, *The State of World Population 1988* (New York: United Nations Population Fund, 1988), p. 12 (recommends introduction of "environmental accounting" in all planning and development activities); and James Gustave Speth, *Environmental Pollution: A Long-Term Perspective* (Washington, DC: World Resources Institute, 1989) (recommends national income accounts that treat depreciation of natural assets as rigorously as depreciation of capital assets).

57.   See WCED, *Our Common Future,* pp. 62-63.

58.   See Ibid., p. 10. In the United States, a revitalized Council on Environmental Quality could also play an important role as a mediator of differences, and as a point of contact for the Executive Branch with nongovernmental organizations, academic centers, and scientific research groups.

59.   See Ibid., p. 64, and William D. Ruckelshaus, "Toward a Sustainable World," *Scientific American,* September 1989, p. 168.

60.   World Resources Institute, *The Crucial Decade: the 1990s and the Global Environmental Challenge* (Washington, DC, 1989), p. iii.

61.   Mathews, "Redefining Security," p. 173.

62.   Repetto, *The Global Possible,* pp. 23-24.

63.   Stavins, "Harnessing Market Forces," *pp.* 6-7; Alfred Endres, "The Search for Effective Pollution Control Policies," in Botkin et al., *Changing the Global Environment,* pp. 439-54.

64.   Endres, "The Search for Effective Pollution Control Policies," pp. 446-49; Stavins, "Harnessing Market Forces," p. 6.

65.   Endres, "The Search for Effective Pollution Control Policies," pp. 448-49.

66.   Stavins, "Harnessing Market Forces," p. 6.

67.   Ibid., p. 7.

68.   Jeffrey A. McNeely, *Economics and Biological Diversity: Developing and Using Economic Incentives and Conserving Biological Resources* (Gland, Switzerland: International Union for the Conservation of Nature and Natural Resources, 1988).

69.   Stavins, "Harnessing Market Forces;" and *Project 88: Harnessing Market Forces to Protect Our Environment: Initiatives for the New President,* A Public Policy Study sponsored by Senator Timothy E. Wirth and Senator John Heinz (Washington, DC, 1988). The study is available from the offices of Senator Wirth and Senator Heinz.

70.   James J. MacKenzie, *Breathing Easier: Taking Action on Climate Change, Air Pollution, and Energy Insecurity* (Washington, DC: World Resources Institute, undated [1988]), p. 21; *Project 88,* pp. 40-46.

71.   Repetto, *The Global Possible,* pp. 28-29.

72.   Ibid., pp. 28-29.

73.   Sadik, *The State of World Population 1988,* p. 15.

74.   Brokensha and Riley, "Managing Natural Resources: The Local Level," *pp.* 341-64.

75.   WCED, *Our Common Future,* Chapter 12. Jessica Mathews has stressed the need for new institutions and regulatory regimes and a new diplomacy to cope with growing environmental interdependence. Mathews, "Redefining Security," p. 174. Regarding the strengths and weaknesses of different kinds of institutions for dealing with different types of large-scale environmental problems, see Clark and Munn, "Sustainable Development of the Biosphere," p. 13.

76.   David Sarokin, "What Are We to Do About the Planet's Health?," *The Washington Post,* August 30, 1988, p. A23.

For recommendations on how international development assistance can be made more effective and more supportive of environmental protection and sustainable development, see WCED, *Our Common Future*, pp. 336-40; and Repetto, *The Global Possible*, pp. 513-14.

On the need for international development assistance agencies to elevate the importance of environmental protection in development projects, see Russell Train, "Bush Faces Major Environmental Challenges," *Conservation Foundation Letter*, 1989, No. 1, p. 7.

77. United Nations Environment Program (UNEP), *UNEP North American News*, Vol. 3, No. 4, August 1988.

78. Benedick, "Environment: An Agenda for the Next G-7 Summit." For recommendations of measures to strengthen UNEP, see WCED, *Our Common Future*, pp. 20, 319-23.

79. Mathews, "Redefining Security," pp. 175-76.

80. William C. Clark, Kennedy School of Government, Harvard University, quoted in Kidder, "Agenda 2000," p. B9. Other proposals offered at the "Agenda 2000" conference (described below) included a suggestion to create an international "interdependent political process that takes account of all levels of society," and a forum for rural people who are generally left out of governmental decision-making; Kidder, "Agenda 2000," p. B10.

81. Vance Martin (ed.), *For the Conservation of Earth* (Golden, CO: International Wilderness Leadership Foundation, 1988), pp. 361-63; Richard D. Lamm and Thomas A. Barron, "The Environmental Agenda for the Next Administration," *Environment*, Vol. 30, No. 4, May 1988, p. 29.

82. Tolba, Statement; *UNEP North American News*, Vol. 4, No. 2, April 1989; and Vol. 4, No. 3, June 1989.

83. Kidder, "Agenda 2000," p. B5. Jessica Mathews has recommended invention of a set of indicators to measure global environmental health, analogous to economic indicators such as unemployment rates and GNP, and demographic indicators such as fertility rates and life expectancy. Mathews, "Redefining Security," pp. 173-74.

84. See Mathews, "Redefining Security," pp. 174-75.

85. WCED, *Our Common Future*, p. 325.

86. See Jessica Tuchman Mathews, "Is There More Risk in the World?," *The Washington Post*, March 27, 1989, p. A25.

87. Mathews, "Redefining Security," p. 176.

88. WCED, *Our Common Future*, p. 21.

89. Ibid., p. 301.

90. Myers, "Environmental Challenges," p. 411.

91. Ibid., pp. 411-12.

92. Ibid., p. 412.

93. Ibid. For additional examples of innovative waste recovery programs in industry, see Repetto, *The Global Possible*, p. 27.

94. Myers, "Environmental Challenges," p. 412. For a list of recommended actions by private business in the area of global resources and environment, see Repetto, *The Global Possible*, pp. 515-17.

95. Repetto, *The Global Possible*, p. 517; Myers, "Environmental Challenges," p. 412.

96. Myers, "Environmental Challenges," p. 412.

97. Ibid.

98. Alan B. Durning, "Mobilizing at the Grassroots," in Lester R. Brown et al., *State of the World 1989*, p. 157.

99. Repetto, *The Global Possible*, p. 29.

100. Myers, "Environmental Challenges," p. 413. For a wide range of case studies involving successful environmental action by NGOs, see Walter V. Reid et al., *Bankrolling Success: A Portfolio of Sustainable Development Projects* (Washington, DC: Environmental Policy Institute and National Wildlife Federation, undated [1988]); McNeely, *Economics and Biological Diversity;* and Gradwohl and Greenberg, *Saving the Tropical Forests*. For recommendations on making NGOs more effective, see WCED, *Our Common Future*, pp. 326-29.

101. *Making Common Cause: A Statement and Action Plan by U.S.-Based International Development, Environment, and Population NGOs* (Washington, DC: World Resources Institute, undated [1986]); *Making Common Cause Internationally*: A Statement and Action Plan by International Development, Environment and Population NGOs (Geneva, Switzerland, April 1988).

102. Myers, "Environmental Challenges," p. 413.

103. See Kidder, "Agenda 2000," p. B10.

104. Repetto, *The Global Possible*, p. 12.

105. Ibid., pp. 12-15, 29. See also Norman Myers, *Not Far Afield: U.S. Interests and the Global Environment* (Washington, DC: World Resources Institute, 1987), pp. 56-57.

106. WCED, *Our Common Future*, p. 65.

107. See Robert Repetto, "The Mechanisms for Improvement: Strategies for Sustainable Development," in Repetto, *The Global Possible*, pp. 16-29.

108. See Peterson, "An Earth Ethic," *pp*. 366-67; Reid, *Bankrolling Successes*, and The Panos Institute, *Towards Sustainable Development* (London, 1987).

109. For case studies of programs representing various aspects of sustainable development, see, for example, Reid, *Bankrolling Successes;* Gradwohl and Greenberg, *Saving the Tropical Forests;* McNeely, *Economics and Biological Diversity;* The Panos Institute, *Towards Sustainable Development;* Worldwatch Institute's *State of the World* series (New York: Norton, published annually since 1984); World Resources Institute, *World Resources* (New York: Basic Books, published annually since 1986); and Renew

America, *State of the States* (Washington, DC, published annually since 1987). A number of organizations publish periodicals that regularly review sustainable programs; a selection of these periodicals is listed under Further Information at the end of this chapter.

110.   *The Renew America Report*, Summer 1989. The 22 environmental categories are: air pollution reduction, drinking water protection, energy pollution control, environmental beautification, environmental education, fish conservation, food safety, forest management, greenhouse gases/ozone, groundwater protection, growth management, hazardous materials reduction, pesticide contamination, public lands and open space, radioactive waste reduction, range conservation, renewable energy/efficiency, soil conservation, solid waste reduction/recycling, surface water protection, transportation efficiency, and wildlife conservation.

111.   For estimates of the global costs of sustainable development programs, see Brown and Wolf, "Reclaiming the Future," summarized in Chapter 15; and WCED, *Our Common Future*, p. 303.

112.   See Repetto, *The Global Possible*, p. 15.

113.   WCED, *Our Common Future*, p. 16.

114.   The estimate for family planning costs is from The World Bank, *World Development Report 1984* (New York: Oxford University Press, 1984), pp. 151-52; see also J. Joseph Speidel, "Population Policies and Programs in Developing Countries—Is the U.S. Doing Its Part?" (Washington, DC: Population Crisis Committee, January 1989). Estimates for agroforestry are from Mathews, "Redefining Security," p. 172; estimates for energy are from Mathews, "Redefining Security," p. 172, and Myers, "Environmental Challenges," p. 412; estimates for fisheries, watershed protection, drinking water, and pollution reduction are from Repetto, *The Global Possible*, pp. 13-15.

115.   *Science*, Vol. 238, December 18, 1987, p. 1634; Myers, "Environmental Challenges," p. 412.

116.   Michael Marien (ed.), *Future Survey Annual 1988-89* (Bethesda, MD: World Future Society, 1989), pp. vi-viii. In the 1988-89 survey, following the environment the other nine top fears, in descending order, were economic concerns, arms races, drug wars, AIDS, energy, poverty, health care costs, traffic congestion, and concerns about children. The assessment was based on 765 books, articles,

and reports reviewed in *Future Survey* from October 1987 through October 1988.

117.   Lamm and Barron, "The Environmental Agenda," p. 28; Riley E. Dunlap, "Polls, Pollution, and Politics Revisited: Public Opinion in the Reagan Era," *Environment*, July-August 1987; and Peter Borelli, "Environmentalism at a Crossroads," in Borelli (ed.), *Crossroads: Environmental Priorities for the Future* (Washington, DC: Island Press, 1988), p. 10.

118.   Clark Norton, "Green Giant," *The Washington Post Magazine*, September 3, 1989, p. 26.

119.   Train, "Bush Faces Major Environmental Challenges," p. 3; *The Sun* (Baltimore), November 27, 1988, p. 5P.

120.   Benedick, "Environment: An Agenda for the Next G-7 Summit;" Richard Morin, "Poll Shows Public Wants Cleanup," *The Washington Post*, June 18, 1989.

121.   Louis Harris and Associates, *Public and Leadership Attitudes to the Environment in Four Continents*, A Report of a Survey in 16 Countries. Conducted for The United Nations Environment Program. Revised draft, August 1989. In addition to the United States, the nations included were Japan, China, and India (Asia); Nigeria, Kenya, Senegal, and Zimbabwe (Africa); Saudi Arabia (Middle East); West Germany, Norway, and Hungary (Europe); Argentina, Brazil, Mexico, and Jamaica (Latin America and the Caribbean). The survey was based on interviews with 8325 members of the public and 750 leaders.

122.   Ibid., Executive Summary, pp. 6-10.

123.   Louis Harris and Associates, Introduction prepared for National Public Radio's "Morning Edition" program on public and leadership attitudes on the environment for August 14, 1989.

124.   See Train, "Bush Faces Major Environmental Challenges;" Lamm and Barron, "The Environmental Agenda;" "What the U.S. Should Do," *Time*, January 2, 1989, p. 65; and Mathews, "Redefining Security," esp. pp. 175, 177.

125.   Lamm and Barron, "The Environmental Agenda," pp. 28-29.

126.   WCED, *Our Common Future*, p. 9.

Chapter 17: What You Can Do

1.   John M. Richardson, Jr., (ed.), *Making It Happen: A Positive Guide to the Future* (Washington, DC: The U.S. Association for The Club of Rome, 1982).

2.   See Russell W. Peterson, "A College of Integrated Studies: Education for the Professional Generalist," *L&S*

Magazine (University of Wisconsin, Madison), Spring 1988.

3.   Kingsley Davis et al. (eds.), *Population and Resources in a Changing World*, Current Readings (Stanford, CA: Morrison Institute for Population and

Resource Studies, 1989). The book is available through the Stanford University bookstore.

4. For information on members of the U.S. Congress and state governors, consult Michael Barone and Grant Ujifusa, *The Almanac of American Politics 1990* (Washington, DC: National Journal, 1989).

5. For example, see National Audubon Society, *The Guide for Citizen Action* (Washington, DC, 1981), available from the Society's Washington office for $2.50; and Sierra Club, *Conservation Action Handbook* (San Francisco, CA, 1987), available from the Club's San Francisco office for $5.00.

Appendix: List of Organizations

This list includes organizations that are referred to in the text of *The Global Ecology Handbook*, including members of the Global Tomorrow Coalition, that can provide information and materials on the topics covered.

(P) denotes GTC Participating Member Organization   (A) denotes GTC Affiliate Member Organization
(F) denotes supplier of films and other audiovisual materials

Acid Rain Foundation (P)
1410 Varsity Drive
Raleigh, North Carolina 27606
919/828-9443

African Wildlife Foundation (A)
1717 Massachusetts Avenue, N.W.
Washington, D.C. 20036
202/265-8394

Agency for International Development
State Building
320 21st Street, N.W.
Washington, D.C. 20523
202/647-1850

Alan Guttmacher Institute (P)
2010 Massachusetts Avenue, N.W./5th floor
Washington, D.C. 20036
202/296-4012

Alliance for the Chesapeake Bay
6600 York Road
Baltimore, Maryland 21212
301/377-6270

Alliance to Save Energy
1725 K Street, N.W./914
Washington, D.C. 20006-1401
202/857-666

American Association for the Advancement
  of Science
1333 H Street, N.W.
Washington, D.C. 20005
202/326-6400

American Cetacean Society (A)
P.O. Box 2639
San Pedro, California 90731 0943
213/548-279

American Committee for International
  Conservation (P)
c/o Michael McCloskey, Sierra Club
408 C Street NE
Washington, D.C. 20002
202/547-1141

American Conservation Association (P)
30 Rockefeller Plaza/5402
New York, New York 10112
212/649-5600

American Council for an Energy-Efficient
  Economy
1001 Connecticut Avenue, N.W./535
Washington, D.C. 20036
202/429-8873

American Council of Life Insurance (A)
1001 Pennsylvania Avenue, N.W.
Washington, D.C. 20004-2599
202/624-2465

American Council for Voluntary
  International Action (InterAction)
200 Park Avenue South
New York, New York 10003
212/777-8210

American Council for Voluntary
 International Action (InterAction)
1815 H Street, N.W./11th floor
Washington, D.C. 20006
202/822-8429

American Farm Foundation (P)
c/o Professor James Austin
Harvard Business School
Boston, Massachusetts 02163
617/495-6497

American Farmland Trust
1920 N Street, N.W./400
Washington, D.C. 20036
202/659-5170

American Forestry Association
1516 P Street, N.W.
Washington, D.C. 20036
202/667-3300

The American Forum for Global Education
(Formerly Global Perspectives in Education)
45 John Street/1200
New York, New York 10038
212/732-8606

American Gas Association
1515 Wilson Boulevard
Arlington, Virginia 22209
703/841-8400

American Geological Institute
4220 King Street
Alexandria, Virginia 22302
703/379-2480

American Institute of Biological Sciences
730 11th Street, N.W.
Washington, D.C. 20001

American Institute of Mining,
 Metallurgical, and Petroleum Engineers
345 East 47th Street, 14th floor
New York, New York 10017
212/705-7695

American Littoral Society
Sandy Hook
Highlands, New Jersey 07732
201/291-0055

American Metal Market
7 East 12th Street
New York, NY 10003

American Mining Congress (F)
1920 N Street, N.W.
Washington, DC 20036
202/861-2800

American Museum of Natural History
79 Central Park West
New York, New York 10023

American Petroleum Institute
1220 L Street, N.W./900
Washington, D.C.
20005
202/682-8000

American Rivers Conservation Council
801 Pennsylvania Avenue, S.E./303
Washington, D.C. 20003
202/547-6900

American Society for the Prevention of
 Cruelty to Animals (P)
441 East 92nd Street
New York, New York 10128
212/876-7700

American Solar Energy Society
850 West Morgan Street
Raleigh, North Carolina 27603
919/832-6303

American Water Resources Association
5410 Grosvenor Lane/220
Bethesda, Maryland 20814
301/493-8600

American Water Works Association
6666 West Quincy Avenue
Denver, Colorado 80235
303/794-7711

American Wind Energy Association
1730 North Lynn Street/610
Arlington, Virginia 22209
703/276 8334

Animal Protection Institute
 of America (P)
2831 Fruitridge Road
Sacramento, California 95820
916/422-1921

Animal Welfare Institute (A)
Post Office Box 3650
Washington, D.C. 20007
202/337-2332

The Antarctica Project (P)
218 D Street, S.E.
Washington, D.C. 20003
202/ 544-2600

Appropriate Technology International
1331 H Street, N.W.
Washington, D.C. 20005
202/879-2900

Arms Control Association
11 Dupont Circle, NW
Washington, D.C. 20036
202/797-6450

Association of State Drinking Water
 Administrators
1911 North Fort Myer Drive
Arlington, Virginia 22209
703/524-2428

Association of State and Territorial
 Waste Management Officials
444 North Capitol Street, N.W.
Washington, D.C. 20001
202/624-5828

Association of State and Interstate Water
 Pollution Control Administration
444 N. Capitol Street, NW
Washington, D.C. 20001
202/624-7782

Audubon Naturalist Society of the
 Central Atlantic States (P)
8940 Jones Mill Road
Chevy Chase, Maryland 20815
202/652-9188

Bat Conservation International
P.O. Box 162603
Austin, Texas 78716
512/327-9721

Better World Society (P)
1140 Connecticut Avenue, N.W./1006
Washington, D.C. 20036
202/331-3770

Biocycle
Box 351
Emmaus, PA 18049

Biosphere Films (F)
4 West 105th Street/6A
New York, New York 10025

Bolton Institute for a
 Sustainable Future (P)
Four Linden Square
Wellesley, Massachusetts 02181-4709
617/235-5320

Bullfrog Films (F)
Oley, Pennsylvania 19547
800/543-3764

Bureau of Mines
U.S. Department of the Interior
2401 E Street, N.W.
Washington, D.C. 20241
202/634-1263

Californians for Population Stabilization
1025 9th Street/217
Sacramento, California 95814
916/446-1033

Carrying Capacity (P)
1325 G Street, N.W./1003
Washington, D.C. 20005
202/879-3045

CARE International (P)
660 First Avenue
New York, New York 10016
212/686-3110

Care Film Unit (F)
660 First Avenue
New York, New York 10016

Center for Clean Air Policy
444 North Capitol Street/526
Washington, D.C. 20001
202/624-7709

Center for Conservation Biology (P)
Department of Biological Sciences
Stanford University
Stanford, California 94305
415/723-5924

Center for Defense Information
1500 Massachusetts Avenue, NW
Washington, D.C. 20005
202/862-0700

Center for Economic Conversion
222 View Street
Mountain View, California 94041
415/968-8798

Center for Innovative Diplomacy
17931 Sky Park Circle/F
Irvine, California 92714
719/250-1296

Center for International Development
and Environment
1709 New York Avenue/700
Washington, D.C. 20006
202/638-6300

Center for Development Policy
418 10th Street, S.E.
Washington, D.C. 20003
202/547-6406

Center for Marine Conservation
1725 DeSales Street, N.W./500
Washington, D.C. 20036
202/429-5609

Center for National Policy
3/7 Massachusetts Avenue, N.E.
Washington, D.C. 20002
202/546-9300

Center for Population Options (A)
1012 14th Street, N.W. /1200
Washington, D.C. 20005
202/347-5700

Center for Religion, Ethics and Social Policy (A)
(Eco-Justice Project)
Anabel Taylor Hall
Cornell University
Ithaca, New York 14853
607/255-4225

Center for Science in the Public Interest
1501 16th Street, N.W.
Washington, D.C. 20036

Center for Teaching International
Relations
University of Denver
Denver, Colorado 80208
303/753-3106

Centre for Our Common Future
Palais Wilson
52, rue des Paquis/CH-1201
Geneva, Switzerland

Chesapeake Bay Foundation
162 Prince George Street
Annapolis, Maryland 21401
301/268-8816

Citizen's Clearinghouse for
Hazardous Wastes
P.O. Box 926
Arlington, Virginia 22216
703/276-7070

Citizens for Ocean Law
1601 Connecticut Avenue, N.W./202
Washington, D.C. 20009
202/462-3737

Clean Sites, Inc.
1199 North Fairfax Street/400
Alexandria, Virginia 22314
703/683-8522

Clean Water Action Project
317 Pennsylvania Avenue, S.E.
Washington, D.C. 20005
202/547-1196

Coalition on Superfund
1730 Pennsylvania Avenue, N.W./200
Washington, D.C.· 20006
202/393-4760

Coast Alliance
1536 16th Street, N.W.
Washington, D.C. 20036
202/265-5518

Colorado Outdoor Education Center (P)
Florissant, Colorado 80816
303/748-3341

Committee for National Security
1601 Connecticut Avenue, N.W./3011
Washington, D.C. 20009
202/745-2450

Common Cause
2030 M Street, N.W.
Washington, D.C. 20036
202/833-1200

Concern, Inc. (P)
1794 Columbia Road, N.W.
Washington, D.C. 20009
202/328-8160

Congressional Clearinghouse on the Future
555 House Annex No. 2
Washington, D.C. 20515
202/226-3434

Conservation Film Service (F)
408 East Main Street
P.O. Box 776
League City, Texas 77574-0776

The Conservation Foundation (P)
1250 24th Street, N.W./500
Washington, D.C. 20037
202/293-4800

Conservation International
1015 18th Street, N.W./1000
Washington, D.C.
202/429-5660

Conservatree Paper Company
10 Lombard Street/250
San Francisco, California 94111
800/522-9200 (outside California)
415/433-1000 (from anywhere)

Coolidge Center for Environmental
  Leadership (A)
1675 Massachusetts Avenue/4
Cambridge, Massachusetts 02138-1836
617/864-5085

Co-op America
2100 M Street, N.W./310
Washington, D.C. 20036
202/872-5307
800/424-2667

Coronet Film and Video (F)
108 Wilmot Road
Deerfield, Illinois 60015
800/621-2131

Council on Economic Priorities
30 Irving Place
New York, New York 10003
212/420-1133

Council on Environmental Quality
722 Jackson Place, N.W.
Washington, D.C. 20006
202/395-5750

Council on International Educational
  Exchange
205 East 42nd Street
New York, New York 10017
212/661-1450

Council on International and
  Public Affairs (P)
777 United Nations Plaza
New York, New York 10017
212/972-9877

Council on Ocean Law
1709 New York Avenue, N.W.
Washington, D.C. 20006
202/347-3766

Council for Solid Waste Solutions
1275 K Street, N.W./400
Washington, D.C. 20005
202/371-5319

Critical Mass Energy Project
215 Pennsylvania Avenue, S.E.
Washington, D.C. 20003
202/546-4996

Defenders of Wildlife
1244 19th Street, N.W.
Washington, D.C. 20036
202/659-9510

Development Group for Alternative
  Policies (GAP)
1400 I Street, N.W./520
Washington, D.C. 20005
202/898-1566

Direct Cinema Ltd. (F)
P.O. Box 69589
Los Angeles, California 90060
213/656-4700

Earth Care Paper Company
325 Beech Lane
Harbor Springs, Michigan 49740
616/526-7003

Earthwatch Expeditions (A)
1228 31st Street, N.W.
Washington, D.C. 20007
202/342-2564

East-West Center
1777 East West Road
Honolulu, Hawaii 96848

Eco-Justice Project (see Center for
Religion, Ethics, and Social Policy)

Education Film and Video Project (F)
1725-B Seabright Avenue
Santa Cruz, California 95062
408/427-2627

Electric Power Research Institute
3412 Hillview Avenue
Palo Alto, California 94303
415/855-2000

Emerging Public Issues Corporation (A)
First Interstate Bank Center
6301 Gaston Avenue/344
Dallas, Texas 75214
214/821-1968

Employment Research Associates
115 West Allegan/810
Lansing, Michigan 48933
517/485-7655

Encyclopedia Brittanica Educational
 Corporation (F)
310 South Michigan Avenue
Chicago, Illinois 60604
312/347-7000

Energy Conservation Coalition
1525 New Hampshire Avenue, N.W.
Washington, D.C. 20036
202/745-4874

Environment Liaison Centre International
P.O. Box 30552
Nairobi Kenya
011/2542-333930

Environmental Action
1525 New Hampshire Avenue, N.W.
Washington, D.C. 20036
202/745-4870

Environmental Coalition for
 North America (P)
1325 G Street, N.W./1003
Washington, D.C. 20005
202/289-5009

Environmental Defense Fund (P)
257 Park Avenue South
New York, New York 10010
212/505-2100
EDF Hotline: 1-800-CALL-EDF

Environmental Defense Fund (EDF) (P)
1616 P Street, N.W./150
Washington, D.C. 20036
202/387-3500

Environmental Education Association
 of Oregon (P)
Post Office Box 40047
Portland, Oregon 97240
1-800/322-3326

Environmental Educators Provincial
 Specialist Association of B.C. (P)
2460 Tanner Road
Victoria
British Columbia, Canada V8Z 5R1
604/828-0026

Environmental and Energy Study Institute
122 C Street, N.W./700
Washington, D.C. 20001
202/628-1400

Environmental Hazards Management
 Institute
10 Newmarket Road
P.O. Box 932
Durham, New Hampshire 03824
603/868-1496

Environmental Policy Institute (P)
218 D Street, S.E.
Washington, D.C. 20003
202/544-2600

Environmental Task Force
1012 14th Street, N.W./15th Floor
Washington, D.C. 20005
202/842-2222

Federation for American Immigration
 Reform (P)
1424 16th Street, N.W./701
Washington, D.C. 20036
202/328-7004

Federation of American Scientists
307 Massachusetts Avenue, N.E.
Washington, D.C. 20002
202/546-3300

Film Distribution Center (F)
13500 N.E. 124th Street/2
Kirkland, Washington 98034-8010
206/820-2592

Food and Agriculture Organization
 of the United Nations
1001 22nd Street, N.W./300
Washington, D.C. 20437
202/653-2398

Food First
(See Institute for Food and
 Development Policy)

Foreign Policy Association
205 Lexington Avenue
New York, New York 10016
212/764-4050

Foundation for Advancements in Science
 and Education (FASE) (P)
Park Mile Plaza
4801 Wilshire Boulevard
Los Angeles, California 90010
213/937-9911

Freshwater Foundation
2500 Shadywood Road, Box 90
Navarre, Minnesota 55392
612/471-8407

Friends Committee on National Legislation
245 2nd Street, N.E.
Washington, D.C. 20002
202/547-6000

Friends of the Earth (P)
218 D Street, S.E.
Washington, D.C. 20003
202/544-2600

Fund for Animals
200 West 57th Street
New York, New York 10019
212/246-2096

The Fund for Peace (A)
345 E. 46th Street/207
New York, New York 10017
212/661-5900

Futures Foundation (A)
4436 280th Street West
Castle Rock, Minnesota 55010
612/463-3987

The Futures Group
76 Eastern Boulevard
Glastonbury, Connecticut 06033
203/633-3501

The Futures Group
1101 14th Street, NW
Washington, D.C. 20005
202/347-8165

Global Action Network
P.O. Box 819
Ketchum, Idaho 83340
208/726-4333

Global Environment Project Institute (P)
Post Office Box 1111
Ketchum, Idaho 83340
208/726-4030

Global Perspectives in Education
(see The American Forum for
  Global Education)

Global Tomorrow Coalition
1325 G Street N.W./915
Washington, D.C. 20005-3014
202/628-4016

Goldman Environmental Foundation (P)
1090 Sansome St. 3rd. Floor
San Francisco, California 94111
415/788-1090

Greater Caribbean Energy and
  Environment Foundation (P)
Post Office Box 490449
Miami, Florida 33149
305/361-7417

Greenhouse Crisis Foundation
1130 17th Street, N.W./630
Washington, D.C. 20036
202/466-2823

Greenpeace USA
Main Office
1436 U Street, N.W.
Washington, D.C. 20009
202/462-1177

Greenpeace USA
Regional Offices:
Fort Mason Center, Building E
San Francisco, California 94123
and
1310 College Avenue/301
Boulder, Colorado 80302

The Human Lactation Center Ltd (P)
666 Sturges Highway
Westport, Connecticut 06880
203/259-5995

Humane Society of the United States (P)
2100 L Street, N.W.
Washington, D.C. 20037
202/452-1100

Hunger and Development Coalition
  of Central Ohio (A)
299 King Avenue
Columbus, Ohio 43201
614/424-6203

The Hunger Project
1 Madison Avenue
New York, New York 10010
212/532-4255

The Hunger Project
1388 Sutter Street
San Francisco, California 94109
415/928-3700

Icarus Films (F)
200 Park Avenue South
New York, New York 10003
212/674-3375

Idaho Conservation League (P)
Post Office Box 844
Boise, Idaho 83701
208/345-6933

INFORM, Inc.
381 Park Avenue South
New York, New York 10016
212/689-4040

Institute for Alternative Agriculture (P)
9200 Edmonston Road/117
Greenbelt, Maryland 20770
301/441-8777

Institute for Alternative Futures (P)
108 North Alfred Street
Alexandria, Virginia 22314
703/684-5880

Institute for Defense and
  Disarmament Studies
2001 Beacon Street
Brookline, Massachusetts 02146

Institute for Development Anthropology
State University of New York
  at Binghamton
Binghamton, New York 13902

Institute for Food and Development Policy
  (Food First)
145 Ninth Street
San Francisco, California 94103
415/864-8555

Institute for Policy Studies
1601 Connecticut Avenue, N.W.
Washington, D.C. 20009
202/234-9382

Institute for Soviet-American Relations
1608 New Hampshire Avenue, N.W.
Washington, D.C. 20009
202/387-3034

Institute for the Study of
  Natural Systems (A)
26 High Street
Rockport, Massachusetts 01966
617/546-7057

Institute for 21st Century Studies (P)
1611 N. Kent Street/610
Arlington, Virginia 22209
703/841-0048

The Institute of the North
  American West (P)
6251 28th. Avenue N.E.
Seattle, Washington 98115-7117

Institute of Scrap Recycling Industries
1627 K Street, N.W./700
Washington, DC 20006
202/466-4050

Inter-American Development Bank
1300 New York Avenue, N.W.
Washington, D.C. 20577
202/623-1000

International Alliance for Sustainable
  Agriculture (P)
1201 University Avenue, S.E./202
Minneapolis, Minnesota 55414
612/331-1099

International Center for
  Development Policy (P)
731 8th Street, S.E.
Washington, D.C. 20003
202/547-3800

International Center for Dynamics
  of Development (P)
4201 South 31st Street/616
Arlington, Virginia 22206
703/578-4627

International Christian Youth Exchange (P)
134 West 26th Street/4th floor
New York, New York 10001
212/206-7307

International Development Conference (A)
1401 New York Avenue/1100
Washington, D.C. 20005
202/638-3111

International Environmental Bureau
61, Route de Chene
CH-1208
Geneva, Switzerland

International Environmental Education
  Foundation (P)
Post Office Box 1092
Estes Park, Colorado 80517
303/586-2019

International Food Policy Research
  Institute
1776 Massachusetts Avenue, N.W./4th floor
Washington, D.C. 20036
202/862-5600

International Friendship League
55 Mount Vernon Street
Boston, Massachusetts 02108

International Fund for Agricultural
  Development
1889 F Street, NW
Washington, DC 20006
202/289-8670

International Marine Life Alliance USA
94 Station Street/1645
Hingham, Massachusetts 02043
617/383-1209

International Planned Parenthood
  Federation
105 Madison Avenue
New York, New York 10016
212/995-8800

International Rivers Network
300 Broadway/28
San Francisco, California 94133
415/788-3666

International Union for the Conservation
  of Nature and Natural Resources
Avenue du Mont-Blanc
CH-1196
Gland, Switzerland
022/64-71-81)

Izaak Walton League
1701 North Ft. Myer Drive/1100
Arlington, Virginia 22209
703/528-1818

Jobs With Peace Campaign
76 Summer Street/3rd Floor
Boston, Massachusetts 02110
617/338-5783

Keep America Beautiful, Inc.
Mill River Plaza
9 West Broad Street
Stamford, Connecticut 06902

League of Women Voters of the
  United States
1730 M Street, N.W.
Washington, D.C. 20036
202/429-1965

Linfield College (P)
McMinnville, Oregon 97128
503/472-4121

Massachusetts Audubon Society (A)
South Great Road
Lincoln, Massachusetts 01773
617/259-9500

Mershon Center
Ohio State University
199 West 10th Street
Columbus, Ohio 43201
614/422-1681

Michigan Media (F)
416 Fourth Street
Ann Arbor, Michigan 48109
313/764-8298

Milwaukee Audubon Society (P)
4759 North Woodburn
Milwaukee, Wisconsin 53211
414/453-5640

Mineral Information Institute, Inc.
1125 17th. Street/2070
Denver, Colorado 80202
303/297-3226

Missouri Botanical Garden
P.O. Box 299
St. Louis, Missouri 63166-0299

Morrison Institute for Population
  and Resource Studies (P)
Box 9281
Stanford University
Stanford, California 95305
415/723-7517

National Academy Press (F)
2101 Constitution Avenue, N.W.
Washington, D.C. 20418
202/334-2665

National Association for Plastic
  Container Recovery (NAPCOR)
5024 Parkway Plaza Boulevard/200
Charlotte, North Carolina 28217
704/357-3250

National Association of Social Workers
7981 Eastern Avenue
Silver Spring, Maryland 20910
301/565-0333

National Association of State Departments
  of Agriculture
1616 H Street, N.W.
Washington, D.C. 20006
202/628-1566

National Audubon Society (P)
950 Third Avenue
New York, New York 10022
212/546-9100

National Audubon Society (P)
801 Pennsylvania Avenue, S.E./301
Washington, D.C. 20003
202/547-9009

National Center for Policy Alternatives
2000 Florida Avenue, N.W.
Washington, D.C. 20009
202/387-6030

National Clean Air Coalition
801 Pennsylvania Avenue, S.E.
Washington, D.C. 20003
202/543-8200

National Coalition Against the
Misuse of Pesticides
530 7th Street, S.E.
Washington, D.C. 20001
202/543-5450

National Coalition for Marine
Conservation
1 Post Office Square
Boston, Massachusetts 02109
617/338-2909

National Commission for Economic
Conversion and Disarmament
Box 15025
Washington, D.C. 20003
202/544-5059

National Committee for World Food Day
1001 22nd Street, N.W./300
Washington, D.C. 20437
202/653-2404

National Family and Reproductive Health
Association
122 C Street,N.W./380
Washington, D.C. 20001
202/628-3535

National Film Board of Canada (F)
111 East Wacker Drive/313
Chicago, Illinois 60601

National Geographic Society
17th and M Streets, N.W.
Washington, D.C. 20036
202/857-7000

National Geographic Society
Educational Services (F)
Department 88
Washington, D.C. 20036
800/368 2728 (to order films)
201/628-9111 (to rent or preview films)

National Parks and Conservation
Association (P)
1015 31st Street, N.W./4th floor
Washington, D.C. 20007
202/944-8530

National Recycling Coalition
P.O. Box 80729
Lincoln, Nebraska 68729

National Recycling Congress
P.O. Box 10540
Portland, Oregon 97210

National Wildlife Federation (P) (F)
1400 16th Street, N.W.
Washington, D.C. 20036
202/637-3700

National Wood Energy Association
1730 North Lynn Street/610
Arlington, Virginia 22209
703/524-6104

Natural Resources Council of America
1015 31st Street, N.W.
Washington, D.C. 20007
202/333-8495

Natural Resources Defense Council (P)
40 West 20th Street
New York, New York 10011
212/727-2700

Natural Resources Defense Council (P)
1350 New York Avenue, N.W./300
Washington, D.C. 20005
202/783-7800

Natural Science for Youth Foundation
130 Azalea Drive
Roswell, Georgia 30075
404/594-9364

National Clean Air Coalition
801 Pennsylvania Avenue, S.E.
Washington, D.C. 20003
202/543-8200

National Environmental Health Association
720 South Colorado Boulevard
Denver, Colorado 80222

National Solid Waste Management
Association
1730 Rhode Island Avenue, N.W.
Washington, D.C. 20036

National Toxics Campaign
29 Temple Place/5th floor
Boston, Massachusetts 02111
617/482-1477
(Regional offices in California, Colorado,
Massachusetts, Maine, Oklahoma, Texas,
and the District of Columbia)

The Nature Conservancy
1815 North Lynn Street
Arlington, Virginia 22209
703/841-4860

Negative Population Growth (P)
210 The Plaza
Teaneck, New Jersey 07666
201/837-3555

New Forests Fund
731 8th Street, S.E.
Washington, D.C. 20003
202/547-3800

New York Botanical Garden (P)
Bronx, New York 10458
212/220-8777

New York Zoological Society (P)
185th Street and Southern Boulevard
Bronx, New York 10460
212/367-1010

Noranda, Inc. (F)
P.O. Box 45
Commerce Court West
Toronto, Ontario M5L 2B6 Canada

North American Association for
  Environmental Education
P.O. Box 400
Troy, New York 45373

North Carolina Botanical Gardens (P)
University of North Carolina
3375 Totten Center
Chapel Hill, North Carolina 27599-3375
919/967-2246

The Jessie Smith Noyes Foundation (A)
16 East 34th Street
New York, New York 10016
212/684-6577

The Oceanic Society (P)
218 D Street, S.E.
Washington, D.C. 20003
202/544-2600

Office of Technology Assessment,
  U.S. Congress
Washington, D.C. 20510
202/226-2115

Organization for Economic Cooperation
  and Development (U.S. Office)
2001 L Street, N.W./700
Washington, D.C. 20036
202/785-6323

Orleans Audubon Society (A)
2720 Octavia Street
New Orleans, Louisiana 70115
504/246-2473

Overseas Development Council
1717 Massachusetts Avenue, N.W./501
Washington, D.C. 20036
202/234-8701

Parliamentarians for Global Action
211 East 43rd Street/1604
New York, New York 10017
212/687-7755

Panos Institute
1405 King Street
Alexandria, Virginia 22314
703/836-1302

Partners of the Americas
1424 K Street, N.W./700
Washington, D.C. 20005

Partridge Films (F)
38 Mill Lane
London NW6 1NR England

Passive Solar Industries Council
2836 Duke Street
Alexandria, Virginia 22314
703/823-3356

The Pathfinder Fund (P)
9 Galen Street
Watertown, Massachusetts 02172
617/731-1700

Peace Net
3228 Sacramento Street
San Francisco, California 94115
415/486-0264

Pennsylvania State University (F)
Audiovisual Services
Division of Media and Learning Resources
Special Services Building
University Park, Pennsylvania 16802

Pergamon Press
395 Saw Mill River Road
Elmsford, New York 10523
914/592-7700

Physicians for Social Responsibility
1000 16th Street, N.W./810
Washington, D.C. 20036
202/785-3777

Planetary Citizens (P)
325 9th Street
San Francisco, California 94103
415/325-2939

Planetary Society (P)
65 North Catalina Avenue
Pasadena, California 91106
818/793-5100

Planet Earth Foundation (P)
2701 First Avenue/ 400
Seattle, Washington 98121
206/547-7000

Planned Parenthood Federation
 of America (P)
810 Seventh Avenue
New York, New York 10019
212/541-7800

Planned Parenthood of New York City
380 Second Avenue
New York, New York 10010
212/777-2002

Planning and Conservation
 League Foundation
909 12th Street/203
Sacramento, California 95814
916/444-8726

Population Communication (P)
1489 East Colorado Boulevard/202
Pasadena, California 91106
213/973-4750

Population Crisis Committee (P)
1120 19th Street, N.W./550
Washington, D.C. 20036
202/659-1833

Population-Environment Balance (P)
1325 G Street, N.W./1003
Washington, D.C. 20005-3104
202/879-3000

Population Food Fund (A)
2017 Hudson Lane
Monroe, Louisiana 71207
318/322-1422

Population Institute
110 Maryland Avenue, N.E.
Washington, D.C. 20036
202/544-3300

Population Reference Bureau (P) (F)
777 14th Street, N.W./800
Washington, D.C. 20005
202/639-8040

Population Resources Center (A)
500 E. 62nd Street
New York, New York 10021
212/888-2820

Population Services International (P)
1120 19th Street, N.W./600
Washington, D.C. 20036
202/785-0072

Portland Audubon Society (P)
2255 N.W. Johnson Street
Portland, Oregon 97210
503/292-6855

Prairie Paper
825 M Street/101
Lincoln, Nebraska 68508
402/477-0825

Public Citizen Critical Mass
 Energy Project
215 Pennsylvania Avenue, S.E.
Washington, D.C. 20003
202/546-4996

Public Media, Inc. (F)
119 West 57th Street/1511
New York, New York 10019
212/247-9050

Rachel Carson Council (P)
8940 Jones Mill Road
Chevy Chase, Maryland 20815
301/652-1877

Rainforest Action Network
301 Broadway/A
San Francisco, California 94133
415/398-4044

Rainforest Alliance
295 Madison Avenue/1804
New York, NY 10017
212/599-5060

Renew America (A)
1400 16th Street/710
Washington, D.C. 20036
202/232-2252

Renewable Fuels Association
201 Massachusetts Avenue, N.W./C-4
Washington, D.C. 20002
202/543-3802

Resources for the Future
1616 P Street, N.W.
Washington, D.C. 20036
202/328-5000

Rocky Mountain Institute
1739 Snowmass Creek Road
Snowmass, Colorado 81654-9199

Rodale Press (P)
33 East Minor Street
Emmaus, Pennsylvania 18049
215/967-5171

Rural America
1346 Connecticut Avenue, N.W.
Washington, D.C. 20036
202/659-2800

Safe Energy Communication Council
1717 Massachusetts Avenue, N.W./LL215
Washington, D.C. 20036
202/483-8491

SANE/FREEZE
711 G Street, S.E.
Washington, D.C. 20003
202/546-7100

Save the Children Federation
54 Wilton Road
Westport, Connecticut 06880
203/226-7271

Scenic Shoreline Preservation
  Conference (P)
4623 More Mesa Drive
Santa Barbara, California 93110
805/964-2492

School for Field Studies (P)
376 Hale Street
Beverly, Massachusetts 01915
617/927-5339

Sierra Club (P)
530 Bush Street
San Francisco, California 94108
415/981-8634

Sierra Club (P)
408 C Street, N.E.
Washington, D.C. 20002
202/547-1141

Smithsonian Institution Traveling
  Exhibition Service (SITES) (F)
Washington, D.C. 20560
202/357-3168

Smithsonian Tropical Research Institute
c/o Office of Assistant Secretary
  for Research
Smithsonian Institution
Washington,D.C. 20560
202/357-2954

Social Issues Resource Series (P)
11 Dixie Boulevard
Delray Beach, Florida 33444
305/994-0079

Society for International Development
Washington D.C. Chapter
1401 New York Avenue, N.W./1100
Washington, D.C. 20005
202/347-1800

Soil and Water Conservation Society
7515 Northeast Ankeny Road
Ankeny, Iowa 50021-9764

Solar Box Cookers International
1724 11th Street
Sacramento, California 95814
916/447-8691

Solar Energy Industries Association
1730 North Lynn Street/610
Arlington, Virginia 22209
703/524-6100

Solar Energy Research Institute
1617 Cole Boulevard
Golden, Colorado 80401
303/231-1000

Southwest Research and
  Information Center
Albuquerque, New Mexico 87106
505/262-1862

State University of New York
  at Stony Brook (F)
Department of Anatomical Sciences
Health Sciences Center
Stony Brook, New York 11794

The Student Conservation Association
P.O. Box 550C
Charlestown, New Hampshire 03603

Survival International USA
2121 Decatur Place, N.W.
Washington, D.C. 20008
202/265-1077

Texas Committee on Natural Resources (P)
4144 Cochran Chapel Road
Dallas, Texas 75209
214/352-8370

Tree People
12601 Mulholland Drive
Beverly Hills, California 90210
818/769-2663

Trust for Public Land (P)
116 New Montgomery/4th floor
San Francisco, California 94105
415/495-4014

Tufts University (P)
  Department of Urban and Environmental Policy
Lincoln Filene Center for Citizenship
  and Public Affairs
Medford, Massachusetts 02155
617/381-3451

Turner Broadcasting System (P)
CNN Center
Post Office Box 105366
Atlanta, Georgia 30348-5366
404/827-1700

Union of Concerned Scientists
26 Church Street
Cambridge, Massachusetts 02238
617/547-5552

UNIPUB
4611-F Assembly Drive
Lanham, Maryland 20706-4391
800/233-0504

United Nations
1 United Nations Plaza
New York, New York 10017
212/963-1234

United Nations Association of the USA
485 Fifth Avenue
New York, New York 10017
212/697-3232
Washington, D.C. Office, 966-5393

United Nations Environment Program
North American Liaison Office
Room DC 2-0803
United Nations, New York 10017
212/963-8093

United Nations Environment Program
1889 F Street, N.W.
Washington, D.C. 20006
202/289-8456

The United Nations Population Fund
  (UNFPA)
220 East 42nd Street
New York, New York 10017
212/850-5600

U.S. Association for The Club of Rome (P)
1325 G Street, N.W./1003
Washington, D.C. 20005-3104
202/879-3000

U.S. Bureau of Land Management
Department of the Interior
18th and C Streets, N.W./3619
Washington, D.C. 20240
202/343-4603

U.S. Bureau of Mines
Minerals Information Office
Department of Interior Building/2647
18th and C Streets, N.W. MS 2647-MIB
Washington, DC 20240

U.S. Bureau of Mines
  Motion Pictures (F)
Cochrans Mill Road
P.O. Box 18070
Pittsburgh, Pennsylvania 15236
412/675-4338

U.S. Department of Agriculture
Independence Avenue
  between 12th and 14th Streets, S.W.
Washington, D.C. 20250
202/477-8732

U.S. Department of Energy
Conservation and Renewable Energy
  Division
1000 Independence Avenue, S.W.
Washington D.C. 20585
202/586-9220

U.S. Department of Energy
Energy Information Administration
Forrestal Building
Washington, D.C. 20585
202/586-8800

U.S. Environmental Protection Agency
401 M Street, S.W.
Washington, D.C. 20460
202/382-2090

U.S. Fish and Wildlife Service
Department of the Interior
Washington, D.C. 20240
202/343-7445

U.S. Forest Service
P.O. Box 96090
Washington, D.C. 20090
202/447-3957

U.S. Geological Survey
12201 Sunrise Valley Drive
Reston, VA. 22092
703/648-4000

U.S. Government Printing Office
Superintendent of Documents
Washington, D.C. 20402
202/783-3238

U.S. Institute of Peace
1550 M Street, NW/700
Washington, D.C. 20005-1708

U.S. Public Interest Research Group (PIRG)
215 Pennsylvania Avenue S.E.
Washington, D.C. 20003
202/546-9707

University of California (F)
Extension Media Center
Berkeley, California 94702

Virginia Polytechnic Institute
 and State University
College of Architecture and
 Urban Studies (A)
202 Cowgill Hall
Blacksburg, Virginia 24061

Volunteers in Technical Assistance (VITA)
1815 North Lynn Street/200
Arlington, Virginia 22209
703/276-1800

Water Pollution Control Federation
601 Wythe Street
Alexandria, Virginia 22314
703/684-2400

Wellman, Inc. (Plastic recycling)
1040 Broad Street/302
Shrewsbury, N.J. 07702
201/542-7300

Western North Carolina Tomorrow (A)
Western Carolina University
Post Office Box 222
Cullowhee, North Carolina 28723
704/227-7492

Wilderness Society (P)
1400 I Street, N.W./10th floor
Washington, D.C. 20005
202/842-3400

Wildlife Conservation International
New York Zoological Society
Bronx, New York 10460
212/220-5090

Windstar Foundation
Post Office Box 286
Snowmass, Colorado 81654
303/927-4777

Winrock International
1611 North Kent Street
Arlington, Virginia 22209
703/525-4430

Wisconsin Department of Natural
 Resources
P.O. Box 7921
Madison, Wisconsin 53707

World Affairs Council
1800 M Street, N.W./299 North
Washington, D.C. 20036
202/293-1051

Work on Waste (F)
82 Judson Street
Canton, New York 13617

Working Group on Community
 Right-To-Know
215 Pennsylvania Avenue S.E.
Washington, D.C. 20003
202/546-9707

World Bank
1818 H Street, N.W.
Washington, D.C. 20433
202/477-1234

World Bank Publications
Department 0552
Washington, D.C. 20073-0552

World Citizens (A)
312 Sutter Street/506
San Francisco, California 94108
415/421-0836

World Environment Center
419 Park Avenue South/1403
New York, New York 10016
212/683-4700

World Federalist Association (P)
418 7th Street, S.E.
Washington, D.C. 20003
202/546-3950

World Future Society
4916 St. Elmo Avenue
Bethesda, Maryland 20814
202/656-8274

World Hunger Education Service
3018 4th Street, N.E.
Washington, D.C. 20017
202/269-1075

World Information Systems
Post Office Box 535
Cambridge, Massachusetts 02238

World Neighbors (F)
5116 North Portland Avenue
Oklahoma City, Oklahoma 73112
800/242-6387

World Population Society (P)
1333 H Street, N.W./760
Washington, D.C. 20005
202/898-1303

World Resources Institute (P)
1709 New York Avenue, N.W./700
Washington, D.C. 20006
202/638-6300

Worldwatch Institute (P)
1776 Massachusetts Avenue N.W.
Washington, D.C. 20036
202/452-1999

WorldWIDE (P)
P.O. Box 40885
Washington, D.C. 20016
202/234-0657

World Wildlife Fund (P)
1250 24th Street, N.W./500
Washington, D.C. 20037
202/293-4800

Xerces Society (A)
10 S.W. Ash Street
Portland, Oregon 97204
503/222-2788

Zero Population Growth (P)
1400 16th Street, N.W./3rd floor
Washington, D.C. 20036
202/332-2200

561 1853